Outstanding Contributions to Logic

Volume 8

More information about this series at http://www.springer.com/series/10033

Katalin Bimbó
Editor

J. Michael Dunn on Information Based Logics

Springer

Editor
Katalin Bimbó
Department of Philosophy
University of Alberta
Edmonton, AB
Canada

ISSN 2211-2758 ISSN 2211-2766 (electronic)
Outstanding Contributions to Logic
ISBN 978-3-319-29298-4 ISBN 978-3-319-29300-4 (eBook)
DOI 10.1007/978-3-319-29300-4

Library of Congress Control Number: 2016930054

Printed on acid-free paper

This Springer imprint is published by SpringerNature
The registered company is Springer International Publishing AG Switzerland

To J. Michael Dunn
on (or near) his 75th birthday

Preface

An obvious excuse for the project that resulted in the present collection of papers is provided by the dedication: J. Michael Dunn turns 75 years old in 2016. This book celebrates his research and his career as a logician, which already spans more than half a century.

Another obvious rationale for this book is that we are aware, more than ever before, of the importance and pervasiveness of information. It is a truism that we live in an information age. The developments in computer technology in the past 20–30 years—including increased storage, transmission, and search capabilities—undoubtedly contribute to our perception of the ubiquity of information. A way to use information is to reason with it. Remarkably, J. M. Dunn was thinking about logic in terms of information well before everybody jumped onto the (i-)bandwagon wheeling along the information superhighway. The papers in this volume evidence that treating logic as an organon for manipulating information is a fruitful approach.

An opportunity to assemble this volume arose because Springer established a new book series. The *Outstanding Contributions to Logic* series provides a different focal point for a collection of papers than some others do. Although the OCL volumes have the flavor of a Festschrift, they support greater flexibility and a narrower theme than what could be achieved by cataloging all the works of a famous logician.

Logic, in general, should interest a wide range of people. The particular approach to logic that is exemplified by this volume will primarily appeal to readers who are involved with disciplines such as mathematics, computer science, the information sciences, and philosophy. Some of the papers include not only new research results, but draw a chronologically faithful picture of the development of certain ideas—these sources will be especially useful for historians of science and philosophers. A reflection on achievements (spanning several decades) motivated some authors to take stock of the accumulated results; such papers are excellent for reference purposes too.

It is expected that the present book will be useful to scholars who are interested in the area that is somewhat vaguely called nonclassical logics. While the papers will definitely be invaluable for researchers, most of them should be accessible to graduate students as well as to researchers working in other fields. Some articles in this collection are written in a style which ensures that anyone who is willing to dabble into a subject (outside their expertise) will enjoy reading them.

Acknowledgments. In a volume of this kind, the person whose work and research results provide the justification for editing the volume is to be thanked first: I am grateful to J. Michael Dunn for allowing me to take on the (somewhat complicated) task of editing this volume and for his continuous help in making the project a success.

I would like to thank the authors of this volume, who not only responded to the initial invitation to contribute to this volume, but have written a paper for this collection. The papers were refereed using the "single-blind" type of refereeing. Thanks to those who refereed a paper, and thereby, contributed to the project.

The series editor, Sven Ove Hansson not only provided a document about how to edit a book for the *Outstanding Contributions to Logic* series, but he was helpful in various other ways from start to finish. I am grateful for his help in the process.

Christi Lue of Springer Science provided forms, guidelines, and advice from the publisher's side. I am thankful for her ongoing support to the project.

This book has been typeset using the program TeX (which was originally designed by D. Knuth) from the source files submitted by the authors. In particular, the volume uses the LaTeX format, a class file provided by Springer as well as several packages that were developed under the auspices of the American Mathematical Society.

Edmonton
September 2015

Katalin Bimbó

Contents

Editor and Contributors

About the Editor

Katalin Bimbó is a professor at the University of Alberta, in Edmonton, Canada. Her main research interests and results are in the areas of relevance logic (especially, ticket entailment), substructural logics (including linear logic) and combinatory logic and λ-calculi. She received her Ph.D. from Indiana University Bloomington (with J. M. Dunn as her thesis supervisor), and then worked at universities in New Zealand, Australia and the U.S.A. She is an author of numerous journal articles and book chapters. She co-authored a book with J. M. Dunn, entitled *Generalized Galois Logics: Relational Semantics of Nonclassical Logical Calculi* (2008), and she wrote two more books, *Combinatory Logic: Pure, Applied and Typed* (2012) and *Proof Theory: Sequent Calculi and Related Formalisms* (2015).

Contributors

Gerard Allwein received his BS in Computer Science from Purdue University in 1979, and his Ph.D. from Indiana University in 1992 with J. Michael Dunn as his advisor. He spent several years as an engineer in the machine tools industry. He is the former Assistant Director of the Visual Inference Laboratory at Indiana University and has taught the Philosophy Department's graduate level logic sequence. Since 2002, Dr. Allwein has worked at the U.S. Naval Research Laboratory in Washington D.C. on theoretical aspects of security. He has also assisted NRL in several engineering projects either analyzing their security or designing and building systems. Dr. Allwein is primarily an algebraic logician when he's not involved with engineering.

Arnon Avron obtained his Ph.D. in Mathematics from Tel Aviv University in 1985. His thesis was devoted to paraconsistent logics in general, and to relevant logics in particular—subjects to which he has continued to contribute since then.

After finishing his thesis he spent his postdoc period at the laboratory for the foundations of computer science at Edinburgh University, where he belonged to the original LF team—the first computerized logical framework. After that he joined the Computer Science Department of Tel Aviv University, where he is a full professor since 1999. Avron's areas of interest and research include nonclassical logics (especially those connected to uncertainty reasoning), proof theory, automated reasoning in general, and the mechanization of mathematics in particular, and foundations of logics and of mathematics, where he is mainly interested in the predicativist approach.

Johan van Benthem is university Professor Emeritus of logic at the University of Amsterdam, Henry Waldgrave Stuart professor of philosophy at Stanford University, and Changjiang Professor of humanities at Tsinghua University Beijing. His main research and publications are in the areas of modal logic, temporal logic, and logic applied to philosophy, natural language, information, computation, agency, and game theory. His most recent books are *Modal Logic for Open Minds* (2010), *Logical Dynamics of Information and Interaction* (2011), and *Logic in Games* (2014). He is a member of the Dutch Royal Academy of Arts and Sciences, the Academia Europaea, and the American Academy of Arts and Sciences.

Ross Brady has held positions in Philosophy at La Trobe University, Melbourne, Australia, for 40 years, ending as Reader and Associate Professor. He retired in 2012 as an Adjunct Reader and Associate Professor. He has devoted his teaching and research career to Formal Logic and closely related fields. His main specialities have been relevant logic and paradox solution. His major works are the books, *Universal Logic*, and *Relevant Logics and their Rivals*, Vol. 2. The *Universal Logic* book contains his main philosophical position in logic, which is the avoidance of the set-theoretic and semantic paradoxes using a weak relevant logic. The book also pursues the goal of introducing a logic based on meaning containment, providing a non-ad hoc solution to the paradoxes. The other book is a continuation of the first volume, updating much of the material with help from a number of scholars in the area of relevant logic.

Janusz Czelakowski is a Polish logician and mathematician. He graduated from the University of Wrocław in 1972. In 1975 he defended his Ph.D. thesis in Mathematics at the University of Warsaw. He worked in the Polish Academy of Sciences till 1992. He is Full Professor at Opole University, Poland, and the head of Chair of Logic and Algebra in the Institute of Mathematics and Informatics. He was Visiting Professor in the Department of Mathematics of the Iowa State University in Ames, U.S.A. (1990/1991 and 1994/1995). His main interests include abstract algebraic logic, universal algebra, action theory, infinitistic methods in the theory of definitions, and algebraic methods in set theory. He is the author of almost 100 publications, including the following monographs: *Protoalgebraic Logics*, Kluwer 2001; *Freedom and Enforcement in Action. Elements of Formal Action Theory*, Springer 2015; *The Equationally Defined Commutator. A Study in Equational Logic and Algebra*, Birkhäuser 2015.

J. Michael Dunn's research has included information based logics, algebraic logic, proof theory, non-standard logics (especially, relevance logic), quantum computation/logic, and relations between logic and computer science. He has an A.B. in Philosophy, Oberlin College (1963), and a Ph.D. in Philosophy (Logic), University of Pittsburgh (1966). His career started at Wayne State University, and he retired from Indiana University Bloomington in 2007 after 38 years, where he is emeritus Oscar Ewing Professor of Philosophy, emeritus Professor of Computer Science and Informatics, and was founding dean of the School of Informatics. He is an author or editor of 6 books and over 100 articles. He has been an editor of *The Journal of Symbolic Logic* and chief editor of *The Journal of Philosophical Logic*. He has received many awards, and is a Fellow of the American Academy of Arts and Sciences.

Robert Goldblatt is Professor of Mathematics at the Victoria University of Wellington. His research has been mainly in algebraic logic and the model theory of modal and other nonclassical logics. His Ph.D. thesis introduced the use of category-theoretic duality to describe the relationship between algebraic semantics and Kripke-style structural semantics. In addition to many articles, he is the author of seven books *Topoi: The Categorical Analysis of Logic*, *Axiomatising the Logic of Computer Programming*, *Logics of Time and Computation*, *Orthogonality and Spacetime Geometry*, *Lectures on the Hyperreals*, and *Quantifiers, Propositions and Identity*. He has been coordinating editor of the *Journal of Symbolic Logic* and was a managing editor of *Studia Logica* for twenty years.

Matt Grice completed his BSc Honors at the Victoria University of Wellington in 2014 and received the University Medal for the top student graduating with that degree across all disciplines. He is now a research student at the University of Waterloo, Ontario.

William Harrison received his BA in Mathematics from Berkeley in 1986 and his doctorate from the University of Illinois at Urbana–Champaign in 2001 in Computer Science. From 2000 to 2003, he was a postdoctoral research associate at the Oregon Graduate Institute in Portland, Oregon, where he was a member of the Programatica project. Dr. Harrison is an associate professor in the Computer Science Department at the University of Missouri, where he has been since 2003. In December 2007, he received the career award from the National Science Foundation. His interests include all aspects of programming languages research (e.g., language-based computer security, semantics, design and implementation), formal methods, and malware analysis.

Chrysafis Hartonas received his Bachelor's degree (1978) in Architecture and Engineering from the Aristotle University of Thessaloniki, Greece. He obtained a D.E.A. in Sociology, a D.E.A. in the History of Sciences in Paris, France (1984), then moved to the U.S.A. (1986) pursuing graduate studies in the History and Philosophy of Science at Indiana University, Bloomington. He switched to Mathematics and Philosophy, obtaining an MSc in Mathematics (1989), an MSc in

Philosophy (1990), and a double Ph.D. in Mathematics and in Philosophy (1994) supervised by J. M. Dunn, J. Barwise, and L. Moss. He worked in the Computer Science Departments of the Universities of Leicester, Manchester, Birmingham, and of the University of Sussex in Brighton, UK. He returned to Greece (1998) and thereafter holds a professorship at the University of Applied Sciences of Thessaly (TEI of Thessaly) in Greece. His research interests include Stone representation and duality, semantics for substructural logics, denotational semantics of programming languages, and computer science logic.

Larisa L. Maksimova is Principal Researcher of Sobolev Institute of Mathematics, Novosibirsk, Russia, and Professor of Novosibirsk State University. Her fields of interest are in Mathematical Logic, especially, Non-classical Logics and the Theory of Algebraic Systems. Maksimova graduated from Novosibirsk State University in 1965. She wrote her Ph.D. Thesis on "Logical calculi of rigorous implication" (1968), under the supervision of Anatoly I. Maltsev. Her Habilitation thesis is entitled "Decidable properties of superintuitionistic and modal logics" (Novosibirsk 1986).

Edwin Mares is a professor of philosophy and founding member of the Centre for Logic, Language and Computation at Victoria University of Wellington, New Zealand. He works mainly in logic and the philosophy of logic, in particular on nonclassical logics such as relevant logic. He is the author of more than sixty articles and three books, including *Relevant Logic: A Philosophical Interpretation* (Cambridge University Press 2004) and *A Priori* (Acumen/Routledge 2011). He is also the managing editor of the *Australasian Journal of Logic.*

Chris Mortensen was born in Rockhampton Queensland. He received his BA in Philosophy from the University of Queensland, his Ph.D. in Philosophy and DSc in Mathematics from the University of Adelaide. Mortensen is the author of the books *Inconsistent Mathematics* (Kluwer 1995) and *Inconsistent Geometry* (College 2010). His research interests include inconsistent mathematics, impossible figures, logic, philosophy of science, metaphysics, Buddhism, and ancient philosophy.

Lawrence S. Moss is Professor of Mathematics at Indiana University, Bloomington. He is also Director of Graduate Studies in Cognitive Science, and Adjunct Professor in Computer Science, Informatics, Linguistics, and Philosophy. His BA and MA degrees were from UCLA, and he earned his Ph.D. degree from there in 1984. He held positions at Stanford University's Center for the Study of Language and Information, the University of Michigan, and the IBM Thomas J. Watson Research Center before moving to Indiana. He served as chair of the Steering Committee of the North American Summer School in Logic, Language, and Information and has chaired several conference program committees. He is (co-)author of over 90 publications. These include the 1996 book *Vicious Circles* (with Jon Barwise), as well as articles and handbook chapters on natural logic, coalgebra, modal logic, dynamic epistemic logic, natural language semantics, recursion, and other topics.

Sebastian Sequoiah-Grayson is a Visiting Fellow at the Sydney Centre for the Foundations of Science at the University of Sydney. He completed his DPhil under Timothy Williamson and Luciano Floridi at Balliol College with the Faculty of Philosophy at the University of Oxford. Sebastian's research reaches across Formal Philosophy, especially Formal Epistemology, Philosophical Logic, Philosophy of Language, and Philosophy of Mathematics. He is especially interested in sub-structural logic approaches to modeling dynamic epistemic actions.

Yaroslav Shramko is currently Professor of Philosophy and Logic at the Kryvyi Rih Pedagogical University, Ukraine. He earned his Ph.D. in Logic from the Lomonosov Moscow State University in 1990, working with Prof. Eugeny Voishvillo. In 1998 he was awarded a Doctor of Science (DSc) degree from the Institute of Philosophy, National Academy of Sciences of Ukraine. He was Research Fellow at the Institute of Philosophy, Humboldt University of Berlin (1996–1998), and at the Institute of Philosophy, University of Uppsala (1997). He was on a Research Fulbright at Indiana University, Department of Philosophy, working with Prof. J. Michael Dunn, 1999–2000. He received the Friedrich Wilhelm Bessel Research Award of the Alexander von Humboldt Foundation for a research stay by the Dresden University of Technology, Institute of Philosophy, 2003–2004, with Prof. Heinrich Wansing. He is a member of the editorial boards of the "European Journal of Mathematics," "Bulletin of the Section of Logic," and "Logic and Logical Philosophy."

Dolph Ulrich ("Ted"), now Professor Emeritus at Purdue University, was born and raised near Tuscarawas, Ohio. He received his A.B. degree in 1963 from Oberlin College, where he and Michael Dunn met as undergraduates and began what became a lifelong friendship. In 1968, he received the first Ph.D. degree granted by the philosophy department at Wayne State University in Detroit. Coincidentally, Dunn started his own academic career at Wayne State in 1966, and consequently served as Ulrich's dissertation director. Ulrich joined the philosophy department at Purdue in 1967. His main research interests concern matrices for sentential calculi, the finite model property, semantics for modal logics and for their implicational fragments, and related open problems posed in the literature. Since 1999, he has been searching as well for short bases and single axioms for various sentential systems.

Alasdair Urquhart is Professor Emeritus of Philosophy and Computer Science at the University of Toronto. His doctoral dissertation "The Semantics of Entailment" (University of Pittsburgh, 1973) was supervised by Nuel D. Belnap. He taught at the University of Toronto from 1970 to 2007 and has published in the areas of nonclassical logic, lattice theory, proof complexity, and history and philosophy of logic. He has served as managing editor of the reviews section of the *Bulletin of Symbolic Logic*, as well as editor of the *Journal of Symbolic Logic* and the *Review of Symbolic Logic*. In 2013, he was elected President of the Association for Symbolic Logic.

Heinrich Wansing is a professor of Logic and Epistemology at the Ruhr-University Bochum (Germany). Before that he was a professor of Philosophy of Science and Logic at Dresden University of Technology (1999–2010). He is the author of *The Logic of Information Structures* (Springer 1993), *Displaying Modal Logic* (Kluwer 1998), *Truth and Falsehood. An Inquiry into Generalized Logical Values* (with Y. Shramko, Springer 2011), *Proof Theory of N4-related Paraconsistent Logics* (with N. Kamide, College Publications, 2015), and numerous articles in professional journals. Heinrich Wansing has been working mainly on philosophical logic, including the semantics and proof theory of modal, constructive, paraconsistent, many-valued, and other nonclassical logics. Moreover, he is the editor-in-chief of the book series *Trends in Logic* (Springer), a managing editor of the journal *Studia Logica*, and a member of a number of other editorial boards of logic and philosophy journals.

Chunlai Zhou is currently associate professor in the Computer Science Department of Renmin University of China, Beijing. He has been working there since 2009. His research interests include Markov processes, probabilistic logic and reasoning, and also other uncertain reasoning. He received his Ph.D. in Mathematics from Indiana University, Bloomington in 2007 (his supervisors were J. Michael Dunn and Larry Moss). From 2007 to 2009, he was a postdoctoral fellow at Computer Science Department of Tsinghua University, Beijing.

An Engineer in Philosopher's Clothing

> I am convinced that the act of thinking logically cannot possibly be natural to the human mind. If it were, then mathematics would be everybody's easiest course at school and our species would not have taken several millennia to figure out the scientific method.
>
> —Neil deGrasse Tyson, *The Sky is Not the Limit: Adventures of an Urban Astrophysicist*

David Brooks, the NY Times columnist, has recently made the distinction between "résumé virtues" and "obituary virtues," saying that the résumé virtues are the skills you bring to the marketplace, whereas the eulogy virtues are the ones that are talked about at your funeral.[1] This autobiography is neither a resume nor an obituary, since I am at a point in my life when (I hope) neither is appropriate. Also since I am a Midwesterner, I am supposed to be humble (as fans of Garrison Keillor's Lake Wobegon will know). But I will do the best I can to tell an interesting story, which I confess will mean bragging, at least a little, about my virtues of both kinds. I will also do a lot of name dropping, because this is my opportunity to in effect thank a lot of people for their support. I of course will accidentally overlook somebody, and if that somebody is you, I deeply apologize.

I will also use this "autobio" a bit as a bully pulpit, to share with you my developing views about logic. The introductory quote from Neil deGrasse Tyson is very relevant to these views. As I shall reveal further on, I started out to become an engineer, changed to science, and ended up in philosophy. There is a saying that has guided me, but which I cannot reference (this is not a criticism of Google). "Philosophy is the art of the sciences, and the science of the arts." I want to be clear that I truly value the arts (which include in this context the humanities). I was indeed once executive associate dean of the College of Arts and Sciences at Indiana University. And I have always had a strong inclination toward the sciences—I don't think I have published a single paper which doesn't have a strong sprinkling of

[1] *The Road to Virtue*, Random House, New York, 2015.

mathematical symbols in it.[2] But the hidden engineer in me is perhaps my best kept secret. We will get back to this later.

Growing Up in Ft. Wayne and Then Lafayette, Indiana: On the Banks of the Wabash. *On the Banks of the Wabash, Far Away* is the official Indiana state song. It begins:

> Round my Indiana homestead wave the cornfields,
> In the distance loom the woodlands clear and cool.
> Oftentimes my thoughts revert to scenes of childhood,
> Where I first received my lessons, nature's school.

I was born in Fort Wayne, Indiana, June 19, 1941. Fort Wayne is not really on the Wabash, although it is close. There were no cornfields or woodlands either in this city, though I enjoyed these on visits to my grandparents' farm. My father had moved off of his family's farm south of Indianapolis to take a job with John Deere's warehouse center in Fort Wayne. Family legend has it that he started as a janitor during the depression, and I know he worked his way up to become the assistant director of that center. My mother had been an accountant but gave up her career to become a "full-time mom" for my two siblings and me. I lived in Ft. Wayne until my father moved to Lafayette, Indiana, to take up the position of territory manager.

There my mother became active in politics and ended up as chair of the Democratic Party of Tippecanoe County. My father was a Republican so I was the product of a mixed marriage. It was also a mixed marriage in the more traditional sense, in that my mother was a devout Catholic and my father seemed pretty much irreligious. I had gone to a parochial school (St. Jude's) in Ft. Wayne, but there was no parochial school in Lafayette when we moved there, and so I went to the public high school Lafayette Jefferson. I think this might have been the best combination I could have had to pursue a career in philosophy, and maybe especially logic.

My first eight years of education, almost entirely with nuns as my teachers, was somewhat stereotypical (even down to the nun marching up and down the aisles to catch students who were daydreaming, occasionally me). But I got a first-rate education, and my philosophical and logical education was largely based in catechism classes. There was lots of room for discussion and argumentation in those classes, and paradoxically I have to say that the catechism classes probably led to my atheism.

My childhood dream was to become an aeronautical engineer. I made countless model planes. My "plan" was to study engineering in college, and then join the Air Force and learn to be a pilot, even a test pilot. But I was saved from engineering by science.

---

[2] I think the only paper without symbols that I tried to publish was a paper titled "Wittgensteinian Scepticism" that I sent to *Mind* while I was in graduate school. I argued that certain arguments from Wittgenstein and his followers (notably Normal Malcolm) were essentially old sceptical arguments in new clothing. I got an immediate rejection from then editor Gilbert Ryle, saying that although my paper contained some admirably pointed barbs, he thought his readers were getting tired of Wittgenstein.

I had the good fortune to take a biology course during my first year in high school from Mr. D.O. Neidigh. He encouraged me to create a Science Fair project, and I was fortunate enough to not only be a winner, but also to attract the attention of Henry Koffler, then chair of the Biology Department at Purdue University. He was a microbiologist and a pioneer in the emerging area of molecular biology—he ended up as President of the University of Arizona. He hired me to work in various labs, most of them associated with a project to study how bacteria flagella move. Henry (I can now call him that) was my mentor and a tremendous influence on my life. I continued to work in his labs throughout high-school summers and after school, and then for two summers after I graduated and went off to Oberlin College. I chose Oberlin at Henry's advice because of its record in turning out graduates who went on to graduate schools and scientific careers.

I also want to acknowledge the great help of Mr. R.W. Levering, physics teacher and academic counsellor, who helped me in various ways on my way to college, and also encouraged me in Science Fair projects. I was a first-generation college student, something I think I have not previously mentioned publicly in my career, and this support and encouragement were very critical. Also my parents' support was very important. As one small example, I remember my father driving me over to Purdue to borrow some equipment for my first Science Fair project.

Going to College: Oberlin, Ohio. When I went to Oberlin I of course took freshmen chemistry since Henry told me a major in chemistry was appropriate for someone intending to become a molecular biologist. But also, to meet a general education requirement, I took an introductory philosophy course.[3] I thus took the first step toward choosing a new career goal, and I took the second step in my sophomore year when I took a course in symbolic logic from Bruce Aune. I remember meeting with Henry to tell him that I was going to major in philosophy and study logic. He suggested to me that Watson and Crick's work on DNA suggested there might be a connection between logic and biology, but I politely (I hope) said that I was interested in pure logic.

I took every course related to logic that Oberlin offered, both in the Philosophy and Mathematics Departments. One of the most memorable of these was taught by Robert Stoll. It was based on a draft of his book *Set Theory and Logic*. The class had maybe a dozen students, but they included the future logicians Peter Woodruff and Jonathan Seldin, and my future wife Sally Dunn (a math major). The draft had a number of errors, as every manuscript does. Ironically, I think we all learned a lot from trying to catch them all.

I also took a number of "private reading" courses in logic with Daniel Merrill, and under his supervision wrote my senior honors thesis on "Logical Behaviorism," i.e., the question of whether logical connectives can be defined by their role in inference. I like to tell the story that my interests at Oberlin developed as I was looking for foundations: biology depends on chemistry, which depends on physics,

[3]Taught by Roger Buck, later to be my colleague at Indiana University.

which depends on math, which depends on philosophy, which involves logic. Something like that is true but I had diversionary interests in political science (with an eye on the law) and poetry. I almost dropped out in the middle of my first year of graduate school to write a novel.

Graduate School: University of Pittsburgh. I was fortunate to be a graduate student in Philosophy at Pitt in the mid-1960s when Pittsburgh was not just the home to the steel industry, but also to relevance logic. Once upon a time the strongest tools were built out of stone (Aristotelean logic), then later came iron (classical logic), and then steel and various substructural alloys (intuitionistic logic, relevance logic, and various other nonclassical logics).

Bruce Aune, a teacher of mine in philosophy of mind at Oberlin, was an important influence in my decision to go to the University of Pittsburgh for graduate study. Bruce went to Pittsburgh as a faculty member in the same year (1963) and he was passing information on to me and a number of other students at Oberlin. Through him we heard the news that both Nuel Belnap and Wilfrid Sellars, along with Jerry Schneewind, were leaving Yale to join the Pitt Philosophy Department (Kurt Baier, Adolph Grünbaum, and Nicholas Rescher were already there). This fit my interests really well since I was somewhat undetermined as to whether I would specialize in logic or philosophy of mind. Philosophy of mind was a strong interest of mine when I was an undergraduate rebelling against behaviorism both in psychology and philosophy. One of the reasons I chose to go to the University of Pittsburgh for my graduate studies was I felt they had the right faculty members no matter which direction I finally chose to go. Not only I, but five other graduates of Oberlin went to do graduate study in Philosophy at Pitt in the fall of 1963.

I was fortunate to have a Woodrow Wilson Fellowship for that first year, and Pitt topped in off with a guarantee of graduate support for several years after that. My interest in logic exploded after I took a course from Nuel Belnap on the logic of questions in my first semester. At that time Philosophy was located in a building that was a remodeling of the old Schenley Hotel. I shared an office with fellow graduate student Tryg Ager, and we had our own bathroom since our office was once a hotel room. The office was right down the hall from Nuel's. He would often stop by, at least several times a week, and ask me to prove something for him. Also I felt free to look in on him and see what he was working on, and he would often stop me as I walked by his door.

Nuel Belnap was a huge positive influence when I was a student, and still to this day. I remember a fellow graduate student Richard Schuldenfrei referring to him as "Nuel Call-Me-Nuel," because of Nuel's practice of wanting students to call him by his first name. To give credit where credit is due, this was also a practice of Nuel's own teacher Alan Ross Anderson. Alan moved from Yale to Pitt in my last year there, and shortly after his arrival invited me to lunch. He told me during lunch that there were two things he didn't like in a student: first, if the student didn't call him by his first name, and second, if the student didn't tell him when he was being stupid. I said something as we parted like "Thanks for lunch Professor Anderson." About a week later he asked me to lunch again, and told me the same thing. I said

"Which would you prefer Professor Anderson, that I call you Alan or that I call you stupid." He replied, "Please call me Alan," which I did.

Another Pitt faculty member important to my career was Storrs McCall. I took his logic seminar in my first semester at Pitt and it was in writing a paper for this course where I was trying to show that the implicational fragment of E was equivalent to the intersections of that of S4 and that of some other logic. (Nuel Belnap had already shown that the other logic was not R.) Storrs told me about the "mingle rule" of Onishi and Matsumoto, and suggested that I add this to E. This ultimately led to my exploring the system obtained by adding this rule to the relevance logic R, and ultimately to my paper showing that every normal extension of R-Mingle had a finite characteristic matrix (Dunn 1970). Bob Meyer and I (1971) were able to show a similar result concerning the intuitionistic-style logic of Dummett's LC.

I was very lucky to have some other excellent fellow students working with Nuel. Besides Bob Meyer (Robert K. Meyer—he always insisted on being called "Bob"), there were Louis Goble and Peter Woodruff, among others. Curiously, both had been students at Oberlin with me.

Another person who clearly helped was Nuel himself. I had a Woodrow Wilson Dissertation Fellowship for my fourth year at Pitt. Later, when I discuss my research career, I will discuss some of what was in my dissertation, but perhaps more important than what my dissertation contained is something that it did not contain. My plan was for the final chapter to show how the logics E and R are decidable, and I was looking forward to showing this in my fourth year. But Nuel told me in my third year that I had done enough. This was a very wise move on his part, since some 18 years later Alasdair Urquhart (another of Nuel's students, who came to Pitt soon after I finished) would show that both E and R (and a wide range of other logics) were undecidable.

First Academic Position: Wayne State University. In 1966 I received a Ph.D. in Philosophy (Logic). That is literally the way it reads. This was back in the "old days," and I am sure with the help of Nuel I got two job offers before I had even defended my dissertation, and I accepted the one from Wayne State University. George Nakhnikian had built a legendary philosophy department there, and I started my career with colleagues such as Hector Castañeda, Edmund Gettier, Richard Cartwright, Alvin Plantinga, Robert Sleigh, Larry Powers (and Arthur Danto, Henry Kyburg, Keith Lehrer, had recently left, I think not because they heard rumors of my coming). My interview there was scary, as were the questions after my talk. Richard Cartwright asked a round of questions, and I tried to answer them one after the other. I wasn't particularly happy with my final answer, and I thought he likely wasn't either. But I was told by someone that in the meeting afterwards he said that he hadn't expected anyone to keep up with his questions far as I had, and I should get the job. Talks at Wayne State were like that, the question period being if anything more interesting than the talk. Roderick Chisholm used to regularly give talks at Wayne as a kind of preparation for the final version of his papers. My fellow junior faculty member Larry Powers was particularly famous for his questions.

I was very happy at Wayne. Echoing Al Plantinga in his "Profiles" book, life at Wayne State was one ongoing philosophical discussion, moving from office to coffee to lunch back to office, etc., interrupted only by teaching. Actually the teaching there was great too, and I had many excellent students, both graduate and undergraduate.

But all good things have to end, and the Wayne State philosophers began to leave, for various individual reasons. Ed Gettier and Bob Sleigh went to University of Massachusetts—Amherst, George Nakhnikian went to Indiana University to rebuild its Philosophy Department and took Hector Castañeda and Nino Cocchiarella with him, Dick Cartwright went to M.I.T. while his wife Helen got a permanent position at Tufts, etc. As luck would have it Yale showed an interest in recruiting me and I took a visiting position in Philosophy there, attracted by the fact that both Nuel and Alan had been at Yale. I guess I thought I was going to sit and watch to see whether U. Mass or IU was the new Wayne State department.

For various reasons I was not real happy with the Philosophy Department at Yale. I was treated well enough, in fact far better than the average junior faculty member. But that was part of the problem. All philosophy faculty were treated as equals at Wayne State, but at Yale I found a very hierarchical system. There was much discussion among junior faculty about who went to a dinner at the chair's home, who was invited to Mory's for drinks with a colloquium speaker, etc. And this was made worse by the fact that almost all of the faculty had their PhDs from Yale. I remember Fred Fitch, who I think was optimistic about my staying at Yale, telling me how he had spent his whole time since he was a student at Yale, and how when he once visited someplace he came to understand that he would never want to leave. For me it was quite the opposite, and when I ended up getting offers from both U. Mass and Indiana and I decided (with some difficulty) to accept the Indiana offer, I never even bothered to try to negotiate with the Philosophy Chair at Yale. But I made some good friends among my colleagues at Yale, including Fred Fitch, Bob Fogelin, Rich Thomason, and Bruce Kuklick—who later wrote "Philosophy at Yale in the Century after Darwin," *History of Philosophy Quarterly*, (July 2004), **21** (3): 313–336.

Second (and Last Academic Position): Back Home Again in Indiana. *Back Home Again in Indiana* is in effect the unofficial state song, and it certainly characterizes me. I went to Indiana University as an associate professor of philosophy in August 1969 (and was promoted to full professor in 1976). Having grown up in Indiana, I never expected to end up living in Indiana again and was pleased to find that southern Indiana had many more woodlands than cornfields, unlike the northern Indiana I was used to, and that Bloomington was a kind of oasis, culturally and politically. The pattern looked like 2 years at Wayne State, 1 year at Yale, etc. But this is where Sally and I have been ever since, even after our retirements.

When I joined the Philosophy Department at IU I not only found my old colleagues from Wayne State there (Castañeda, Cocchiarella, and Nakhnikian), but also Bob Meyer had been recruited there. As fellow logicians I not only had Bob and Nino, but soon also in the new Computer Science Department there were

George Epstein, Dan Friedman, Stu Shapiro, and Mitch Wand, who all had logic-related aspects to their work. This turned out to be very important to my development. George was an important figure in the multiple-valued logic community, Dan and Mitch's work on programming languages had a strong influence from the λ-calculus, and Stu was involved with AI and knowledge representation. And several years later Doug Hofstadter (AI, soon then to publish *Gödel, Escher, Bach*) and Ed Robertson (database theorist) joined the Computer Science Department.

George was the chair for the Fifth International Symposium on Multiple Valued Logic (1975), and he arranged for me to be program chair. I learned a secret thereby, which I have shared with a number of faculty involved in organizing a conference. You know everything that goes wrong, even if everyone else is ignorant and enjoying themselves immensely. I also should mention George for introducing me to Paul Erdős. Sally and I had invited George and his wife for dinner at our house, and at almost the last minute Erdős (I don't dare refer to him by his first name) called George and asked if he could give a lecture at IU. George of course said yes, and Sally and I included Erdős in our dinner. It was a wonderful evening. Erdős told us that he was funded by the Hungarian NSF to travel and stay out of Hungary. He clearly was a practiced guest. He had clearly read the local newspaper and discussed politics in Indiana, did a couple of magic tricks for our children, and finally would not let Sally and me near the kitchen as he cleaned up after dinner. I remember mentioning a couple of open problems in relevance logic, but he didn't bite. He clearly had larger fish to try. Therefore I missed my chance to have the Erdős number one, and had to content myself with three (Erdős, Joel H. Spencer, Belnap, Dunn, or Erdős, Marcel Erné, Mai Gehrke, Dunn).

I have been involved in multidisciplinary activities, possibly because logic is one of those areas that cannot be neatly pigeonholed into the usual academic departments. In fact, when I was a visiting assistant professor at Yale, my position was funded by the Departments of Philosophy, Electrical Engineering, and Linguistics, and I presented several lectures in a Mathematics seminar led by the famous logician Abraham Robinson.

When the computer science department installed its first time-sharing computer system (VAX 11/780—I believe in 1978), I was given a Unix account by Ed Robertson, a database theorist and then chair of the Computer Science Department. A few weeks later, when Ed asked how I liked Unix, eyes sparkling, I responded: "All my career I've studied formal systems, and now at last I have one that is truly responsive."

I had a number of opportunities to leave IU, but my principle was to pursue them only if I was prepared to leave. I believe I took only two offers seriously. One of these occurred in the late 1980s, and the dean of Arts and Sciences said, when I met with him, "I don't think I have ever seen one like this before." Although I was fully prepared to leave, IU responded in the most generous way possible and I stayed. This had something to do with the schools, housing, and traffic near the competing institution, but it is not a coincidence that the year 1989 was particularly eventful, in titles. I became the Oscar Ewing Professor of Philosophy and a

Professor of Computer Science. And the generous response allowed me to recruit a number of excellent logicians to IU including Jon Barwise, Anil Gupta, and Alice ter Meulen. One good thing leads to another and Jon was able to recruit Larry Moss and Slawomir Solecki in Mathematics. We already had Bill Wheeler in Mathematics, Daniel Leivant in Computer Science, and in Philosophy Nino Cocchiarella, David McCarty, and Raymond Smullyan, not to mention Geoffrey Hellman (philosophy of mathematics), Ed Martin (Frege), and Paul Spade (medieval logic). So IUB had at that time one of the best programs in logic anywhere. Jon was the first director of what is now the Program in Pure and Applied Logic (the director is now Larry Moss).

I somehow have been successful in combining research, teaching, and service. My research is reflected in more than 100 publications, including five coauthored books and over 150 talks at conferences and universities. But when I formally retired from Indiana University in 2007 I was amazed to count back and find that I had spent over half of my 38 years there as an administrator at the chair level or above. I was twice chair of the Department of Philosophy, and in the early 1990s served as the first executive associate dean of the College of Arts and Sciences. And I ended my official career at IU as the founding dean of the School of Informatics (now the School of Informatics and Computing), then the first completely new school at IU in a quarter of a century. I must like to start things since I was also involved in the creation of the Cognitive Science Program, and also the Program in Pure and Applied Logic.

Besides my official administrative service, I led or served on numerous faculty committees at all levels (over 75 campus and university committees alone). In particular I have been on practically every committee having to do with computing at IU, from chairing the campus word processing committee 1985 to chairing the university's Information Technology Committee, which put together the strategic plan for IT at IU in 1999.

Perhaps the most frustrating committee I have ever served on was the Campus Calendar Committee. This was early in my career and it took me a while to figure out that I was probably the only one on the committee without a vested interest. The other members represented athletics, student housing, the laboratory sciences, anything where schedules really mattered. My wife Sally later served on the same committee and had much the same experience.

My academic service outside of IU included being an editor of the *Journal of Symbolic Logic* and the *Journal of Philosophical Logic* (and on the editorial boards of a number of other journals). Also, I was president of the Society for Exact Philosophy and vice chair of the Computing Research Association's IT Deans Group. I accepted the Mira Award from TechPoint (the Indiana state IT association) in 2002 for the School of Informatics. I also received the i-School Caucus's "Bookends Award" for "vision and pioneering leadership in the formation of the i-Schools community." I have been on external review committees for a number of universities, and served on academic advisory boards for Carnegie Mellon University, Spelman College, the University of Cincinnati, and the University of Dubai.

I was elected (2010) a fellow of the American Academy of Arts and Sciences. My other favorite honors include being awarded the IUB Provost's Medal and being appointed as a Sagamore of the Wabash by the Governor of Indiana. (From Wikipedia: "Sagamore was the term used by Algonquian-speaking American Indian tribes of the northeastern United States for the tribal chiefs.")

Occasionally campus conversations became confused at the mention of "Dean Dunn," a confusion that arises because my wife, Sally Dunn was dean of the freshman division. The confusion continues with "Jon Dunn" as our son, Jon William Butcher Dunn, is director of technology and an assistant dean for the IUB Library. Our daughter, Jennifer Knight Dunn, works as a senior geographic information specialist, and both of our grandchildren love math—so there must be something in the genes.

Somewhat ironically, given my interest in technology, my wife and I live in a 100-year old house, and I have been active in historic and neighborhood preservation. I have also been involved in other forms of civic service, for example, serving on strategic planning task forces of the Indiana Chamber of Commerce and the Indiana Health Industry Forum. Most recently, I serve on the Board of HealthLINC (appropriately for my academic interests, the regional health information exchange) and was president of the board for 3 years.

I directed 17 Ph.D. dissertations. It is difficult to talk about my graduate students because I love each and every one of them, just like my children (but of course there were many more graduate students). Some of these did work not really related to my own, and I was really just their "supervisor." Others did work closely related, even joint with my own. Some of these were invited to contribute to this volume, and I greatly appreciate their contributions and will let them speak in their own words. The dissertations I directed were mostly in philosophy but also in computer science and mathematics. All but three of these were in logic and the other three in philosophy of mind (reflecting my own ambivalence, as indicated earlier, between these two subjects). 14 of those now hold permanent positions at universities, two have equivalent positions in government labs, and 1 is an IT entrepreneur (who also published a book on philosophy of mind with Oxford University Press). I am very proud of them all.

Around the World. We have been residents of Indiana since 1969, but this has not prevented Sally and I from travelling the world together, and with our two children when they were young. I have been a visitor at a number of different universities. In 1975–1976 I went on a research Fulbright with my family to the Australian National University. This was my first of a number of visits to Australia, and it was a tremendous experience. Bob Meyer, Richard Routley (later Sylvan), and Val Routley (later Plumwood) were there, and Len Goddard and Nuel Belnap also were visitors (Nuel just for a month). Errol Martin and Michael McRobbie were among the postgraduate students. Michael now claims to be a Hoosier, and indeed he is President of Indiana University.

It was an absolutely incredible environment for doing relevance logic. I returned to Australia in 1983 to teach a seminar at the University of Melbourne, and again found an incredible environment for relevance logic, with both Len Goddard and Michael McRobbie at Uni Melbourne, Ross Brady at LaTrobe Uni, and Lloyd Humberstone at Monash Uni. In between I spent close to a year (1978) in Oxford with my family, and courtesy of Dana Scott visited the Mathematical Institute. Among the Oxford faculty in logic, besides Dana, were Robin Gandy, Michael Dummett, and Dan Isaacson, and visitors that year included Dov Gabbay, Rob Goldblatt, Saul Kripke, and Jonathan Seldin. I have also visited the University of Pittsburgh twice, once as a visiting fellow at the Center for Philosophy of Science and most recently (after I retired) as a visiting professor. It was a real privilege to teach in the seminar room that was sadly graced with photographs of many of my deceased teachers, and to rekindle by friendships with Nuel Belnap, Anil Gupta, and others.

In 1985, with the support of a fellowship from the American Council of Learned Societies, I visited the University of Massachusetts—Amherst. It was great to see my former teacher Bruce Aune, and my former colleagues Ed Gettier and Bob Sleigh, but the purpose of my visit was to spend time with Gary Hardegree to finish the book *Algebraic Methods in Philosophical Logic* which we had begun when Gary visited Indiana a couple of years earlier.

Since my formal retirement in 2007 I have continued to attend conferences and speak at universities, often combining this with some tourism. In the first 6 months of 2015 for example Sally and I travelled to India, England, and Russia.

Research: Pencil, Paper, and a Wastebasket. There is the old joke that mathematics is cheap to fund because all you need is a pencil, paper, and a wastepaper basket. Only one field is cheaper to fund, namely philosophy. Because there you can do without the wastepaper basket. Since logic has a part of mathematics and a part of philosophy in it, I am not sure whether I needed the wastepaper basket. But I do know that it would have been hard to combine my research and administrative careers (teaching too) if I had to manage a laboratory, visit distant archives, etc. Being able to do my work with pencil and paper (and more lately a computer, which fortunately has a delete key) was a huge plus. Also it helped with my multitasking—I could be in a meeting, seeming to take scrupulous notes, when in fact I was trying to prove a new theorem.

I had some early publication problems. The first was that Nuel Belnap suggested that I submit my dissertation to the North-Holland series: Studies in Logic and the Foundations of Mathematics. This was very famous at the time, and so I did so quite willingly. After considering it a year or two they replied that they were no longer publishing dissertations. I think this policy changed again a year or two later. Murphy's Law! In the meantime I had not done the usual thing of submitting various chapters (rewritten of course) as articles to journals. Instead, I accepted the opportunity to have some of these chapters published as a "contributing author" to Anderson and Belnap's *Entailment: The Logic of Relevance and Necessity*, vol. 1. This was a great opportunity, but I am afraid that it meant that many of my original contributions got a bit lost in the literature. I was more than paid back when I became a "first class" author of Anderson, Belnap, and Dunn's *Entailment: The Logic of Relevance and Necessity*, vol. 2.

My dissertation (Dunn 1966), *The Algebra of Intensional Logics*, was about the algebraic treatment of relevance logics. The name itself was a marketing error (the word "intensional" had not yet become proprietary to modal logic). It would have been better to call it simply "The Algebra of Relevance Logics." The two major themes in the dissertation were algebraic foundations for the algebras of the relevance logics E and R, and in particular their first-degree entailments (entailments between truth-functional formulas). The first rested on viewing implication as residuation on an underlying monoid (in the case of R what I labeled as a De Morgan monoid) and anticipated algebraic treatments by others of various so-called substructural logics. Ultimately this led to my super-generalization called generalized Galois logics (or Gaggles). More on these later.

The second theme came from viewing first-degree entailments as corresponding to quasi-Boolean algebras (Białynicki-Birula and Rasiowa) or equivalently De Morgan lattices (Monteiro). Białynicki-Birula and Rasiowa (1957)[4] gave an interesting representation of these lattices, and I gave an equivalent representation together with an interpretation of that representation. These, and other representations, were published by me only as an abstract (Dunn 1967b). As it turns out the representation of Białynicki-Birula and Rasiowa was in effect published by Richard and Valerie Routley in 1972 as a semantics for first-degree entailments (with no reference to Białynicki-Birula, Rasiowa, or myself).[5]

The story of my own representation is more complicated. In my dissertation (1966) I had a result where each element of a De Morgan lattice was to be viewed as a pair of sets ($X+$, $X-$), and thus indirectly this was an assignment to a sentence. $X+$ was the "topics" that the sentence gave positive information about, and $X-$ was the set of "topics" that the sentence gave negative information about. You can see from this the beginnings of my interest in informational semantics for logics.

I pointed out that a sentence could give both positive and negative information about the same topic, as well as giving neither. But I did not have the nerve, though it crossed my mind, to speak publicly of a sentence being both true and false (or neither, though this was less controversial because of the Łukasiewicz 3-valued logic). I finally took this public position in my talk in a joint symposium of the American Philosophical Association in 1969 and the Association for Symbolic Logic on "Natural Language vs. Formal Language." Because of the nature of the symposium, I gave this talk and wrote the accompanying paper (unpublished, available online at www.philosophy.indiana.edu/people/papers/natvsformal.pdf) in a very philosophical style—almost no symbols and certainly no mention of the 4-valued lattice that was so much a centerpiece of my dissertation. I thought this was a good way to help "sell" the ideas, but perversely it got in the way of

[4]Białynicki-Birula, A. and Rasiowa H. (1957). On the representation of quasi-Boolean algebras, *Bulletin de l'Académie Polonaise des Sciences* **5**: 259–261.

[5]Routley, R. and Routley, V. (1972). The semantics of first degree entailment, *Noûs* **6**: 335–359.

connecting those ideas to the algebraic way of looking at things to be made popular by Nuel Belnap.

For no sooner had I done this than Nuel Belnap published his own version of the 4-valued semantics in two separate venues ("How a Computer Should Think," and "A Useful Four-valued Logic"). In each of these he carefully cited my work, but almost no one seemed to notice and there was much talk of Belnap's 4-valued logic. Over the years this has changed, and now I often see mention of the "Belnap–Dunn logic," or even the "Dunn–Belnap logic." I do not mean to suggest that the name Belnap not be linked to this logic. Nuel did many interesting things in his two papers on the 4-valued logic, both technically and promotionally, e.g., his connecting it to bilattices and his emphasizing its usefulness to computer applications. Years later, when Yaroslav Shramko and Tatsutoshi Takenaka were visitors at IU, we worked on 8- and 16-valued trilattices.

I had of course hoped that the 4-valued semantics could be extended beyond first-degree entailments to encompass the whole of the systems R and E. I even managed to do a 3-valued version of this for the logic R-mingle allowing a sentence to take both truth values, but not neither), essentially modifying the Kripke/Grzegorczyk semantics for the intuitionistic sentential calculus, so as to allow that a sentence might be both affirmed and denied in a given "evidential situation." The binary accessibility relation was interpreted as one evidential situation extending another.

I think it was the 4-valued approach, and the Białynicki-Birula and Rasiowa/Routleys approach to first-degree entailments that led me to be a bit obsessed with negation. I just did a search and found that the word "negation" occurs 16 times in my CV, including talks as well as publications. Besides the work I did in my dissertation and early on regarding negation as De Morgan complement, I discovered that the Routleys' * operator can be replaced with a binary relation of "incompatibility" (much as in the modeling of quantum logic), and that a^* can be viewed as the weakest information state compatible with a. Various properties can be put upon the incompatibility relation to get various logics. I also learned that when negation is viewed this way, the representation of both negation and implication can be viewed as falling under a common abstraction. This gives an algebraic approach to these, and other connectives and their semantics via representations of their underlying algebraic structure. This can be viewed as growing out of the work relating representations and semantics begun in my dissertation, but vastly generalized. It was also motivated as a generalization of Jónsson and Tarski's (1951–1952) work on "Boolean algebras with operators" so as to apply to relevance logic and other substructural logics.

Thus I began in Dunn, 1991, to publish on what I called generalized Galois logics, or GGLs, which I insisted should be pronounced "Gaggles," not "Giggles." (You are encouraged though to giggle here. :) This is probably the abstraction I created/discovered of which I am the most proud and it resulted in perhaps my longest thread of publications, some joint with Gerry Allwein, Kata Bimbó, Mai Gehrke, Gary Hardegree, Alessandra Palmigiano, and Chunlai Zhou.

One of the most exciting, and also frustrating, of my achievements was providing a Gentzen system for the positive fragment of the logic R of relevant implication. I had the idea of modifying the usual Gentzen sequents so as to allow for two kinds of "commas" (actually using a semi-colon for one of these), so as to mimic the two kinds of conjunctions available in R (intensional as well as the usual extensional). Anyone who has ever proved a cut elimination theorem knows the difficulty in getting every case to work, and the excitement when they all click.[6] I thought I was only another day (and maybe a night) away from adding negation, and then another day or so away from proving decidability. (Again this was before Alasdair Urquhart obtained his undecidability result.) Regarding negation, a number of other researchers worked on extending the Gentzen system for R_+ so as to include it, and certainly the most persistent of these was Ross Brady. Nuel Belnap ultimately came up with his Display Logic, which allowed for adding De Morgan negation to R_+, but there was a price to pay for this—he also had Boolean negation as part of his basic structure.

Perhaps the single publication that pleased me the most was a joint publication with Bob Meyer (Meyer and Dunn 1969) proving the admissibility of Ackermann's rule γ for the systems R and E. In his paper in this volume Alasdair Urquhart shares some things I told him, so I will not tell the stories again about how Bob and I produced the proof (working first independently and then together). Bob and I were once students together, later colleagues, and always friends, at least until his regrettable death. I think I do not reveal any secrets if I say that Bob was somewhat of a character, albeit a loving and much loved one. After Bob moved to Australia he would often visit the US, driving a rental car to see his son and his family who lived near Bloomington. Usually Bob would call us ahead of time with an hour or two's notice, and then he would drop by our house all set to do logic. Bob and I, working with Hugues Leblanc extended the admissibility of γ to first-order versions of R and E.

Writing this autobio has led me to reflect on my CV, searching for patterns. One thing I noticed is that I have done almost no work on first-order logic, classical or nonclassical. The exceptions are Dunn and Belnap (1968b), Meyer et al. (1974), Dunn (1976c), and my papers on relevant predication (Dunn 1987b, 1990a, b, c). This was at least partly a conscious decision on my part. There are so many different choices to make, e.g., constant domain, expanding domain, individuals or individual concepts, infinite meets/joins, cylindric, polyadic algebras, etc., I wanted to stick with the basics.

One of my most cited papers was a survey of relevance logic (Dunn 1986), updated in (Dunn and Restall 2002).

I have to confess that this search for patterns in my CV confirms my antecedent view that I had no grand research program. I have followed where the paths have

---

[6]I must mention that Grigori Mints did the same thing at roughly the same time, and when we met each other for the first time many years later he agreed with me about both the excitement and the frustration.

led me. I was very fortunate to have had the teachers, colleagues, and students I have had. It is only with reflection, when asked what the title should be for this volume, that I suggested "Information Based Logics." Information has been a common theme throughout much of my research, but it was never intended as a programmatic theme. That the concept of information turned out to be so useful in itself proves its importance, at least to me.

I have also worked on quantum logic (Dunn 1981, 1988), and particularly in recent years on the relationship of quantum logic to quantum computation (Dunn et al. 2005; Dunn et al. 2013). Representations play an important role again in these last two. Zhenghan Wang was a faculty member in the IU Mathematics Department, who is now a lead researcher in Microsoft Research's "Station Q" project to build a topological quantum computer.

I have done a little to advance the actual application of relevance logics (you would think they would be "relevant" to something :)), particularly the system R. Perhaps my most sustained attempt (four papers around 1990) was to use relevance logic as a means of defining relevant predication and then using that to define intrinsic properties, essential properties, and internal relations. I also got involved in the idea, pushed by Bob Meyer, of basing mathematics, particularly arithmetic, on relevance logic (Dunn 1979b, c). But despite many interesting results by Bob Meyer, Ross Brady, Chris Mortensen, Zach Weber, myself, and others, it still seems to me to be a not very well-developed area.

My dream application would be to have some version of relevance logic undergirding searches on the Web (where notoriously one can find any side of a question that one wishes). I have written about this in my recent paper (Dunn 2010) "Contradictory Information: Too Much of a Good Thing," and offered my own preliminary thoughts about how this might be done.

Another small theme in my work has to do with relation algebras. In Dunn (1982b) I showed how De Morgan lattices (I called them "quasi-Boolean algebras" following Białynicki-Birula and Rasiowa) could be represented as relations closed under intersection, union, and complement of converse. In Dunn (2001b) I in effect extended this using the idea of the Routley–Meyer ternary accessibility relation to show how the relation algebras that satisfy Tarski's equations for relation algebras can be represented using this relation. The representation ends up showing that relation algebras can be represented as sets of relations, not as relations themselves, since a well-known result of Lyndon showed that relation algebras cannot be represented in the natural way as relations. In Dunn (2014b) I related this to work by Johan van Benthem and Yde Venema on what they dubbed as "Arrow Logics." I had the nice occasion to work with Kata Bimbó and Roger Maddux on a paper (Bimbó et al. 2009) that contains a series of results relating relation algebras and relevance logic.

I was fortunate to have Katalin Bimbó as a student and later as a research associate at Indiana University when she and I wrote *Generalized Galois Logics: Relational Semantics of Nonclassical Logical Calculi* (2008), using a draft as a text for a graduate seminar we co-taught. Kata and I have published a number of things together. I want to thank Kata for including me as a collaborator on a couple of grants she has received. She and I have been working on several projects together, and developed some novel Gentzen systems. Using these we were able to solve the problem of the decidability of the implicational fragment of the logic of Ticket Entailment $T_{\rightarrow}$ (a problem that had been open since circa 1960), and hence by the Curry–Howard Isomorphism, the decidability of the inhabitation problem for simple types by the corresponding set of combinators. Another use of them has been to show a certain logic decidable that had been thought to have been proved undecidable. Our paper is still in the refereeing process so I should not say anything more.

Our most recent project is to work on a conceptual history of the development of the ternary relational semantics for relevance logic and some other logics. This is often referred to as the "Routley–Meyer semantics." In the context of that project I have been working on various intuitive interpretations of the ternary accessibility relation. I was up to an even dozen on the last talk I gave on the subject. Among these interpretations are of course the informational interpretations due to Urquhart and Kit Fine: the information *a* when combined with the information *b* equals (or is included in) the information *c*. But two are "dynamic" informational interpretations based on von Neumann's idea of a "stored program," where one or both of the information states *a* and *b* can be thought of as standing for an action (binary relation). (Dunn 2001a, c, 2003). Dunn and Meyer (1997) show how this can be applied so as to give a semantics of various combinatory logics (the combinators replacing the structural rules), and Bimbó and Dunn (2005) show how to apply it to Kleene Logic and other "action logics."

The latest to appear on my list was something that should have been there for a long time, but wasn't. Dunn (2015) actually gives an interpretation of the ternary relation in terms of (contextual) relevance: information state *b* is relevant to information state *c* in the context of information state *a*.

Another way I have been spending my time since I "retired" is as an affiliate (and member of the advisory board) of the Info-Metrics Institute of American University. Its founding director Amos Golan defines "info-metrics" as "the science and practice of inference and quantitative information processing." What could be nicer? The first time I participated in an Info-Metrics workshop I seemed to hear John Denver singing, "Coming home to a place he had never been before." Amos and I are involved in a joint project to understand, perhaps define, the value of information.

Ruminations. I was first drawn to logic as an undergraduate because of its certainty. I was interested in deductive logic, and not probability and statistics, because they were tools for dealing with uncertainty. Classical logic was the one true logic. Already as an undergraduate I had inklings that there were matters of choice in logic. What set theory did one use, how are numbers to be represented, why not limit proofs to the constructive ones, etc. And when I went to graduate school and was exposed to relevance logic I never truly accepted the system E of entailment as the one true logic, as Alan Anderson is supposed to have done. I quickly came to notice that there were various systems of relevance logic to choose from, and that they could not do everything, and in particular it seemed that they could not do their own metatheory.

Arthur Prior in his *Past, Present and Future* (1967) said: "The logician must be like a lawyer … in the sense that he is there to provide the metaphysician, perhaps even the physicist, the tense-logic he wants, provided that it be consistent." I only came across this quote recently, but I once had the privilege of being introduced by Timothy Smiley when I gave a lecture at Cambridge, and Tim said that I was a "lawyer of logics." Just as a lawyer might draw up various legal documents for you according to your specs, so I might draw up various logics. Do you want excluded middle, do you want distribution, etc.? Here are several logics to choose from. I was somewhat surprised at his description of me, but quickly came to accept its main point. Except I wouldn't describe myself so much as a lawyer of logics, but more as an engineer of logics—a maker of tools. As the inventor of the World Wide Web Tim Berners-Lee said: "We are not analyzing a world, we are building it. We are not experimental philosophers, we are philosophical engineers." See https://www.academia.edu/5222185/An_interview_with_Tim_Berners-Lee.

Man has been defined in the Aristotelean tradition as a "rational animal." Benjamin Franklin defined man as "a tool making animal." Both of these definitions seem to deny contemporary evidence that rationality and tool making, in various degrees, extend to other animals (and I add that they are not always found in humans). But there is no doubt that these are important characteristics of humans. Primitive man had primitive tools, and primitive man also had primitive rationality. As humanity developed, it developed more sophisticated tools, and these included tools, even specialized tools, for reasoning. Now, as we turn more and more of our reasoning over to "the machines," it is important that we outfit them with not just a general-purpose logic, but also the appropriate specialized logics that they will need to solve more and more of our intellectual problems.

Let me close these ramblings with a related observation about definitions, and in particular about the definition of "information." Although I have written much about information, I have never really defined what it was. Perhaps the closest I came was when I said in (Dunn 2008, p. 581):

> I like to think of information, at least as a first approximation, as what is left from knowledge when you subtract, justification, truth, belief, and any other ingredients such as reliability that relate to justification. Information is, as it were, a mere "idle thought." Oh, one other thing, I want to subtract the thinker.

Or, to put it dually, I have used a number of different definitions as I have discussed information from both the classical Shannon, Carnap, Bar-Hillel framework and the nonclassical 4-valued framework. This is the way I think it should be. Our language typically has all the precision of a hunk of rock. But if it is a potentially useful rock then it is important to shape it and sharpen it for the purpose at hand. There are analogies in physics. A hunk of matter can be assigned either a weight or a mass—its weight becoming meaningless without gravity. The rate of its movement can either be speed or velocity (the latter adding a vector for direction). Its momentum can be…. The definition depends upon the purpose. Oops, I just revealed that against all of my platonic instincts when I first entered philosophy, I have somewhere along the way become a pragmatist. I now worship the trinity of Charles Sanders Peirce, William James, and John Dewey.

I close by emphasizing something very important. The fact that I take a pragmatic attitude toward both logics and definitions does not mean that all is conventional or arbitrary. I believe strongly in an underlying reality. Rocks have underlying properties, and that it is because of these that flint can be shaped into a useful knife for say cutting flesh, whereas pumice makes a useful tool to remove dead or dry skin without cutting. This is not a mere matter of convention—try to reverse these if you don't believe me. This was nicely stated by Frank Herbert in his Book three of *Dune*: "Deep in the human unconscious is a pervasive need for a logical universe that makes sense. But the real universe is always one step beyond logic."

OK—sounds like I need to stop writing and get back to work. :) However I do want to mention one new thing that is relevant to the title of my autobio. The Indiana University School of Informatics and Computing is adding a program in Intelligent Systems Engineering in 2015–2016. I stepped down as dean too soon. :)

J. Michael Dunn
Indiana University
e-mail: dunn@indiana.edu
URLs: www.philosophy.indiana.edu/people/dunn.shtml
and www.soic.indiana.edu/all-people/profile.html?profile_id=194

J. Michael Dunn's Publications

Dissertation

1. Dunn, J. M. (1966). *The Algebra of Intensional Logics*, PhD thesis, University of Pittsburgh, Ann Arbor (UMI).

Books

1. Dunn, J. M., Epstein, G., Cocchiarella, N. and Shapiro, S. (eds.) (1975). *Proceedings of the 1975 International Symposium on Multiple-Valued Logic*, (Indiana University, Bloomington, IN, May 13–16, 1975), IEEE Computer Society, Long Beach, CA.
2. Dunn, J. M. and Epstein, G. (eds.) (1977). *Modern Uses of Multiple-Valued Logic*, D. Reidel, Dordrecht.
3. Dunn, J. M. and Gupta, A. (eds.) (1990). *Truth or Consequences: Essays in Honor of Nuel Belnap*, Kluwer, Amsterdam.
4. Anderson, A. R., Belnap, N. D. and Dunn, J. M. (1992). *Entailment: The Logic of Relevance and Necessity*, Vol. II, Princeton University Press, Princeton, NJ.
5. Dunn, J. M. and Hardegree, G. M. (2001). *Algebraic Methods in Philosophical Logic*, Vol. 41 of *Oxford Logic Guides*, Oxford University Press, Oxford, UK.
6. Bimbó, K. and Dunn, J. M. (2008). *Generalized Galois Logics: Relational Semantics of Nonclassical Logical Calculi*, Vol. 188 of *CSLI Lecture Notes*, CSLI Publications, Stanford, CA.

Articles in Journals and Books

1. Dunn, J. M. (1967). Drange's paradox lost, *Philosophical Studies* **18**: 94–95.
2. Dunn, J. M. and Belnap, N. D. (1968a). Homomorphisms of intensionally complemented distributive lattices, *Mathematische Annalen* **176**: 28–38.
3. Dunn, J. M. and Belnap, N. D. (1968b). The substitution interpretation of the quantifiers, *Noûs* **2**: 177–185.
4. Meyer, R. K. and Dunn, J. M. (1969). E, R and γ, *Journal of Symbolic Logic* **34**: 460–474. Reprinted in Anderson, A. R. and Belnap, N. D., *Entailment: The*

Logic of Relevance and Necessity, Vol. I, Princeton University Press, Princeton, NJ, 1975, §25.2, pp. 300–314.

5. Dunn, J. M. (1970). Algebraic completeness results for R-mingle and its extensions, *Journal of Symbolic Logic* **35**: 1–13.
6. Dunn, J. M. and Meyer, R. K. (1971). Algebraic completeness results for Dummett's LC and its extensions, *Zeitschrift für Mathematische Logik und Grundlagen der Mathematik* **17**: 225–230.
7. Dunn, J. M. (1972). A modification of Parry's analytic implication, *Notre Dame Journal of Formal Logic* **13**: 195–205.
8. Dunn, J. M. (1973). A truth value semantics for modal logic, *in* H. Leblanc (ed.), *Truth, Syntax and Modality. Proceedings of the Temple University Conference on Alternative Semantics*, North-Holland, Amsterdam, pp. 87–100.
9. Meyer, R. K., Dunn, J. M. and Leblanc, H. (1974). Completeness of relevant quantification theories, *Notre Dame Journal of Formal Logic* **15**: 97–121.
10. Dunn, J. M. (1975a). Axiomatizing Belnap's conditional assertion, *Journal of Philosophical Logic* **4**: 383–397.
11. Dunn, J. M. (1975b). The algebra of **R**, §28.2, *in* Anderson, A. R. and Belnap, N. D., *Entailment: The Logic of Relevance and Necessity*, Vol. I, Princeton University Press, Princeton, NJ, pp. 352–371.
12. Dunn, J. M. (1975c). Consecution formulation of positive **R** with co-tenability and **t**, §28.5, *in* Anderson, A. R. and Belnap, N. D., *Entailment: The Logic of Relevance and Necessity*, Vol. I, Princeton University Press, Princeton, NJ, pp. 381–391.
13. Dunn, J. M. (1975d). Extensions of **RM**, §29.3, *in* Anderson, A. R. and Belnap, N. D., *Entailment: The Logic of Relevance and Necessity*, Vol. I, Princeton University Press, Princeton, NJ, pp. 420–429.
14. Dunn, J. M. (1975e). Intensional algebras, §18, *in* Anderson, A. R. and Belnap, N. D., *Entailment: The Logic of Relevance and Necessity*, Vol. I, Princeton University Press, Princeton, NJ, pp. 180–206.
15. Dunn, J. M. (1976a). Intuitive semantics for first-degree entailments and 'coupled trees', *Philosophical Studies* **29**: 149–168.
16. Dunn, J. M. (1976b). A Kripke-style semantics for R-mingle using a binary accessibility relation, *Studia Logica* **35**: 163–172.
17. Dunn, J. M. (1976c). Quantification and ***RM***, *Studia Logica* **35**: 315–322.
18. Dunn, J. M. (1976d). A variation on the binary semantics for R-Mingle, *Relevance Logic Newsletter* **1**: 56–67.
19. Dunn, J. M. (1979a). **R**-mingle and beneath. Extensions of the Routley–Meyer semantics for **R**, *Notre Dame Journal of Formal Logic* **20**: 369–376.
20. Dunn, J. M. (1979b). Relevant Robinson's arithmetic, *Studia Logica* **38**: 407–418.
21. Dunn, J. M. (1979c). A theorem in 3-valued model theory with connections to number theory, type theory, and relevant logic, *Studia Logica* **38**: 149–169.
22. Meyer, R. K., Routley, R. and Dunn, J. M. (1979). Curry's paradox, *Analysis (n.s.)* **39**: 124–128.

23. Dunn, J. M. (1980). A sieve for entailments, *Journal of Philosophical Logic* **9**: 41–57.
24. Belnap, N. D., Gupta, A. and Dunn, J. M. (1980). A consecutive calculus for positive relevant implication with necessity, *Journal of Philosophical Logic* **9**: 343–362.
25. Belnap, N. D. and Dunn, J. M. (1981). Entailment and the disjunctive syllogism, *in* G. Fløistad and G. H. von Wright (eds.), *Contemporary Philosophy: A New Survey*, Vol. I, Philosophy of Language/Philosophical Logic, Martinus Nijhoff, The Hague, pp. 337–366.
26. Dunn, J. M. (1981). Quantum mathematics, *in* P. D. Asquith and R. N. Giere (eds.), *PSA 1980: Proceedings of the 1980 Biennial Meeting of the Philosophy of Science Association*, Vol. 2, Philosophy of Science Association, East Lansing, MI, pp. 512–531.
27. Dunn, J. M. (1982a). Anderson and Belnap, and Lewy on entailment, *in* L. J. Cohen, J. Łoś, H. Pfeiffer and K.-P. Podewski (eds.), *Logic, Methodology and Philosophy of Science, VI. Proceedings of the Sixth International Congress of Logic, Methodology and Philosophy of Science, Hannover 1979*, North-Holland and PWN—Polish Scientific Publishers, Amsterdam and Warszawa, pp. 291–297.
28. Dunn, J. M. (1982b). A relational representation of quasi-Boolean algebras, *Notre Dame Journal of Formal Logic* **23**: 353–357.
29. Dunn, J. M. (1986). Relevance logic and entailment, *in* D. Gabbay and F. Guenthner (eds.), *Handbook of Philosophical Logic*, 1st edn, Vol. 3, D. Reidel, Dordrecht, pp. 117–224.
30. Dunn, J. M. and Hellman, G. (1986). Dualling: A critique of an argument of Popper and Miller, *The British Journal for the Philosophy of Science* **37**: 220–223.
31. Dunn, J. M. (1987a). Incompleteness of the bibinary semantics for *R*, *Bulletin of the Section of Logic of the Polish Academy of Sciences* **16**: 107–110.
32. Dunn, J. M. (1987b). Relevant predication 1: The formal theory, *Journal of Philosophical Logic* **16**: 347–381.
33. Dunn, J. M. (1988). The impossibility of certain higher-order non-classical logics with extensionality, *in* D. F. Austin (ed.), *Philosophical Analysis: A Defense by Example. (A Festschrift for Edmund Gettier)*, Kluwer, Dordrecht, pp. 261–279.
34. Dunn, J. M. and Meyer, R. K. (1989). Gentzen's cut and Ackermann's gamma, *in* J. Norman and R. Sylvan (eds.), *Directions in Relevant Logic*, Kluwer, Dordrecht, pp. 229–240.
35. Dunn, J. M. (1990a). The frame problem and relevant predication, *in* H. E. Kyburg, R. P. Loui and G. N. Carlson (eds.), *Knowledge Representation and Defeasible Reasoning*, Kluwer, Dordrecht, pp. 89–95.
36. Dunn, J. M. (1990b). Relevant predication 2: Intrinsic properties and internal relations, *Philosophical Studies* **60**: 177–206.

37. Dunn, J. M. (1990c). Relevant predication 3: Essential properties, *in* J. M. Dunn and A. Gupta (eds.), *Truth or Consequences: Essays in Honor of Nuel Belnap*, Kluwer, Amsterdam, pp. 77–95.
38. Dunn, J. M. (1991). Gaggle theory: An abstraction of Galois connections and residuation, with applications to negation, implication, and various logical operators, *in* J. van Eijck (ed.), *Logics in AI: European Workshop JELIA '90, Amsterdam, The Netherlands, September 10–14, 1990*, number 478 in *Lecture Notes in Computer Science*, Springer, Berlin, pp. 31–51.
39. Franco, J., Dunn, J. M. and Wheeler, W. (1992). Recent work at the interface of logic, combinatorics, and computer science, *Annals of Mathematics and Artificial Intelligence* **6**: 1–16.
40. Dunn, J. M. (1993a). Partial gaggles applied to logics with restricted structural rules, *in* K. Došen and P. Schroeder-Heister (eds.), *Substructural Logics*, Clarendon and Oxford University Press, Oxford, UK, pp. 63–108.
41. Dunn, J. M. (1993b). Star and perp: Two treatments of negation, *Philosophical Perspectives* **7**: 331–357. (Language and Logic, 1993, J. E. Tomberlin (ed.)).
42. Allwein, G. and Dunn, J. M. (1993). Kripke models for linear logic, *Journal of Symbolic Logic* **58**: 514–545.
43. Dunn, J. M. (1995a). Gaggle theory applied to intuitionistic, modal and relevance logics, *in* I. Max and W. Stelzner (eds.), *Logik und Mathematik. Frege-Kolloquium Jena 1993*, W. de Gruyter, Berlin, pp. 335–368.
44. Dunn, J. M. (1995b). Positive modal logic, *Studia Logica* **55**: 301–317.
45. Dunn, J. M. (1996a). Generalized ortho negation, *in* H. Wansing (ed.), *Negation: A Notion in Focus*, W. de Gruyter, New York, NY, pp. 3–26.
46. Dunn, J. M. (1996b). Is existence a (relevant) predicate?, *Philosophical Topics* **24**: 1–34.
47. Dunn, J. M. (1996c). A logical framework for the notion of *natural property*, *in* J. Earman and J. D. Norton (eds.), *The Cosmos of Science: Essays of exploration*, number 6 in *Pittsburgh-Konstanz Series in the Philosophy and History of Science*, University of Pittsburgh Press and Universitäts-Verlag Konstanz, Pittsburgh and Konstanz, pp. 458–497.
48. Hartonas, C.and Dunn, J. M. (1997). Stone duality for lattices, *Algebra Universalis* **37**: 391–401. [Preliminary version: Duality theorems for partial orders, semilattices, Galois connections and lattices, *Indiana University Logic Group Preprint Series* IULG–93–26, 1993.]
49. Dunn, J. M. and Meyer, R. K. (1997). Combinators and structurally free logic, *Logic Journal of the IGPL* **5**: 505–537.
50. Bimbó, K. and Dunn, J. M. (1998). Two extensions of the structurally free logic *LC*, *Logic Journal of the IGPL* **6**: 403–424.
51. Dunn, J. M. (1999). A comparative study of various model-theoretic treatments of negation: A history of formal negation, *in* D. M. Gabbay and H. Wansing (eds.), *What is Negation?*, Kluwer, Dordrecht, pp. 23–51.
52. Dunn, J. M. (2000). Partiality and its dual, *Studia Logica* **65**: 5–40.
53. Dunn, J. M. (2001a). The concept of information and the development of modern logic, *in* W. Stelzner and M. Stöckler (eds.), *Zwischen traditioneller

und moderner Logik: Nichtklassische Ansätze, (Perspektiven der Analytischen Philosophie), Mentis-Verlag, Paderborn, pp. 423–447.

54. Dunn, J. M. (2001b). A representation of relation algebras using Routley–Meyer frames, *in* C. A. Anderson and M. Zelëny (eds.), *Logic, Meaning and Computation. Essays in Memory of Alonzo Church*, Kluwer, Dordrecht, pp. 77–108. [Preliminary version: A representation of relation algebras using Routley–Meyer frames, *Indiana University Logic Group Preprint Series* IULG–93–28, 1993.]
55. Dunn, J. M. (2001c). Ternary relational semantics and beyond: Programs as arguments (data) and programs as functions (programs), *Logical Studies* **7**: 1–20.
56. Bimbó, K. and Dunn, J. M. (2001). Four-valued logic, *Notre Dame Journal of Formal Logic* **42**: 171–192.
57. Shramko, Y., Dunn, J. M. and Takenaka, T. (2001). The trilattice of constructive truth values, *Journal of Logic and Computation* **11**: 761–788.
58. Dunn, J. M. and Restall, G. (2002). Relevance logic, *in* D. Gabbay and F. Guenthner (eds.), *Handbook of Philosophical Logic*, 2nd edn, Vol. 6, Kluwer, Amsterdam, pp. 1–128.
59. Bimbó, K. and Dunn, J. M. (2005). Relational semantics for Kleene logic and action logic, *Notre Dame Journal of Formal Logic* **46**: 461–490.
60. Dunn, J. M., Gehrke, and M. Palmigiano, A. (2005). Canonical extensions and relational completeness of some substructural logics, *Journal of Symbolic Logic* **70**: 713–740.
61. Dunn, J. M., Hagge, T. J., Moss, L. S. and Wang, Z. (2005). Quantum logic as motivated by quantum computing, *Journal of Symbolic Logic* **70**: 353–359.
62. Dunn, J. M. and Zhou, C. (2005). Negation in the context of gaggle theory, *Studia Logica* **80**: 235–264.
63. Dunn, J. M. (2008). Information in computer science, *in* P. Adriaans and J. van Benthem (eds.), *Philosophy of Information*, Vol. 8 of *Handbook of the Philosophy of Science*, (D. M. Gabbay, P. Thagard, J. Woods (eds.)), Elsevier, Amsterdam, pp. 581–608.
64. Bimbó, K. and Dunn, J. M. (2009). Symmetric generalized Galois logics, *Logica Universalis* **3**: 125–152.
65. Bimbó, K., Dunn, J. M. and Maddux, R. D. (2009). Relevance logics and relation algebras, *Review of Symbolic Logic* **2**: 102–131.
66. Dunn, J. M. (2010). Contradictory information: Too much of a good thing, *Journal of Philosophical Logic* **39**: 425–452.
67. Bimbó, K. and Dunn, J. M. (2010). Calculi for symmetric generalized Galois logics, *in* J. van Benthem and M. Moortgat (eds.), *Festschrift for Joachim Lambek*, Vol. 36 of *Linguistic Analysis*, Linguistic Analysis, Vashon, WA, pp. 307–343.
68. Beall, J., Brady, R., Dunn, J. M., Hazen, A. P., Mares, E., Meyer, R. K., Priest, G., Restall, G., Ripley, D., Slaney, J. and Sylvan, R. (2012). On the ternary relation and conditionality, *Journal of Philosophical Logic* **41**: 595–612.

69. Bimbó, K. and Dunn, J. M. (2012). New consecution calculi for $R^t_{\to}$, *Notre Dame Journal of Formal Logic* **53**: 491–509.
70. Bimbó, K. and Dunn, J. M. (2013). On the decidability of implicational ticket entailment, *Journal of Symbolic Logic* **78**: 214–236.
71. Dunn, J. M., Moss, L. S. and Wang, Z. (2013). The third life of quantum logic: Quantum logic inspired by quantum computing, *Journal of Philosophical Logic* **42**: 443–459.
72. Dunn, J. M. (2014a). Some stories and theorems inspired by Raymond Smullyan, *in* J. Rosenhouse (ed.), *Four Lives: A Celebration of Raymond Smullyan*, Dover Publications, Mineola, NY, pp. 65–75.
73. Dunn, J. M. (2014b). Arrows pointing at arrows: Arrow logic, relevance logic, and relation algebras, *in* A. Baltag and S. Smets (eds.), *Johan van Benthem on Logic and Information Dynamics*, Outstanding Contributions to Logic, Springer, New York, NY, pp. 881–894.
74. Bimbó, K. and Dunn, J. M. (2014). Extracting BB'IW inhabitants of simple types from proofs in the sequent calculus $LT^t_{\to}$ for implicational ticket entailment, *Logica Universalis* **8**: 141–164.
75. Dunn, J. M. and Eisenberg, P. (2014). Chorus: Hector-Neri Castañeda. A conversation about Hector by two of his colleagues, *in* A. Palma (ed.), *Castañeda and his Guises: Essays on the Work of Hector-Neri Castañeda*, W. de Gruyter, Boston, MA, pp. 15–18.
76. Dunn, J. M. (2015). The relevance of relevance to relevance logic, *in* M. Banerjee and S. N. Krishna (eds.), *Logic and its Applications: 6th Indian Conference, ICLA 2015, Mumbai, India, January 8–10, 2015*, number 8923 in *Lecture Notes in Computer Science*, Springer, Heidelberg, pp. 11–29.
77. Bimbó, K. and Dunn, J. M. (2016). Larisa Maksimova's early contributions to relevance logic, *in* S. Odintsov (ed.), *L. Maksimova on Implication, Interpolation and Definability*, (Outstanding Contributions to Logic), Springer, New York, NY, (25 pages, accepted).
78. Bimbó, K. and Dunn, J. M. (2015). Modalities in lattice-*R*. (37 pages, submitted).

Abstracts

1. Dunn, J. M. (1967a). An algebraic completeness proof for the first degree fragment of entailment, *Abstracts of Papers: Third International Congress for Logic, Methodology and Philosophy of Science, Amsterdam*, p. 9.
2. Dunn, J. M. (1967b). The effective equivalence of certain propositions about de Morgan lattices, *Journal of Symbolic Logic* **32**: 433–434.
3. Dunn, J. M. and Belnap, N. D. (1967). Homomorphisms of intensionally complemented distributive lattices, *Journal of Symbolic Logic* **32**: 446.
4. Dunn, J. M. (1968). Representation theory of normal R-mingle matrices, *Journal of Symbolic Logic* **33**: 637.
5. Meyer, R. K. and Dunn, J. M. (1968). Entailment logics and material implication, *Notices of the American Mathematical Society* **15**: 1021–1022.

6. Dunn, J. M. (1970a). Comments on N. D. Belnap, Jr.'s 'Conditional assertion and restricted quantification', *Noûs* **4**: 13.
7. Dunn, J. M. (1970b). Extensions of RM and LC, *Journal of Symbolic Logic* **35**: 360.
8. Dunn, J. M. (1971). An intuitive semantics for first degree relevant implications, *Journal of Symbolic Logic* **36**: 362–363.
9. Dunn, J. M. (1973). A 'Gentzen system' for positive relevant implication, *Journal of Symbolic Logic* **38**: 356–357.
10. Dunn, J. M. (1976). A natural family of sentential calculi intermediate between **R** and **R-mingle**, *Journal of Symbolic Logic* **41**: 553.
11. Dunn, J. M. (1978). A theorem in 3-valued model theory with connections to number theory, type theory, and relevant logic, *Journal of Symbolic Logic* **43**: 615–616.
12. Dunn, J. M. (1979). A sieve for entailments, *Abstracts of Papers: Sixth International Congress of Logic, Methodology and Philosophy of Science (§5.7)*, pp. 1–5.
13. Meyer, R. K., Bimbó, K. and Dunn, J. M. (1998). Dual combinators bite the dust, *Bulletin of Symbolic Logic* **4**: 463–464.
14. Meyer, R. K., Dezani, M., Dunn, J. M., Motohama, Y. and Sylvan, R. (2000). The key to the universe—intersection types, lambda, CL and relevant entailment, *Bulletin of Symbolic Logic* **6**: 252–253.
15. Dunn, J. M. (2003). Ternary semantics for dynamic logic, Hoare logic, action logic, etc., *Smirnov's Readings 4*, Institute of Logic, Russian Academy of Sciences, Moscow, pp. 70–71.
16. Bimbó, K. and Dunn, J. M. (2009). Canonicity in symmetric generalized Galois logics, *Bulletin of Symbolic Logic* **15**: 240.
17. Bimbó, K. and Dunn, J. M. (2012). From relevant implication to ticket entailment, *Bulletin of Symbolic Logic* **18**: 288.
18. Bimbó, K. and Dunn, J. M. (2013). The decision problem of $T_{\to}$, *Bulletin of Symbolic Logic* **19**: 226–227.
19. Bimbó, K. and Dunn, J. M. (2014a). Inhabitants of $T^{t}_{\to}$ theorems, *Bulletin of Symbolic Logic* **20**: 123.
20. Bimbó, K. and Dunn, J. M. (2014b). Combinatory inhabitants of $R_{\to}$ theorems extracted from sequent calculus proofs, *Bulletin of Symbolic Logic* **20**: 258–259.
21. Bimbó, K. and Dunn, J. M. (2015). On the decidability of classical linear logic, *Bulletin of Symbolic Logic* **21**: 358.
22. Dunn, J. M. (2015). Various interpretations of the ternary accessibility relation, *Smirnov's Readings 9*, Sovremennye Tetradi, Moscow, pp.10–13.

Book Reviews

1. Dunn, J. M. (1971). Review of Jaakko Hintikka (ed.), *Philosophy of Mathematics*, *American Mathematical Monthly* **78**: 91.
2. Dunn, J. M. (1978). Review of Hugues Leblanc, *Truth-value Semantics*, *Journal of Symbolic Logic* **43**: 376–377.

3. Dunn, J. M. (2013). Guide to the Floridi keys: Essay review of Luciano Floridi's *The Philosophy of Information*, *MetaScience* **22**: 93–98.

Article Reviews in the *Mathematical Reviews*

1. MR0292640 (45 #1725). Leblanc, H. (1971). Truth-value semantics for a logic of existence, *Notre Dame Journal of Formal Logic* **12**: 153–168.
2. MR0292649 (45 #1734). Bayart, A. (1970). On truth-tables for M, B, S4 and S5, *Logique et Analyse (n.s.)* **13**: 335–375.
3. MR0295887 (45 #4949). Parks, R. Z. (1972). A note on **R**-Mingle and Sobociński's three-valued logic, *Notre Dame Journal of Formal Logic* **13**: 227–228.
4. MR0295888 (45 #4950). Tichý, P. (1969). Intension in terms of Turing machines, *Studia Logica* **24**: 7–25.
5. MR0323532 (48 #1888). Urquhart, A. (1972). Semantics for relevant logics, *Journal of Symbolic Logic* **37**: 159–169.
6. MR0327515 (48 #5857). Meyer, R. K. and Routley, R. (1972). Algebraic analysis of entailment. I, *Logique et Analyse (n.s.)* **15**: 407–428.
7. MR0329865 (48 #8205). Ono, H. (1973). Incompleteness of semantics for intermediate predicate logics. I. Kripke's semantics, *Proceeding of the Japan Academy* **49**: 711–713.
8. MR0337500 (49 #2269). Stalnaker, R. C. and Thomason, R. H. (1968). Abstraction in first-order modal logic, *Theoria* **34**: 203–207.
9. MR0347556 (50 #59). Fitting, M. (1972). An ε-calculus system for first-order *S4*, *Conference in Mathematical Logic, London '70 (Bedford College, London, (1970)*, number 255 in *Lecture Notes in Mathematics*, Springer, Berlin, pp. 103–110.
10. MR0351748 (50 #4236). Prucnal, T. (1974). Interpretations of classical implicational sentential calculus in nonclassical implicational calculi, *Studia Logica* **33**: 59–64.
11. MR0351759 (50 #4247). Czermak, J. (1974). Matrix calculi *SS1M* and *SS1I* compared with axiomatic systems, *Notre Dame Journal of Formal Logic* **15**: 312–316.
12. MR0351763 (50 #4251). Satre, T. W. (1972). Natural deduction rules for $S1^0 - S4^0$, *Notre Dame Journal of Formal Logic* **13**: 565–568.
13. MR0363789 (51 #44). Meyer, R. K. and Routley, R. (1973 and 1974). Classical relevant logics. I and II, *Studia Logica* **32** and **33**: 51–68 and 183–194.
14. MR0441695 (56 #94). Manor, R. (1974). A semantic analysis of conditional assertion, *Journal of Philosophical Logic* **3**: 37–52.
15. MR0441726 (56 #122). Weese, M. (1976). The isomorphism problem of superatomic Boolean algebras, *Zeitschrift für Mathematische Logik und Grundlagen der Mathematik* **22**: 439–440.
16. MR0446909 (56 #5227). Urquhart, A. (1973). A semantical theory of analytic implication, *Journal of Philosophical Logic* **2**: 212–219.

17. MR0497933 (58 #16145). Davidson, B. and Jackson, F. C. and Pargetter, R. (1977). Modal trees for T and S5, *Notre Dame Journal of Formal Logic* **18**: 602–606.
18. MR0497955 (58 #16158). Siemens, D. F., Jr. (1977). Fitch-style rules for many modal logics, *Notre Dame Journal of Formal Logic* **18**: 631–636.
19. MR0497983 (58 #16180). da Costa, N. C. A. and Alves, E. H. (1977). A semantical analysis of the calculi $\mathbf{C}_n$, *Notre Dame Journal of Formal Logic* **18**: 621–630.
20. MR0497988 (58 #16183). Leschine, T. M. (1978). Propositional logic for topology-like matrices: A calculus with restricted substitution, *Studia Logica* **37**: 161–165.
21. MR0505276 (58 #21478). Meyer, R. K. (1976). Two questions from Anderson and Belnap, *Reports on Mathematical Logic* **7**: 71–86.
22. MR500772 (80a:03085). Yutani, H. (1978). Congruences on the products of BCK-algebras, *Kobe University, Mathematics Seminar Notes* **6**: 169–176.
23. MR509688 (81b:03033). Dywan, Z. (1978). Decidability of structural completeness for strongly finite propositional calculi, *Polish Academy of Sciences, Institute of Philosophy and Sociology, Bulletin of the Section of Logic* **7**: 129–132.
24. MR524111 (81b:03016). Bellissima, F. (1978). On the modal logic corresponding to diagonalizable algebra theory, *Unione Matematica Italiana, Bolletino B (5)* **15**: 915–930.
25. MR527438 (81b:03018). Fitting, M. (1978). Subformula results in some propositional modal logics, *Studia Logica* **37**: 387–391.
26. MR530140 (81b:03025). Ursini, A. (1979). Two remarks on models in modal logics, *Unione Matematica Italiana, Bolletino A (5)* **16**: 124–127.

Introduction: From Information at Large to Semantics of Logics

Katalin Bimbó

Abstract We take stock of various views and approaches to information, in general. Then we explicate some relationships between logics and information—to provide an explanation (even a justification) for the title of the volume. The multifaceted character of information based logics leads to a bewildering assortment of linkages between the papers in this volume. We highlight touching points between consecutive papers in the rest of the volume.

Keywords Gaggle theory · Information theories · Relational semantics · Relevance logic · Situation

1 Information on Information

This introduction has several aims. First, it connects logic and various approaches to information. Second, we delineate information based logics and how they feature in J. Michael Dunn's research. Lastly, the order in which the papers appear in this volume is explained via the connections between the papers in this collection as well as between the papers and Dunn's work.

"Information" is a fashionable term, even a buzz-word nowadays. But perhaps we have to ask first: *What is information?* This section offers some answers to this question with the intention to situate the connection between logic and information within the landscape of various approaches to information. Then, in the next section, we elaborate on the fundamental relationship between information and logic, that is, we answer the question: *Which logics are information based?*

The *New Oxford American Dictionary* defines one of the meanings of "information" as "what is conveyed or represented by a particular arrangement or sequence of things." Of course, the usage of a natural language term is rarely (if ever) unambiguous. "Information" is no exception. The *NOAD* provides a second sense, in which "facts" are mentioned, which may be thought of to be correct descriptions. Some

K. Bimbó (✉)
University of Alberta, Edmonton, Canada
e-mail: bimbo@ualberta.ca
URL: http://www.ualberta.ca/~bimbo

K. Bimbó (ed.), *J. Michael Dunn on Information Based Logics*,
Outstanding Contributions to Logic 8, DOI 10.1007/978-3-319-29300-4_1

further exploration, this time within the *Oxford American Writer's Thesaurus*, reveals that "[i]nformation may be no more than a collection of data or facts ... gathered through observation, reading, or hearsay, with no guarantee of their validity ...." The absolution from the burden of establishing truth is a welcome ramification, because it is notoriously difficult to determine what is true and what is not. Moreover, the latter task, most of the time, might go well beyond what is covered by a logician's job description.

There are situations where it is reasonable to expect that the information received is true, at least to a certain extent. At a public transit station, we might expect to obtain information about the arrival and departure times of buses or trains. In the summer, it is part of our expectation that the scheduled buses ran on schedule earlier in the day, and they will continue to run on schedule for the rest of the day. On the other hand, after the first big snow fall, it is also part of our expectation that there might be delays. In other words, the degree of expected truthfulness varies depending on pragmatic considerations such as the season of the year or the weather. A whole philosophical debate could ensue about how precise a schedule of trains or a GPS map should or could be. Barring clarification on such questions, the problem of the truth of information in any practical situations would not make much sense, because almost all the information almost all the time will turn out to be false, if we can demand greater and greater precision.

Dunn in his (2001) and (2008) describes a somewhat similar sense of "information" which does not include the requirement of truthfulness; nor does it include other characteristics that are usually associated to a concept of knowledge. Every philosopher heard the adage that knowledge is justified true belief. Given that, we could say that information is what remains when the three listed features are subtracted from knowledge. First it may appear that nothing remains from knowledge once it is neither justified nor true, and may not be believed by anybody. However, unjustified unbelievable falsehoods can be communicated, without anybody believing them. Malicious gossip that is spread with the intent to harm is an example where perhaps not even the speaker, and we hope not the listener, believes the information being conveyed.

In order to get a grasp of various aspects of information that emerge when we start to think about it, let us start with a concrete example.

Example 1.1 The following 21-character sequence of letters is nothing more—at first glance—than 21 letters from the English (or a similar) alphabet.

$$\text{bpqajmkiumqvnwzuibqwv} \tag{1}$$

On the other hand, if I am in the middle of a lecture on (simple) substitution ciphers while I write the above sequence of letters on the white board, and perhaps even mention "shift +9," then the string (with some segmentation) suddenly conveys information. The string now can be seen as:

$$\text{thisbecameinformation.} \tag{2}$$

Some might contend that the gibberish in (1) contains the same information as (2) does, independently of the situation in which it appears. The information could be recovered from the string in (1) by a statistical attack relying on the frequencies by which various letters occur in English texts. Alternatively, a brute-force attack will work too, and it will require checking 25 shifts in the worst case. Even then, the situation could matter of course, since there are other languages (beyond English) that use the Latin alphabet.

The example hints toward two areas that also deal with information in some sense, however, they will not concern us at all in this book. *Pragmatics* (as part of linguistics) deals with the use of language in context. The broader context such as a lecture during which the string in (1) is written would be subject to pragmatical analysis. (Other disciplines that incorporate considerations of pragmatics or take interpretations of signs above and beyond spoken languages are anthropology and semiotics. In these areas the notion of a "language" is expanded and attempts are made to extract meaning or information from a wide variety of contextualized signs.)

Another area hinted at by Example 1.1 is where the goal is to *measure information*. A quantitative approach to information was initiated by Claude Shannon. The information conveyed by a string turns out to be measurable, eventually, by the length of the string. We could continue with our example to explain how the correlation between information and the length of a string can be derived, but first we glance at binary strings. Digital information is often stored and manipulated in the form of bits (i.e., 0's and 1's), and considering binary strings affords certain simplifications.

Example 1.2 Let us assume that we consider a binary string $w = 0101101$ of 7 bits. It is easy to see that there are 2^7 possible binary strings of length 7. Our concrete string w is one among all those strings, and this can be expressed using probabilistic terminology by saying that $p(w) = \frac{1}{128}$, that is, w has probability $\frac{1}{128}$ with respect to the collection of similar strings (i.e., binary strings of length 7). The "inverse probability" of w is the total pool of strings of length 7, that is, 128. Taking the base-2 logarithm of the inverse probability of w is simply 7, and that is defined to be the *amount of information* carried by w.

It is easy to see that for the string s in (1), we could proceed with a similar reasoning in the context of strings over a 26-letter alphabet. $\log_{26}(\frac{1}{p(s)}) = \log_{26}(21^{26}) = 21$, which is the length of the string s. However, $\log_{26}(21^{26}) = \log_2(21^{26})/\log_2 26$ which means that we could use $\log_2$ to measure the information carried by strings over the English alphabet too. The information carried by a string of 28 characters viewed as $\log_{26}(28^{26})$ is proportional to $\log_2(28^{26})$ by a constant factor c (in particular, c is $\frac{1}{\log_2 26}$).

An alternative way to measure the information in a string is to look at its *internal complexity*. It might appear that the string 1111111 is simpler than 1101101. Algorithmic complexity theory measures the information in a string by its Kolmogorov complexity, and as the label "algorithmic" suggests, complexity is characterized via algorithms or programs. The Kolmogorov complexity of a string is the length of the shortest program plus the length of the input that will produce the string. We may expect that to print 7 1's, we can specify a program that prints 1's, and a counter that goes to 0 in 7 steps. On the other hand, to print our second sample string, a program should count to 2 while printing 1's, then print a 0, repeat this block of steps (or this subroutine) once, and finish with a 1. Another example may be constructed using the widely known fact that there is a Turing machine that duplicates a string of 1's. That is, the size of the duplication program does not depend on the length of the input string. Then we can expect that a string comprising 256 1's is simpler than a string having one fewer 1's (i.e., 255 1's) or one more 1's (i.e., 257 1's).

The cryptographic example above and the quantitative theories of information presuppose a very fine-grained anatomy of information carriers: their units are bite-sized or even bit-sized. For our purposes larger components such as sentences are important.

In everyday situations the primary way to convey information is by sentences—whether spoken or written. Although many languages may be written using letters and utterances may be analyzed into syllables, letters are usually not considered to be capable of conveying information by themselves. It seems sensible to concentrate on sentences when we are interested in how information is used in reasoning.

The definition quoted above from the NOAD describes information as represented and as conveyed. In a similar vein, we can view sentences as *static* or as *dynamic* carriers of information. Admittedly, analyzing objects (of any kind) without their interactions is likely to pose a less challenging problem. Most logical systems do not deal with dynamic aspects of reasoning. For example, while we might talk about "drawing a conclusion from a set of premises" in an introductory logic course, we are likely to move on quickly to the notion of logical consequence, where the set of premises and the conclusion are already given. Perhaps a more dynamic part of classical logic is a proof system, where even if there is a target formula that is to be proved, some intermediate formulas have to be derived (as conclusions) from the axioms or premises. On the semantic side, we could think of building models via games as discovering the information that is contained in formulas. The intuitive connection between logical consequence and information is nicely formulated by Barwise (1993, pp. 6–7):

> The intuition is that the notion of logical consequence has to do with information containment, and that inference has to do with the extraction of logically implicit information.

The syntactic and semantic processes by which we to tease out some implicit information as described above belong to the practice of logic, rather than the logical systems themselves. Yes, we may use a sequent or natural deduction calculus to prove a theorem, but a formula is provable (or it isn't) in an eternal (or atemporal) sense—even if we haven't found a proof, and perhaps we can't or we won't, because of a shortage in our supply of pens and paper, or time. Similarly, when we build a model, we may gain insight and understanding of the meaning of the formulas, but that activity is auxiliary to logic (as a theory).

We talked about bits as carriers of information and the quoted definition from NOAD emphasized that information is communicated or conveyed; both stress the *dynamic* aspects of information. The problem of change is an ancient philosophical problem, and from time to time, logicians pitched in to provide a formal framework to describe change. In connection to information specifically, we have to mention at least four approaches, even though not all of them are represented in this volume.

In the first approach known as *dynamic semantics*, it is assumed that the basic information carriers are formulas which are usual entities in a logic, and the change that is brought about by a formula can be represented by an ordered pair. In the form of a slogan, the meaning of a sentence is the modification it causes to the hypothetical situation. In the area of natural language semantics a prototypical dynamic phenomenon is an anaphoric pronoun. The effect of linking a pronoun to its antecedent is an update in the current information state.[1] There are certain similarities between updating states and switching to other possible worlds as a result of applications of modal operators. However, anaphoric updates to a situation are more restricted than those that result from considerations about possibilities and necessities.

The semantic properties of modalities in a natural language can be captured at the level of the meaning of sentences. However, some linguistic phenomena emerge only when multiple sentences are considered at once. In particular, (short) sequences of declarative sentences crystallize into manageable and interesting constituents of texts; these are called discourses. The dynamics within a discourse are not limited to building up an information state by a sequence of sentences, but it includes connections between sentences that could not exist outside a discourse. Certain particles of a natural language, for instance, pronouns hold much less meaning than others do. The thin meaning of a pronoun can be filled out by tracing connections that arise in a sequence of sentences: pronouns that gain meaning via links to other parts of a discourse are called anaphoric (when they are backward looking) and cataphoric (when they are forward looking). In English, for instance, pronouns are predominantly anaphoric (when they are not deictic). Pronouns have a variable denotation in an obvious way, however, other syntactic elements of a natural language sentence can behave alike. Definite noun phrases and tenses also can function as if they would include a hidden parameter the meaning of which is determined as a discourse proceeds. Languages differ a lot not only in their vocabulary but also in their syntax. Thus the role of logic in the description of a discourse is to provide a sufficiently rich formalism, which inevitably has to be complemented with empirical considerations for each concrete natural language.

The second approach that we mention takes into consideration that there can be other actions than uttering sentences one after another. Some of the actions that can be described formally are programs, and some of the states (or certain aspects thereof) that can be characterized by formulas being true or false (or unknown) are states of a computer. *Dynamic logic* incorporates programs as labels for modalities together with operations on programs such as program composition. Preconditions and postconditions can be expressed by formulas, which leads to straightforward applications in computer science, because it is possible to state what will hold after the termination of a possibly complex program.

[1] See, for instance, Groenendijk and Stokhof (1991).

Epistemic logics were developed independently from dynamic logic as an application of some of the well-known normal modal logics. The **S1**, …, **S5** systems were invented by Clarence I. Lewis with some sort of necessity being part of his strict implication, but soon other interpretations of the non-truth functional components were suggested, for example, provability instead of necessity, by Kurt Gödel. It is not difficult to see that some of the most often used axioms can be interpreted in terms of belief and knowledge too. For example, $\Box\varphi \supset \neg\Box\neg\varphi$ can be rendered as "If φ is necessary, then not-φ is not necessary." Then if we change $\Box$ to K_i, then $\mathsf{K}_i\varphi \supset \neg\mathsf{K}_i\neg\varphi$ can be read as "If agent i knows that φ, then i does not know that not-φ." Depending of which properties of knowledge and belief are deemed to be desirable (or even plausible), it is possible to formulate a range of *epistemic logics* using the machinery of normal modal logics.

To reiterate the connection between logic and information on one hand, and the differences between knowledge and information, on the other, let us quote Barwise again (1989, p. 203):

> Information travels at the speed of logic, genuine knowledge travels only at the speed of cognition and inference. Put another way, I would argue that much of the work in logic of knowledge is best understood in terms of the logic of information.

We may interpret what is said here to mean that logic, including epistemic logic, is primarily about information or the transfer of information. While an agent (surely) does not know all the logical consequences of what he or she already knows (i.e., he or she is not logically omniscient), it is plausible to say that an agent has all the information (including implied information) that is contained in what she or he has been informed about.

Dynamic logic and epistemic logics are naturally multi-modal logics, that is, they contain several modal operators, which are not definable from each other. However, the various modalities encompass orthogonal dimensions so to speak. Although agents can be grouped, and indeed, for the investigations of common knowledge some agents are considered together, agents are not subject to the same sort of operations that can be performed on programs. We will surely reach a point when we will want to reason about the mental (or epistemic) states of robots and the consequences of their actions in light of their epistemic states, but at the moment, a more straightforward combination of the latter two approaches is what is called *dynamic epistemic logic*. The agents remain (idealized) people, and the actions are limited to those that are capable of changing the epistemic state of an agent. Moreover, the changes in the beliefs or knowledge of an agent can be expected to be systematic with respect to their previous epistemic state and the action.

The third and slightly different approach we mention here focuses on ways in which beliefs in a belief set (or in a belief base) change when new information is received. An easy operation is expansion, which adds new beliefs to the stock of already believed sentences. Contraction and revision are less straightforward operations, because they require the exclusion of some beliefs from the belief set. The whole approach gets its name "*belief revision*" from the revision operation, which is perhaps the most complicated among the three belief change operations that we listed.[2] The removal of certain beliefs would be unproblematic if all the beliefs would be mutually independent, however, when they are not—as they are in any interesting belief set—the exclusion of certain beliefs must respect the logic that serves as the inference engine in the belief set.

Lastly, the fourth approach we mention is about communicating information. The flow of a river has been a metaphor for change since Heraclitus, which bestows the impression of elusiveness on the changing object. Thus the term "information flow" seems a particularly well-chosen phrase when information is perceived as being shaped by communication. It would be possible to select a concrete device or method that aids the transmission of information and investigate its characteristics. Although we do not consider any of those here, we rush to say that such considerations are practically important. From a more abstract point of view, information flows through channels between sites, which may support certain types (i.e., formulas). Channels may be typed by constraints, which are compound formulas, and channels themselves may have operations performed on them (not completely unlike how programs can be manipulated). The regularities that sites, channels, types and constraints must obey can be described in a formal theory called *channel theory*, that is somewhat similar to a typed natural deduction system.[3]

In this section we have already seen certain connections between logic and information. Now we proceed to explicate the more specific relationship that is behind information based logics.

2 Information in Logic

According to an old sentiment tautologies do not say anything about the world. This idea goes back (at least) to Wittgenstein's *Tractatus* (1999, p. 29, 1–1.2). Roughly speaking, if the world is the sum of some facts, and a tautology such as $A \vee \neg A$ ("A or not A") is always true no matter what the facts are, then $A \vee \neg A$ cannot distinguish between those states of the world in which certain facts are present and those states in which the same facts are absent. This is contrasted with other sentences (or sets thereof), let us say, $\{ B, \neg C \wedge (D \vee E) \}$. If the latter set of sentences is satisfied by the current state of the world, then a fact that supports B must be the case in the world. Similarly, there is a fact that $\neg C$, and a fact that $D \vee E$. (We tacitly assumed that the letters stand for different independent propositions, and the whole set is not contradictory.) Were a set contradictory, it could not be considered to be satisfied by any state of the world, because there can be no fact or facts that support both F and $\neg F$.

The idea that the world is a collection of facts may be a good first approximation to describe a situation using classical propositional logic, but such a description hardly seems plausible outside an informal explanation of a completeness proof for

[2] See for example Gärdenfors and Rott (1995) for a survey.

[3] See for example Barwise (1993).

some proof system. Both logic and physics made enormous steps forward since the early part of the 20th century, and the idea that the structure of reality adheres to theories which can be presented in classical (propositional) logic looks completely untenable now. Indeed, it is difficult to see why logic (of any kind) would be suitable to emulate the structure of the world, rather than to function merely as a framework for descriptions of the world.

The connection between completeness proofs and facts about the world can be depicted as a historical development. Rudolf Carnap loosened the tight connection between factual statements and the world by talking about state-descriptions, which can be thought to be complete, that is, they can provide full descriptions of how the world is or could be. (Of course, the fullness of the description is modulated by the choice of the language.) Through the work of several logicians, including A. Lindenbaum, Leon Henkin, Stig Kanger, Jaakko Hintikka and Saul Kripke, maximally consistent sets of sentences became a staple of canonical models that in model theory and in the semantics of modal, intuitionistic and other intensional logics are frequently called possible worlds. Carnap's experience with the completeness of the system **S5** shows that it is useful to consider state-descriptions or possible worlds without much metaphysical weight.

Once logic has been successfully freed from the looming confusion stemming from so-called metaphysical commitments, we can see that formal semantics for logics are a suitable tool to make explicit the information that is contained in a sentence and what can be conveyed when the sentence is communicated.

Defining a *semantics* for a logic can start with algebraizing a logic, that is, with the Lindenbaum algebra of the logic. This process is lossy, because certain elements of the information about formulas are discarded, and it is an abstraction in a philosophical sense of the word. Metaphorically speaking, algebraization sharpens our picture of a logic. Some logics can be given various semantics, for instance, algebraic semantics, a semantics utilizing finitely many truth values, a semantics built from elements of proofs, or a category-theoretic semantics. A sort of semantics that proved to be very fruitful since the pioneering work of Tarski on the semantics of classical logic, is a semantics that comprises sets together with relations and functions. For decades, a semantics that was couched in anything else than in terms of sets, was not considered "truly mathematical." In this sense, the so-called term semantics for λ-calculus and for combinatory logic were unsatisfactory compared to Dana Scott's work where functions are interpreted as continuous functions in a topology defined on a complete lattice.

In a *set-theoretic semantics* a formula is interpreted—roughly speaking—as a set of situations, which is the information content of the sentence. If a logic cannot distinguish between its theorems, on one hand, and between its contradictions, on the other, then theorems are maximally informative, whereas contradictions are void of information. (Alternatively, theorems are empty of information and contradictions provide overabundant information, if we size up information by the converse of $\subseteq$.)

For *non-classical logics*, these kinds of semantics are often labeled by Kripke's name, and for relevance logics in particular, by those of Meyer and Routley. Dunn's gaggle theory both encompasses and generalizes the standard Kripke-style and

Routley–Meyer-style semantics for a whole range of non-classical logics. *Gaggle theory* ties together the starting point for the definition of a semantics and the set-theoretic semantics: gaggles are various algebras motivated by logics and gaggle theory constructs set-theoretic representations of gaggles. The soundness and completeness theorems for logics become special instances of mathematical representation theorems.

Finding a suitable semantics for various non-classical logics was a formidable challenge in the 1960s, especially so for relevance logics. Dunn had the idea that expanding the concept of possible worlds might alleviate the difficulties. He first considered pairs of sets in a *semantics for* **FDE** which has no nested implications; he called these pairs proposition surrogates. The first element of the pair can be viewed as the set of situations that carry positive information about a formula, whereas the second element of the pair adds negative information. Dunn then was able to extend a slight modification of this semantics to allow for implications in the context of the logic **R**-mingle (or **RM**, that is, **R** with the mingle axiom added). He stipulated that the set of situations is linearly ordered and he required that each situation provide either positive or negative information for a given proposition surrogate. (A linear order is simpler than the partial order in Kripke's semantics for intuitionistic logic; in the semantics for **FDE**, a situation may provide neither positive nor negative information.) Thus, Dunn defined the first *set-theoretic semantics for a relevance logic*, but this semantics turned out not to be completely general in the sense of gaggle theory, because the implication of **RM** is modeled from a binary relation rather than from a ternary one. Having created the general framework—*gaggle theory*—to provide set-theoretic semantics for non-classical logics, Dunn gave semantics for several other non-classical logics including the Lambek calculi, negation-free modal logics, linear logic and structurally free logics. He has also given information based interpretations of the ternary relation in the Routley–Meyer semantics.

Syntactic calculi are usually provided with sound and complete semantics, which suggests that proof systems also encompass the information that can be manipulated in a logic. It is perhaps less clear how the various components of axiomatic or natural deduction systems contribute to all the information that can be used in the process of proving theorems in a calculus. However, it should be mentioned that Dunn invented not only *tableaux for* **FDE**, but he also introduced a *new type of sequent calculi*, originally for the negation-free fragment of the relevance logic **R**. These sequent calculi turned out to be very successful in formalizing positive logics with distributive conjunction and disjunction (and later on were further developed by Belnap into display logics). Instead of going into speculations as to how we should think about the two kinds of structures in these sequent calculi in terms of information, we simply point out that Dunn's research yielded powerful new results and influential ideas not only in the area of information based logics but also in proof theory and in algebraic studies of logic.

3 Threads of Ideas

What follows after this introduction is a series of papers together with a response to them. The papers are connected to the work of Dunn in various ways, and there are other thematic links between the papers. In a collection, papers are inevitably linearly ordered, and in the case of this particular volume it seemed that splitting the whole set of the papers into groups would not be helpful for the reader. Instead, the papers are arranged so that the adjacent papers have some connections between them, thereby, creating a path in the landscape of ideas, so to speak.

The paper by Avron deals with the logic **RM** (**R**-*mingle*), which grew out of some early research of Dunn (with S. McCall). Relevance logics, including the logic called entailment, are often motivated by the desire to have a connective (e.g., $\rightarrow$), which is better aligned with a relevant consequence relation (that mandates the use of all the premises) than the conditional connective of two-valued logic. Avron proves several metatheorems about **RM** (including some earlier results) and gives rigorous criteria to position **RM** as a member of the family of relevance logics.

The usefulness of **RM**, more precisely of **RM3**, the *3-valued version of* **RM**, is shown in the next paper. Mortensen considers descriptions of mathematical structures that result from identifying elements of two (distinct) structures. The prevalent mathematical approach is to form a quotient structure with a congruence relation, that is, to "typelift" the whole combined structure. A typical example is modular arithmetic, where numbers with the same remainder (mod n) are the elements of an equivalence class. Mortensen uses a topological example, in which the disjoint union of two spaces is formed save a pair of points, which are identified. **RM3** allows for theories which include both a formula and its negation—without the theory becoming trivial. Mortensen argues that merging the (consistent) theories of the component structures into a inconsistent **RM3** theory is the natural counterpart of the operation that combines the mathematical structures themselves.

Ulrich's paper continues the theme of considering **RM**, however, from an axiomatic point of view. The economy of an axiom system may be measured along different lines: the number of axioms, the (total) length of the axioms, the number of rules or perhaps even the length of the proofs of select theorems. This paper is concerned with implicational fragments of logics; hence, comparisons by length and by the number of axioms are forthright. Ulrich shows that $\mathbf{RM}_{\rightarrow}$ can be *axiomatized by two axioms*, and even *by a single one*. Additionally, he provides numerous small axiomatic systems for implicational fragments of other logics—from strict implication, entailment, **BCK**, **BCI** to the implicational fragments of intuitionistic and classical logics.

RM is the starting point of Maksimova's paper too. **RM** itself is not a finitely-valued logic, however, by a result of Dunn, all its proper extensions are, that is, **RM** is *pretabular*. Maksimova considers similar logics—except that they are extensions of intuitionistic logic. Then she turns to the positive fragment of intuitionistic logic and minimal logic, as well as to modal and relevance logics. Maksimova also notes the complexity of questions such as the tabularity problem of superintuitionistic logics.

Most of these problems turn out to be computationally intractable. Intuitionistic, minimal and certain normal modal logics have a handful of pretabular extensions, whereas **RM** is only one of infinitely many pretabular extensions of **R** (the logic of relevant implication).

Urquhart's paper is about proofs of the *admissibility of the* γ *rule for* **E** and **R**. The result was first proved by Dunn and Meyer. Urquhart uncovers—with some help from Dunn—that the original proof had a strong connection to the completeness theorem of **RM**, because a key trick in the proof of the admissibility of γ is to duplicate elements in the algebra of **R** (or **E**, respectively) on which negation has a fixed point. Urquhart outlines two further proofs, both of which rely (eventually) on the ternary relational semantics of relevance logics.

Mares interprets the ternary relation in the semantics of logics in terms of *agents manipulating sources of information*. It might seem that there is no need to interpret a semantics, however, formal semantics sometimes benefit from informal interpretations, especially, when it comes to their wider acceptance. Mares's interpretations—the productive interpretation and the functional interpretation—are adequate for multiplicative–additive linear logic (**MALL**) and contraction-free relevance logic (**RW**), and for the logics of lattice-**R** (**LR**) and relevant implication (**R**), respectively. Sources of information are typed, and rules govern how combinations of sources of information result in changes in their types.

Sequoiah-Grayson's paper is also about interpreting the ternary accessibility relation R. He develops a theme from some recent work by Dunn. Sequoiah-Grayson takes the *relevance interpretation of* R to mean epistemic relevance for an agent. Then the ternary relation connects information states. R is decomposed into an epistemic action and a binary relation between information states. Several epistemic actions can be imagined, and accordingly, Sequoiah-Grayson considers which properties of an epistemic action appear to be plausible when the action combines pieces of data, a pair of programs, or applies a program to data. Although contemporary computer science treats programs as data on purpose, a careful analysis of these three kinds of epistemic operations may reveal differences in the logics that are motivated by the different epistemic operations.

Brady in his paper connects the semantics of relevance logics with information differently than the previous two authors. Brady reserves the term "information" for true information and focuses on content (which means here what is called information elsewhere in the volume). He considers the *logic of meaning containment* (**MC**), which is a contraction- and distribution-free logic. In the interpretation, situations are taken to be pieces of content with certain operations on them, whereas the ternary accessibility relation is replaced by a combination of a binary relation and a unary operation. Brady also discusses the connections between his semantic approach and the work of Dunn and Mares who draw distinctions in the semantics of relevance logics, for instance, between prime and non-prime information, and true and non-true information.

Wansing's paper focuses on how *positive and negative information* may be used in the interpretation of substructural logics. Negation applied to a formula creates negative information, and various kinds of negation create various kinds of negative information. Dunn investigated different negations, especially, in the framework of gaggle theory. Wansing scrutinizes split negations (i.e., Galois negations) in the case where they are defined from a falsity constant by implications, similarly to how negation is definable in intuitionistic logic. It is often thought that facts support certain statements, but the justification of negative statements may require more than the mere lack of support for their positive counterparts. Similarly, in certain interpretations of intuitionistic logic, there is an asymmetry between what is true and what is not true. Wansing argues that it is possible to view positive and negative information in a symmetric fashion by appeal to direct verification and direct falsification, or by considering the four-valued logic **FDE** and the logic of the sixteen-element tri-lattice.

The *American Plan* occurring in the title of Shramko's paper, refers to the idea that the four truth values that can be used in an interpretation of **FDE** may be reconstructed as subsets of the set of "usual truth values," that is, $\{ T, F \}$. Dunn worked out several interpretations of **FDE**, each rooted in this idea, and he defined a version of truth trees as well as a sieve for **FDE**; the latter are syntactic implementations of the semantic insights. Shramko considers the bi-lattice **4** in which the two orders are the truth and the information orders. A natural step is to consider a tri-lattice with sixteen elements where falsity becomes the third order of the tri-lattice. Shramko presents a further generalization in which a multi-lattice is considered with multiple order relations that in turn allow several consequence relations to be defined.

Zhou shows in his paper how to combine **FDE**, and in general, *De Morgan lattices with belief functions*. Belief functions—compared to multiple truth values—can be seen as a different aspect of the relation between a sentence and reality, which involves (tacitly) an agent too. The paper also mentions the complexity of validity and satisfiability: they turn out to be the same as those for two-valued propositional logic. That is, the replacement of orthonegation (of two-valued logic) and the addition of belief functions does not increase the complexity of the resulting logics. Zhou also proves a representation theorem for finite De Morgan lattices using join-irreducible elements with an order-inverting operation of period two defined on sets of join-irreducible elements.

Czelakowski deals with probabilities per se, without (explicitly) stipulating an agent or even beliefs held by an agent. He considers *probabilistic interpretations of predicates*: an n-place predicate may hold of an n-tuple of objects with probability p, where $p \in [0,1]$ (rather than simply 0 or 1). The characteristic functions of predicates are cumulative distribution functions, which give rise to a De Morgan lattice. Czelakowski shows that further operations are definable in a natural way, for instance, convolution, bounded addition and strong conjunction. These operations may be added to the De Morgan lattice and the augmented algebra can be bounded (by a 0 and a 1). The operations are analogs of Łukasiewicz's logic (with uncountably many values) in the sense that the consequence relation on the algebra of cumulative distribution functions (without convolution) coincides with the consequence relation in Łukasiewicz's logic.

De Morgan lattices are distributive lattices in which the negation operation obeys the so-called De Morgan laws. Hartonas investigates *representations of negation operations* from Dunn's kite of negations when they are added to a lattice (which is not stipulated to be distributive). A difficulty in a set-theoretic representation of a lattice is that there is no obvious operation on sets for $\vee$ once $\cap$ (or $\cup$) has been chosen to stand for $\wedge$ (or vice versa). Hartonas's representation of lattices uses a neighborhood of information sites, where the neighborhood is understood in a similar sense as in the semantics of certain non-normal modal logics: the neighborhood of an information site is a set of information sites. Hartonas also considers the addition of a possibility operator to the language, which together with negation allows for impossibility to be expressed; impossibility implies more than the lack of evidence of possibility at an information site. Both negation and possibility are unary operations and they are represented using binary relations along the lines of the generalized Galois logics approach of Dunn.

The paper by Goldblatt and Grice proves a *categorial duality theorem* between Boolean contact algebras and mereotopological spaces. A special feature of this representation result is that a Boolean contact algebra is a *BA* with a binary relation which can be viewed as extensive connection between regions. In the context of logic, the typical representations are those of Lindenbaum algebras which usually have only an order relation (perhaps tacitly). On the topological side, a mereotopological space has a subalgebra of the closed regular sets of the topology which is a *BA*. Moreover, the sets that make up the selected subalgebra constitute a closed basis for the topology. The authors define morphisms between Boolean contact algebras, on one hand, and between mereotopological spaces, on the other, and they prove full duality between the respective categories.

The paper by Allwein and Harrison deals with distributed modal logics, the components of which are normal modal logics that have a *BA* reduct in their algebra. Multi-modal logics typically intertwine several modalities within one logic—possibly including axioms characterizing the interactions between different modalities. Distributed logics do not merge the separate modal logics, rather the connections between the logics are regulated by modalities. A motivation for distributed logic is the aim to *model information access* in systems, where certain parts of information contained in one component have to be shielded when another component attempts to access the information. Allwein and Harrison provide a set of axiom schemas, and algebraize distributed modal logic using heterogenous algebras. A semantics for the logic is defined based on neighborhood frames. The authors also investigate the properties of implication operations that are definable. Non-interference in a compound system means that a user with restricted access cannot distinguish filtered output from an unrestricted source from unfiltered output from the same source. The paper shows how non-interference can be modeled in distributed logic by a special simulation relation.

Van Benthem's paper continues the idea of handling information on different levels, though here the levels are thought be models that have fewer or more details. The level chosen in this paper is *propositional epistemic logic* enriched with operators that go beyond the standard *K* (knowledge) and *B* (belief) operators. Announcement and radical update are epistemic actions, which can be viewed as restrictions on the model (as a set of possible worlds). Having investigated the relationship between epistemic and plausibility models, van Benthem turns to evidence models, in which an evidence set is a set of possible worlds. He shows axiomatizability for conditional addition of evidence and conditional belief with evidence addition. An operator f can be traced on the level of evidence models by g when f and g commute with the transformation that turns an evidence model into a plausibility model. The paper shows that deletion of evidence cannot be traced, which supports the claim of the paper that information can be used and modeled in several ways at various levels in logic.

The paper by Moss shows another way how a logic allows us to manipulate information about a situation without having comprehensive information about all aspects of a situation. He extends a *syllogistic logic* with (generalized) quantifiers "at least as many as" and "(strictly) more than" (assuming finite sets of objects). The logic allows straightforward formalizations of English sentences that contain these expressions or the traditional "all" and "some" quantifiers. Moss provides two axiom systems—one with and the other without negation. These syllogistic logics are decidable, and the author wrote a program that generates a proof or a counterexample for a given inference. The interpretation of the quantifiers is unproblematic, so is the soundness of the axiom system. However, the language of syllogistic logic does not allow us to talk about individuals, which means that a Henkin-style completeness proof that relies on a model constructed from pieces of the language cannot be carried out. The bulk of the paper is the completeness proof itself, which shows that it is not an easy task to recover in a different form the information that can be reasoned with in these syllogistic logics.

While the above descriptions aim at providing a path, I should emphasize that the connections between these papers and Dunn's work as well as between the papers themselves are much richer than what could be squeezed into this section. I hope that the reading of these papers will be an enjoyable and worthwhile experience.

References

Barwise, J. (ed.) (1989). *The situation in logic*. Volume 17 of *CSLI Lecture Notes*. Stanford, CA: CSLI Publications.

Barwise, K. J. (1993). Constraints, channels and the flow of information. In P. Aczel, D. Israel, Y. Katagiri & S. Peters (Eds.), *Situation theory and its applications (v. 3)*, Volume 37 of *CSLI Lecture Notes* (pp. 3–27). Stanford, CA: CSLI.

Dunn, J. M. (2001). The concept of information and the development of modern logic. In W. Stelzner & M. Stöckler (Eds.), *Zwischen traditioneller und moderner Logik: Nichtklassische Ansatze* (pp. 423–447). Paderborn: Mentis-Verlag.

Dunn, J. M. (2008). Information in computer science. In P. Adriaans & J. van Benthem (Eds.), *Philosophy of information*, Volume 8 of *Handbook of the philosophy of science (D. M. Gabbay, P. Thagard, J. Woods (eds.))* (pp. 581–608). Amsterdam: Elsevier

Gärdenfors, P., & Rott, H. (1995). *Belief revision* (pp. 35–132). Handbook of Logic in Artificial Intelligence and Logic Programming. Oxford, UK: Oxford University Press

Groenendijk, J., & Stokhof, M. (1991). Dynamic predicate logic. *Linguistics and Philosophy, 14*, 39–100.

Wittgenstein, L. (1999). *Tractatus logico-philosophicus*. Mineola, NY: Dover.

RM and its Nice Properties

Arnon Avron

Abstract Dunn–McCall logic **RM** is by far the best understood and the most well-behaved logic in the family of logics developed by the school of Anderson and Belnap. However, it is not considered to be a relevant logic by the relevant logicians, since it fails to have the variable-sharing property. Instead, **RM** is usually characterized as being "semi-relevant," without explaining what this notion means. In this paper we suggest a plausible definition of semi-relevance, and show that according to it, **RM** is a strongly maximal semi-relevant logic having a conjunction, a disjunction, and an implication. We also review and prove the most important nice properties of **RM**, especially strong completeness results about it (the full proofs of which are difficult to find in the literature).

Keywords Degrees of truth · Fuzzy logics · Paraconsistent logics · Relevant logics · Semi-relevance

1 Introduction

The central idea behind the design of $\mathbf{R}_{\rightarrow}$, the basic, purely implicational fragment of the relevant logic **R**, is that $\varphi \rightarrow \psi$ should relevantly follow from $\mathcal{T}$ iff there is a proof of ψ from $\mathcal{T}, \varphi$ in which φ is actually *used*.[1] But what exactly is meant by '$\mathcal{T}, \varphi$' in this formulation? In textbooks on logics, this is usually just an abbreviation for $\mathcal{T} \cup \{\varphi\}$, where $\mathcal{T}$ is a *set* of formulas. However, this interpretation is problematic from the point of view of $\mathbf{R}_{\rightarrow}$. To see why, consider the question whether $\varphi \rightarrow \varphi$ should relevantly follow from the assumption φ. According to the above criterion, this is the case iff there is a proof of φ from φ, φ that actually uses φ. By the standard interpretation, this means that there is a proof of φ from $\{\varphi\} \cup \{\varphi\}$ that uses φ, i.e., there is a proof of φ from $\{\varphi\}$ that uses φ. This is certainly the case, and so we should

[1] This principle is practically abandoned in the full system **R**.

A. Avron (✉)
School of Computer Science, Tel Aviv University, Tel Aviv, Israel
e-mail: aa@cs.tau.ac.il
URL: http://www.cs.tau.ac.il/~aa

K. Bimbó (ed.), *J. Michael Dunn on Information Based Logics*,
Outstanding Contributions to Logic 8, DOI 10.1007/978-3-319-29300-4_2

conclude that indeed $\varphi \to \varphi$ relevantly follows from φ, implying that $\varphi \to (\varphi \to \varphi)$ should be provable. Unfortunately, this formula is *not* provable in $\mathbf{R}_{\to}$. The reason is that the above criterion leads to $\mathbf{R}_{\to}$ only if the term '$\mathcal{T}, \varphi$' in its formulation is understood as the *multiset* which is obtained by adding (a copy of) φ to the *multiset* $\mathcal{T}$.

It somewhat looks strange to take relevant entailment as a relation between *multisets* of formulas and formulas, rather than between *sets* of formulas and formulas (as consequence relations are usually and most naturally taken to be). This observation motivated J.M. Dunn and S. McCall in investigating the results of adding to **R** and its fragments the mingle axiom $\varphi \to (\varphi \to \varphi)$ considered above. In the case of $\mathbf{R}_{\to}$, this addition yields $\mathbf{RM0}_{\to}$, which is the minimal system in which the above criterion for relevant entailment is met, with the latter taken as a relation between *sets* of formulas and formulas. In the case of the full system **R**, it yields a very interesting system called **RM** ("**R**-mingle"). As noted in Dunn and Restall (2002), Dunn–McCall logic **RM** "is by far the best understood of the Anderson–Belnap style systems." However, it is not considered to be a relevant logic by the relevant logicians, since it fails to have the variable-sharing property. Instead, **RM** is usually characterized as being "semi-relevant," without explaining what this notion means. In this paper we suggest a plausible definition of semi-relevance, and show that according to it, **RM** is a strongly maximal semi-relevant logic having a conjunction, a disjunction, and an implication. We also review and prove known important properties of **RM**, especially *strong* completeness results whose full proofs are difficult to find in the literature.

2 Preliminaries

2.1 Propositional Logics

In the sequel, $\mathcal{L}$ denotes a propositional language. The set of well-formed formulas of $\mathcal{L}$ is denoted by $\mathcal{W}(\mathcal{L})$, and φ, ψ, σ vary over its elements. $\mathcal{T}$, $\mathcal{S}$ vary over theories of $\mathcal{L}$ (where by a 'theory' we simply mean here a subset of $\mathcal{W}(\mathcal{L})$), and Γ, Δ vary over *finite* sets of formulas. We denote by $\mathsf{Atoms}(\varphi)$ ($\mathsf{Atoms}(\mathcal{T})$) the set of atomic formulas that appear in φ (in formulas of $\mathcal{T}$).

Definition 2.1 A (Tarskian) *consequence relation* (tcr) for a language $\mathcal{L}$ is a binary relation $\vdash$ between theories in $\mathcal{W}(\mathcal{L})$ and formulas in $\mathcal{W}(\mathcal{L})$, satisfying the following three conditions.

[R] *Reflexivity*: $\psi \vdash \psi$ (i.e., $\{\psi\} \vdash \psi$).
[M] *Monotonicity*: If $\mathcal{T} \vdash \psi$ and $\mathcal{T} \subseteq \mathcal{T}'$, then $\mathcal{T}' \vdash \psi$.
[C] *Cut (Transitivity)*: If $\mathcal{T} \vdash \psi$ and $\mathcal{T}', \psi \vdash \varphi$, then $\mathcal{T} \cup \mathcal{T}' \vdash \varphi$

Definition 2.2 Let $\vdash$ be a Tarskian consequence relation for $\mathcal{L}$.

- $\vdash$ is *structural*, if for every $\mathcal{L}$-substitution θ and every $\mathcal{T}$ and ψ, if $\mathcal{T} \vdash \psi$, then $\theta(\mathcal{T}) \vdash \theta(\psi)$.
- $\vdash$ is *non-trivial*, if $p \nvdash q$ for distinct atoms $p, q \in \mathsf{Atoms}(\mathcal{L})$.
- $\vdash$ is *finitary*, if for every theory $\mathcal{T}$ and every formula ψ such that $\mathcal{T} \vdash \psi$, there is a *finite* theory $\Gamma \subseteq \mathcal{T}$ such that $\Gamma \vdash \psi$.

Note 2.1 The condition of non-triviality is strictly stronger than the more familiar condition of consistency used by Dunn in Sect. 29.4 of Anderson and Belnap (1975), which says that $\nvdash q$ for $q \in \mathsf{Atoms}(\mathcal{L})$. Thus the tcr $\vdash$ for which $\mathcal{T} \vdash \varphi$ iff $\mathcal{T} \neq \emptyset$ is structural, finitary, and consistent, but not non-trivial.

Definition 2.3

- A (propositional) *logic* is a pair $\mathbf{L} = \langle \mathcal{L}, \vdash_{\mathbf{L}} \rangle$, where $\mathcal{L}$ is a propositional language, and $\vdash_{\mathbf{L}}$ is a structural and non-trivial tcr for $\mathcal{L}$.[2]
- A logic $\langle \mathcal{L}, \vdash_{\mathbf{L}} \rangle$ is *finitary* if $\vdash_{\mathbf{L}}$ is finitary.

Definition 2.4 Let $\mathbf{L}_1 = \langle \mathcal{L}_1, \vdash_{\mathbf{L}_1} \rangle$ and $\mathbf{L}_2 = \langle \mathcal{L}_2, \vdash_{\mathbf{L}_2} \rangle$ be propositional logics.

- $\mathbf{L}_1$ is an *extension* of $\mathbf{L}_2$, if $\mathcal{L}_2 \subseteq \mathcal{L}_1$ and $\vdash_{\mathbf{L}_2} \subseteq \vdash_{\mathbf{L}_1}$.
- $\mathbf{L}_1$ is a *simple extension* of $\mathbf{L}_2$, if $\mathcal{L}_2 = \mathcal{L}_1$ and $\vdash_{\mathbf{L}_2} \subseteq \vdash_{\mathbf{L}_1}$.
- $\mathbf{L}_1$ is a *proper extension* of $\mathbf{L}_2$, if $\mathcal{L}_2 \subseteq \mathcal{L}_1$ and $\vdash_{\mathbf{L}_2} \subsetneq \vdash_{\mathbf{L}_1}$.
- $\mathbf{L}_1$ is a *strongly proper* extension of $\mathbf{L}_2$, if $\mathcal{L}_2 \subseteq \mathcal{L}_1$, and there is a sentence φ of $\mathcal{L}_2$ such that $\vdash_{\mathbf{L}_1} \varphi$ but $\nvdash_{\mathbf{L}_2} \varphi$.
- $\mathbf{L}_1$ is a *conservative extension* of $\mathbf{L}_2$, if $\mathcal{L}_2 \subseteq \mathcal{L}_1$, and $\mathcal{T} \vdash_{\mathbf{L}_1} \psi$ iff $\mathcal{T} \vdash_{\mathbf{L}_2} \psi$ whenever $\mathcal{T} \cup \{\psi\} \in 2^{\mathcal{W}(\mathcal{L}_2)}$.
- $\mathbf{L}_1$ is a *weakly conservative extension* of $\mathbf{L}_2$, if $\mathcal{L}_2 \subseteq \mathcal{L}_1$, and $\vdash_{\mathbf{L}_1} \psi$ iff $\vdash_{\mathbf{L}_2} \psi$ whenever $\psi \in \mathcal{W}(\mathcal{L}_2)$.
- $\mathbf{L}_1$ is an *axiomatic extension* of $\mathbf{L}_2$, if $\mathcal{L}_2 \subseteq \mathcal{L}_1$, and there is a set $\mathcal{S}$ of sentences in $\mathcal{L}_1$ such that $\vdash_{\mathbf{L}_1}$ is the minimal structural tcr $\vdash$ on $\mathcal{L}_1$ which satisfies the following conditions: $\vdash_{\mathbf{L}_2} \subseteq \vdash$, and $\vdash \varphi$ for every $\varphi \in \mathcal{S}$.

Definition 2.5 Let $\mathbf{L} = \langle \mathcal{L}, \vdash_{\mathbf{L}} \rangle$ be a propositional logic.

- A binary connective $\supset$ of $\mathcal{L}$ is called an *implication for* $\mathbf{L}$ if the classical deduction theorem holds for $\supset$ and $\vdash_{\mathbf{L}}$. That is,

$$\mathcal{T}, \varphi \vdash_{\mathbf{L}} \psi \quad \text{iff} \quad \mathcal{T} \vdash_{\mathbf{L}} \varphi \supset \psi.$$

- A binary connective $\wedge$ of $\mathcal{L}$ is called a *conjunction for* $\mathbf{L}$ if it satisfies the following condition:

$$\mathcal{T} \vdash_{\mathbf{L}} \psi \wedge \varphi \quad \text{iff} \quad \mathcal{T} \vdash_{\mathbf{L}} \psi \text{ and } \mathcal{T} \vdash_{\mathbf{L}} \varphi.$$

[2]The condition of non-triviality is not always explicitly demanded, but we have found it (here and elsewhere) convenient to include it in order to avoid uninteresting pathological cases.

- A binary connective $\vee$ of $\mathcal{L}$ is called a *disjunction for* **L** if it satisfies the following condition:

$$\mathcal{T}, \psi \vee \varphi \vdash_{\mathbf{L}} \sigma \quad \text{iff} \quad \mathcal{T}, \psi \vdash_{\mathbf{L}} \sigma \text{ and } \mathcal{T}, \varphi \vdash_{\mathbf{L}} \sigma.$$

Definition 2.6 We call a logic *normal* if it has all the basic connectives above (conjunction, disjunction, implication).[3]

Definition 2.7 Let $\mathcal{L}$ be a propositional language.

- A *matrix* for $\mathcal{L}$ is a triple $\mathcal{M} = \langle \mathcal{V}, \mathcal{D}, \mathcal{O} \rangle$, where
  (1) $\mathcal{V}$ is a non-empty set of truth values;
  (2) $\mathcal{D}$ is a non-empty proper subset of $\mathcal{V}$ (the *designated* elements of $\mathcal{V}$);
  (3) $\mathcal{O}$ is a function that associates an n-ary function $\tilde{\diamond}_{\mathcal{M}}: \mathcal{V}^n \to \mathcal{V}$ with every n-ary connective $\diamond$ of $\mathcal{L}$.

  We say that $\mathcal{M}$ is *(in)finite*, if so is $\mathcal{V}$.
- Let $\mathcal{M} = \langle \mathcal{V}, \mathcal{D}, \mathcal{O} \rangle$ be a matrix for $\mathcal{L}$. An $\mathcal{M}$*-valuation* for $\mathcal{L}$ is a function $\nu: \mathcal{W}(\mathcal{L}) \to \mathcal{V}$ such that $\nu(\diamond(\psi_1, \ldots, \psi_n)) = \tilde{\diamond}_{\mathcal{M}}(\nu(\psi_1), \ldots, \nu(\psi_n))$ for every n-ary connective $\diamond$ of $\mathcal{L}$ and every $\psi_1, \ldots, \psi_n$ in $\mathcal{W}(\mathcal{L})$.
- An $\mathcal{M}$-valuation ν is an $\mathcal{M}$*-model* of a formula ψ, or ν $\mathcal{M}$*-satisfies* ψ (notation: $\nu \models_{\mathcal{M}} \psi$), if $\nu(\psi) \in \mathcal{D}$. We say that ν is an $\mathcal{M}$*-model* of a theory $\mathcal{T}$ (notation: $\nu \models_{\mathcal{M}} \mathcal{T}$), if it is an $\mathcal{M}$-model of every element of $\mathcal{T}$.
- Let $\mathcal{M}$ be a matrix for $\mathcal{M}$. $\vdash_{\mathcal{M}}$, the consequence relation that is induced by $\mathcal{M}$, is defined by: $\mathcal{T} \vdash_{\mathcal{M}} \psi$ if every $\mathcal{M}$-model of $\mathcal{T}$ is an $\mathcal{M}$-model of ψ. We shall denote by $\mathbf{L}_{\mathcal{M}}$ the logic $\langle \mathcal{L}, \vdash_{\mathcal{M}} \rangle$ which is induced by $\mathcal{M}$.

Definition 2.8 Let $\mathbf{L} = \langle \mathcal{L}, \vdash_{\mathbf{L}} \rangle$ be a propositional logic, and let $\mathcal{M}$ be a matrix for $\mathcal{L}$.

- If $\mathbf{L}_{\mathcal{M}}$ is an extension of **L**, we say that **L** is *sound* for $\mathcal{M}$.
- If **L** is an extension of $\mathbf{L}_{\mathcal{M}}$, we say that **L** is *complete* for $\mathcal{M}$.
- $\mathcal{M}$ is a *characteristic* matrix for **L**, if $\mathbf{L} = \mathbf{L}_{\mathcal{M}}$ (that is, if **L** is both sound and complete for $\mathbf{L}_{\mathcal{M}}$).
- $\mathbf{L}_{\mathcal{M}}$ is *weakly sound* for **L**, if for every $\psi \in \mathcal{W}(\mathcal{L})$, $\vdash_{\mathcal{M}} \psi$ implies that $\vdash_{\mathbf{L}} \psi$. $\mathbf{L}_{\mathcal{M}}$ is *weakly complete* for **L**, if $\vdash_{\mathbf{L}} \psi$ implies that $\vdash_{\mathcal{M}} \psi$.
- $\mathcal{M}$ is a *weakly characteristic* matrix for **L**, if **L** is both weakly sound and weakly complete for $\mathbf{L}_{\mathcal{M}}$ (that is, $\vdash_{\mathcal{M}} \psi$ iff $\vdash_{\mathbf{L}} \psi$).

2.2 *Some Basic Relevant Logics*

In this section, we shortly review some basic relevant logics, together with their properties that will be used later in our study of **RM**. (**RM** itself will be introduced in Sect. 4.) We start with the central relevant logic **R**.

---

[3]Our notion of normality should not be confused with the notion of normality used in modal logics, or the notion of normal theory used in Anderson and Belnap (1975).

Definition 2.9 Let $\mathcal{L}_R \equiv \{\wedge, \vee, \rightarrow, \neg\}$.

Definition 2.10 **R** is the logic in $\mathcal{L}_R$ which is induced by the system *HR* that is presented in Fig. 1.

For our purposes, the most important property of **R** is the following theorem, an (implicit) proof of which can be found, e.g., in Anderson and Belnap (1975, p. 301).

Theorem 2.11 $\vee$ *is a disjunction for any axiomatic extension of* **R**.

A particularly important fragment of **R** is its intensional fragment.

Definition 2.12 Let $HR_{\overrightarrow{\neg}}$ be the Hilbert-type systems in $\{\neg, \rightarrow\}$ whose axioms and rule are those axioms and rule of *HR* which do not mention $\wedge$ or $\vee$ (i.e., [Id], [Tr], [Pe], [Ct], [N1], [N2], and [MP]). $\mathbf{R}_{\overrightarrow{\neg}}$ is the logic in $\{\neg, \rightarrow\}$ which is induced by $HR_{\overrightarrow{\neg}}$.

The following theorem has been proved by Meyer (see Anderson and Belnap 1975).

Proposition 2.13 **R** *is a conservative extension of* $\mathbf{R}_{\overrightarrow{\neg}}$. *In other words,* $HR_{\overrightarrow{\neg}}$ *axiomatizes the* $\{\neg, \rightarrow\}$*-fragment of* **R**.

The most significant property of $\mathbf{R}_{\overrightarrow{\neg}}$ is that very natural relevant deduction theorems obtain for it. The simplest one is the following proposition from Avron (2014) (originally due to Church).

Fig. 1 The proof system *HR*

Axioms:

[Id]	$\varphi \rightarrow \varphi$
[Tr]	$(\varphi \rightarrow \psi) \rightarrow ((\psi \rightarrow \sigma) \rightarrow (\varphi \rightarrow \sigma))$
[Pe]	$(\varphi \rightarrow (\psi \rightarrow \sigma)) \rightarrow (\psi \rightarrow (\varphi \rightarrow \sigma))$
[Ct]	$(\varphi \rightarrow (\varphi \rightarrow \psi)) \rightarrow (\varphi \rightarrow \psi)$
[N1]	$(\varphi \rightarrow \neg\psi) \rightarrow (\psi \rightarrow \neg\varphi)$
[N2]	$\neg\neg\varphi \rightarrow \varphi$
[ConE1]	$\varphi \wedge \psi \rightarrow \varphi$
[ConE2]	$\varphi \wedge \psi \rightarrow \psi$
[ConI])	$(\sigma \rightarrow \varphi) \wedge (\sigma \rightarrow \psi) \rightarrow (\sigma \rightarrow \varphi \wedge \psi)$
[DisI1]	$\varphi \rightarrow \varphi \vee \psi$
[DisI2]	$\psi \rightarrow \varphi \vee \psi$
[DisE]	$(\varphi \rightarrow \sigma) \wedge (\psi \rightarrow \sigma) \rightarrow (\varphi \vee \psi \rightarrow \sigma)$
[Dist]	$\varphi \wedge (\psi \vee \sigma) \rightarrow (\varphi \wedge \psi) \vee (\varphi \wedge \sigma)$

Rules:

[MP] $\dfrac{\varphi \quad \varphi \rightarrow \psi}{\psi}$ [Ad] $\dfrac{\varphi \quad \psi}{\varphi \wedge \psi}$

Proposition 2.14 *Let* **L** *be an axiomatic extension of* $\mathbf{R}_{\overrightarrow{\neg}}$. *Then* **L** *satisfies the following relevant deduction theorem:*

$$\mathcal{T}, \varphi \vdash_{\mathbf{L}} \psi \ \textit{ iff } \ \mathcal{T} \vdash_{\mathbf{L}} \psi \textit{ or } \mathcal{T} \vdash_{\mathbf{L}} \varphi \to \psi.$$

Another important property of $\mathbf{R}_{\overrightarrow{\neg}}$ (see Anderson and Belnap 1975; Dunn and Restall 2002) is the fact that it has a corresponding cut-free Gentzen-type calculus, which can be used for a decision procedure. That system can also be used for an easy proof of the next lemma.

Definition 2.15

- $\varphi + \psi =_{Df} \neg\varphi \to \psi$
- $\varphi \otimes \psi =_{Df} \neg(\varphi \to \neg\psi)$
- $\varphi \leftrightarrow \psi =_{Df} (\varphi \to \psi) \otimes (\psi \to \varphi)$

Lemma 2.16 *All instances of the following formulas are provable in* $HR_{\overrightarrow{\neg}}$:

1. $(\varphi \leftrightarrow \psi) \to (\varphi \to \psi)$ *and* $(\varphi \leftrightarrow \psi) \to (\psi \to \varphi)$
2. $(\varphi + \psi) \leftrightarrow (\psi + \varphi)$ *and* $((\varphi + \psi) + \sigma) \leftrightarrow (\varphi + (\psi + \sigma))$
3. $(\varphi + \varphi) \to \varphi$
4. $(\varphi_1 \to \psi_1) \to ((\varphi_2 \to \psi_2) \to ((\varphi_1 + \varphi_2) \to (\psi_1 + \psi_2)))$
5. $(\varphi \to \sigma) \to ((\psi \to \sigma) \to ((\varphi + \psi) \to \sigma))$
6. $\neg\varphi + \varphi$
7. $\neg\neg\varphi \leftrightarrow \varphi$
8. $(\neg\psi \to \sigma) \to ((\psi \to \sigma) \to \sigma)$
9. $(\psi \to \neg\sigma) \to ((\psi \to \sigma) \to \neg\psi)$
10. $(\varphi \to (\varphi + \varphi)) \leftrightarrow (\neg\varphi \to (\neg\varphi \to \neg\varphi))$
11. $((\varphi \to \varphi) + (\psi \to \psi)) \leftrightarrow (\varphi \to (\psi \to (\varphi + \psi))) \leftrightarrow ((\varphi \to \psi) + (\psi \to \varphi))$

With the help of [DisI1] and [DisI2], item 5 of Lemma 2.16 entails

Lemma 2.17 $\vdash_{\mathbf{R}} (\varphi + \psi) \to \varphi \vee \psi$.

One more important property of **R** and $\mathbf{R}_{\overrightarrow{\neg}}$ that we will need is given in the next proposition.

Proposition 2.18 *Every simple extension* **L** *of either* **R** *or* $\mathbf{R}_{\overrightarrow{\neg}}$ *has the* replacement property, *that is, if* $\vdash_{\mathbf{L}} \psi \leftrightarrow \varphi$, *then* $\vdash_{\mathbf{L}} \sigma\{\varphi/p\} \leftrightarrow \sigma\{\psi/p\}$ *for every sentence* σ *and atom* p.

Another central purely intensional relevant logic is the following logic, which can easily be seen to be a simple axiomatic extension of $\mathbf{R}_{\overrightarrow{\neg}}$.

Definition 2.19 Let $HRMI_{\overset{\neg}{\to}}$ be the Hilbert-type system in $\{\neg, \to\}$ that is obtained from $HR_{\overset{\neg}{\to}}$ by replacing the identity axiom [Id] by the mingle axiom:

$$[\mathrm{Mi}] \quad \varphi \to (\varphi \to \varphi)$$

$\mathbf{RMI}_{\overset{\neg}{\to}}$ is the logic in $\{\neg, \to\}$ which is induced by $HRMI_{\overset{\neg}{\to}}$.

$\mathbf{RMI}_{\overset{\neg}{\to}}$ has been investigated in Avron (1984). It is shown there that it has the following properties.

- the variable-sharing property[4];
- a weakly characteristic infinite matrix that provides a decision procedure;
- Scroggs' property (which **RM** has as well—see Theorem 6.9);
- an associated cut-free Gentzen-type system $GRMI_{\overset{\neg}{\to}}$ which provides a decision procedure too.

$GRMI_{\overset{\neg}{\to}}$ can be used for verifying the next lemma. Alternatively, the lemma can easily be proved with the help of Lemma 2.16, the definition of $+$, and the mingle axiom.

Lemma 2.20 *All instances of the following formulas are theorems of* $\mathbf{RMI}_{\overset{\neg}{\to}}$.

1. $\psi + \psi \leftrightarrow \psi$ *(and so* $\psi \leftrightarrow (\neg\psi \to \psi)$*)*
2. $(\psi \to \sigma) \to (\sigma \to (\psi \to \sigma))$
3. $(\psi \to \sigma) \to (\neg\psi \to (\psi \to \sigma))$
4. $\neg((\varphi \to \psi) \to (\varphi \to \psi)) \to \psi$
5. $\neg(\psi \to \sigma) \to ((\psi \to \sigma) \to \sigma)$
6. $\neg(\psi \to \sigma) \to ((\psi \to \sigma) \to \neg\psi)$
7. $(\neg\psi \to \sigma) \to (\neg\psi \to (\psi \to \sigma))$
8. $(\sigma \to \neg\psi) \to (\sigma \to (\psi \to \sigma))$

3 Semi-relevance

In Avron (2014) we have tried to characterize the notion of a relevant logic. A central part in that characterization was the presence of an implication $\to$ with certain properties, including the famous variable-sharing property of Anderson and Belnap (see Anderson and Belnap 1975). Now we turn to the problem of characterizing "semi-relevance." Naturally, this should be a weaker notion, for which the notion of relevance is still relevant. Our idea is to look for general conditions, not depending on the properties of any particular connective, which seem relevant. One such condition that seems absolutely necessary was already given in Avron (2014):

---

[4]This was observed already in Parks (1972). See also (Anderson and Belnap 1975, p. 148).

Definition 3.1 A logic $\mathbf{L} = \langle \mathcal{L}, \vdash_{\mathbf{L}} \rangle$ satisfies the *basic relevance criterion* if for every two theories $\mathcal{T}_1, \mathcal{T}_2$ and formula ψ, we have that $\mathcal{T}_1 \vdash_{\mathbf{L}} \psi$ whenever $\mathcal{T}_1 \cup \mathcal{T}_2 \vdash_{\mathbf{L}} \psi$ and $\mathcal{T}_2$ has no atomic formulas in common with $\mathcal{T}_1 \cup \{\psi\}$.

Note 3.1 As explained in Avron (2014), the idea behind the basic relevance criterion is that if a theory $\mathcal{T}_2$ shares no content with $\mathcal{T}_1 \cup \{\psi\}$ then it should not be *relevant* to the question whether $\mathcal{T}_1 \vdash \psi$ or not. These idea and criterion were already implicit in the claim denoted by RM87, on p. 418 of Anderson and Belnap (1975), and almost explicit in the discussion that follows it. It is argued there that this criterion is in fact stronger than the usual relevance criterion (i.e., the variable-sharing property). RM87 (actually, the discussion that follows it) claims that **RM** and **R** satisfy it. Though these claims are correct, their proofs in Anderson and Belnap (1975) are not: they use a false deduction theorem for those logics.[5] Below we provide a correct proof in the case of **RM** (see Proposition 6.5).

The following proposition is an immediate consequence of Definition 3.1.

Proposition 3.2 *Suppose* $\mathbf{L} = \langle \mathcal{L}, \vdash_{\mathbf{L}} \rangle$ *is a logic that satisfies the basic relevance criterion. Then:*

1. *if* $\mathcal{T} \vdash_{\mathbf{L}} \psi$, *then either* $\vdash_{\mathbf{L}} \psi$, *or* $\mathcal{T}$ *and* ψ *share an atomic formula;*
2. $\mathcal{T} \nvdash_{\mathbf{L}} q$ *whenever* q *is an atom that does not occur in any formula of* $\mathcal{T}$*;*
3. $\mathbf{L}$ *is* paraconsistent *with respect to any (primitive or defined) unary connective* $\neg$ *of* $\mathcal{L}$*, i.e.,* $\neg p, p \nvdash_{\mathbf{L}} q$ *in case* p *and* q *are distinct atoms.*

Example 3.2 1. Since q follows from $\{p, \neg p\}$ in classical logic and in intuitionistic logic, these logics do not satisfy the basic relevance criterion. However, their *positive fragments* are easily seen to satisfy it.

2. Let $\mathcal{M} = \langle \{t, \top, f\}, \{t, \top\}, \mathcal{O} \rangle$ be a three-valued logic. Assume that all operations of $\mathcal{O}$ are $\{\top\}$-closed (i.e., that $\tilde{\diamond}(\top, \top, \ldots, \top) = \top$ for every connective $\diamond$ of the language). Then $\mathbf{L}_{\mathcal{M}}$ satisfies the basic relevance criterion. That is, if $\mathsf{Atoms}(\mathcal{T}_2) \cap \mathsf{Atoms}(\mathcal{T}_1 \cup \{\psi\}) = \emptyset$, then by assigning $\top$ to any p in $\mathsf{Atoms}(\mathcal{T}_2)$ we can turn any countermodel of $\mathcal{T}_1 \vdash_{\mathcal{M}} \psi$ into a countermodel of $\mathcal{T}_1, \mathcal{T}_2 \vdash_{\mathcal{M}} \psi$.

Proposition 3.3 *Suppose* $\mathbf{L} = \langle \mathcal{L}, \vdash_{\mathbf{L}} \rangle$ *is a finitary logic that satisfies the basic relevance criterion. Then* $\mathbf{L}$ *has a characteristic matrix.*

Proof A logic which satisfies the basic relevance criterion is by definition *uniform* (Urquhart 2001), while according to Łoś–Suszko's Theorem (see Łoś and Suszko 1958; Urquhart 2001), a uniform finitary propositional logic has a single characteristic matrix. □

[5]Thus if $\mathcal{T}$ is $\{p\}$, and ψ is $p \wedge (q \to q)$, then ψ follows from $\mathcal{T}$ in **RM**, but there is no 'appropriate form of the deduction theorem' for either **R** or **RM** that would justify the argument outlined in those proofs.

Example 3.2 shows that we cannot be satisfied with the basic relevance criterion. Thus both of the positive logics mentioned in its first item have $q \supset (p \supset q)$ as a valid formula, while the rejection of this "paradox of material implication" has been one of the main motivations for developing relevant logics. Proposition 3.3 suggests a natural direction for going beyond the basic relevance criterion: to impose appropriate constraints on the characteristic matrices of the logics which satisfy it (whose existence is guaranteed by that proposition). Next is an analysis which leads to a reasonable constraint of this sort.

Let **L** be one of the three-valued logics mentioned in Example 3.2. Then any two paradoxical formulas are necessarily indistinguishable in **L**. (Formally, this is reflected by the fact that $p, \neg p, q, \neg q, \psi[p/r] \vdash_{\mathbf{L}} \psi[q/r]$ for every p, q, r and ψ.) Intuitively, this state of affairs is in a direct conflict with principles of relevance. More generally, if a logic is induced by a finite matrix with n elements, then in any state of affairs any set of $n + 1$ formulas necessarily includes two different formulas which are absolutely indistinguishable in that state of affairs. (Formally, if ν is a valuation, and $\psi_1, \ldots, \psi_{n+1}$ are formulas, then there are $1 \leq i < j \leq n + 1$ such that for any formula φ and any atom p, $\nu(\varphi[\psi_i/p]) = \nu(\varphi[\psi_j/p])$.) This again is in conflict with the idea of relevance. It seems counterintuitive that there is an a priori, logically dictated, fixed finite bound on the number of distinct propositions (or even just distinct paradoxical propositions). According to this intuition, any characteristic matrix for a relevant logic should necessarily be infinite. Actually, it seems reasonable to make a little bit stronger demand.

Definition 3.4 (*Minimal semantic relevance criterion*) A logic **L** satisfies the *minimal semantic relevance criterion* if it does not have a finite weakly characteristic matrix.

Note 3.3 The main reason that the minimal semantic relevance criterion forbids a relevant logic **L** to have even a finite *weakly* characteristic matrix is that the existence of finite weakly characteristic matrix is often reflected in the validity of counter-intuitive (from a relevance point of view) formulas. Thus, the existence of a 3-valued weakly characteristic matrix is frequently reflected by a formula of the form

$$(p_1 \leftrightarrow p_2) \vee (p_1 \leftrightarrow p_3) \vee (p_1 \leftrightarrow p_4) \vee (p_2 \leftrightarrow p_3) \vee (p_2 \leftrightarrow p_4) \vee (p_3 \leftrightarrow p_4),$$

where $\leftrightarrow$ and $\vee$ are appropriate equivalence and disjunction connectives, respectively, available in the logic.

Note 3.4 To the best of our knowledge, our minimal semantic relevance criterion has never been suggested before as a criterion for relevance (not even in Avron (2014)). Nevertheless, all the main systems that have been designed to be relevant logics do satisfy it (see Anderson and Belnap 1975).

The two criteria suggested above do not seem sufficient for characterizing relevant logics. However, we believe that they do suffice for characterizing *semi*-relevance.

Definition 3.5 A logic **L** which satisfies both the basic relevance criterion and the minimal semantic relevance criterion is called *semi-relevant*.

4 Introducing RM and $\mathbf{RM}_{\overrightarrow{\neg}}$

Now we turn at last to the subject of this paper, the logic **RM**.

Definition 4.1

1. *HRM* is the Hilbert-type system which is obtained from *HR* by replacing the identity axiom $\varphi \to \varphi$ by the mingle axiom [Mi] (Definition 2.19).
2. **RM** is the logic in $\mathcal{L}_R$ which is induced by *HRM*.
3. $\mathbf{RM}_{\overrightarrow{\neg}}$ is the $\{\neg, \to\}$-fragment of **RM**.

The next proposition lists some of the most characteristic properties of **RM**.

Proposition 4.2

1. $\vdash_{\mathbf{RM}} \varphi + \varphi \leftrightarrow \varphi$
2. $\vdash_{\mathbf{RM}} \varphi \wedge \psi \to \varphi + \psi$
3. *If* $\vdash_{\mathbf{RM}} \varphi$*, and* $\vdash_{\mathbf{RM}} \psi$*, then* $\vdash_{\mathbf{RM}} \varphi + \psi$.
4. *Each of the three equivalent formulas in the last item of Lemma 2.16 is provable in* **RM**.

Proof 1. Immediate from item 1 of Lemma 2.20.
2. First substitute in item 4 of Lemma 2.16 $\varphi \wedge \psi$ for φ_1 and φ_2, φ for ψ_1, and ψ for ψ_2. Then by using the conjunction axioms of *HRM*, we get that $\vdash_{\mathbf{RM}} (\varphi \wedge \psi + \varphi \wedge \psi) \to \varphi + \psi$. Hence the claim follows from the first part.
3. Immediate from item 2 and the adjunction rule [Ad].
4. From item 3 it follows that $\vdash_{\mathbf{RM}} (\varphi \to \varphi) + (\psi \to \psi)$. Now apply item 11 of Lemma 2.16. □

It was observed by Parks (1972) (see also Anderson and Belnap 1975, p. 148) that $\mathbf{RM}_{\overrightarrow{\neg}}$ is not identical with $\mathbf{RMI}_{\overrightarrow{\neg}}$. Indeed, item 3 (or 4) of Proposition 4.2 implies that unlike $\mathbf{RMI}_{\overrightarrow{\neg}}$, $\mathbf{RM}_{\overrightarrow{\neg}}$ does not have the variable-sharing property for $\to$. Accordingly, our first task is to provide an axiomatization of $\mathbf{RM}_{\overrightarrow{\neg}}$. This is what we do next.

Definition 4.3 1. $HRM_{\overrightarrow{\neg}}$ is the Hilbert-type system that is obtained from $HR_{\overrightarrow{\neg}}$ by replacing the identity axiom $\varphi \to \varphi$ by axiom [++] below.

$$[++] \quad (\varphi \to \varphi) + (\psi \to \psi)$$

2. $\mathbf{L}_{HRM_{\overrightarrow{\neg}}}$ is the logic induced by $HRM_{\overrightarrow{\neg}}$.

Proposition 4.4 $\mathbf{RMI}_{\overrightarrow{\neg}} \subseteq \mathbf{L}_{HRM_{\overrightarrow{\neg}}}$.

Proof By substituting φ for ψ in [++] and in the last item of Lemma 2.16, we get that $\vdash_{HRM_{\overrightarrow{\neg}}} \varphi \to (\varphi \to \varphi + \varphi)$. Using contraction, it follows that $\vdash_{HRM_{\overrightarrow{\neg}}} \varphi \to \varphi + \varphi$. By item 10 of Lemma 2.16 (using item 7 of that lemma), this implies that the mingle axiom [Mi] is provable in $HRM_{\overrightarrow{\neg}}$. □

Proposition 4.5 *If* $\mathcal{T} \cup \{\varphi\}$ *is in the language* $\{\neg, \rightarrow\}$, *and* $\mathcal{T} \vdash_{HRM_{\overset{\neg}{\rightarrow}}} \varphi$, *then* $\mathcal{T} \vdash_{\mathbf{RM}_{\overset{\neg}{\rightarrow}}} \varphi$.

Proof Immediate from the last item of Proposition 4.2. □

That the converse of Proposition 4.5 also holds (and so $\mathbf{RM}_{\overset{\neg}{\rightarrow}} = \mathbf{L}_{HRM_{\overset{\neg}{\rightarrow}}}$) will be shown in Theorem 5.11.

5 Semantics of RM

In this section, we introduce a semantics for **RM** (and $\mathbf{RM}_{\overset{\neg}{\rightarrow}}$) for which it is (strongly) complete.

Definition 5.1 (*Sugihara chains*) A *Sugihara chain* is a triple $\langle \mathcal{V}, \leq, - \rangle$ such that $\mathcal{V}$ has at least two elements, $\leq$ is a linear order on $\mathcal{V}$, and $-$ is an involution for $\leq$ on $\mathcal{V}$ (i.e., for every $a, b \in \mathcal{V}$, $- - a = a$, and $-b \leq -a$ whenever $a \leq b$).

Example 5.1 There are plenty of examples of Sugihara chains in all areas of mathematics. The most important for our needs are the following.

- $S_{\mathbb{R}} = \langle \mathbb{R}, \leq, - \rangle$, $S_{\mathbb{Z}} = \langle \mathbb{Z}, \leq, - \rangle$, $S_{\mathbb{Z}^*} = \langle \mathbb{Z} - \{0\}, \leq, - \rangle$, $S_{\mathbb{Q}} = \langle \mathbb{Q}, \leq, - \rangle$, and $S_{\mathbb{Q}^*} = \langle \mathbb{Q} - \{0\}, \leq, - \rangle$, where $\mathbb{R}$ is the set of real numbers, $\mathbb{Z}$ is the set of integers, $\mathbb{Q}$ is the set of rationals, $\leq$ is the usual order relation on $\mathbb{R}$, and $-a$ is the usual additive inverse of a.
- The *finite* substructures $S_{\mathbb{Z}_n} = \langle \mathbb{Z}_n, \leq, - \rangle$ and $S_{\mathbb{Z}_n^*} = \langle \mathbb{Z}_n^*, \leq, - \rangle$ of $S_{\mathbb{Z}}$, where for $n > 0$ $\mathbb{Z}_n = \{ z \in \mathbb{Z}\colon\ -n \leq z \leq n \}$, and $\mathbb{Z}_n^* = \mathbb{Z}_n - \{0\}$.
- $S_{[0,1]} = \langle [0, 1], \leq, \lambda x.\, 1 - x \rangle$, where $\leq$ is again the usual order relation. Note that here the underlying ordered set is *bounded and complete*.

The next two lemmas about ordered sets will be useful in the sequel.

Lemma 5.2 *Let* $n > 0$ *be a natural number. Every finite Sugihara chain which has* $2n + 1$ *elements is isomorphic to* $S_{\mathbb{Z}_n}$, *and every finite Sugihara chain which has* $2n$ *elements is isomorphic to* $S_{\mathbb{Z}_n^*}$.

Proof By an easy induction on n. □

Lemma 5.3 *Every countable Sugihara chain can be embedded in* $S_{[0,1]}$.

Proof It is well known that every countable linearly ordered set can be embedded in any closed interval $[a, b]$ of $\mathbb{R}$, so that a is assigned to the minimal element of the set (if such exists), and b is assigned to the maximal element of the set (if such exists). Now let $\langle \mathcal{V}, \leq, - \rangle$ be a countable Sugihara chain, and let $D = \{a \in \mathcal{V}\colon\ -a \leq a\}$. First, suppose that there is $a \in \mathcal{V}$ such that $-a = a$. It is easy to prove that in such a case a is unique, and it is the minimal element of D. Let f be an embedding of

D into $[1/2, 1]$ such that $f(a) = 1/2$, and extend f to the whole of $\mathcal{V}$ by letting $f(x) = -f(-x)$ in case $x \notin D$. (Note that if $x \notin D$ then $-x \in D$, because $\leq$ is linear, and $--x = x$.) If there is no $a \in \mathcal{V}$ such that $-a = a$, then we let f be any embedding of D into $[2/3, 1]$ (say), and we again extend f to the whole of $\mathcal{V}$ by letting $f(x) = -f(-x)$ in case $x \notin D$. In both cases, f is easily seen to be an embedding of $\langle \mathcal{V}, \leq, - \rangle$ into $[0, 1]$. □

Definition 5.4 Let $S = \langle \mathcal{V}, \leq, - \rangle$ be a Sugihara chain, and let $a, b \in \mathcal{V}$.

- $a < b$ if $a \leq b$ and $a \neq b$.
- $|a| = \max(-a, a)$.
- $a \preceq_+ b$ iff either $|a| < |b|$, or $|a| = |b|$ and $a < b$.

The following lemma is easily verified.

Lemma 5.5 *If $\langle \mathcal{V}, \leq, - \rangle$ is a Sugihara chain, then $\preceq_+$ linearly orders $\mathcal{V}$.*

Definition 5.6 (*Sugihara matrix*) Let $S = \langle \mathcal{V}, \leq, - \rangle$ be a Sugihara chain.

- The *multiplicative Sugihara matrix* based on S is the matrix $\mathcal{M}_m(S) = \langle \mathcal{V}, \mathcal{D}, \mathcal{O} \rangle$ for $\{\neg, \rightarrow\}$ in which $\mathcal{D} = \{a \in \mathcal{V} \colon -a \leq a\}$ (equivalently, $\mathcal{D} = \{a \in \mathcal{V} \colon |a| = a\}$), $\tilde{\neg} a = -a$, and $a \tilde{\rightarrow} b = \max_{\preceq_+}(-a, b)$.
- The *Sugihara matrix* $\mathcal{M}(S)$ based on S is the extension of $\mathcal{M}_m(S)$ to $\mathcal{L}_R$ in which $a \tilde{\wedge} b = \min(a, b)$ and $a \tilde{\vee} b = \max(a, b)$.
- A matrix $\mathcal{M}$ for $\mathcal{L}_R$ (for $\{\neg, \rightarrow\}$) is a *(multiplicative) Sugihara matrix* if for some Sugihara chain S, $\mathcal{M}$ is the (multiplicative) Sugihara matrix which is based on S.

Note 5.2 Obviously, we have that in a (multiplicative) Sugihara matrix $a \tilde{+} b = \max_{\preceq_+}(a, b)$. It is also easy to see that the above definition of $\tilde{\rightarrow}$ in Sugihara matrices is equivalent to the following original definition from Sugihara (1955):

$$a \tilde{\rightarrow} b = \begin{cases} \max(-a, b) & \text{if } a \leq b, \\ \min(-a, b) & \text{if } a > b. \end{cases}$$

It easily follows that $a \tilde{\rightarrow} b \in \mathcal{D}$ iff $a \leq b$.

Note 5.3 It is easy to see that the set $\mathcal{D}$ is upward closed in a Sugihara matrix $\mathcal{M}$. That is, if $a \in \mathcal{D}$ and $a \leq b$ (where $\leq$ is the order relation of the Sugihara chain which underlies $\mathcal{M}$), then $b \in \mathcal{D}$.

The following observation will be useful in the sequel.

Proposition 5.7 *Let $S = \langle \mathcal{V}, \leq, - \rangle$ be a Sugihara chain, and suppose that $\mathcal{V}'$ is a subset of $\mathcal{V}$ which is closed under $-$, and has at least two elements. Then $S' = \langle \mathcal{V}', \leq, - \rangle$ is also a Sugihara chain, and $\mathcal{M}(S')$ ($\mathcal{M}_m(S')$) is a submatrix of $\mathcal{M}(S)$ ($\mathcal{M}_m(S)$).*

Proof The definitions of the operations immediately imply that if $\mathcal{V}'$ is closed under $-$, then it is closed under $\tilde{\to}$, $\tilde{\wedge}$, and $\tilde{\vee}$ as well. The proposition easily follows from this fact and the definition of the set $\mathcal{D}$ of designated elements in Sugihara matrices. □

Notation For $A \in \{\mathbb{R}, \mathbb{Z}, \mathbb{Z}^*, \mathbb{Q}, \mathbb{Q}^*, [0,1], \mathbb{Z}_n, \mathbb{Z}_n^*\}$ we shall henceforth write just $\mathcal{M}(A)$ instead of $\mathcal{M}(S_A)$, and $\mathcal{M}_m(A)$ instead of $\mathcal{M}_m(S_A)$.

Next we prove a *strong* soundness and completeness theorem for $HRM_{\overset{\neg}{\to}}$.

Theorem 5.8 (Strong soundness and completeness of $HRM_{\overset{\neg}{\to}}$)

1. *$HRM_{\overset{\neg}{\to}}$ is strongly sound and complete for the class of multiplicative Sugihara matrices.*
2. *$HRM_{\overset{\neg}{\to}}$ is strongly sound and complete for* $\mathcal{M}_m([0,1])$.

Proof 1. For the soundness part we need to prove that the axioms and rule of $HRM_{\overset{\neg}{\to}}$ are all valid in any Sugihara matrix. We leave the straightforward but tedious details of this to the reader.

For completeness, assume $\mathcal{T} \nvdash_{HRM_{\overset{\neg}{\to}}} \varphi$. Extend $\mathcal{T}$ to a maximal theory $\mathcal{T}^*$ such that $\mathcal{T}^* \nvdash_{HRM_{\overset{\neg}{\to}}} \varphi$. Then the relevant deduction theorem of $HRM_{\overset{\neg}{\to}}$ (Proposition 2.14) implies that for every sentence ψ, $\psi \notin \mathcal{T}^*$ iff $\psi \to \varphi \in \mathcal{T}^*$. By item 5 of Lemma 2.16, this in turn implies:

(1) If $\psi + \sigma \in \mathcal{T}^*$, then either $\psi \in \mathcal{T}^*$ or $\sigma \in \mathcal{T}^*$.

(1) together with [++] and items 11 and 6 of Lemma 2.16 imply:

(2) For every ψ, σ, either $\psi \to \sigma \in \mathcal{T}^*$ or $\sigma \to \psi \in \mathcal{T}^*$.
(3) For every sentence ψ, either $\psi \in \mathcal{T}^*$ or $\neg\psi \in \mathcal{T}^*$.

- Now construct the Lindenbaum Algebra $\mathcal{M}_{\mathcal{T}^*}$ of $\mathcal{T}^*$ in the usual way. We define that $\psi \equiv \sigma$ iff $\psi \leftrightarrow \sigma \in \mathcal{T}^*$ (and so both $\psi \to \sigma \in \mathcal{T}^*$ and $\sigma \to \psi \in \mathcal{T}^*$, by item 1 of Lemma 2.16). By Proposition 2.18, this is obviously a congruence relation. Let $\mathcal{V}$ be the set of equivalence classes, and let $\mathcal{D} = \{[\psi] : \psi \in \mathcal{T}^*\}$. Define the operations $\neg$ and $\to$ on $\mathcal{V}$ as $[\psi] \to [\sigma] = [\psi \to \sigma]$ and $\neg[\psi] = [\neg\psi]$. To show that the resulting matrix is a multiplicative Sugihara matrix, we let $[\psi] \le [\sigma]$ iff $\psi \to \sigma \in \mathcal{T}^*$. These are all legitimate definitions because $\equiv$ is a congruence relation. It is a standard matter to show that $\le$ is a partial order on $\mathcal{V}$ and that the negation axioms of $\mathbf{R}_{\overset{\neg}{\to}}$ ensure that $\neg$ is an involution on $\langle \mathcal{V}, \le \rangle$. (2) above implies that $\le$ is also linear. It follows that $S = \langle \mathcal{V}, \le, \neg \rangle$ is a Sugihara chain. Next we show that $\mathcal{M}_{\mathcal{T}^*} = \mathcal{M}_m(S)$. That $[\psi] \in \mathcal{D}$ iff $\neg[\psi] \le [\psi]$ easily follows from the definitions of $\mathcal{D}$ and $\le$, and the fact that both $\psi \to (\neg\psi \to \psi)$ and $(\neg\psi \to \psi) \to \psi$ are theorems of $\mathbf{RMI}_{\overset{\neg}{\to}}$ (Lemma 2.20, 1). It remains to show that the operation $\to$ of $\mathcal{M}_{\mathcal{T}^*}$ is identical to that of $\mathcal{M}_m(S)$. We use for that the characterization of $\mathcal{M}_m(S)$ given in Note 5.2.

- Suppose $[\psi] \leq [\sigma]$. Then $\psi \to \sigma \in \mathcal{T}^*$. By items 3 and 2 of Lemma 2.20, it follows that both $\neg\psi \to (\psi \to \sigma)$ and $\sigma \to (\psi \to \sigma)$ are in $\mathcal{T}^*$. Hence $[\psi] \to [\sigma] \geq \max(\neg[\psi], [\sigma])$. To prove the converse, note that since $\leq$ is linear, $\max(\neg[\psi], [\sigma])$ is either $[\sigma]$ or $[\neg\psi]$. In the first case $\neg[\psi] \leq [\sigma]$, and so $\neg\psi \to \sigma \in \mathcal{T}^*$. By item 8 of Lemma 2.16, we get that in this case $[\psi \to \sigma] \leq [\sigma]$. In the second case, $[\sigma] \leq \neg[\psi]$, and so $[\psi] \leq \neg[\sigma]$, implying that $\psi \to \neg\sigma \in \mathcal{T}^*$. By item 9 of Lemma 2.16, we get that in this case $[\psi \to \sigma] \leq \neg[\psi]$. In both cases, we have that $[\psi] \to [\sigma] = [\psi \to \sigma] \leq \max(\neg[\psi], [\sigma])$.
- Suppose $[\psi] \not\leq [\sigma]$. Then $\psi \to \sigma \notin \mathcal{T}^*$. Hence (3) implies that $\neg(\psi \to \sigma) \in \mathcal{T}^*$. By items 5 and 6 of Lemma 2.20, it follows that both $(\psi \to \sigma) \to \sigma$ and $(\psi \to \sigma) \to \neg\psi$ are in $\mathcal{T}^*$. Hence $[\psi] \to [\sigma] \leq \min(\neg[\psi], [\sigma])$. To prove the converse, note that since $\leq$ is linear, $\min(\neg[\psi], [\sigma])$ is either $[\sigma]$ or $[\neg\psi]$. In the first case, $\neg[\psi] \leq [\sigma]$, and so $\neg\psi \to \sigma \in \mathcal{T}^*$. By item 7 of Lemma 2.20, we get that $\neg[\psi] \leq [\psi \to \sigma]$ in this case. In the second case, $[\sigma] \leq \neg[\psi]$, and so $\sigma \to \neg\psi \in \mathcal{T}^*$. By item 8 of Lemma 2.20, we get that $[\sigma] \leq [\psi \to \sigma]$ in this case. In both cases, we have that $[\psi] \to [\sigma] = [\psi \to \sigma] \geq \min(\neg[\psi], [\sigma])$.

The end of the proof is now standard. Let $\nu(\psi) = [\psi]$. This is easily seen to be a legitimate valuation (the canonical one) in $\mathcal{M}_{\mathcal{T}^*}$. Obviously, ν is a model of ψ iff $\psi \in \mathcal{T}^*$. Hence ν is a model of $\mathcal{T}$ in the Sugihara matrix $\mathcal{M}_{\mathcal{T}^*}$ which is not a model of φ.

2. The multiplicative Sugihara matrix constructed in the proof of the first part is countable. Hence the second part follows from the first (and its proof) by Lemma 5.3 and Proposition 5.7. □

Proposition 5.9 *$\mathcal{M}_m(\mathbb{Z}_1)$ is weakly characteristic for $\mathbf{L}_{HRM_{\vec{\neg}}}$, but it is not strongly characteristic for it.*

Proof From the first part of Theorem 5.8 it follows that if $\vdash_{HRM_{\vec{\neg}}} \varphi$, then $\vdash_{\mathcal{M}_m(\mathbb{Z}_1)} \varphi$. For the converse, assume that $\nvdash_{HRM_{\vec{\neg}}} \varphi$. By the second part of Theorem 5.8, it follows that there is an assignment ν in $\mathcal{M}_m([0,1])$ such that $\nu(\varphi) < 1/2$. Let $\nu(\varphi) = a$. Without loss of generality, we may assume that $\nu(p) \in [a, 1-a]$ for every $p \notin \mathsf{Atoms}(\varphi)$, while the definitions of the operations in $\mathcal{M}_m([0,1])$ imply that necessarily $\nu(p) \in [a, 1-a]$ also for every $p \in \mathsf{Atoms}(\varphi)$. Hence $\nu(\psi) \in [a, 1-a]$ for every ψ. Define $f : [a, 1-a] \to \{-1, 0, 1\}$ as

$$f(x) = \begin{cases} 1 & \text{if } x = 1-a, \\ 0 & \text{if } a < x < 1-a, \\ -1 & \text{if } x = a. \end{cases}$$

It is easy to verify that $\nu^* = f \circ \nu$ is an assignment in $\mathcal{M}_m(\mathbb{Z}_1)$ such that $\nu^*(\psi) = f(\nu(\psi))$ for every formula ψ. In particular, $\nu^*(\varphi) = -1$, and so $\nvdash_{\mathcal{M}_m(\mathbb{Z}_1)} \varphi$.

To see that $\mathcal{M}_m(\mathbb{Z}_1)$ is not strongly characteristic for $\mathbf{L}_{HRM_{\vec{\neg}}}$, it suffices to note that $\varphi \otimes \psi \vdash_{\mathcal{M}_m(\mathbb{Z}_1)} \varphi$, but $\varphi \otimes \psi \nvdash_{\mathcal{M}_m([0,1])} \varphi$. □

Note 5.4 That $\mathcal{M}_m(\mathbb{Z}_1)$ is weakly characteristic for $\mathbf{L}_{HRM_{\vec{\neg}}}$ and $\mathbf{RM}_{\vec{\neg}}$ was (essentially) shown in Parks (1972). The fact that it is not strongly characteristic for them was (to my best knowledge) first shown in Avron (1997).

Our next goal is to prove a counterpart of Theorem 5.8 for the whole of **RM**. The main obstacle in doing that is that the relevant deduction theorem, which was used in the proof of Theorem 5.8 for showing the crucial property that was denoted by (1) in that proof, fails for **RM**. Therefore, we shall use instead for that purpose the fact that $\vee$ is a disjunction for **RM**.

Theorem 5.10 (Strong soundness and completeness of **RM**)

1. **RM** *is strongly sound and complete for the class of Sugihara matrices.*
2. **RM** *is strongly sound and complete for* $\mathcal{M}([0, 1])$.

Proof

1. Given the strong soundness of $HRM_{\vec{\neg}}$ for multiplicative Sugihara matrices (Theorem 5.8), the proof of the strong soundness of **RM** for Sugihara matrices is straightforward, and is left to the reader.
   For completeness, assume $\mathcal{T} \nvdash_{\mathbf{RM}} \varphi$. Extend $\mathcal{T}$ to a maximal theory $\mathcal{T}^*$ such that $\mathcal{T}^* \nvdash_{\mathbf{RM}} \varphi$. Then $\psi \notin \mathcal{T}^*$ iff $\mathcal{T}^*, \psi \vdash_{\mathbf{RM}} \varphi$. Hence Theorem 2.11 implies that $\mathcal{T}^*$ is *prime*, i.e., if $\psi \vee \sigma \in \mathcal{T}^*$, then either $\psi \in \mathcal{T}^*$ or $\sigma \in \mathcal{T}^*$. Therefore, it follows from Lemma 2.17 that (1) from the proof of Theorem 5.8 holds for $\mathcal{T}^*$. From this point on, the proof is almost identical to the proof of the first part of Theorem 5.8, except that we show that $\mathcal{M}_{\mathcal{T}^*} = \mathcal{M}(S)$ (where S is defined like in that proof), rather than that $\mathcal{M}_{\mathcal{T}^*} = \mathcal{M}_m(S)$. For this, all we have to add to the proof of Theorem 5.8 is that $[\psi \wedge \sigma] = \min([\psi], [\sigma])$ and $[\psi \vee \sigma] = \max([\psi], [\sigma])$. This is obvious from the axioms concerning $\wedge$ and $\vee$ of **RM**, and the linearity of $\leq$.
2. The proof is identical to that of the second part of Theorem 5.8. □

Note 5.5 Theorem 5.10 is essentially due to Dunn (1970). However, Dunn used the countable matrix $\mathcal{M}(\mathbb{Q})$ for strongly characterizing **RM**, rather than the uncountable $\mathcal{M}([0, 1]$ used by us here.[6]

Now we can at last prove the following theorem.

Theorem 5.11 $\mathbf{RM}_{\vec{\neg}} = \mathbf{L}_{HRM_{\vec{\neg}}}$.

Proof Immediate from Proposition 4.5, and the second parts of Theorems 5.8 and 5.10. □

[6]An advantage of choosing $\mathcal{M}([0, 1])$ is that its use allows us to view **RM** as a fuzzy logic. (See Sect. 7.) Another advantage is that it can be expanded very naturally to provide semantics for first-order **RM**, as well as for the logic that is obtained from **RM** by adding to its language the propositional constants **T** and **F**, together with the axioms $\mathbf{F} \to \varphi$ and $\varphi \to \mathbf{T}$.

Next we show that for *weak* completeness the set of finite Sugihara matrices and each of the countable Sugihara matrices $\mathcal{M}(\mathbb{Z})$ and $\mathcal{M}(\mathbb{Z}^*)$ suffice.

Definition 5.12 (*The matrices $\mathcal{RM}_n$*) For $k = 1, 2, \ldots$, we let $\mathcal{RM}_{2k} = \mathcal{M}(\mathbb{Z}_k^*)$ and $\mathcal{RM}_{2k+1} = \mathcal{M}(\mathbb{Z}_k)$.

Proposition 5.13 *Every finite Sugihara matrix which has n elements is isomorphic to $\mathcal{RM}_n$. Hence every such matrix is isomorphic to some finite submatrix of $\mathcal{M}(\mathbb{Z})$.*

Proof This is an easy corollary of Lemma 5.2 and Proposition 5.7. □

Theorem 5.14 *Suppose that* $\mathsf{Atoms}(\mathcal{T} \cup \{\varphi\})$ *is finite.*

1. *Let n be the number of atomic variables which occur in $\mathcal{T} \cup \{\varphi\}$. Then $\mathcal{T} \vdash_{\mathbf{RM}} \varphi$ iff $\mathcal{T} \vdash_{\mathcal{RM}_k} \varphi$ for every $2 \leq k \leq 2n$.*
2. *$\mathcal{T} \vdash_{\mathbf{RM}} \varphi$ iff $\mathcal{T} \vdash_{\mathcal{M}(\mathbb{Z})} \varphi$.*

Proof From the soundness of **RM** for Sugihara matrices, it follows that if $\mathcal{T} \vdash_{\mathbf{RM}} \varphi$, then $\mathcal{T} \vdash_{\mathcal{M}(\mathbb{Z})} \varphi$, and $\mathcal{T} \vdash_{\mathcal{RM}_k} \varphi$ for every $k \geq 2$. For the converse, assume $\mathcal{T} \nvdash_{\mathbf{RM}} \varphi$. By Theorem 5.10, there is a Sugihara chain $S = \langle \mathcal{V}, \leq, - \rangle$ and a valuation ν in $\mathcal{M}(S)$ which is a model of $\mathcal{T}$ but not of φ. Suppose $\mathsf{Atoms}(\mathcal{T} \cup \{\varphi\}) = \{p_1, \ldots, p_n\}$, and let $\mathcal{V}' = \{\nu(p_1), -\nu(p_1), \ldots, \nu(p_n), -\nu(p_n)\}$. An easy induction on the complexity of a sentence ψ shows that $\nu(\psi) \in \mathcal{V}'$ for every ψ such that $\mathsf{Atoms}(\psi) \subseteq \{p_1, \ldots, p_n\}$. Since ν is not a model of φ, this implies that $\mathcal{V}'$ has at least two elements (and of course not more than $2n$). Hence Proposition 5.7 implies that $S' = \langle \mathcal{V}', \leq, - \rangle$ is also a Sugihara chain, and $\mathcal{M}(S')$ is a submatrix of $\mathcal{M}(S)$. Let ν' be any valuation in $\mathcal{M}(S')$ such that $\nu'(p_i) = \nu(p_i)$ for $1 \leq i \leq n$. Then $\nu'(\psi) = \nu(\psi)$ for every ψ such that $\mathsf{Atoms}(\psi) \subseteq \{p_1, \ldots, p_n\}$. It follows that ν' is a model of $\mathcal{T}$ in $\mathcal{M}(S')$ which is not a model of φ. Hence $\mathcal{T} \nvdash_{\mathcal{M}(S')} \varphi$. By Proposition 5.13, this implies that $\mathcal{T} \nvdash_{\mathcal{RM}_k} \varphi$ for some $2 \leq k \leq 2n$, and that $\mathcal{T} \nvdash_{\mathcal{M}(\mathbb{Z})} \varphi$. □

Corollary 5.15 *If $\mathcal{T}$ is a finite theory, then $\mathcal{T} \vdash_{\mathbf{RM}} \varphi$ iff $\mathcal{T} \vdash_{\mathcal{M}(\mathbb{Z})} \varphi$. In particular, $\mathcal{M}(\mathbb{Z})$ is weakly characteristic for* **RM**.

In contrast we have the following.

Proposition 5.16 *$\mathcal{M}(\mathbb{Z})$ is not strongly characteristic for* **RM**.

Proof Let $\mathcal{T} = \{p_i : i \geq 1\} \cup \{(p_i \to p_{i+1}) \to p_0 : i \geq 1\}$, and let $S = \langle \mathcal{V}, \leq, - \rangle$ be a Sugihara chain. It is not difficult to check that a valuation ν in $\mathcal{M}(S)$ can be a model of $\mathcal{T}$ which is not a model of p_0 iff $\nu(p_0) < -\nu(p_0)$, while for $i > 0$, $\nu(p_i) \geq -\nu(p_i)$ and $\nu(p_i) > \nu(p_{i+1})$. Such ν does not exist in $\mathcal{M}(\mathbb{Z})$, but it does in $\mathcal{M}([0, 1])$. Hence $\mathcal{T} \vdash_{\mathcal{M}(\mathbb{Z})} p_0$, while $\mathcal{T} \nvdash_{\mathbf{RM}} p_0$. □

The characterization of **RM** in terms of finite matrices that is given in Theorem 5.14 can in fact be improved using the next proposition.

Proposition 5.17 *For every $n \geq 2$, if $\vdash_{\mathcal{RM}_{n+1}} \varphi$, then also $\vdash_{\mathcal{RM}_n} \varphi$.*

Proof The claim is obvious in case n is even, since $\mathcal{RM}_{2k}$ is a submatrix of $\mathcal{RM}_{2k+1}$ for every $k \geq 1$.

Now suppose that $n = 2k + 1$ for some $k \geq 1$, and that $\nvdash_{\mathcal{RM}_{2k+1}} \varphi$. We show that also $\nvdash_{\mathcal{RM}_{2k+2}} \varphi$. Let ν be a valuation in $\mathcal{RM}_{2k+1}$ such that $\nu(\varphi) < 0$. Define a valuation ν^* in $\mathcal{RM}_{2k+2}$ by letting $\nu^*(p) = \nu(p) + 1$ in case $\nu(p) \geq 0$, and $\nu^*(p) = \nu(p) - 1$ in case $\nu(p) < 0$. By induction on the complexity of ψ, it is not difficult to show that for every sentence ψ we have the following.

- If $\nu(\psi) > 0$, then $\nu^*(\psi) = \nu(\psi) + 1$.
- If $\nu(\psi) = 0$, then $\nu^*(\psi) \in \{-1, 1\}$.
- If $\nu(\psi) < 0$, then $\nu^*(\psi) = \nu(\psi) - 1$.

It follows in particular that $\nu^*(\varphi) < 0$. Hence $\nvdash_{\mathcal{RM}_{2k+2}} \varphi$. □

Corollary 5.18 *If $n \geq 2$ and $\vdash_{\mathcal{RM}_n} \varphi$, then $\vdash_{\mathcal{RM}_m} \varphi$ for every $2 \leq m \leq n$.*

Proposition 5.19 *Suppose $|\mathsf{Atoms}(\varphi)| = n$. Then $\vdash_{\mathbf{RM}} \varphi$ iff $\vdash_{\mathcal{RM}_{2n}} \varphi$.*

Proof Immediate from Part 1 of Theorem 5.14 and Corollary 5.18. □

Proposition 5.20 *$\mathcal{M}(\mathbb{Z}^*)$ is weakly characteristic for* **RM**.

Proof That if $\vdash_{\mathbf{RM}} \varphi$ then $\vdash_{\mathcal{M}(\mathbb{Z}^*)} \varphi$ follows from the soundness of **RM** for Sugihara matrices. For the converse, assume $\nvdash_{\mathbf{RM}} \varphi$. Then by Proposition 5.19, there is n such that $\nvdash_{\mathcal{RM}_{2n}} \varphi$. Since $\mathcal{RM}_{2n}$ is a submatrix of $\mathcal{M}(\mathbb{Z}^*)$, this implies that $\nvdash_{\mathcal{M}(\mathbb{Z}^*)} \varphi$. □

Note 5.6 The second part of Corollary 5.15, and Propositions 5.16, 5.19, and 5.20 are due to Meyer (see Anderson and Belnap (1975, Sect. 29.3)). Corollary 5.18 and Proposition 5.19 are due to Dunn (see Anderson and Belnap (1975, Sect. 29.4)).

Corollary 5.21 *If γ is a finite set of sentences, and $\gamma \vdash_{\mathcal{M}(\mathbb{Z}^*)} \varphi$, then the rule γ/φ is admissible in* **RM**.

Proof Let θ be a substitution such that $\vdash_{\mathbf{RM}} \theta(\psi)$ for every $\psi \in \gamma$. By Proposition 5.20, $\vdash_{\mathcal{M}(\mathbb{Z}^*)} \theta(\psi)$ for every $\psi \in \gamma$. Since $\gamma \vdash_{\mathcal{M}(\mathbb{Z}^*)} \varphi$, it follows that $\vdash_{\mathcal{M}(\mathbb{Z}^*)} \theta(\varphi)$ as well. Hence $\vdash_{\mathbf{RM}} \theta(\varphi)$, by Proposition 5.20 again. □

Note 5.7 Since $\neg p, p \vee q \vdash_{\mathcal{M}(\mathbb{Z}^*)} q$, Corollary 5.21 entails that the disjunctive syllogism is admissible in **RM**. That is, if $\vdash_{\mathbf{RM}} \neg\varphi$, and $\vdash_{\mathbf{RM}} \varphi \vee \psi$, then $\vdash_{\mathbf{RM}} \psi$.[7] On the other hand, it is easy to see that $\neg p, p \vee q \nvdash_{\mathcal{M}(\mathbb{Z})} q$. By Theorem 5.14, this implies that $\neg p, p \vee q \nvdash_{\mathbf{RM}} q$. It follows that the analogue of Theorem 5.14 does *not* hold for $\mathcal{M}(\mathbb{Z}^*)$.

[7] This is another famous result of Meyer and Dunn. See Meyer and Dunn (1969) and Sect. 25 of Anderson and Belnap (1975). In the latter, two different proofs of this theorem (for the main relevant and semi-relevant logics) are presented.

6 The Nice Properties of RM

RM has several nice properties. The first we present in this section is one that according to a famous theorem of Urquhart (1984), the main logics developed by Anderson and Belnap's school lack.[8]

Theorem 6.1 **RM** *is decidable.*[9]

Proof Immediate from Theorem 5.14. (See also Proposition 5.19 for the special case of theoremhood in **RM**.) □

Our next goal is to show that **RM** is normal.

Definition 6.2 $\varphi \supset \psi =_{Df} (\varphi \rightarrow \psi) \vee \psi$

Note 6.1 It is easy to see that in any Sugihara matrix we have that

$$a \tilde{\supset} b = \begin{cases} -a & \text{if } a \leq b \leq -a, \\ b & \text{otherwise.} \end{cases}$$

Proposition 6.3 $\supset$ *is an implication for* **RM**.

Proof By Theorem 2.11, $\vee$ is a disjunction for **RM**. Given the definition of $\supset$, this easily implies that $\varphi, \varphi \supset \psi \vdash_{\mathbf{RM}} \psi$. It follows that if $\mathcal{T} \vdash_{\mathbf{RM}} \varphi \supset \psi$, then $\mathcal{T}, \varphi \vdash_{\mathbf{RM}} \psi$.

For the converse, assume $\mathcal{T} \nvdash_{\mathbf{RM}} \varphi \supset \psi$. By Theorem 5.10, this implies that there is a valuation ν in $\mathcal{M}([0, 1])$ such that $\nu(\sigma) \geq 1/2$ for every $\sigma \in \mathcal{T}$, while $\nu(\varphi \supset \psi) < 1/2$. The latter means that $\nu(\psi) < 1/2$ and $\nu(\varphi) > \nu(\psi)$. If $\nu(\varphi) \geq 1/2$, then ν is a model in $\mathcal{M}([0, 1])$ of $\mathcal{T} \cup \{\varphi\}$ which is not a model of ψ, and so $\mathcal{T}, \varphi \nvdash_{\mathbf{RM}} \psi$. So assume $1/2 > \nu(\varphi) > \nu(\psi)$. Define a new valuation ν^* in $\mathcal{M}([0, 1])$ as follows.

$$\nu^*(\sigma) = \begin{cases} 1/2 & \text{if } \nu(\varphi) \leq \nu(\sigma) \leq 1 - \nu(\varphi), \\ \nu(\sigma) & \text{otherwise.} \end{cases}$$

It is easy to verify that ν^* is indeed a legitimate valuation. Now $\nu^*(\sigma) \in \{\nu(\sigma), 1/2\}$ for every $\sigma \in \mathcal{T}$, $\nu^*(\varphi) = 1/2$, while $\nu^*(\psi) = \nu(\psi) < 1/2$. Since $\nu(\sigma) \geq 1/2$ for every $\sigma \in \mathcal{T}$, this implies that ν^* is a model in $\mathcal{M}([0, 1])$ of $\mathcal{T} \cup \{\varphi\}$ which is not a model of ψ, and so again $\mathcal{T}, \varphi \nvdash_{\mathbf{RM}} \psi$. □

---

[8]Here it should be noted that there are many contraction-free logics which are closely related to Anderson and Belnap's relevant logics, and *are* decidable (like **RW** (Brady 1990) or the multiplicative-additive fragment of Girard's linear logic). However, logics without contraction are not relevant logics according to our understanding of this notion (see Avron 2014).

[9]This result too is due to Meyer. See Anderson and Belnap (1975, Sect. 29.3).

Note 6.2 It is also possible to prove Proposition 6.3 purely syntactically using the standard inductive method of converting a proof in **RM** of ψ from $\mathcal{T} \cup \{\varphi\}$ into a proof in **RM** of $\varphi \supset \psi$ from $\mathcal{T}$. In addition to the validity of [MP] for $\supset$ in **RM** (which was shown above purely syntactically), one should only provide derivations of the following four formulas in **RM**: $\varphi \supset \varphi$, $\varphi \supset (\psi \supset \varphi)$, $(\varphi \supset (\psi \to \sigma)) \supset ((\varphi \supset \psi) \supset (\varphi \supset \sigma))$, and $(\varphi \supset \psi) \wedge (\varphi \supset \sigma) \supset (\varphi \supset \psi \wedge \sigma)$. None of these tasks is difficult.

Note 6.3 The above formulation of Definition 6.2 and Proposition 6.3 are due to Avron (1986). An equivalent definition and proposition have already been given in Dunn and Meyer (1971). However, they were given there only for $\mathbf{RM^t}$, a conservative extension of **RM** which is obtained from **RM** by adding to its language the propositional constant $\mathbf{t}$ together with the axioms $\mathbf{t}$ and $\mathbf{t} \to (\varphi \to \varphi)$. In $\mathbf{RM^t}$ $\varphi \supset \psi$ is equivalent to $\varphi \wedge \mathbf{t} \to \psi$, and this was the definition used in Dunn and Meyer (1971).

Proposition 6.4 **RM** *is normal.*

Proof The axioms [ConE1], [ConE2], and the adjunction rule [Ad] ensure that $\wedge$ is a conjunction for every extension of **R**, including **RM**. Hence the proposition follows from Theorem 2.11 and Proposition 6.3. □

Proposition 6.5 **RM** *satisfies the basic relevance criterion.*[10]

Proof Suppose $\mathcal{T}_1, \mathcal{T}_2 \vdash_{\mathbf{RM}} \psi$ and $\mathcal{T}_2$ has no atomic formulas in common with $\mathcal{T}_1 \cup \{\psi\}$. We show that $\mathcal{T}_1 \vdash_{\mathbf{RM}} \psi$. Suppose otherwise. Then, by Theorem 5.10, there is a valuation ν in $\mathcal{M}([0, 1])$ such that $\nu(\varphi) \geq 1/2$ for every $\varphi \in \mathcal{T}_1$, while $\nu(\psi) < 1/2$. Since $\mathcal{T}_2$ has no atomic formulas in common with $\mathcal{T}_1 \cup \{\psi\}$, we may assume without loss of generality that $\nu(p) = 1/2$ for every atom p which occurs in $\mathcal{T}_2$. But then $\nu(\varphi) = 1/2$ for every $\varphi \in \mathcal{T}_2$, and so ν is a model in $\mathcal{M}([0, 1])$ of $\mathcal{T}_1 \cup \mathcal{T}_2$ that is not model of ψ. By Theorem 5.10 again, this contradicts our assumption that $\mathcal{T}_1, \mathcal{T}_2 \vdash_{\mathbf{RM}} \psi$. □

Next we show that **RM** is a semi-relevant logic (Definition 3.5). For this we need to show that it does not have a weakly characteristic matrix. Actually, we prove something significantly stronger.

Proposition 6.6 **RM** *has no finite weakly characteristic non-deterministic matrix (Nmatrix).*[11] *In particular, it satisfies the minimal semantic criterion.*

Proof Assume for contradiction that **RM** has a weakly characteristic Nmatrix $\mathcal{M} = \langle \mathcal{V}, \mathcal{D}, \mathcal{O} \rangle$, where the number of elements in $\mathcal{V}$ is a natural number $n > 1$. Let $p_1, \ldots, p_{n+1}$ be $n + 1$ distinct atomic formulas. Define

[10] As pointed out in Note 3.1, this was first claimed by Meyer, but with a wrong proof, in Anderson and Belnap (1975).

[11] See Avron and Zamansky (2011) about this generalization of the notion of a matrix for a logic, including a lot of examples of logics which do not have a finite weakly characteristic matrix, but do have a finite weakly characteristic Nmatrix.

$$\varphi_n =_{Df} (p_1 \to p_2) \vee (p_2 \to p_3) \vee \cdots \vee (p_n \to p_{n+1}).$$

By assigning $-i$ to p_i, we see that φ_n is not valid in $\mathcal{M}(\mathbb{Z})$. Hence $\nvdash_{\mathbf{RM}} \varphi_n$, and so φ_n is not valid in $\mathcal{M}$. It follows that there is a valuation ν in $\mathcal{M}$ such that $\nu(\varphi) \notin \mathcal{D}$. Now by the pigeonhole principle, there are $1 \leq i < j \leq n+1$ such that $\nu(p_i) = \nu(p_j)$. Obtain ψ_n from φ_n by replacing the first occurrence of p_j in φ_n by p_i, and define a valuation ν' in $\mathcal{M}$ by letting $\nu'(\sigma) = \nu(\sigma')$, where σ' is the formula obtained from σ by replacing in σ each subformula of the form $p_{j-1} \to p_i$ by $p_{j-1} \to p_j$. Since $\nu(p_i) = \nu(p_j)$, ν' is easily seen to be a legitimate valuation in $\mathcal{M}$. Now $\nu'(\psi_n) = \nu(\varphi_n)$. Hence ψ_n is not valid in $\mathcal{M}$, and so $\nvdash_{\mathbf{RM}} \psi_n$. On the other hand, ψ_n is valid in $\mathcal{M}(\mathbb{Z})$ (since $(p_i \to p_{i+1}) \vee \cdots \vee (p_{j-2} \to p_{j-1}) \vee (p_{j-1} \to p_i)$ is easily seen to be valid in $\mathcal{M}(\mathbb{Z})$), and so $\vdash_{\mathbf{RM}} \psi_n$. A contradiction. □

Note 6.4 That **RM** has no finite weakly characteristic *deterministic* (i.e., ordinary) matrix was first observed by Dunn in (1970).

Theorem 6.7 **RM** *is a normal semi-relevant logic.*

Proof This follows from Propositions 6.4–6.6. □

Here is another well-known way in which the logic **RM** is "semi-relevant."

Proposition 6.8 *1.* **RM** *does not have the variable-sharing property.*

2. *If* $\vdash_{\mathbf{RM}} \varphi \to \psi$ *then either* φ *and* ψ *share an atomic formula, or* both $\neg\varphi$ *and* ψ *are theorems of* **RM**.

Proof 1. $\neg(p \to p) \to (q \to q)$ is a theorem of $HRM_{\vec{\neg}}$, and so also of **RM**.
2. Suppose that $\vdash_{\mathbf{RM}} \varphi \to \psi$, but φ and ψ share no atomic formula. We show that both $\neg\varphi$ and ψ are theorems of **RM**. Suppose, for example, that $\neg\varphi$ is not a theorem of **RM**. (The argument in the case where ψ is not a theorem of **RM** is similar.) Then, by Theorem 5.14, there is valuation ν in $\mathcal{M}(\mathbb{Z})$ such that $\nu(\neg\varphi) < 0$, and so $\nu(\varphi) > 0$. Without a loss of generality, we may assume that $\nu(q) = 0$, for every atom $q \notin \mathsf{Atoms}(\varphi)$. Since φ and ψ share no atomic formula, this implies that $\nu(q) = 0$ for every atom $q \in \mathsf{Atoms}(\psi)$. But then $\nu(\psi) = 0$. Since $\nu(\varphi) > 0$ this implies that $\nu(\varphi \to \psi) < 0$, contradicting the assumption that $\vdash_{\mathbf{RM}} \varphi \to \psi$. □

Note 6.5 Relevant logics like **R** have the *variable-sharing property*. This means that if $\varphi \to \psi$ is a tautology, then φ and ψ share an atomic formula. On the other hand, in classical logic there are two other possibilities in such a case: first, that $\neg\varphi$ is a tautology, and second, that ψ is a tautology. Proposition 6.8 shows that **RM** is intermediate in this respect between relevant logics and classical logic. Intuitively, this provides an additional justification for seeing **RM** as a "semi-relevant" logic. Another one is provided by the following strong, "semi-relevant" version of the Craig interpolation theorem that was shown in Avron (1986) for **RM**: if $\vdash_{\mathbf{RM}} \varphi \supset \psi$ (where $\supset$ is the implication for **RM** given in Definition 6.2), then either $\vdash_{\mathbf{RM}} \psi$, or there is an interpolant σ such that $\mathsf{Atoms}(\sigma) \subseteq \mathsf{Atoms}(\varphi) \cap \mathsf{Atoms}(\psi)$, and both $\varphi \supset \sigma$ and $\sigma \supset \psi$ are theorems of **RM**. (In classical logic there is a third possibility:

that $\vdash \neg\varphi$.) In connection to this, it is worth mentioning that Meyer has presented in Anderson and Belnap (1975, Sect. 29.3) an example of a case in which the Craig interpolation theorem fails in **RM** for $\rightarrow$.

Our next goal is to study the set of simple extensions of **RM**.
Notation Let **L** be a logic. $Th(\mathbf{L}) =_{Df} \{\varphi\colon\ \vdash_{\mathbf{L}} \varphi\}$.

Theorem 6.9 *Let* **L** *be a simple strongly proper extension of* **RM**. *Then there is a natural number* $n \geq 2$ *such that* $Th(\mathbf{L}) = Th(\mathcal{RM}_n)$, *i.e.,* $\mathcal{RM}_n$ *is weakly characteristic for* **L**.

Proof First we prove that all theorems of **L** are valid in $\mathcal{RM}_2$. Suppose for contradiction that there is a theorem φ of **L** which is not valid in $\mathcal{RM}_2$. Then there is a valuation ν_0 in $\mathcal{RM}_2$ such that $\nu_0(\varphi) = -1$. By substituting $p_0 \rightarrow p_0$ for every atom p such that $\nu_0(p) = 1$, and $\neg(p_0 \rightarrow p_0)$ for every atom p such that $\nu_0(p) = -1$, we obtain from φ a theorem ψ of **L** such that $\mathsf{Atoms}(\psi) = \{p_0\}$, and $\nu(\psi) = -1$ for any valuation ν in $\mathcal{RM}_2$. It follows that $\neg\psi$ is valid in $\mathcal{RM}_2$. Therefore, Proposition 5.19 implies that $\vdash_{\mathbf{RM}} \neg\psi$. Hence both ψ and $\neg\psi$ are theorems of **L**. But because $\mathsf{Atoms}(\psi) = \{p_0\}$, the first part of Theorem 5.14 implies that $\neg\psi, \psi \vdash_{\mathbf{RM}} p_0$. It follows that $\vdash_{\mathbf{L}} p_0$, contradicting the condition of non-triviality in our definition of a logic.

Now let A be the set of all natural numbers n such that all theorems of **L** are valid in $\mathcal{RM}_n$. By what we have just proved, $2 \in A$, and so A is not empty. On the other hand, the fact that **L** is a simple strongly proper extension of **RM** means that there is a sentence φ_0 of $\mathcal{L}_R$ such that $\vdash_{\mathbf{L}} \varphi_0$, but $\nvdash_{\mathbf{RM}} \varphi_0$. Therefore, Proposition 5.19 implies that there is $n_0 \geq 2$ such that φ_0 is not valid in $\mathcal{RM}_{n_0}$, and so $n_0 \notin A$. It follows, by Corollary 5.18, that A has a maximal element $k \geq 2$. Then by Corollary 5.18 again, every theorem of **L** is valid in $\mathcal{RM}_j$ for every $2 \leq j \leq k$, and there is a theorem of **L** which is not valid in $\mathcal{RM}_j$ for $j > k$. We end the proof by showing that $\mathcal{RM}_k$ is weakly characteristic for **L**. Since $k \in A$, it suffices to show that if $\nvdash_{\mathbf{L}} \varphi$, then φ is not valid in $\mathcal{RM}_k$.

So suppose that $\nvdash_{\mathbf{L}} \varphi$, and let $\mathsf{Atoms}(\varphi) = \{p_1, \ldots, p_n\}$. Define

$$\mathcal{T} = \{\sigma : \mathsf{Atoms}(\sigma) \subseteq \{p_1, \ldots, p_n\} \text{ and } \vdash_{\mathbf{L}} \sigma\}$$

Since $\nvdash_{\mathbf{L}} \varphi$, also $\mathcal{T} \nvdash_{\mathbf{RM}} \varphi$. Therefore, Theorem 5.14 and its proof imply that there is an l and a valuation ν_0 in $\mathcal{RM}_l$ such that ν_0 is a model of $\mathcal{T}$ in $\mathcal{RM}_l$ which is not a model of φ, and for every element a of $\mathcal{RM}_l$ there is $1 \leq i \leq n$ such that either $a = \nu_0(p_i)$ or $a = -\nu_0(p_i) = \nu_0(\neg p_i)$. We show that $l \in A$. So let σ be a theorem of **L**, and let ν be a valuation in $\mathcal{RM}_l$. Let θ be a substitution that assigns to any atomic formula q an element τ of $\{ p_1, \neg p_1, \ldots, p_n, \neg p_n \}$ such that $\nu(q) = \nu_0(\tau)$. Then for any atomic formula q, $\nu(q) = \nu_0 \circ \theta(q)$. This easily implies that $\nu = \nu_0 \circ \theta$, and so $\nu(\sigma) = \nu_0(\theta(\sigma))$. But since **L** is a logic, $\theta(\sigma)$ is also a theorem of **L**, and by definition of θ, this implies that $\theta(\sigma) \in \mathcal{T}$. Since ν_0 is a model of $\mathcal{T}$, $\nu_0(\theta(\sigma))$ is designated, and so $\nu(\sigma)$ is designated. This was shown for every valuation ν in

$\mathcal{RM}_l$ and any theorem σ of **L**, and so it follows that indeed $l \in A$. Hence $l \leq k$. Since φ is not valid in $\mathcal{RM}_l$ (because $v_0(\varphi)$ is not designated), φ is not valid in $\mathcal{RM}_k$ either. □

Theorem 6.10 **RM** *has the Scroggs' property, that is, it does not have a finite weakly characteristic matrix, but every strongly proper extension of it does.*

Proof This follows from Proposition 6.6 and Theorem 6.9. □

Note 6.6 Theorems 6.9 and 6.10 are due to Dunn (see Dunn (1970) and Anderson and Belnap (1975, Sect. 29.4)).

By Theorem 6.9, if **L** is a simple extension of **RM**, then $Th(\mathbf{L})$ belongs to the sequence $\{\, Th(\mathcal{RM}_n)\,\}_{n=2}^{\infty}$. Next we axiomatize each of the elements in this sequence, and show that they are all different from each other.

Definition 6.11 HRM_n is the simple axiomatic extension of **RM** which is obtained by adding φ_n (from the proof of Proposition 6.6) to *HRM* as an extra axiom schema (i.e., by adding to *HRM* all instances of φ_n as new axioms).

Theorem 6.12

1. *For every* $n \geq 2$ *and* $\varphi \in \mathcal{L}_R$, φ *is valid in* $\mathcal{RM}_n$ *iff* $\vdash_{HRM_n} \varphi$. *(In other words,* $Th(\mathcal{RM}_n) = Th(HRM_n)$ *for every* $n \geq 2$.*)*
2. *The sequence* $\{\, Th(\mathcal{RM}_n)\,\}_{n=2}^{\infty}$ *is strictly decreasing, and includes* $Th(\mathbf{L})$ *whenever* **L** *is a simple strongly proper extension of* **RM**.

Proof Let φ_n be like in the proof of Proposition 6.6. It is straightforward to check that for every $n \geq 2$, φ_n is valid in $\mathcal{RM}_n$, but not in $\mathcal{RM}_{n+1}$. Hence n is the maximal number k such that φ_n is valid in $\mathcal{RM}_k$. Hence the first part follows from the proof of Theorem 6.9. That theorem implies also that the sequence $\{\, Th(\mathcal{RM}_n)\,\}_{n=2}^{\infty}$ includes every set of the form $Th(\mathbf{L})$ such that **L** is a simple strongly proper extension of **RM**. That this sequence is decreasing follows from Proposition 5.17. That it is strictly decreasing again follows from the fact that φ_n is valid in $\mathcal{RM}_n$, but not in $\mathcal{RM}_{n+1}$. □

Now we turn to what is perhaps the most important property of **RM** (and certainly the main new result in this paper).

Theorem 6.13 **RM** *is a maximal finitary logic which is both normal and semi-relevant. In other words, every proper simple finitary extension of* **RM** *is either not normal or not semi-relevant.*

Proof Let **L** be a simple finitary extension of **RM** which is both normal and semi-relevant. We show that $\mathbf{L} = \mathbf{RM}$. Now by Theorem 6.9, no strongly proper extension of **RM** can be semi-relevant. It follows that $Th(\mathbf{L}) = Th(\mathbf{RM})$. Let $\Rightarrow$ be a defined connective of $\mathcal{L}_R$ which is an implication for **L**. Then $\mathcal{T}, \varphi \vdash_{\mathbf{L}} \psi$ iff $\mathcal{T} \vdash_{\mathbf{L}} \varphi \Rightarrow \psi$, for every $\mathcal{T}$, φ and ψ.

In the sequel, we denote by $\tilde{\neg}$, $\tilde{\vee}$, $\tilde{\wedge}$, $\tilde{\to}$, and $\tilde{\Rightarrow}$ the interpretations in $\mathcal{M}(\mathbb{Z})$ of $\neg$, $\vee$, $\wedge$, $\to$, and $\Rightarrow$, respectively; and we extensively use the following property of these operations:

(*) If $f : \mathbb{Z}^n \to \mathbb{Z}$ is obtained from $\tilde{\neg}$, $\tilde{\vee}$, $\tilde{\wedge}$, $\tilde{\to}$, and $\tilde{\Rightarrow}$ using compositions, then $f(a_1, \ldots, a_n) \in \{a_1, -a_1, a_2, -a_2, \ldots, a_n, -a_n\}$ for every $a_1, \ldots, a_n \in \mathbb{Z}$.

Next we prove some properties of $\tilde{\Rightarrow}$.

1. For every $n \geq 0$, $n \tilde{\Rightarrow} n = -n \tilde{\Rightarrow} n = -n \tilde{\Rightarrow} -n = n$, while $n \tilde{\Rightarrow} -n = -n$.

Proof Since $p \vdash_{\mathbf{L}} p$, $\vdash_{\mathbf{L}} p \Rightarrow p$. Hence $\vdash_{\mathbf{RM}} p \Rightarrow p$, and so $a \tilde{\Rightarrow} a \geq 0$ for every $a \in \mathbb{Z}$. By (*) this implies that $n \tilde{\Rightarrow} n = -n \tilde{\Rightarrow} -n = n$ for every $n \geq 0$. Now the fact that $\vdash_{\mathbf{L}} p \Rightarrow p$ implies that $p \vdash_{\mathbf{L}} p \Rightarrow p$, and so $\vdash_{\mathbf{L}} p \Rightarrow (p \Rightarrow p)$. It follows that $\vdash_{\mathbf{RM}} p \Rightarrow (p \Rightarrow p)$. Hence for every $n \geq 0$, $-n \tilde{\Rightarrow} (-n \tilde{\Rightarrow} -n) \geq 0$, and so $-n \tilde{\Rightarrow} n \geq 0$. By (*), this implies that $-n \tilde{\Rightarrow} n = n$. Next we note that since $\mathbf{L}$ is a logic (i.e., non-trivial), the fact that $\vdash_{\mathbf{L}} p \Rightarrow p$ implies that $p \Rightarrow p \nvdash_{\mathbf{L}} p$, and so $\nvdash_{\mathbf{L}} (p \Rightarrow p) \Rightarrow p$. It follows that there is some $a \in \mathbb{Z}$ such that $(a \tilde{\Rightarrow} a) \tilde{\Rightarrow} a < 0$. By what we have already shown, if $a \geq 0$ then $(a \tilde{\Rightarrow} a) \tilde{\Rightarrow} a = a \geq 0$. Hence necessarily $a < 0$. So let $a = -k$ for some $k > 0$. Then $(-k \tilde{\Rightarrow} -k) \tilde{\Rightarrow} -k < 0$, and so $k \tilde{\Rightarrow} -k < 0$. By (*), this implies that $k \tilde{\Rightarrow} -k = -k$. Now by Lemma 5.2, for every $n > 0$ the submatrix of $\mathcal{M}(\mathbb{Z})$ induced by $\{-n, n\}$ is isomorphic to the submatrix of $\mathcal{M}(\mathbb{Z})$ induced by $\{-k, k\}$. It follows that $n \tilde{\Rightarrow} -n = -n$ for every $n > 0$, and (*) implies that $n \tilde{\Rightarrow} -n = -n$ also when $n = 0$. □

2. $a \tilde{\Rightarrow} k \in \{|a|, k\}$ for every $a \in \mathbb{Z}$ and $k \geq 0$.

Proof Since $\vdash_{\mathbf{L}} p \Rightarrow p$, also $q \vdash_{\mathbf{L}} p \Rightarrow p$. Hence $\vdash_{\mathbf{L}} q \Rightarrow (p \Rightarrow p)$, and so $\vdash_{\mathbf{RM}} q \Rightarrow (p \Rightarrow p)$. Hence $a \tilde{\Rightarrow} (k \tilde{\Rightarrow} k) \geq 0$ for every $a \in \mathbb{Z}$ and $k \geq 0$. By item 1 above, this means that $a \tilde{\Rightarrow} k \geq 0$ for every $a \in \mathbb{Z}$ and $k \geq 0$. Hence (*) implies that $a \tilde{\Rightarrow} k \in \{|a|, k\}$ for every $a \in \mathbb{Z}$ and $k \geq 0$. □

3. For every $a \in \mathbb{Z}$ and $k \geq 0$, if $|a| \leq k$, then $-k \tilde{\Rightarrow} a \in \{|a|, k\}$.

Proof Using $\mathcal{RM}_4$ it is easy to see that $\vdash_{\mathbf{RM}} \neg((p \to q) \to (p \to q)) \to p$. This entails that $\neg((p \to q) \to (p \to q)) \vdash_{\mathbf{L}} p$. Hence $\vdash_{\mathbf{L}} \neg((p \to q) \to (p \to q)) \Rightarrow p$, and so $\vdash_{\mathbf{RM}} \neg((p \to q) \to (p \to q)) \Rightarrow p$. It follows that if $a \in \mathbb{Z}$ and $k \geq 0$, then $-((a \tilde{\to} k) \tilde{\to} (a \tilde{\to} k)) \tilde{\Rightarrow} a \geq 0$. Now if $|a| \leq k$, then $-((a \tilde{\to} k) \tilde{\to} (a \tilde{\to} k)) = -k$, and so we get that $-k \tilde{\Rightarrow} a \geq 0$ in such a case. By (*), this is equivalent to $-k \tilde{\Rightarrow} a \in \{|a|, k\}$. □

4. If $0 \leq k \leq n$, then $k \tilde{\Rightarrow} -n = -n$.

Proof Since $\mathbf{L}$ is semi-relevant, $\neg(p \to p), (p \to p) \nvdash_{\mathbf{L}} q$. Hence $\neg(p \to p) \nvdash_{\mathbf{L}} (p \to p) \Rightarrow q$, and so $\neg(p \to p) \nvdash_{\mathbf{RM}} (p \to p) \Rightarrow q$ as well. By Corollary 5.15, this implies that there is a valuation ν in $\mathcal{M}(\mathbb{Z})$ which is a model of $\neg(p \to p)$, but not of $(p \to p) \Rightarrow q$. The first fact implies that $\nu(p) = 0$, and so the second one implies that $0 \tilde{\Rightarrow} \nu(q) < 0$. By item 2, this is possible only if $\nu(q) = -n$ for some $n > 0$. But in such a case it easily follows from Proposition 5.13 that $0 \tilde{\Rightarrow} -n < 0$ for *every* $n > 0$. By (*) and item 1, it follows that $0 \tilde{\Rightarrow} -n = -n$ for every n.

From the fact shown above that $\neg(p \to p) \nvdash_{\mathbf{L}} (p \to p) \Rightarrow q$, it follows that $\nvdash_{\mathbf{L}} \neg(p \to p) \Rightarrow ((p \to p) \Rightarrow q)$. Hence $\nvdash_{\mathbf{RM}} \neg(p \to p) \Rightarrow ((p \to p) \Rightarrow q)$. Therefore, Proposition 5.19 implies that there is a valuation ν in $\mathcal{RM}_4$ such that $\nu(\neg(p \to p) \Rightarrow ((p \to p) \Rightarrow q)) < 0$. By item 3, it cannot be the case that $\nu(\neg(p \to p)) = -2$. Hence $|\nu(p)| = 1$, and we get that $-1 \mathrel{\tilde{\Rightarrow}} (1 \mathrel{\tilde{\Rightarrow}} \nu(q)) < 0$. By items 1 and 2, this is impossible if $\nu(q) \in \{-1, 1, 2\}$. It follows that $\nu(q) = -2$, and so $-1 \mathrel{\tilde{\Rightarrow}} (1 \mathrel{\tilde{\Rightarrow}} -2) < 0$. This in turn implies (by items 1 and 2 again) that $1 \mathrel{\tilde{\Rightarrow}} -2 = -2$. As usual, by Proposition 5.13 this means that $k \mathrel{\tilde{\Rightarrow}} -n = -n$ in case $0 < k < n$. By item 1 and what we have shown above about $0 \mathrel{\tilde{\Rightarrow}} -n$, this equation holds also in the cases where $k = n$ or $k = 0$. Hence $k \mathrel{\tilde{\Rightarrow}} -n = -n$ whenever $0 \leq k \leq n$. □

5. If $0 < n < k$, then $k \mathrel{\tilde{\Rightarrow}} -n < 0$.

Proof $(p \wedge \neg p) \vee (p \wedge \neg p \to q)$ is not a tautology of **RM** in case $p \neq q$. (Take $\nu(p) = 1$ and $\nu(q) = -2$ in $\mathcal{M}(\mathbb{Z})$.) Hence it is not provable in **L** either, and so also $q \to q \nvdash_{\mathbf{L}} (p \wedge \neg p) \vee (p \wedge \neg p \to q)$. It follows that $\nvdash_{\mathbf{L}} (q \to q) \Rightarrow (p \wedge \neg p) \vee (p \wedge \neg p \to q)$, and so $\nvdash_{\mathbf{RM}} (q \to q) \Rightarrow (p \wedge \neg p) \vee (p \wedge \neg p \to q)$. Therefore, Proposition 5.19 implies that there is a valuation ν in $\mathcal{RM}_4$ such that $\nu((q \to q) \Rightarrow (p \wedge \neg p) \vee (p \wedge \neg p \to q)) < 0$. By item 2, this is possible only if $\nu((p \wedge \neg p) \vee (p \wedge \neg p \to q)) < 0$. An easy check shows that this is the case only if $\nu(q) = -2$ and $|\nu(p)| = 1$. Hence the fact that $\nu((q \to q) \Rightarrow (p \wedge \neg p) \vee (p \wedge \neg p \to q)) < 0$ means that $2 \mathrel{\tilde{\Rightarrow}} -1 < 0$. By Proposition 5.13 again, it follows that $k \mathrel{\tilde{\Rightarrow}} -n < 0$ whenever $0 < n < k$.

Next we show that $[MP]$ for $\Rightarrow$ is valid in **RM**, i.e., $\varphi, \varphi \Rightarrow \psi \vdash_{\mathbf{RM}} \psi$ for every φ and ψ. Suppose otherwise. Then from Corollary 5.15 it follows that there is a valuation ν in $\mathcal{M}(\mathbb{Z})$ such that $\nu(\varphi) \geq 0$, $\nu(\psi) < 0$, and $\nu(\varphi \Rightarrow \psi) \geq 0$. But this is impossible, by items 4 and 5 of the above list of properties of $\tilde{\Rightarrow}$.

Finally, we prove that $\mathbf{L} = \mathbf{RM}$. Since **L** is an extension of **RM**, it suffices to show that if $\mathcal{T} \vdash_{\mathbf{L}} \varphi$, then $\mathcal{T} \vdash_{\mathbf{RM}} \varphi$. So assume that $\mathcal{T} \vdash_{\mathbf{L}} \varphi$. Since **L** is finitary, there are $\psi_1, \ldots, \psi_n \in \mathcal{T}$ such that $\{\psi_1, \ldots, \psi_n\} \vdash_{\mathbf{L}} \varphi$. It follows that $\vdash_{\mathbf{L}} \psi_1 \Rightarrow (\psi_2 \Rightarrow \cdots (\psi_n \Rightarrow \varphi) \cdots)$. This in turn implies that $\vdash_{\mathbf{RM}} \psi_1 \Rightarrow (\psi_2 \Rightarrow \cdots (\psi_n \Rightarrow \varphi) \cdots)$. But we have shown that $[MP]$ for $\tilde{\Rightarrow}$ is valid in **RM**. Therefore $\{\psi_1, \ldots, \psi_n\} \vdash_{\mathbf{RM}} \varphi$, and so $\mathcal{T} \vdash_{\mathbf{RM}} \varphi$. □

Note 6.7 The fact that **L** is semi-relevant was used in the last proof only for deriving 4. Since $\neg p, p \vdash_{\mathbf{RM}} \neg(p \to p)$, while $\neg(p \to p) \vdash_{\mathbf{RM}} p$ and $\neg(p \to p) \vdash_{\mathbf{RM}} \neg p$, an almost identical proof shows that if **L** is a finitary proper simple extension of **RM** which is both normal and paraconsistent, then **L** has a finite weakly characteristic matrix. In other words, **RM** is a maximal normal paraconsistent logic that satisfies the minimal semantic relevance criterion.

Note 6.8 It is worth noting that in addition to its nice semantic properties and maximality properties as described in this section, **RM** is nice also from a proof-theoretical point of view, since it has a corresponding cut-free Gentzen-type system *GRM* with the subformula property. *GRM* employs *hypersequents*, rather than ordinary sequents, and its logical rules are identical to those used in classical logic (with caution about the chosen form of each rule, namely, whether the rule is multiplicative or additive). See Avron (1987) for details.

7 RM as a fuzzy logic

Fuzzy logics are logics that are designed to deal with propositions that involve imprecise concepts, like "tall" or "old." Their semantics is based on the idea of *degrees of truth*, according to which the truth-value assigned to a proposition of this sort might not be one of the two classical values 0 and 1, but any real number between them. Now, in all the standard fuzzy logics investigated in the literature (see Cintula et al. (2011) for an extensive survey), the consequence relation is based on preserving absolute truth, i.e., 1 is taken as the only designated value. This choice implies that none of these logics is paraconsistent. Therefore, the obvious way to develop useful paraconsistent fuzzy logics is to use a more comprehensive set of designated values. This is precisely what is done in the semantics of **RM** as given in the second part of Theorem 5.10 (i.e., the matrix $\mathcal{M}([0,1])$). Hence **RM** can serve as an excellent candidate for paraconsistent fuzzy logic.[12] However, to view and use **RM** as a fuzzy logic it would be better to take $\supset$ (rather than $\rightarrow$) as a primitive connective. This is possible, since by the next proposition this choice does not affect the expressive power of the language.

Proposition 7.1 *The connective* $\rightarrow$ *of* **RM** *is definable in* $\{\neg, \supset, \wedge, \vee\}$.

Proof By using $\mathcal{M}([0,1])$, it is easy to check that $\varphi \rightarrow \psi$ is equivalent in **RM** to $(\varphi \supset \psi) \wedge (\neg\psi \supset \neg\varphi)$.[13] □

Next we show that not only is **RM** a fuzzy logic according to the above characterization of this notion, but it (more exactly, its natural conservative extension $\mathbf{RM}^{\mathsf{F}}$ defined below) is in fact a conservative extension of one of the three most basic *standard* fuzzy logics (Cintula et al. 2011), namely, of the Gödel–Dummett logic $\mathbf{G}_\infty$.

Definition 7.2 Let $\mathcal{L}_R^{\mathsf{F}} = \mathcal{L}_R \cup \{\mathsf{F}\}$. HRM^{F} is the extension of *HRM* by the axiom $\mathsf{F} \rightarrow \varphi$. $\mathbf{RM}^{\mathsf{F}}$ is the logic in $\mathcal{L}_R^{\mathsf{F}}$ that is induced by HRM^{F}.

[12] Slaney's logic **F** (Slaney 2010) is another recent work on substructural fuzzy logics.

[13] In Avron (1986), it is noted that $\varphi \rightarrow \psi$ is equivalent in **RM** also to $\neg(\varphi \supset \psi) \supset \neg(\psi \supset \varphi)$, so it is definable in terms of just $\neg$ and $\supset$.

Definition 7.3

- A Sugihara chain $\langle \mathcal{V}, \leq, - \rangle$ is *bounded* if $\langle \mathcal{V}, \leq \rangle$ has a minimal element.[14]
- A *bounded Sugihara matrix* for $\mathcal{L}_R^{\mathsf{F}}$ is a Sugihara matrix which is based on a bounded Sugihara chain $\langle \mathcal{V}, \leq, - \rangle$, and in which the interpretation $\tilde{\mathsf{F}}$ of F is the minimal element of $\langle \mathcal{V}, \leq \rangle$.

Here is a particularly important example of a bounded Sugihara matrix.

Definition 7.4 $\mathcal{M}^{\mathsf{F}}([0, 1])$ is the extension of $\mathcal{M}([0, 1])$ to $\mathcal{L}_R^{\mathsf{F}}$ that is obtained by letting $\tilde{\mathsf{F}}$ (the interpretation of F) be 0.

Theorem 7.5

1. $\mathbf{RM}^{\mathsf{F}}$ *is strongly sound and complete for bounded Sugihara matrices.*
2. $\mathbf{RM}^{\mathsf{F}}$ *is strongly sound and complete for* $\mathcal{M}^{\mathsf{F}}([0, 1])$.

Proof A straightforward extension of the proof of Theorem 5.10. □

Corollary 7.6 $\mathbf{RM}^{\mathsf{F}}$ *is a conservative extension of* **RM**.

Definition 7.7 (*Gödel–Dummett logic* $\mathbf{G}_\infty$) Let $\mathcal{IL} = \{\supset, \wedge, \vee, \mathsf{F}\}$, and let HIL be some standard Hilbert-type system in $\mathcal{IL}$ for intuitionistic logic. HG_∞ is the extension of HIL by the following linearity axiom.

$$[\mathrm{Li}] \qquad (\varphi \supset \psi) \vee (\psi \supset \varphi)$$

$\mathbf{G}_\infty$ is the logic in $\mathcal{IL}$ which is induced by HG_∞, and $\mathbf{G}_\infty^+$ is its positive (i.e., F-free) fragment.

Theorem 7.8 $\mathbf{RM}^{\mathsf{F}}$ *is a conservative extension of* $\mathbf{G}_\infty$*, and* **RM** *is a conservative extension of* $\mathbf{G}_\infty^+$.

Proof We show the first part. The proof of the second part is almost identical.

Using Proposition 6.3 (and the fact that $\wedge$ and $\vee$ are, respectively, conjunction and disjunction for **RM**), it is easy to show that HIL is included in $\mathbf{RM}^{\mathsf{F}}$. It is also easy to verify that the extra axiom [Li] of HG_∞ is a theorem of $\mathbf{RM}^{\mathsf{F}}$ too. Hence $\mathbf{RM}^{\mathsf{F}}$ is an extension of $\mathbf{G}_\infty$.

To show that $\mathbf{RM}^{\mathsf{F}}$ *conservatively* extends $\mathbf{G}_\infty$, assume that $\mathcal{T} \nvdash_{HG_\infty} \psi$, where both $\mathcal{T}$ and ψ are in $\mathcal{IL}$. Like in the proof of Theorem 5.10, we get an extension $\mathcal{T}^*$ of $\mathcal{T}$ such that

1. $\mathcal{T}^* \nvdash_{HG_\infty} \psi$;
2. for every φ and τ, $\mathcal{T}^* \vdash_{HG_\infty} \varphi \wedge \tau$ iff both $\mathcal{T}^* \vdash_H \varphi$ and $\mathcal{T}^* \vdash_{HG_\infty} \tau$;
3. for every φ and τ, $\mathcal{T}^* \vdash_{HG_\infty} \varphi \vee \tau$ iff either $\mathcal{T}^* \vdash_{HG_\infty} \varphi$ or $\mathcal{T}^* \vdash_{HG_\infty} \tau$.

[14]Obviously, if a is a minimal element then $-a$ is a maximal one. Hence a Sugihara chain is bounded according to Definition 7.3 iff it is bounded in the usual sense of having both a minimal element and a maximal one.

Now define $\psi \equiv \sigma$ iff both $\vdash_{HG_\infty} \psi \supset \sigma$ and $\vdash_{HG_\infty} \sigma \supset \psi$. Since HG_∞ is an (axiomatic simple) extension of HIL, $\equiv$ is an equivalence relation (indeed, a congruence relation). Let $\mathcal{V}$ be the set of equivalence classes, and define $\leq$ on $\mathcal{V}$ by letting $[\tau] \leq [\sigma]$ iff $\vdash_{HG_\infty} \tau \supset \sigma$. The fact that HG_∞ is an extension of HIL easily implies this time that $\leq$ is well defined, and is a partial order on $\mathcal{V}$. In addition, the $\vee$-primeness of $\mathcal{T}^*$ (item 3 above) and the special axiom [Li] of HG_∞ entail that $\leq$ is a *linear* order. Obviously, $[\mathsf{F}]$ is the minimal element of $\mathcal{V}$ according to this linear order, while axiom [$\supset$ 1] of HIL ensures that $\{\varphi\colon \mathcal{T}^* \vdash_{HG_\infty} \varphi\}$ is its maximal element. Since $\mathcal{V}$ is countable, these facts imply (see the beginning of the proof of Lemma 5.3) that there is a function $e : \mathcal{V} \to [0, 1/2]$ such that e is order preserving, $e([\mathsf{F}]) = 0$, and $e(\{\varphi\colon \mathcal{T}^* \vdash_{HG_\infty} \varphi\}) = 1/2$. Define a valuation ν in $\mathcal{M}([0, 1])$ by letting $\nu(p) = e([p])$ for every atom p. We show that the following is true for every formula φ of $\mathcal{IL}$:

(a) If $\mathcal{T}^* \vdash_{HG_\infty} \varphi$, then $\nu(\varphi) \geq 1/2$.
(b) If $\mathcal{T}^* \nvdash_{HG_\infty} \varphi$, then $\nu(\varphi) = e([\varphi])$ (and so $\nu(\varphi) < 1/2$).

Since $\mathcal{T} \subseteq \mathcal{T}^*$ and $\mathcal{T}^* \nvdash_{HG_\infty} \psi$, these two facts imply that ν is a model of $\mathcal{T}$ in $\mathcal{M}([0, 1])$ that is not a model of ψ. Hence Theorem 7.5 entails that $\mathcal{T}^* \nvdash_{\mathbf{RM}^{\mathsf{F}}} \psi$, which is what we wanted to prove.

We prove (a) and (b) by induction on the complexity of φ.

- The case where φ is an atomic variable or the constant F easily follows from the definition of ν, and the properties of e mentioned above.
- Suppose that $\varphi = \tau \supset \sigma$.
  (a) Suppose $\mathcal{T}^* \vdash_{HG_\infty} \varphi$. Then $[\tau] \leq [\sigma]$, and so $e([\tau]) \leq e([\sigma])$. If $\nu(\sigma) \geq 1/2$, then $\nu(\varphi) \geq 1/2$ (see Note 6.1). If not, then $\mathcal{T}^* \nvdash_{HG_\infty} \sigma$ by (a) of the induction hypothesis, and so $\mathcal{T}^* \nvdash_{HG_\infty} \tau$. Hence (b) of the induction hypothesis implies that $\nu(\tau) = e([\tau])$ and $\nu(\sigma) = e([\sigma])$. Therefore $\nu(\tau) \leq \nu(\sigma)$, and so $\nu(\varphi) \geq 1/2$.
  (b) Suppose $\mathcal{T}^* \nvdash_{HG_\infty} \varphi$. Because of Axiom [$\supset$ 1], this implies that also $\mathcal{T}^* \nvdash_{HG_\infty} \sigma$, and so $\nu(\sigma) = e([\sigma]) < 1/2$ by (a). The assumption also implies that $[\tau] \not\leq [\sigma]$, and so $e([\sigma]) < e([\tau])$. Since by (a) and (b) $\nu(\tau) \geq 1/2$ or $\nu(\tau) = e([\tau])$, it follows that $\nu(\sigma) < \nu(\tau)$, and so (see Note 6.1) $\nu(\varphi) = \nu(\sigma) = e([\sigma])$. It remains to show that $e([\varphi]) = e([\sigma])$ in this case, i.e., that $\varphi \equiv \sigma$. That $\mathcal{T}^* \vdash_{HG_\infty} \sigma \supset \varphi$ is immediate from Axiom [$\supset$ 1]. For the converse implication, note that since $\tau \supset (\tau \supset \sigma) \vdash_{HIL} \tau \supset \sigma$ (immediate from the deduction theorem of HIL), our assumption implies that $\mathcal{T}^* \nvdash_{HG_\infty} \tau \supset (\tau \supset \sigma)$. Hence Axiom [Li] and the $\vee$-primeness of $\mathcal{T}^*$ entail that $\mathcal{T}^* \vdash_{HG_\infty} (\tau \supset \sigma) \supset \tau$. But $\vdash_{HIL} ((\tau \supset \sigma) \supset \tau) \supset ((\tau \supset \sigma) \supset \sigma)$. It follows that $\mathcal{T}^* \vdash_{HG_\infty} (\tau \supset \sigma) \supset \sigma$, i.e., $\mathcal{T}^* \vdash_{HG_\infty} \varphi \supset \sigma$.
- Suppose that $\varphi = \tau \vee \sigma$.
  (a) Suppose $\mathcal{T}^* \vdash_{HG_\infty} \varphi$. Then the $\vee$-primeness of $\mathcal{T}^*$ implies that either $\mathcal{T}^* \vdash_{HG_\infty} \tau$ or $\mathcal{T}^* \vdash_{HG_\infty} \sigma$. It follows by (a) of the induction hypothesis that either $\nu(\tau) \geq 1/2$ or $\nu(\sigma) \geq 1/2$. In both cases, also $\nu(\varphi) \geq 1/2$.
  (b) Suppose that $\mathcal{T}^* \nvdash_{HG_\infty} \varphi$. Then property 3 of $\mathcal{T}^*$ implies that $\mathcal{T}^* \nvdash_{HG_\infty} \tau$ and $\mathcal{T}^* \nvdash_{HG_\infty} \sigma$. It follows by (b) of the induction hypothesis that $\nu(\tau) = e([\tau])$

and $\nu(\sigma) = e([\sigma])$. Assume, without loss of generality, that $[\sigma] \leq [\tau]$. Then $\mathcal{T}^* \vdash_{HG_\infty} \sigma \supset \tau$, and $e([\sigma]) \leq e([\tau])$. The former fact implies (with the help of the Axioms $[\supset \vee]$ and $[\vee \supset]$) that $\varphi \equiv \tau$, and so $e([\varphi]) = e([\tau])$. The latter fact implies that $\nu(\varphi) = e([\tau])$, hence $\nu(\varphi) = e([\varphi])$.
- We leave the case where $\varphi = \tau \wedge \sigma$ to the reader.

This ends the proof of (a) and (b), and so of the theorem. □

Note 7.1 The connection between **RM** and $\mathbf{G}_\infty$ was first observed by Dunn and Meyer in (1971), where it was proved that $\mathbf{RM}^{\mathbf{t}}$ (see Note 6.3) is a *weakly* conservative extension of the positive fragment of $\mathbf{G}_\infty$.

Note 7.2 The standard semantics of Gödel–Dummett logic $\mathbf{G}_\infty$, as described in the literature on fuzzy logics, is provided by the matrix $\langle [0,1], 1, \mathcal{O} \rangle$, where the interpretations in $\mathcal{O}$ of $\vee$, $\wedge$, and F are like in $\mathcal{M}([0,1])$ (the strongly characteristic matrix for $\mathbf{RM}^{\mathsf{F}}$), while $a \tilde{\supset} b$ is 1 if $a \leq b$, and b otherwise. However, the last theorem shows that when we use $\mathbf{G}_\infty$, it is not essential at all to take 1 as the only designated value. It is also interesting to note that the interpretation of $\neg$ in $\mathcal{M}([0,1])$ is identical to that used in the most famous fuzzy logic (except perhaps $\mathbf{G}_\infty$): Łukasiewicz's logic. (In $\mathbf{G}_\infty$ itself $\neg\varphi$ is usually taken as an abbreviation for $\varphi \supset \mathsf{F}$.)

Note 7.3 A Hilbert-type system $HRM^{\supset}$ in $\{\neg, \vee, \wedge, \supset\}$ which is strongly sound and complete for **RM** has been given in Avron (1986). $HRM^{\supset}$ is obtained from HG_∞, by adding to it axioms connected with $\neg$. By adding $\mathsf{F} \supset \varphi$ and $\varphi \supset \neg\mathsf{F}$ as axioms to $HRM^{\supset}$, we get a Hilbert-type system in $\mathcal{IL} \cup \{\neg\}$ that is strongly sound and complete for $\mathbf{RM}^{\mathsf{F}}$.

References

Anderson, A. R., & Belnap, N. D. (1975). *Entailment: The logic of relevance and necessity* (Vol. I). Princeton: Princeton University Press.

Avron, A. (1984). Relevant entailment—semantics and formal systems. *Journal of Symbolic Logic*, *49*, 334–342.

Avron, A. (1986). On an implication connective of RM. *Notre Dame Journal of Formal Logic*, *27*, 201–209.

Avron, A. (1987). A constructive analysis of RM. *Journal of Symbolic Logic*, *52*, 939–951.

Avron, A. (1997). Multiplicative conjunction as an extensional conjunction. *Logic Journal of the IGPL*, *5*, 181–208.

Avron, A. (2014). What is relevance logic? *Annals of Pure and Applied Logic*, *165*, 26–48.

Avron, A. & Zamansky, A. (2011). Non-deterministic semantics for logical systems —a survey, In D. Gabbay & F. Guenthner (Eds.), *Handbook of philosophical logic* (2nd ed., Vol. 16, pp. 227–304). Springer.

Brady, R. T. (1990). The gentzenization and decidability of RW. *Journal of Philosophical Logic*, *19*, 35–73.

Cintula, P., Hájek, P. & Noguera, C. (2011). *Handbook of mathematical fuzzy logic*, Volume 37–38 of Studies in Logic, College Publications.

Dunn, J. M. (1970). Algebraic completeness results for R-Mingle and its extensions. *Journal of Symbolic Logic, 35*, 1–13.
Dunn, J. M., & Meyer, R. K. (1971). Algebraic completeness results for Dummett's LC and its extensions. *Zeitschrift für mathematische Logik und Grundlagen der Mathematik, 17*, 225–230.
Dunn, J. M., & Restall, G. (2002). Relevance logic. In D. Gabbay & F. Guenthner (Eds.), *Handbook of philosophical logic* (2nd ed., Vol. 6, pp. 1–128). Amsterdam: Kluwer.
Łoś, J., & Suszko, R. (1958). Remarks on sentential logics. *Indagationes Mathematicae, 20*, 177–183.
Meyer, R. K., & Dunn, J. M. (1969). E, R and γ. *Journal of Symbolic Logic, 34*, 460–474.
Parks, R. Z. (1972). A note on R-Mingle and Sobociński's three-valued logic. *Notre Dame Journal of Formal Logic, 13*, 227–228.
Slaney, J. K. (2010). A logic for vagueness. *Australian Journal of Logic, 8*, 100–134.
Sugihara, T. (1955). Strict implication free from implicational paradoxes. *Memoirs of the Faculty of Liberal Arts, Fukui University, Series, 1*(4), 55–59.
Urquhart, A. (1984). The undecidability of entailment and relevant implication. *Journal of Symbolic Logic, 49*, 1059–1073.
Urquhart, A. (2001). Many-valued logic, In D. Gabbay & F. Guenthner (Eds.), *Handbook of philosophical logic* (Vol. 11, pp. 249–295). Kluwer

Wedge Sum, Merge and Inconsistency

Chris Mortensen

Abstract This paper investigates the topological construction of Wedge Sum, with the aim of showing that it can be done mathematically, via a quotient construction, or logically, via Merge. Consistent and Inconsistent versions are given, while noting that the natural outcome of Merging is an inconsistent theory. Finally it is observed that algebraic constructions can also be treated via Merge, where the extra functionality makes for various triviality and non-triviality results.

Keywords Inconsistent theory · Leibniz law · Logical theories · Merge · Wedge sum

1 Wedge Sum

The outstanding work on three-valued paraconsistent model theory by J. M. Dunn (1979), and preceding him R. K. Meyer (1976), proved an inspiration to the present author to construct inconsistent mathematical theories (e.g., Mortensen 1995). In particular, these two authors studied and applied the logic *RM*3. This is a three-valued logic which lends itself naturally to an informational interpretation in virtue of the inconsistency-tolerance of the logic. In this paper these model-theoretic methods are applied to study various inconsistent topological and algebraic theories. It is noted that other simple paraconsistent logics, such as *LP* and *P*3, allow for similar results, since results on inconsistent theories tend to be invariant over large classes of background logics. Even so, we will be working with *RM*3 in honour of Meyer and Dunn's brilliant example.

Mathematics and logic approach identification in characteristically different ways. A typical mathematical construction involves equivalence classes which preserve structure, for example topological structure or algebraic structure. The equivalence classes, which "identify" those things in the same equivalence classes, are then the

C. Mortensen (✉)
Department of Philosophy, University of Adelaide, Adelaide, Australia
e-mail: Chris.Mortensen@adelaide.edu.au
URL: http://www.adelaide.edu/directory/chris.mortensen

K. Bimbó (ed.), *J. Michael Dunn on Information Based Logics*,
Outstanding Contributions to Logic 8, DOI 10.1007/978-3-319-29300-4_3

elements of the collapsed space. This construction can then be turned naturally into consistent and inconsistent logical theories, as we will see. By contrast, logic does the work utilizing logical techniques. This technique begins with consistent theories and applies logical operations (specifically Merge, see below) to produce inconsistent theories. The differences of approach are instructive for comparison of inconsistent mathematical structures.

We begin with mathematics. A useful topological construction is Wedge Sum.

Definition 1.1 Let X, Y, be topological spaces, with points $x \in X$, $y \in Y$. Form the disjoint union of X and Y, $X \sqcup Y$, and let a, b be two points with $a \in X$, $b \in Y$. Let $\sim$ be the equivalence relation determined by membership of $\{a, b\}$, that is $x \sim y$ iff either $x = y$ or both x and y are members of $\{a, b\}$. Then the *wedge sum* of X and Y, written $X \vee Y$, is defined as the quotient $X \sqcup Y / \sim$.

This amounts to saying that the wedge sum of two spaces is their union, except for having just one pair of overlapping points identified. So for example for the circle S^1, the wedge sum $S^1 \vee S^1$ is homeomorphic to the numeral 8, while $S^2 \vee S^2$ is homeomorphic to a pair of spheres joined at a single point (see e.g., Hatcher 2002, p. 10).

Clearly, this definition can be extended to more than two spaces, and more than one pair of overlapping points, but we do not need that here. In this paper, we directly construct consistent logical theories which describe wedge sums, and inconsistent theories which extend them. Then we utilize the technique Merge to show how to obtain inconsistent theories of the wedge sum in a different way. Finally, it is seen that this technique can be used to study inconsistent structures other than topological spaces.

2 Consistent and Inconsistent Theories

There is a natural way to produce consistent theories, and extensions to inconsistent theories, which take into account the construction above, (following Mortensen 1995, pp. 93–95). This is necessary for comparison between the mathematical approach above, and the logical approach.

We take a language $\mathcal{L}$ with:

(i) terms for members of $X \sqcup Y$, with metalinguistic variables $t, t_1, t_2, \ldots$ ranging over these terms; terms for subsets of $X \sqcup Y$, with metalinguistic variables $S, S_1, S_2, \ldots$ ranging over them; and a constant term O for the set of open sets of $X \sqcup Y$. For convenience we can take members of $X \sqcup Y$ as naming themselves.
(ii) For binary relations we take identity $=$ and set membership $\in$.
(iii) Atomic sentences are of the form $t_1 = t_2$, $S_1 = S_2$, $t \in S$ and $S \in O$.

An interpretation I can now be defined, which interprets terms in the wedge sum of X and Y, and sentences in the values $\{T, F\}$ of 2-valued Boolean logic:

(1) $I(t) = P(t) = [t]$,
where P is the projection operator taking t to its equivalence class, and $[t]$ is the equivalence class of t;
(2) $I(S) = P(S) = \{[t] : t \in S\}$;
(3) $I(O) = \{P(S) : S \in O\}$;
(4) $I(t_1 = t_2) = T$ *if* $I(t_1) = I(t_2)$, *else* $I(t_1 = t_2) = F$;
(5) $I(t \in S) = T$ *if* $I(t)$ *is in* $I(S)$, *else* $I(t \in S) = F$;
(6) $I(S_1 = S_2) = T$ *if* $I(S_1) = I(S_2)$, *else* $I(S_1 = S_2) = F$;
(7) $I(S \in O) = T$ *if* $I(S)$ *is in* $I(O)$, *else* $I(S \in O) = F$.

This induces an interpretation on non-atomic sentences in accordance with 2-valued Boolean logic. The resulting interpretation determines a theory *Th* by the definition $Th := \{A : I(A) = T\}$.

Definition 2.1 A theory is *inconsistent* if it contains some sentence and its negation, else *consistent*; and *trivial* if it contains every sentence, else *non-trivial*. A theory is *incomplete* if it lacks some sentence (of its language) and its negation, else *complete*.

Hence we may observe that the theory generated by the above interpretation (1)–(7) is consistent and complete, since sentences take exactly one of the two truth values, T or F. Note also that $I(a = b) = T$, that is $a = b$ holds but $\neg(a = b)$ does not hold.

Of interest are theories determined by assigning sentences one of the three values (T, B, F) of the paraconsistent logic *RM*3 (see e.g., Dunn 1979).

Definition 2.2 A sentence is said to *hold* in an *RM*3 interpretation if it takes the value T or the value B. Given an *RM*3 interpretation, an associated theory *Th* can be then defined as being the set of sentences that hold in the interpretation, $Th := \{A : I(A) = T \text{ or } I(A) = B\}$.

It is straightforward to modify the interpretation above to produce an inconsistent theory: simply change T to B in (4)–(7) above. This implies that $I(a = b) = B = I(\neg a = b)$. Call the associated *RM*3-theory $Th(X \vee Y)$. Both $a = b$ and $\neg a = b$ hold in $Th(X \vee Y)$. The theory is non-trivial, since for any pair of points c, d other than a, b, where c is in one of the original two spaces and d is in the other, then $c = d$ does not hold. The theory is complete, since if any sentence is assigned F then its negation is assigned T.

Definition 2.3 A theory is *functional* if, whenever any equation $t_1 = t_2$ holds, then Ft_1 holds iff Ft_2 holds, where Ft_1 is any atomic sentence containing t_1 and Ft_2 is like Ft_1 except for replacing t_1 by t_2 in one or more places. A theory is *transparent* if the same conditions hold except that Ft_1 and Ft_2 can be any sentence (not restricted to atomic).

We now observe:

Theorem 2.4 *The inconsistent theory* $Th(X \vee Y)$ *is functional, and indeed transparent.*

Proof For functionality, we just have to check all the atomic contexts. If $t_1 = t_2$ holds then $I(t_1) = I(t_2)$. Hence $t_1 = t_3$ holds iff $I(t_1) = I(t_3)$ iff $I(t_2) = I(t_3)$ iff $t_2 = t_3$ holds. Similarly, $t_1 \in S$ holds iff $I(t_1) \in I(S)$ iff $I(t_2) \in I(S)$ iff $t_2 \in S$ holds. If $S_1 = S_2$ then by similar arguments $t \in S_1$ holds iff $t \in S_2$ holds, and $S_1 \in O$ holds iff $S_2 \in O$ holds.

Transparency is proved by a straightforward induction on the logical complexity of sentences, of which functionality is the base clause. □

We also note that there are other ways that a model and an associated inconsistent theory can be constructed. For example, define I by (1)–(3) plus:

(4′) $I(t_1 = t_2) = T$ *if* $t_1 = t_2$, *else* $I(t_1 = t_2) = B$ *if* $I(t_1) = I(t_2)$; *otherwise* $I(t_1 = t_2) = F$.
(5′) $I(t \in S) = T$ *if* t *is in* S, *else* $I(t \in S) = B$ *if* $I(t)$ *is in* $I(S)$; *otherwise* $I(t \in S) = F$.
(6′) $I(S_1 = S_2) = T$ *if* $S_1 = S_2$, *else* $I(S_1 = S_2) = B$ *if* $I(S_1) = I(S_2)$; *otherwise* $I(S_1 = S_2) = F$.
(7′) $I(S \in O) = T$ *if* $I(S)$ *is in* $I(O)$, *else* $I(S \in O) = F$.

This theory is inconsistent, since again both $a = b$ and $\neg(a = b)$ hold. Note that here in (7′) the concept of openness is treated consistently. Note that this consistency is optional, but it illustrates the point that inconsistency can be isolated in various ways.

It is easily seen that this theory is functional, by an argument similar to that for $Th(X \vee Y)$ above. It is, however, not transparent: since both $a = b$ and $\neg(a = b)$ hold, then if it were transparent, so would $\neg(a = a)$ hold, contradicting (4′). Failure of transparency is not such a burden, it is an epiphenomenon of logic rather than mathematics, though Dunn and Meyer both liked transparency. (On transparency and functionality, see Mortensen 1995, Chap. 2.)

It is convenient to refer to this theory as **Wedge**. The point of having inconsistency in such theories, especially Wedge, is that the inconsistency "keeps track of" the difference in origin of the identified items, by preserving their original disidentity while still registering the identification where the mathematical (functional) work is done.

3 Merge

There is another way to approach the construction of an inconsistent theory of wedge sum, a proof-theoretic or logical way.

Definition 3.1 If Th_1, Th_2 are two theories (of the same logic $\vdash$), then Merge $(Th_1, Th_2) = (Th_1 \cup Th_2)^{\vdash\&}$.

That is, the Merge of two theories is the theory which is the deductive and conjunctive closure of their union (see Mortensen 2011, Sect. 4). In the present case, the logic in question is $RM3$.

The Merge operation can be used to produce an alternative inconsistent theory of the wedge sum, as follows.

Begin by constructing a consistent and complete theory of one of the topological spaces X. First form the usual Boolean interpretation I by:

$I(t_1 = t_2) = T$ iff both t_1 and t_2 are in X, and $t_1 = t_2$, else $I(t_1 = t_2) = F$.

$I(t \in S) = T$ iff t is in S, else $I(t \in S) = F$.

$I(S_1 = S_2) = T$ iff $S_1 = S_2$, else $I(S_1 = S_2) = F$.

$I(S \in O) = T$ iff S is in O, else $I(S \in O) = F$.

This interpretation forms a consistent and complete theory. Extend I to an interpretation I^{ω} by adding an additional "dummy" constant ω and setting $I^{\omega}(\omega) = I^{\omega}(a) = a$. This ensures that $I^{\omega}(a = \omega) = T$. Call the generated theory $Th^{\omega}(X)$, that is $Th^{\omega}(X) = \{A\colon I^{\omega}(A) = T\}$. Note that $Th^{\omega}(X)$ remains consistent and complete, and that $a = \omega$ and $\neg(a = b)$ are in it, and that $\neg(a = \omega)$ and $a = b$ are not.

Now do the same for Y, save that ω is interpreted to be the same as b. Call this theory $Th^{\omega}(Y)$. Note that $Th^{\omega}(Y)$ is consistent and complete, and that $b = \omega$ and $\neg(a = b)$ are in it, but that $\neg(b = \omega)$ and $a = b$ are not.

Finally, Merge these two theories, obtaining Merge($Th^{\omega}(X)$, $Th^{\omega}(Y)$), or **Merge** for short.

Note that Merging requires a background logic, whose deductive rules are common to the two theories being merged. As above we take $RM3$. It is required here to make one further assumption, specifically that the background logic is closed under *Leibniz Law* as one of its rules. It suffices to assume a weak form of Leibniz Law, namely the substitutivity of identicals in all *atomic* contexts. That is, if $t_1 = t_2$ holds, then Ft_1 holds iff Ft_2 holds where F is atomic. This gives:

Theorem 3.2 *Merge is inconsistent, but non-trivial.*

Proof Since $a = \omega$ holds and $b = \omega$ holds, then substituting b for the ω in the first equation gives by Leibniz Law that $a = b$ holds. However, $\neg(a = b)$ also holds since it holds in (both) the Th^{ω} theories. Also both $\omega = \omega$ and $\neg(\omega = \omega)$ hold. Many sentences continue not to hold, however, such as any identity between points of X other than a, and points of Y other than b; so the theory is non-trivial. □

Theorem 3.3 *Merge is complete and functional.*

Proof If the Merged theory were incomplete then for some A, neither A nor $\neg A$ would be consequences of the union of the two Th^{ω} theories. But this is impossible since these two theories, in the same language, are complete. The requirement of closure under Leibniz Law is the same as functionality. □

We now see that:

Theorem 3.4 *Wedge and Merge have identical atomic sentences in their common language.*

Proof First, every atomic sentence that holds in Wedge, holds in Merge. Any atomic sentence taking the value T in Wedge is true in one of the Boolean theories from which Merge is constructed, and so holds in Merge. Thus we must consider those atomic sentences which are B in Wedge, and show that they all hold in the Merge theory. These are:

(i) $t_1 \neq t_2$ but $I(t_1) = I(t_2)$.

(ii) $t \notin S$ but $I(t) \in I(S)$.

(iii) $S_1 \neq S_2$ but $I(S_1) = I(S_2)$.

On (i), one of the t must be a and the other must be b, since otherwise either $t_1 = t_2$ or $I(t_1) \neq I(t_2)$. But then, in the Merged theory, $a = b$ holds.

On (ii), two cases. First case: t is a and S contains b but not a. Thus, $S = S' \cup \{b\}$ where S' does not include b. By construction of Merge, $S = S' \cup \{b\}$ holds. By Leibniz Law, $S = S' \cup \{a\}$ holds, so that $t \in S$ holds. Second case: t is b and S contains a but not b. Similar argument.

On (iii), two cases. First case: $S_1 = S \cup \{a\}$ and $S_2 = S \cup \{b\}$ where S contains neither a nor b. By construction, both hold. By Leibniz Law, $S_1 = S \cup \{b\}$ holds. Hence $S_1 = S_2$ holds. Second case: S_1 and S_2 are reversed. Similar argument.

Second, every atomic sentence that holds in Merge, holds in Wedge. Both the Boolean theories to be Merged are classically true, so all their sentences take the value T in Wedge. Similarly $a = b$ which holds in Merge, holds by construction in Wedge. So we just have to assure ourselves that the deductive rules of Merge are preserved in Wedge. But the rules of Merge are the rules of $RM3$ and are certainly preserved in Wedge which is an $RM3$ model. In particular, Leibniz Law is preserved in Wedge, since as observed above, Wedge is functional. □

In short, we see that the same effects on the functionality of theories can be obtained by a typically mathematical approach, as in the quotient construction Wedge Sum, or by a more logical approach, as in Merge.

4 Merging Algebras

The technique of Merge is applicable to theories other than those of topological spaces. For instance, we can consider finite additive groups such as the integers with varying moduli. These generate consistent theories, and also inconsistent theories (see Meyer and Mortensen 1984). Bringing in algebraic operations allows for a richer range of properties, and suggests a natural mathematical generalization of Wedge Sum.

For example, consistent Z mod 6 can be merged with consistent Z mod 9. The additive structure provides for zero as a surrogate for the dummy constant ω. Thus we can write $0 = 6$ in mod 6, and $0 = 9$ in mod 9. Closing under Leibniz Law gives $9 = 6$. Now Leibniz Law must respect the additional functional (arithmetic) structure. Hence, from $0 - 0 = 9 - 6$, we get $0 = 3$. The number 3 is the highest common factor of 6 and 9. This theory is inconsistent as either (both) theories also contain $\neg(0 = 3)$. Leibniz Law does not afford further arithmetical reduction, so we have a non-triviality result for Merge: merging integer arithmetics mod 6 and mod 9 does not allow the deduction of $0 = 1$. Merging of higher numbers of theories also resist trivialization as long as the moduli all have a common factor.

Contrast with the case where the two moduli are relatively prime, where repeated applications of Leibniz Law give $0 = 1$. Thus, we cannot automatically expect that Merging will result in a non-trivial theory: it depends on the functional properties of the operators in the theories.

References

Dunn, J. M. (1979). A theorem in 3-valued model theory with connections to number theory, type theory, and relevant logic. *Studia Logica, 38*, 149–169.

Hatcher, A. (2002). *Algebraic topology*. Cambridge: Cambridge University Press.

Meyer, R. K. (1976). Relevant arithmetic. *Bulletin of the Section of Logic of the Polish Academy of Science, 5*, 133–137.

Meyer, R. K., & Mortensen, C. (1984). Inconsistent models for relevant arithmetics. *Journal of Symbolic Logic, 49*, 917–929.

Mortensen, C. (1995). *Inconsistent mathematics*. Dordrecht: Kluwer.

Mortensen, C. (2011). Merge. *Australasian Journal of Logic, 9*, 135–141.

Single Axioms and Axiom-Pairs for the Implicational Fragments of R, R-Mingle, and Some Related Systems

Dolph Ulrich

Abstract Various axiom sets for the implicational fragments $\mathbf{R}_{\rightarrow}$ and $\mathbf{RM}_{\rightarrow}$ of **R** and of **R-Mingle** have appeared in the literature over the last six-and-a-half decades, some of them in other guises well before the full systems with $\sim$, &, and $\vee$ were even introduced. Most such sets are comprised of three or four axioms. For other logics of pure implication, the historical progression has typically been from longer axiom sets to the discovery of deductively equivalent two- and one-axiom bases. This paper continues in that pattern, presenting such bases for $\mathbf{R}_{\rightarrow}$ and $\mathbf{RM}_{\rightarrow}$. Along the way, new axiom pairs and new single axioms are given for a number of other implicational logics as well, some in the paper itself and many in the Appendix attached to it. Prominent among these is C.A. Meredith's system **BCI**. Though single axioms for **BCI** are of independent interest, one of them in particular also plays an invaluable role in the construction of those provided here for implicational **R** and **R-Mingle**.

Keywords Axiom-pairs · BCI · Implicational fragment · R · R-Mingle · Single axioms

1 Implicational Logics: Early Work

Our main concern is with the development of axiom sets for systems of pure implication, that is, sentential logics with a single binary connective and the rules *modus ponens* (equivalently, *detachment*) and *uniform substitution* of formulas for sentence letters. Formulas are displayed throughout using standard infix notation with $\rightarrow$ as the implication connective, though many of the publications cited herein instead use C and the parentheses-free prefix notation of Łukasiewicz.

The first sustained work on axiomatizing implicational logics began in Poland. In 1921, Tarski axiomatized the implicational fragment **IF** of the classical sentential calculus using the axioms $(p \rightarrow q) \rightarrow ((q \rightarrow r) \rightarrow (p \rightarrow r))$, $p \rightarrow (q \rightarrow p)$,

D. Ulrich (✉)
Department of Philosophy, Purdue University, West Lafayette, USA
e-mail: dulrich@purdue.edu
URL: http://www.cla.purdue.edu/philosophy/directory/?p=Dolph_Ulrich

K. Bimbó (ed.), *J. Michael Dunn on Information Based Logics*,
Outstanding Contributions to Logic 8, DOI 10.1007/978-3-319-29300-4_4

and $((p \to q) \to r) \to ((p \to r) \to r)$ (Łukasiewicz and Tarski 1930). In 1928, Bernays (ibid.) shortened the third one, replacing it with Peirce's $((p \to q) \to p) \to p$,[1] and the resulting set below has come to be known as the Tarski–Bernays axioms.

$(p \to q) \to ((q \to r) \to (p \to r))$	**B′/Syll/Syl**
$p \to (q \to p)$	**K/Simp**
$((p \to q) \to p) \to p$	**Peirce**

(The single letters **B′** and **K** used here to label the axioms are taken from the names of the combinators for which they are the principle types, a common modern practice. The others—**Syll**, **Simp**, and so on—are from the older tradition; they were the "nicknames" given those formulas by the early researchers, starting in the 1920s and 1930s.)

As alternate axiomatizations of **IF** emerged, concerns with the lengths of individual axioms became accompanied by interest in reducing the total number of axioms in the set. Wajsberg (1932) (but circa 1925–1926; cf. Prior (1962, p. 302)), produced the first *two-bases*. Each is twenty-two symbols in total length (not counting parentheses—there were none in the notation he was using), with ten occurrences of the implication connective as in the Tarski–Bernays set.

(a) $p \to (q \to (r \to p))$, $((p \to q) \to r) \to ((s \to r) \to ((p \to r) \to r))$
(b) $(p \to q) \to p) \to p$, $(((p \to q) \to r) \to s) \to ((q \to r) \to (p \to s))$

Discovery of shorter, 20-symbol pairs, using just nine occurrences of the connective, soon followed. Łukasiewicz (cf. Prior 1961) found

(c) $p \to (q \to p)$, $(((p \to q) \to r) \to s) \to ((q \to s) \to (p \to s))$,

and (Wajsberg 1939) has two more of the same total length.

(d) $((p \to q) \to p) \to p$, $(p \to q) \to (s \to ((q \to r) \to (p \to r)))$
(e) $((p \to q) \to p) \to (r \to p)$, $(p \to q) \to ((q \to r) \to (p \to r))$

Looking at such pairs some years later, Prior observed that the distribution of occurrences of the connective between the two axioms in (e) is 4/5, in others—cf. (d)—is 3/6, and in (c) is 2/7. Prior (1961) reports asking his colleague C.A. Meredith if there exists also a nine-connective pair in which the distribution is 1/8, that is, if there is a formula containing eight occurrences of the connective which, with $p \to p$, provides an independent two-base for **IF**. Meredith did not answer that question, but in 1960 he did find an even shorter, 18-symbol pair with the distribution 1/7.

(f) $p \to p$, $((p \to q) \to r) \to ((r \to p) \to (s \to (t \to p)))$

[1] This formula appeared originally in Peirce (1885), alongside $(p \to q) \to ((q \to r) \to (p \to r))$, $(p \to (q \to r)) \to (q \to (p \to r))$, and $p \to p$. It turns out that those four together also suffice to axiomatize **IF**.

Beyond (short) two-bases of course lies the ultimate goal of discovering equivalent *single* axioms. Tarski himself devised methods for producing such axioms by the (carefully considered) placement of one or more theorems inside certain others, and was consequently able to construct the first single axioms for **IF** (Leśniewski 1929). Those first axioms were quite long, so Łukasiewicz and Tarski (1930) display instead two such axioms of length 25 found later by others, one from Wajsberg and one from Łukasiewicz.

$$((p \to q) \to ((r \to s) \to t)) \to ((u \to ((r \to s) \to t)) \to ((p \to u) \to (s \to t)))$$
$$((p \to (q \to p)) \to (((((r \to s) \to t) \to u) \to ((s \to u) \to (r \to u))) \to v)) \to v$$

Wajsberg's contains no theorems of **IF** as proper parts and so, following Leśniewski (1929), is said to be *organic*. Łukasiewicz's is clearly *inorganic* containing, as it does, $p \to (q \to p)$. In fact, it is "doubly" inorganic, because it contains $((((r \to s) \to t) \to u) \to ((s \to u) \to (r \to u)))$ as well.

Indeed, $p \to (q \to p)$ and $(((r \to s) \to t) \to u) \to ((s \to u) \to (r \to u))$ together are an alphabetic variant of Łukasiewicz's two-base (c) above. Since $(((r \to s) \to t) \to u) \to ((s \to u) \to (r \to u))$ is an **IF** thesis, then so also is $((((r \to s) \to t) \to u) \to ((s \to u) \to (r \to u)) \to v) \to v$, and Łukasiewicz's axiom thus illustrates the Tarski-style approach to constructing single axioms by inserting one inside another.

Łukasiewicz went on thereafter to produce even shorter single axioms, culminating finally (Łukasiewicz 1948) in the 13-symbol

$$((p \to q) \to r) \to ((r \to p) \to (s \to p)).$$

He showed as well that no shorter theorem of **IF** is a single axiom for it, so that his axiom is *shortest possible* for that system. Later, Tursman (1968), with an assist from Thomas (1970) for one troublesome candidate, went on to prove that Łukasiewicz's axiom is unique among the 13-symbol formulas in this respect: no other **IF** theorem of *equal* length can do the job either.

With Łukasiewicz's axiom at hand, we can take up Prior's question about the existence of an eight-arrow **IF** thesis which, with $p \to p$, axiomatizes **IF**. He himself retained interest in it, concluding Prior (1961) by saying that as far as he knows "the problem of axiomatizing [classical implication] in the way indicated has not yet been either solved or shown to have no solution."

For any sentence letter L, $p \to p$ and $(L \to L) \to ((p \to (q \to r)) \to ((r \to p) \to (s \to p)))$ would provide a trivial solution: the two are independent, each alone having as consequences only its own substitution instances, but detaching the first from the second immediately delivers Łukasiewicz's axiom. Since it is unlikely that either Meredith or Prior overlooked this simple example, it seems reasonable to suppose that what Prior sought was an *organic* pair (as with all those above).

With that understanding, consider then the pair

(g) $p \to p$, $((p \to q) \to r) \to (((q \to s) \to t) \to ((t \to q) \to r))$.

These two are also independent. As before, $p \to p$ alone gives only instances of itself. For the other, $((p \to q) \to r) \to (((q \to s) \to t) \to ((t \to q) \to r))$ takes, according to the following matrix, the value 1, for every assignment of values of the matrix to the letters occurring in it,

$$\boldsymbol{M1} = \begin{array}{c|ccc} \to & 1 & 2 & 3 \\ \hline {}^{*}1 & 1 & 2 & 3 \\ 2 & 1 & 3 & 1 \\ 3 & 1 & 1 & 2, \end{array}$$

and thus is a *tautology* of ***M1***. *Modus ponens* and uniform substitution preserve this feature, leading from tautologies of the matrix only to other such tautologies. But $p \to p$ is not an ***M1*** tautology: it takes the value 3 when p is assigned the value 2.

Together, however, these two also immediately yield Łukasiewicz's single axiom for **IF**. When presenting proofs from axioms in this paper, we will employ the rule of **condensed detachment** invented by Meredith in the 1950s (its first appearance in the literature was in Prior 1956), and will annotate our proofs as he does his. Meredith writes "D$x.y$" as short for the most general formula obtainable (when such exists) by using formula x, or some substitution instance of it, as *major* premise for an application of detachment—that is, *modus ponens*—and formula y, or some substitution instance of it, as *minor* premise. (When such a formula does exist, it is unique up to the renaming of sentence letters.)

In the present case, for example, substituting $s \to p$ for p in $p \to p$ gives $(s \to p) \to (s \to p)$. Then putting s for p, p for q, $s \to p$ for r, q for s, and r for t throughout the longer axiom produces $((s \to p) \to (s \to p)) \to (((p \to q) \to r) \to ((r \to p) \to (s \to p)))$. Detaching the former from the latter delivers Łukasiewicz's $((p \to q) \to r) \to ((r \to p) \to (s \to p))$.

Meredith's annotation method suppresses explicit display of the substitutions involved (one can work them out for oneself, though not always easily). He would simply write, as shall we, the following.

$$\begin{array}{rl} 1. & ((p \to q) \to r) \to (((q \to s) \to t) \to ((t \to q) \to r)) \\ 2. & p \to p \\ \text{D}1.2 = 3. & ((p \to q) \to r) \to ((r \to p) \to (s \to p)) \end{array}$$

The pair (g) of axioms appearing here as the first two lines of our proof is not the only solution to Prior's problem. The 8-arrow formula can be replaced at least with any of $(p \to q) \to (((p \to r) \to s) \to ((s \to q) \to (t \to q)))$, $(p \to q) \to (((q \to r) \to s) \to ((s \to p) \to (t \to q)))$, $(p \to q) \to (((r \to s) \to p) \to ((r \to p) \to (t \to q)))$, and $(p \to q) \to (((r \to s) \to q) \to ((r \to p) \to (t \to q)))$, though the last two lead to Łukasiewicz's axiom much more slowly.

With Meredith's entry into the game, shorter bases and short single axioms for other implicational logics besides **IF** were sought and found. In the 1950s and 1960s, he himself discovered single axioms and short two-bases for a variety of such calculi. For the implicational fragment $\mathbf{H}_{\rightarrow}$ of the intuitionistic sentential calculus, for example, Meredith (1953) presented $((p \rightarrow q) \rightarrow r) \rightarrow (s \rightarrow ((q \rightarrow (r \rightarrow t)) \rightarrow (q \rightarrow t)))$ and Meredith and Prior (1963) add $p \rightarrow ((q \rightarrow r) \rightarrow (((s \rightarrow q) \rightarrow (r \rightarrow t)) \rightarrow (q \rightarrow t)))$. For the strict-implicational fragment **C5** of the modal logic **S5**, Meredith's $((((p \rightarrow p) \rightarrow q) \rightarrow r) \rightarrow (s \rightarrow t)) \rightarrow ((t \rightarrow q) \rightarrow (u \rightarrow (s \rightarrow q)))$ is in Lemmon et al. (1969) along with a couple of two-bases, and Prior also had one of the latter, $p \rightarrow p$ with $(((p \rightarrow q) \rightarrow r) \rightarrow s) \rightarrow ((s \rightarrow p) \rightarrow (u \rightarrow (r \rightarrow p)))$. For his own pure implicational calculus **BCI**, whose axioms are $\mathbf{B} = (p \rightarrow q) \rightarrow ((r \rightarrow p) \rightarrow (r \rightarrow q))$, $\mathbf{C} = (p \rightarrow (q \rightarrow r)) \rightarrow (q \rightarrow (p \rightarrow r))$, and $\mathbf{I} = p \rightarrow p$, Meredith provided $(p \rightarrow (q \rightarrow r)) \rightarrow (((s \rightarrow s) \rightarrow (t \rightarrow q)) \rightarrow (t \rightarrow (p \rightarrow r)))$ and Prior gave $((p \rightarrow p) \rightarrow (q \rightarrow r)) \rightarrow ((s \rightarrow (r \rightarrow t)) \rightarrow (q \rightarrow (s \rightarrow t)))$, while for his **BCK**, in which the axiom **I** is replaced with **K**, Meredith produced $((p \rightarrow q) \rightarrow r) \rightarrow ((s \rightarrow (r \rightarrow t)) \rightarrow (q \rightarrow (s \rightarrow t)))$—see Meredith and Prior (1963) again for these last four.

In more recent times, others have taken up and expanded the project. Most of Meredith's single axioms have been shown to be shortest possible, and additional such axioms have been added to his. For example, in Ernst et al. (2002), Ernst, Fitelson, Harris and Wos present six more axioms for **C5** of the same length as Meredith's and report that they are shortest possible. But they also found an even shorter two-base, $p \rightarrow p$ and $(p \rightarrow q) \rightarrow (((q \rightarrow r) \rightarrow s) \rightarrow r) \rightarrow (p \rightarrow r)$, just 18 symbols and proved minimal in total length among all bases for this system. In addition, they found the first single axiom for the strict-implicational fragment **C4** of **S4**, $(p \rightarrow ((q \rightarrow (r \rightarrow r)) \rightarrow (p \rightarrow q))) \rightarrow ((s \rightarrow t) \rightarrow (u \rightarrow (p \rightarrow t)))$, and showed that no other theorem of **C4** of lesser or even of the *same* length will do.

Meredith and Prior's two single axioms for $\mathbf{H}_{\rightarrow}$ have been extended to twelve, and a proof that these are shortest possible is nearly complete with only four shorter theorems of undetermined status remaining (see Ulrich (1999, 2001), and the Appendix to the present paper). For **BCK** and **BCI**, the author has shown that Meredith's single axioms for each are the shortest possible. The list of those for **BCK** currently stands at thirteen, and for **BCI**—perhaps now of renewed interest since the latter system has re-emerged as the implicational fragment of Girard's ***linear logic*** (Girard 1987)—has grown to eighty single axioms. All of these and more are listed in the Appendix below, and one of the new single axioms for **BCI** to be found there, $((p \rightarrow p) \rightarrow (q \rightarrow (r \rightarrow s))) \rightarrow ((t \rightarrow q) \rightarrow (r \rightarrow (t \rightarrow s)))$, will be especially useful when we turn, as we now do, to the system $\mathbf{R}_{\rightarrow}$ of relevant implication, and to $\mathbf{RM}_{\rightarrow}$, the implicational fragment of Dunn's own **R-Mingle**, later on.

2 Compact Bases and a Single Axiom for $\mathbf{R}_{\rightarrow}$, the Implicational Fragment of R

Anderson and Belnap (1975, p. 89) ask, among other related questions, whether there exists a single axiom for $\mathbf{R}_{\rightarrow}$, the implicational fragment of the relevance logic **R**.

They themselves list a number of four-axiom bases for $\mathbf{R}_{\rightarrow}$, including (p. 20)

$$\begin{array}{ll} (p \to q) \to ((q \to r) \to (p \to r)) & \mathbf{B}' \\ (p \to (q \to r)) \to (q \to (p \to r)) & \mathbf{C} \\ p \to p & \mathbf{I} \\ (p \to (q \to r)) \to ((p \to q) \to (p \to r)) & \mathbf{S/Frege}, \end{array}$$

and their axiom set $\mathbf{R}_{\rightarrow 3}$ (p. 88)

$$\begin{array}{ll} (p \to q) \to ((q \to r) \to (p \to r)) & \mathbf{B}' \\ (p \to (q \to r)) \to (q \to (p \to r)) & \mathbf{C} \\ p \to p & \mathbf{I} \\ (p \to (p \to q)) \to (p \to q) & \mathbf{W}. \end{array}$$

They include also two equivalent sets that appeared in the literature well before full **R** was formulated. The earliest was Moh (1950)'s

$$\begin{array}{ll} (p \to q) \to ((q \to r) \to (p \to r)) & \mathbf{B}' \\ p \to ((p \to q) \to q) & \mathbf{Pon} \\ p \to p & \mathbf{I} \\ (p \to (p \to q)) \to (p \to q) & \mathbf{W}, \end{array}$$

but only a year later Church (1951) independently introduced what has now become, perhaps, the most widely known set and the one that has given rise to the alternate name "**BCIW**" for $\mathbf{R}_{\rightarrow}$:

$$\begin{array}{ll} (p \to q) \to ((r \to p) \to (r \to q)) & \mathbf{B} \\ (p \to (q \to r)) \to (q \to (p \to r)) & \mathbf{C} \\ p \to p & \mathbf{I} \\ (p \to (p \to q)) \to (p \to q) & \mathbf{W} \end{array}$$

New bases using fewer axioms are, of course, possible. Rezus (1982), for example, gives the following 27-symbol three-base, as short as any the author knows of for $\mathbf{R}_{\rightarrow}$:

$$\begin{array}{ll} (p \to (q \to r)) \to ((s \to p) \to (q \to (s \to r))) & \\ p \to p & \mathbf{I} \\ (p \to (p \to q)) \to (p \to q) & \mathbf{W} \end{array}$$

And here's another of the same length:

$$\begin{array}{ll} (((p \to q) \to q) \to r) \to ((r \to s) \to (p \to s)) & \\ p \to p & \mathbf{I} \\ (p \to (p \to q)) \to (p \to q) & \mathbf{W} \end{array}$$

The odd first member of this base can be gotten from three familiar $\mathbf{R}_{\rightarrow}$ theorems as follows:

1.	$(p \rightarrow q) \rightarrow ((r \rightarrow p) \rightarrow (r \rightarrow q))$	**B**
2.	$(p \rightarrow q) \rightarrow ((q \rightarrow r) \rightarrow (p \rightarrow r))$	**B′**
3.	$p \rightarrow ((p \rightarrow q) \rightarrow q)$	**Pon**
D1.2 = 4.	$(p \rightarrow (q \rightarrow r)) \rightarrow (p \rightarrow ((r \rightarrow s) \rightarrow (q \rightarrow s)))$	
D2.3 = 5.	$(((p \rightarrow q) \rightarrow r) \rightarrow r) \rightarrow (p \rightarrow r)$	
D4.5 = 6.	$\mathbf{(((p \rightarrow q) \rightarrow q) \rightarrow r) \rightarrow ((r \rightarrow s) \rightarrow (p \rightarrow s))}$	

To see that this axiom does give full $\mathbf{R}_{\rightarrow}$ with **I** and **W**, note that it suffices all by itself for getting **B**, **B′**, and **C** (it, and its permuted variant, $(p \rightarrow q) \rightarrow ((((r \rightarrow s) \rightarrow s) \rightarrow p) \rightarrow (r \rightarrow q))$, are the shortest theorems of $\mathbf{R}_{\rightarrow}$—indeed, of **IF**—that do so).

1.	$(((p \rightarrow q) \rightarrow q) \rightarrow r) \rightarrow ((r \rightarrow s) \rightarrow (p \rightarrow s))$	
D1.1 = 2.	$(((p \rightarrow q) \rightarrow (r \rightarrow q)) \rightarrow s) \rightarrow ((r \rightarrow p) \rightarrow s)$	
D2.1 = 3.	$(p \rightarrow (q \rightarrow r)) \rightarrow (((p \rightarrow r) \rightarrow s) \rightarrow (q \rightarrow s))$	
D1.2 = 4.	$(((p \rightarrow q) \rightarrow (p \rightarrow r)) \rightarrow s) \rightarrow ((q \rightarrow r) \rightarrow s)$	
D3.1 = 5.	$(((((p \rightarrow q) \rightarrow q) \rightarrow r) \rightarrow (p \rightarrow s)) \rightarrow t) \rightarrow ((r \rightarrow s) \rightarrow t)$	
D1.3 = 6.	$(((((p \rightarrow (q \rightarrow r)) \rightarrow r) \rightarrow s) \rightarrow (q \rightarrow s)) \rightarrow t) \rightarrow (p \rightarrow t)$	
D3.2 = 7.	$(((((p \rightarrow q) \rightarrow (r \rightarrow q)) \rightarrow s) \rightarrow s) \rightarrow t) \rightarrow ((r \rightarrow p) \rightarrow t)$	
D3.4 = 8.	$(((((p \rightarrow q) \rightarrow (p \rightarrow r)) \rightarrow s) \rightarrow s) \rightarrow t) \rightarrow ((q \rightarrow r) \rightarrow t)$	
D5.2 = 9.	$((p \rightarrow q) \rightarrow r) \rightarrow ((p \rightarrow (s \rightarrow q)) \rightarrow (s \rightarrow r))$	
D6.6 = 10.	$(p \rightarrow (q \rightarrow (r \rightarrow s))) \rightarrow (p \rightarrow (r \rightarrow (q \rightarrow s)))$	
D6.7 = 11.	$\mathbf{(p \rightarrow q) \rightarrow ((r \rightarrow p) \rightarrow (r \rightarrow q))}$	**B**
D6.8 = 12.	$\mathbf{(p \rightarrow q) \rightarrow ((q \rightarrow r) \rightarrow (p \rightarrow r))}$	**B′**
D1.9 = 13.	$(((p \rightarrow (q \rightarrow r)) \rightarrow (q \rightarrow r)) \rightarrow s) \rightarrow (p \rightarrow s)$	
D13.13 = 14.	$(p \rightarrow (q \rightarrow r)) \rightarrow (p \rightarrow (q \rightarrow r))$	
D10.14 = 15.	$\mathbf{(p \rightarrow (q \rightarrow r)) \rightarrow (q \rightarrow (p \rightarrow r))}$	**C**

Two-bases can be obtained by adding **W** to any 19-symbol single axiom for **BCI**, but those are 28 symbols long.

Anderson and Belnap's question about the existence of a *single* axiom for $\mathbf{R}_{\rightarrow}$ was answered by Rezus (1982), who devised a method (which he attributes to Tarski) involving the insertion of certain bases for $\mathbf{R}_{\rightarrow}$ into a general template. The axioms produced tend to be quite long and he does not actually display any of them but only gives instructions for their construction. The shortest such single axiom the author has found is 93 symbols long. To aid readability a bit, the members of the three-base used, also due to Rezus, are underlined below. Notice that alphabetic variants of two of them appear twice each:

$$((p \to (((q \to q) \to ((r \to r) \to ((s \to s) \to ((t \to t) \to (p \to u))))) \to u)) \to$$
$$(((((v \to w) \to ((w \to x) \to (v \to x))) \to (((((y \to (y \to z)) \to (y \to z)) \to$$
$$(((a \to b) \to ((b \to c) \to (a \to c))) \to d)) \to d) \to e)) \to e) \to$$
$$((f \to (((g \to g) \to ((h \to h) \to ((i \to i) \to ((j \to j) \to (f \to k))))) \to k))$$
$$\to l))) \to l$$

On the way to a shorter single axiom, it will be helpful to employ some of the derived rules of inference that all bases for $\mathbf{R}_{\to}$ provide: *prefixing*, which allows the inference of $(C \to A) \to (C \to B)$ from $A \to B$, *transitivity*, which lets one move from $A \to B$ and $B \to C$ to $A \to C$, and $\mathbf{R}_{\to}$'s distinctive deduction theorem, which assures the provability of $A \to B$ whenever there exists a deduction of B from A in which A is actually used. With their help, a proof for the following result is straightforward.

Theorem 2.1 *For each theorem T of $\mathbf{R}_{\to}$, $(T \to ((s \to s) \to (t \to (u \to v)))) \to ((w \to t) \to (u \to (w \to v)))$ is a theorem of $\mathbf{R}_{\to}$.*

Proof (Asterisks indicate dependence on the hypothesis.)

1.	$T \to ((s \to s) \to (t \to (u \to v)))$	*	hyp
2.	T		theorem of $\mathbf{R}_{\to}$
3.	$(s \to s) \to (t \to (u \to v))$	*	1, 2 modus ponens
4.	$s \to s$		theorem of $\mathbf{R}_{\to}$
5.	$t \to (u \to v)$	*	3, 4 modus ponens
6.	$(w \to t) \to (w \to (u \to v))$	*	5 prefixing
7.	$(w \to (u \to v)) \to (u \to (w \to v))$		theorem of $\mathbf{R}_{\to}$
8.	$(w \to t) \to (u \to (w \to v))$	*	6, 7 transitivity
9.	$(T \to ((s \to s) \to (t \to (u \to v)))) \to ((w \to t) \to (u \to (w \to v)))$		1–8, deduction theorem

Now take T to be the formula $((p \to q) \to (r \to p)) \to ((p \to q) \to (r \to q))$, which is certainly a theorem of $\mathbf{R}_{\to}$: it is the result of using Frege to distribute the antecedent, $p \to q$, of $\mathbf{B}$ over its consequent. With that choice, Theorem 2.1 provides us with a single 35-symbol theorem from which, it turns out, all of $\mathbf{R}_{\to}$'s theses follow.

Theorem 2.2 $((((p \to q) \to (r \to p)) \to ((p \to q) \to (r \to q))) \to ((s \to s) \to (t \to (u \to v)))) \to ((w \to t) \to (u \to (w \to v)))$ *is a single axiom for $\mathbf{R}_{\to}$.*

Proof By Theorem 2.1, and the fact that $(((p \to q) \to (r \to p)) \to ((p \to q) \to (r \to q)))$ is provable in $\mathbf{R}_{\to}$, the formula of our present Theorem is also provable. To show that it is in fact a single axiom for that system, we derive Moh's base, $\mathbf{B}'$, **Pon**, **I**, and **W**, from it.

1.	$((((p \to q) \to (r \to p)) \to ((p \to q) \to (r \to q))) \to$ $((s \to s) \to (t \to (u \to v)))) \to ((w \to t) \to (u \to (w \to v)))$	
D1.1 = 2.	$(p \to q) \to ((q \to q) \to (p \to q))$	
D1.2 = 3.	$(p \to ((q \to r) \to (s \to q))) \to ((q \to r) \to (p \to (s \to r)))$	
D1.3 = 4.	$(p \to ((q \to q) \to (r \to q))) \to (r \to (p \to q))$	
D4.2 = 5.	$\boldsymbol{p \to ((p \to q) \to q)}$	**Pon**
D2.5 = 6.	$(((p \to q) \to q) \to ((p \to q) \to q)) \to (p \to ((p \to q) \to q))$	
D5.6 = 7.	$(((((p \to q) \to q) \to ((p \to q) \to q)) \to$ $(p \to ((p \to q) \to q))) \to r) \to r$	
D1.7 = 8.	$\boldsymbol{(p \to q) \to ((q \to r) \to (p \to r))}$	**B′**
D1.4 = 9.	$(p \to (((q \to r) \to (q \to r)) \to ((s \to s) \to$ $(q \to r)))) \to (q \to (p \to r))$	
D9.2 = 10.	$p \to (((q \to q) \to (p \to r)) \to r)$	
D10.5 = 11.	$((p \to p) \to ((q \to ((q \to r) \to r)) \to s)) \to s$	
D11.10 = 12.	$\boldsymbol{p \to p}$	**I**
D8.6 = 13.	$((p \to ((p \to q) \to q)) \to r) \to ((((p \to q) \to q) \to$ $((p \to q) \to q)) \to r)$	
D8.8 = 14.	$(((p \to q) \to (r \to q)) \to s) \to ((r \to p) \to s)$	
D1.13 = 15.	$(p \to (q \to r)) \to (((q \to r) \to q) \to (p \to r))$	
D15.5 = 16.	$(((p \to q) \to q) \to (p \to q)) \to (p \to q)$	
D14.16 = 17.	$\boldsymbol{(p \to (p \to q)) \to (p \to q)}$	**W**

3 RM$_\to$, the Implicational Fragment of R-Mingle

Sobociński (1952) introduced a 3-valued logic designed to avoid most of the paradoxes of material implication.

M2 =

$\to$	1	2	3	$\sim$
*1	1	3	3	3
*2	1	2	3	2
3	1	1	1	1

The tautologies of ***M2*** are of course the formulas in $\to$ and $\sim$ that take only values designated here with asterisks regardless of how values of the matrix are assigned to the letters occurring in them, and Sobociński axiomatized the set of all ***M2***-tautologies with the following five formulas.

$(p \to q) \to ((q \to r) \to (p \to r))$	**B′**
$p \to ((p \to q) \to q)$	**Pon**
$(p \to (p \to q)) \to (p \to q)$	**W**
$p \to (q \to (\sim q \to p))$	
$(\sim p \to \sim q) \to (q \to p)$	

He asked explicitly if the implicational fragment of his system could be axiomatized as well, and that question was first answered in Rose (1956) using twenty-one different axioms (some of which were over 100 symbols in length).

Shortly after the introduction of **R-Mingle** in Dunn (1970), Parks (1972) showed that the theorems of its implication–negation fragment were exactly the theorems of Sobociński's system (noting that this result was obtained independently by Robert K. Meyer as well) so that the implicational fragment of **R-Mingle**, $\mathbf{RM}_{\rightarrow}$, is also characterized by the (implicational part of the) matrix ***M2***. Meyer and Parks (1972) then provided an elegant set of axioms answering Sobociński's question, and simultaneously axiomatizing $\mathbf{RM}_{\rightarrow}$ as follows.

$$(p \rightarrow q) \rightarrow ((q \rightarrow r) \rightarrow (p \rightarrow r)) \quad \mathbf{B}'$$
$$p \rightarrow ((p \rightarrow q) \rightarrow q) \quad \mathbf{Pon}$$
$$(p \rightarrow (p \rightarrow q)) \rightarrow (p \rightarrow q) \quad \mathbf{W}$$
$$((((p \rightarrow q) \rightarrow q) \rightarrow p) \rightarrow r) \rightarrow (((((q \rightarrow p) \rightarrow p) \rightarrow q) \rightarrow r) \rightarrow r)$$

In the tradition of moving to reduce the number of axioms in the base for any logic of interest, Ernst et al. (2001) presented a pair of three-bases for $\mathbf{RM}_{\rightarrow}$. Each includes $\mathbf{B}'$ and **Pon** together with either **RM1** or **RM2**:

$$(p \rightarrow q) \rightarrow ((q \rightarrow r) \rightarrow (p \rightarrow r)) \quad \mathbf{B}'$$
$$p \rightarrow ((p \rightarrow q) \rightarrow q) \quad \mathbf{Pon}$$
$$((p \rightarrow (((q \rightarrow p) \rightarrow r) \rightarrow q)) \rightarrow r) \rightarrow r \quad \mathbf{RM1}$$
$$((((p \rightarrow q) \rightarrow r) \rightarrow (q \rightarrow p)) \rightarrow r) \rightarrow r \quad \mathbf{RM2}$$

An additional three-base can be obtained by replacing the third axiom in either set with the following slightly longer theorem, which will prove useful shortly:

$$((((p \rightarrow q) \rightarrow r) \rightarrow (q \rightarrow p)) \rightarrow (s \rightarrow r)) \rightarrow (s \rightarrow r) \quad \mathbf{RM3}$$

RM3 is readily shown to be a tautology of ***M2*** so it is a theorem of $\mathbf{RM}_{\rightarrow}$. To see that it can replace the third axiom in either of the three-bases above, observe that D **RM3**.**Pon** = **RM2**: substitute the antecedent, $(((p \rightarrow q) \rightarrow r) \rightarrow (q \rightarrow p)) \rightarrow r$, of **RM2** for each occurrence of s in **RM3**, put that same formula for p, and put r for q in **Pon**; then, detach the latter from the former.

After the three-bases were discovered, Wos and Pieper (2003) posed two open questions:

OQ29.RM. Does there exist a 2-basis for $\mathbf{RM}_{\rightarrow}$?
OQ30.RM. Does there exist a single axiom for $\mathbf{RM}_{\rightarrow}$?

To answer the first of these questions, a two-base 28 symbols in length results (cf. Ulrich 2009) from pairing the powerful first axiom from the 27-symbol three-base for $\mathbf{R}_{\rightarrow}$ above with either **RM1** or **RM2** from Ernst et al. (2001).

$(((p \to q) \to q) \to r) \to ((r \to s) \to (p \to s))$
$((p \to (((q \to p) \to r) \to q)) \to r) \to r$ **RM1**
$((((p \to q) \to r) \to (q \to p)) \to r) \to r$ **RM2**

The first axiom, as shown above when considering the three-base for $\mathbf{R}_{\to}$, is provable in $\mathbf{R}_{\to}$ and so in $\mathbf{RM}_{\to}$, and gives **B′**, **B**, and **C**. Taking **RM1** as the second axiom (with no loss since the two are interdeducible in this setting), it suffices then to derive the remaining member of Ernst et al. (2001)s three-base, **Pon**, and that can be done quickly when the two-base is supplemented with those three known consequences of the first axiom.

1.	$(((p \to q) \to q) \to r) \to ((r \to s) \to (p \to s))$	1st member of 2-base
2.	$((p \to (((q \to p) \to r) \to q)) \to r) \to r$	**RM1**
3.	$(p \to q) \to ((q \to r) \to (p \to r))$	**B′**
4.	$(p \to q) \to ((r \to p) \to (r \to q))$	**B**
5.	$(p \to (q \to r)) \to (q \to (p \to r))$	**C**
D3.1 = 6.	$(((p \to q) \to (r \to q)) \to s) \to ((((r \to t) \to t) \to p) \to s)$	
D1.4 = 7.	$(((p \to (q \to r)) \to (p \to r)) \to s) \to (q \to s)$	
D6.5 = 8.	$(((p \to q) \to q) \to r) \to (p \to ((r \to s) \to s))$	
D7.2 = 9.	$((p \to q) \to (q \to p)) \to (q \to p)$	
D9.8 = 10.	$\boldsymbol{p \to ((p \to q) \to q)}$	**Pon**

Turning finally to the search for a single axiom for $\mathbf{RM}_{\to}$, Theorem 2.1 above suggests a place to look. That result provides for the construction of theorems of $\mathbf{R}_{\to}$ by inserting various theorems of that system into the antecedent of the formula $((p \to p) \to (q \to (r \to s))) \to ((t \to q) \to (r \to (t \to s)))$. This formula itself is a theorem of both $\mathbf{R}_{\to}$ and $\mathbf{RM}_{\to}$. In fact, it is one of the many single axioms for Meredith's system **BCI** listed in the Appendix, where it is shown as **BCI-22**.

BCI-22. $((p \to p) \to (q \to (r \to s))) \to ((t \to q) \to (r \to (t \to s)))$

B, **C**, and **I** can be seen to give **BCI-22** (on line 13) as follows.

1.	$(p \to q) \to ((r \to p) \to (r \to q))$	**B**
2.	$(p \to (q \to r)) \to (q \to (p \to r))$	**C**
3.	$p \to p$	**I**
D1.2 = 4.	$(p \to (q \to (r \to s))) \to (p \to (r \to (q \to s)))$	
D2.1 = 5.	$(p \to q) \to ((q \to r) \to (p \to r))$	
D2.3 = 6.	$p \to ((p \to q) \to q)$	
D5.2 = 7.	$((p \to (q \to r)) \to s) \to ((q \to (p \to r)) \to s)$	
D5.5 = 8.	$(((p \to q) \to (r \to q)) \to s) \to ((r \to p) \to s)$	
D6.3 = 9.	$((p \to p) \to q) \to q$	
D5.9 = 10.	$(p \to q) \to (((r \to r) \to p) \to q)$	
D7.10 = 11.	$(p \to (q \to r)) \to (((s \to s) \to q) \to (p \to r))$	
D8.4 = 12.	$(p \to q) \to ((q \to (r \to s)) \to (r \to (p \to s)))$	
D11.12 = 13.	$\boldsymbol{((p \to p) \to (q \to (r \to s))) \to ((t \to q) \to (r \to (t \to s)))}$	

That **BCI-22** (line 13 above), in turn, gives **B**, **C**, and **I** takes a bit longer.

$$\begin{array}{rll}
1. & ((p \to p) \to (q \to (r \to s))) \to ((t \to q) \to (r \to (t \to s))) & \\
\text{D1.1} = 2. & (p \to (q \to (r \to s))) \to (r \to (p \to (q \to s))) & \\
\text{D2.1} = 3. & p \to (((q \to q) \to (r \to (p \to s))) \to ((t \to r) \to (t \to s))) & \\
\text{D1.2} = 4. & (p \to q) \to ((r \to (q \to s)) \to (p \to (r \to s))) & \\
\text{D1.3} = 5. & (p \to ((q \to q) \to (r \to ((s \to s) \to t)))) \to & \\
 & \qquad ((u \to r) \to (p \to (u \to t))) & \\
\text{D5.1} = 6. & (p \to q) \to (((r \to r) \to ((s \to s) \to (q \to t))) \to (p \to t)) & \\
\text{D5.3} = 7. & (p \to ((q \to q) \to (r \to s))) \to (r \to (p \to s)) & \\
\text{D5.4} = 8. & (p \to q) \to ((q \to r) \to (p \to r)) & \\
\mathbf{D7.3 = 9.} & \mathbf{(p \to (q \to r)) \to (q \to (p \to r))} & \mathbf{C} \\
\text{D7.6} = 10. & p \to ((p \to q) \to q) & \\
\text{D9.9} = 11. & p \to ((q \to (p \to r)) \to (q \to r)) & \\
\text{D7.9} = 12. & p \to ((p \to ((q \to q) \to r)) \to r) & \\
\text{D11.10} = 13. & (p \to ((q \to ((q \to r) \to r)) \to s)) \to (p \to s) & \\
\mathbf{D9.8 = 14.} & \mathbf{(p \to q) \to ((r \to p) \to (r \to q))} & \mathbf{B} \\
\mathbf{D13.12 = 15.} & \mathbf{p \to p} & \mathbf{I}
\end{array}$$

Unfortunately, taking T in Theorem 2.1 to be either **RM1** or **RM2** is unproductive, because the resulting theorems of $\mathbf{RM}_{\to}$ turn out to have no consequences whatever (apart from their own substitution instances). But with **RM3** (cf. Ulrich 2009), the results are different.

Theorem 3.1 $((((((p \to q) \to r) \to (q \to p)) \to (s \to r)) \to (s \to r)) \to ((t \to t) \to (u \to (v \to w)))) \to ((x \to u) \to (v \to (x \to w)))$ *is a single axiom for* $\mathbf{RM}_{\to}$.

Proof Since **RM3** is a theorem of $\mathbf{RM}_{\to}$, it follows by Theorem 2.1 that the formula here is a theorem as well. It remains only to show that a known base for $\mathbf{RM}_{\to}$ can be inferred from it. The Ernst et al. (2002) base, but with **RM3** in place of either of their third axioms, is a convenient choice.

$$\begin{array}{rll}
1. & ((((((p \to q) \to r) \to (q \to p)) \to (s \to r)) \to (s \to r)) \to & \\
 & ((t \to t) \to (u \to (v \to w)))) \to ((x \to u) \to (v \to (x \to w))) & \\
\mathbf{D1.1 = 2.} & \mathbf{(p \to q) \to ((q \to r) \to (p \to r))} & \mathbf{B'} \\
\text{D1.2} = 3. & (p \to ((((q \to r) \to s) \to (r \to q)) \to (t \to s))) \to (t \to (p \to s)) & \\
\text{D2.2} = 4. & (((p \to q) \to (r \to q)) \to s) \to ((r \to p) \to s) & \\
\text{D3.2} = 5. & p \to ((p \to ((q \to r) \to (r \to q))) \to (r \to q)) & \\
\text{D5.4} = 6. & (((((p \to q) \to (r \to q)) \to s) \to ((r \to p) \to s)) \to & \\
 & ((t \to u) \to (u \to t))) \to (u \to t) & \\
\mathbf{D1.6 = 7.} & \mathbf{(p \to (q \to r)) \to (q \to (p \to r))} & \mathbf{C} \\
\text{D2.7} = 8. & ((p \to (q \to r)) \to s) \to ((q \to (p \to r)) \to s) & \\
\text{D3.7} = 9. & p \to ((p \to ((((q \to r) \to s) \to (r \to q)) \to s)) \to s) & \\
\text{D7.9} = 10. & (p \to ((((q \to r) \to s) \to (r \to q)) \to s)) \to (p \to s) & \\
\mathbf{D8.10 = 11.} & \mathbf{((((p \to q) \to r) \to (q \to p)) \to (s \to r)) \to (s \to r)} & \mathbf{RM3} \\
\mathbf{D6.11 = 12.} & \mathbf{p \to p} & \mathbf{I}
\end{array}$$

This is already enough since **B′**, **C**, **I**, and **RM3** is also a base for $\mathbf{RM}_{\rightarrow}$, but **Pon** is just a step away.

D7.12 = 13. $p \rightarrow ((p \rightarrow q) \rightarrow q)$ **Pon**

Other bases are also within easy reach. **RM1** is D10.8, for example, and **RM2** is DD4.6D3.3.[2]

The existence of the 28-symbol two-bases for $\mathbf{RM}_{\rightarrow}$ and $\mathbf{R}_{\rightarrow}$ encourages the conjectures that shorter single axioms of length 29 or less exist for both. But the author closes on a cautionary note by quoting a remark once made by Dunn himself: "Conjectures, it should be remembered, are cheap."

Appendix: Axiom Sets for Some Implicational Logics

Modus ponens (i.e., detachment) and uniform substitution of formulas for sentence letters are the rules of inference throughout all sections. Names for axioms are handled as follows. When an axiom involved in any of the systems below first appears, what the author takes to be the most commonly used name for it (if such exists) is listed first, with alternate names appended thereafter. When such an axiom is also used in axiom sets farther down the list, he is occasionally inconsistent: usually only the name taken to be most common is used, but sometimes, when a reminder seems to be in order, the formula it names is repeated.

Axiom sets discovered before 1961 for several of the logics appearing here are given in the Appendix to Prior's *Formal Logic* (Prior 1962). There is some overlap, but the author has tried to minimize the duplication whence readers interested in seeing more axiomatizations are encouraged to consult Prior. Several of the early results, especially those from Tarski, Łukasiewicz, Wajsberg, and C.A. Meredith, were obtained considerably earlier than the date of the reference cited below in which they first appeared in the literature. Prior's Appendix gives approximate years in several such cases.

Results not attributed below to others are, barring accidental omissions, believed by the author to be new.

A.1 IF: The implicational fragment of the classical sentential calculus

Four-base: **B′/Syll/Syl** $= (p \rightarrow q) \rightarrow ((q \rightarrow r) \rightarrow (p \rightarrow r))$, **C/Com** $= (p \rightarrow (q \rightarrow r)) \rightarrow (q \rightarrow (p \rightarrow r))$, **I/Id** $= p \rightarrow p$, **Peirce** $= ((p \rightarrow q) \rightarrow p) \rightarrow p$ [in Peirce (1885)]

---

[2]The author has found that the axiom shown in the Appendix as **BCI-33** $= ((p \rightarrow p) \rightarrow (q \rightarrow r)) \rightarrow (q \rightarrow ((s \rightarrow (r \rightarrow t)) \rightarrow (s \rightarrow t)))$ can play the role played in Sects. 2 and 3 above by **BCI-22**. An analog for **BCI-33** of Theorem 2.1 is easily established, whence $((((p \rightarrow q) \rightarrow (r \rightarrow p)) \rightarrow ((p \rightarrow q) \rightarrow (r \rightarrow q))) \rightarrow ((s \rightarrow s) \rightarrow (t \rightarrow u))) \rightarrow (t \rightarrow ((v \rightarrow (u \rightarrow w)) \rightarrow (v \rightarrow w)))$ is another single axiom for $\mathbf{R}_{\rightarrow}$, as is $((((((p \rightarrow q) \rightarrow r) \rightarrow (q \rightarrow p)) \rightarrow (s \rightarrow r)) \rightarrow (s \rightarrow r)) \rightarrow ((t \rightarrow t) \rightarrow (u \rightarrow v))) \rightarrow (u \rightarrow ((w \rightarrow (v \rightarrow x)) \rightarrow (w \rightarrow x)))$ for $\mathbf{RM}_{\rightarrow}$. However, his best current proofs that these axiomatize the systems in question currently run over fifty steps each.

Three-bases: **B**′, **K/Simp** = $p \to (q \to p)$, **Tarski** = $((p \to q) \to r) \to ((p \to r) \to r)$ [Tarski, Bernays circa 1921; cf. Łukasiewicz and Tarski (1930)]
B′, **K**, **Peirce** [circa 1926; ibid.]

Two-bases: $p \to (q \to (r \to p))$, $((p \to q) \to r) \to ((s \to r) \to ((p \to r) \to r))$ [Wajsberg (1932) but circa 1925–1926; cf. Prior (1962, p. 302)]
Peirce, $(((p \to q) \to r) \to s) \to ((q \to r) \to (p \to s))$ [Wajsberg circa 1925–1926; ibid.]
K, $(((p \to q) \to r) \to s) \to ((q \to s) \to (p \to s))$ [Łukasiewicz circa 1926; cf. Łukasiewicz and Tarski (1930)]
Peirce, $(p \to q) \to (s \to ((q \to r) \to (p \to r)))$ [Wajsberg (1939)]
$((p \to q) \to p) \to (r \to p)$, **B**′ [Wajsberg (1939)]
K, $((p \to q) \to r) \to ((r \to p) \to (s \to (t \to p)))$ [Meredith circa 1956, in Lemmon et al. (1969)]
K, **M2** = $(((p \to q) \to r) \to q) \to ((q \to s) \to (p \to s))$ [Prior circa 1960; cf. Prior (1961)]
I, $((p \to q) \to r) \to (((q \to s) \to t) \to ((t \to q) \to r))$
I with any one of $(p \to q) \to (((p \to r) \to s) \to ((s \to q) \to (t \to q)))$, $(p \to q) \to (((q \to r) \to s) \to ((s \to p) \to (t \to q)))$, $(p \to q) \to (((r \to s) \to p) \to ((r \to p) \to (t \to q)))$, or $(p \to q) \to (((r \to s) \to q) \to ((r \to p) \to (t \to q)))$, and probably others, will also do.
I, $((p \to q) \to r) \to ((r \to p) \to (s \to (t \to p)))$ [Meredith; see Prior (1961).]

One-bases: $((p \to q) \to ((r \to s) \to t)) \to ((u \to ((r \to s) \to t)) \to ((p \to u) \to (s \to t)))$ [Wajsberg circa 1926; cf. Łukasiewicz and Tarski (1930).]
$((p \to (q \to p)) \to (((((r \to s) \to t) \to u) \to ((s \to u) \to (r \to u))) \to v)) \to v$ [Łukasiewicz; ibid.]
$((p \to q) \to (r \to s)) \to (t \to ((s \to p) \to (r \to p)))$ [in Łukasiewicz (1948), but discovered earlier.]
$((p \to q) \to (r \to s)) \to ((s \to p) \to (t \to (r \to p)))$ [also in Łukasiewicz (1948), and also discovered earlier.]
$(p \to (q \to r)) \to ((r \to p) \to (s \to p))$ [Łukasiewicz (1948), but discovered in 1936.]

A.2 BCI

Three-bases: **B** = $(p \to q) \to ((r \to p) \to (r \to q))$, **C** = $(p \to (q \to r)) \to (q \to (p \to r))$, **I** [Meredith; cf. Prior (1962, p. 302). **BCI** turns out to be the implicational fragment of linear logic from Girard (1987).]
B′, **C**, **I** [An obvious variant of the preceding base.]
B′, **Pon/Assertion/T** = $p \to ((p \to q) \to q)$, **I** [Another variant.]

Two-bases: **BB′C-1** = $(((p \to q) \to q) \to r) \to ((r \to s) \to (p \to s))$, **I**
BB′C-2 = $(p \to q) \to ((((r \to s) \to s) \to p) \to (r \to q))$, **I**
$(p \to q) \to ((((q \to r) \to r) \to s) \to (p \to s))$, **I**
$(p \to (q \to r)) \to ((t \to q) \to (t \to (p \to r)))$, **I**

BB′C-1 and **BB′C-2** are the only theorems of **BCI** (or even of **IF**) of length 15 or less that give all of **B**, **B′**, and **C**. Each is a single axiom for the system (also new) whose axioms are **B** and either $(((p \to q) \to q) \to r) \to ((r \to s) \to (p \to s))$ or $(((p \to q) \to q) \to r) \to (p \to ((r \to s) \to s))$ (equivalently, **B′** and either; **C** and either).

One-bases: **BCI-1** = $(p \to (q \to r)) \to (((s \to s) \to (t \to q)) \to (t \to (p \to r)))$
BCI-2 = $((p \to p) \to (q \to r)) \to ((s \to (r \to t)) \to (q \to (s \to t)))$

BCI-1 is Meredith's circa 1956, and **BCI-2** is Prior's; see Meredith and Prior (1963). The 78 additional single axioms for **BCI** below are new.

BCI-3. $(p \to q) \to (((r \to r) \to (((q \to s) \to s) \to t)) \to (p \to t))$
BCI-4. $(p \to q) \to ((q \to ((r \to r) \to (s \to t))) \to (s \to (p \to t)))$
BCI-5. $((((p \to p) \to (q \to r)) \to r) \to s) \to ((s \to t) \to (q \to t))$
BCI-6. $(((p \to q) \to q) \to r) \to (((s \to s) \to (r \to t)) \to (p \to t))$
BCI-7. $(p \to (q \to r)) \to (((((s \to s) \to p) \to r) \to t) \to (q \to t))$
BCI-8. $(p \to (q \to r)) \to (((s \to s) \to (r \to t)) \to (q \to (p \to t)))$
BCI-9. $((p \to p) \to (((q \to r) \to r) \to s)) \to ((s \to t) \to (q \to t))$
BCI-10. $((p \to p) \to (q \to r)) \to (s \to ((r \to (s \to t)) \to (q \to t)))$
BCI-11. $((p \to p) \to (q \to r)) \to ((((r \to s) \to s) \to t) \to (q \to t))$
BCI-12. $((p \to p) \to (q \to r)) \to ((r \to (s \to t)) \to (s \to (q \to t)))$
BCI-13. $(p \to ((q \to q) \to (r \to s))) \to ((s \to t) \to (r \to (p \to t)))$
BCI-14. $(p \to ((q \to q) \to r)) \to (s \to ((r \to (s \to t)) \to (p \to t)))$
BCI-15. $(p \to q) \to (((r \to r) \to (q \to (s \to t))) \to (s \to (p \to t)))$
BCI-16. $(p \to (q \to r)) \to (q \to (((s \to s) \to (r \to t)) \to (p \to t)))$
BCI-17. $p \to (((q \to q) \to (r \to s)) \to ((s \to (p \to t)) \to (r \to t)))$
BCI-18. $p \to ((q \to (p \to r)) \to (((s \to s) \to (r \to t)) \to (q \to t)))$
BCI-19. $(p \to q) \to (r \to (((s \to s) \to (q \to (r \to t))) \to (p \to t)))$
BCI-20. $((p \to p) \to (q \to r)) \to ((((s \to t) \to t) \to q) \to (s \to r))$
BCI-21. $p \to (((q \to q) \to (r \to (p \to s))) \to ((t \to r) \to (t \to s)))$
BCI-22. $((p \to p) \to (q \to (r \to s))) \to ((t \to q) \to (r \to (t \to s)))$
BCI-23. $(p \to q) \to (((r \to r) \to (((s \to p) \to q) \to t)) \to (s \to t))$
BCI-24. $(p \to ((q \to q) \to (r \to s))) \to ((t \to p) \to (r \to (t \to s)))$
BCI-25. $((p \to p) \to (q \to r)) \to ((((s \to q) \to r) \to t) \to (s \to t))$
BCI-26. $((p \to p) \to (q \to r)) \to ((s \to (t \to q)) \to (t \to (s \to r)))$
BCI-27. $(p \to ((q \to q) \to r)) \to ((r \to (s \to t)) \to (s \to (p \to t)))$
BCI-28. $(p \to (q \to r)) \to (((s \to s) \to (t \to p)) \to (q \to (t \to r)))$
BCI-29. $((p \to p) \to (q \to (r \to s))) \to ((s \to t) \to (r \to (q \to t)))$
BCI-30. $((p \to p) \to (q \to (r \to s))) \to ((t \to r) \to (t \to (q \to s)))$
BCI-31. $((p \to p) \to (q \to (r \to s))) \to (r \to ((s \to t) \to (q \to t)))$
BCI-32. $((p \to p) \to (q \to (r \to s))) \to (r \to ((t \to q) \to (t \to s)))$
BCI-33. $((p \to p) \to (q \to r)) \to (q \to ((s \to (r \to t)) \to (s \to t)))$
BCI-34. $(p \to ((q \to q) \to (r \to s))) \to ((t \to r) \to (t \to (p \to s)))$
BCI-35. $(p \to (q \to r)) \to ((s \to ((t \to t) \to p)) \to (q \to (s \to r)))$
BCI-36. $(p \to (q \to r)) \to (q \to (((s \to s) \to (t \to p)) \to (t \to r)))$
BCI-37. $(p \to q) \to (((((r \to r) \to (s \to t)) \to t) \to p) \to (s \to q))$

BCI-38. $(p \to q) \to (((r \to r) \to (s \to (q \to t))) \to (p \to (s \to t)))$
BCI-39. $(p \to q) \to (((r \to r) \to (s \to (t \to p))) \to (t \to (s \to q)))$
BCI-40. $(p \to q) \to ((r \to ((s \to s) \to (q \to t))) \to (p \to (r \to t)))$
BCI-41. $(p \to q) \to (r \to (((s \to s) \to (t \to (r \to p))) \to (t \to q)))$
BCI-42. $p \to ((q \to ((r \to r) \to (p \to s))) \to ((s \to t) \to (q \to t)))$
BCI-43. $p \to (((q \to q) \to (r \to s)) \to ((t \to (p \to r)) \to (t \to s)))$
BCI-44. $p \to ((q \to ((r \to r) \to (p \to s))) \to ((t \to q) \to (t \to s)))$
BCI-45. $p \to ((q \to ((r \to r) \to s)) \to ((s \to (p \to t)) \to (q \to t)))$
BCI-46. $(p \to (q \to r)) \to (s \to (((t \to t) \to (s \to q)) \to (p \to r)))$
BCI-47. $(p \to q) \to ((r \to ((s \to s) \to (t \to p))) \to (t \to (r \to q)))$
BCI-48. $p \to ((q \to (r \to s)) \to (((t \to t) \to (p \to r)) \to (q \to s)))$
BCI-49. $p \to ((q \to r) \to (((s \to s) \to (r \to (p \to t))) \to (q \to t)))$
BCI-50. $((((p \to q) \to q) \to r) \to s) \to ((q \to (t \to r)) \to (t \to s))$
BCI-51. $(((p \to q) \to q) \to r) \to (((s \to s) \to (t \to p)) \to (t \to r))$
BCI-52. $(((p \to q) \to r) \to s) \to (((t \to t) \to (q \to r)) \to (p \to s))$
BCI-53. $((p \to p) \to (((q \to r) \to r) \to s)) \to ((t \to q) \to (t \to s))$
BCI-54. $((p \to p) \to (((q \to r) \to s) \to t)) \to ((r \to s) \to (q \to t))$
BCI-55. $(p \to q) \to (p \to (((r \to r) \to (s \to (q \to t))) \to (s \to t)))$
BCI-56. $p \to ((q \to (p \to r)) \to (((s \to s) \to (t \to q)) \to (t \to r)))$
BCI-57. $p \to ((p \to q) \to (((r \to r) \to (s \to (q \to t))) \to (s \to t)))$
BCI-58. $p \to ((q \to r) \to ((r \to ((s \to s) \to (p \to t))) \to (q \to t)))$
BCI-59. $((((p \to p) \to (q \to r)) \to r) \to s) \to ((t \to q) \to (t \to s))$
BCI-60. $(((p \to (q \to r)) \to r) \to s) \to (((t \to t) \to p) \to (q \to s))$
BCI-61. $((p \to p) \to q) \to ((((q \to (r \to s)) \to s) \to t) \to (r \to t))$
BCI-62. $(p \to (q \to r)) \to (q \to ((s \to ((t \to t) \to p)) \to (s \to r)))$
BCI-63. $((p \to p) \to (q \to (r \to s))) \to (t \to ((t \to r) \to (q \to s)))$
BCI-64. $((p \to p) \to (((q \to r) \to ((s \to q) \to r)) \to t)) \to (s \to t)$
BCI-65. $(p \to q) \to ((((q \to r) \to r) \to ((s \to s) \to t)) \to (p \to t))$
BCI-66. $(p \to q) \to ((((r \to p) \to q) \to ((s \to s) \to t)) \to (r \to t))$
BCI-67. $p \to (((q \to q) \to (r \to (p \to s))) \to ((s \to t) \to (r \to t)))$
BCI-68. $p \to (((q \to q) \to (r \to (s \to t))) \to ((p \to s) \to (r \to t)))$
BCI-69. $p \to ((q \to ((r \to r) \to (s \to t))) \to ((p \to s) \to (q \to t)))$
BCI-70. $p \to ((q \to (p \to r)) \to ((s \to ((t \to t) \to q)) \to (s \to r)))$
BCI-71. $p \to ((q \to r) \to ((s \to ((t \to t) \to (p \to q))) \to (s \to r)))$
BCI-72. $p \to ((p \to q) \to ((r \to ((s \to s) \to (q \to t))) \to (r \to t)))$
BCI-73. $((p \to p) \to (q \to r)) \to (s \to ((t \to (s \to q)) \to (t \to r)))$
BCI-74. $(p \to q) \to (((((r \to r) \to (q \to s)) \to s) \to t) \to (p \to t))$
BCI-75. $(p \to q) \to ((((r \to s) \to s) \to ((t \to t) \to p)) \to (r \to q))$
BCI-76. $(((p \to q) \to ((r \to p) \to q)) \to ((s \to s) \to t)) \to (r \to t)$
BCI-77. $((((p \to p) \to (q \to r)) \to ((s \to q) \to r)) \to t) \to (s \to t)$
BCI-78. $p \to ((q \to r) \to (((s \to s) \to (t \to (p \to q))) \to (t \to r)))$
BCI-79. $((((p \to p) \to (q \to r)) \to s) \to t) \to ((r \to s) \to (q \to t))$
BCI-80. $(p \to q) \to (((((r \to r) \to (s \to p)) \to q) \to t) \to (s \to t))$

Working from a list of **BCI** candidates circulated by the author starting in 2007, John Halleck and Larry Wos found the first proof for **BCI-29**. Halleck also provided the first proofs for **BCI-73** and **BCI-74**, Wos gave proofs for **BCI-75**, **BCI-76**, and **BCI-77**, and Mark Stickel discovered proofs for **BCI-78** through **BCI-80**.

It is well known that every theorem of any substitution-detachment system is a substitution instance of at least one formula derivable from the axioms by condensed detachment. Axiom sets from which every theorem can be derived by that method are said to be D-*complete*. None of the bases for **BCI** shown above is D-complete. Each of them has Belnap's two-property (Belnap 1976), that is, each axiom appearing in any of them contains each sentence letter occurring in it exactly twice, and Hindley (1993) shows that, consequently, each theorem derivable from any such set of axioms by condensed detachment alone will have the two-property as well. So, for example, though $p \to p$ is obtainable in all of the systems above when condensed detachment is used, instances of it such as $(p \to p) \to (p \to p)$ that lack the two-property cannot be so derived.

An example of a D-complete axiom set for **BCI**—in fact, a single axiom—is given in Ulrich (2005a): $((((((p \to (q \to r)) \to (((s \to s) \to (t \to q)) \to (t \to (p \to r)))) \to s) \to s) \to s) \to s)$.

There are fifteen length-19 theorems of **BCI** whose status remains an open question, having been neither ruled out nor confirmed as being single axioms:

BCI-Q1. $(p \to ((q \to q) \to r)) \to ((((s \to t) \to t) \to p) \to (s \to r))$
BCI-Q2. $(((p \to q) \to q) \to ((r \to r) \to s)) \to ((s \to t) \to (p \to t))$
BCI-Q3. $(((p \to q) \to q) \to ((r \to r) \to s)) \to ((t \to p) \to (t \to s))$
BCI-Q4. $(((p \to q) \to r) \to ((s \to s) \to t)) \to ((q \to r) \to (p \to t))$
BCI-Q5. $(((p \to q) \to q) \to r) \to ((r \to ((s \to s) \to t)) \to (p \to t))$
BCI-Q6. $(((p \to q) \to q)) \to r) \to ((r \to ((s \to s) \to s)) \to (p \to s)$
BCI-Q7. $(p \to ((q \to q) \to (q \to r))) \to ((s \to p) \to (q \to (s \to r)))$
BCI-Q9. $(p \to (q \to r)) \to (((s \to s) \to (r \to s)) \to (q \to (p \to s)))$
BCI-Q10. $(p \to (q \to r)) \to (((s \to s) \to (s \to p)) \to (q \to (s \to r)))$
BCI-Q11. $(p \to (q \to r)) \to (q \to (((s \to s) \to (r \to s)) \to (p \to s)))$
BCI-Q13. $(p \to q) \to ((((r \to p) \to q) \to ((s \to s) \to s)) \to (r \to s))$
BCI-Q16. $p \to (((q \to q) \to (q \to r)) \to ((r \to (p \to s)) \to (q \to s)))$
BCI-Q17. $p \to (((q \to q) \to (r \to q)) \to ((s \to (p \to r)) \to (s \to q)))$
BCI-Q18. $p \to ((q \to ((p \to p) \to (p \to r))) \to ((s \to q) \to (s \to r)))$
BCI-Q20. $p \to ((q \to r) \to ((r \to ((p \to p) \to (p \to s))) \to (q \to s)))$

The formula which appeared as **BCI-Q14** when the author originally sent out the list was ruled out by John Halleck. Those formerly appearing as **BCI-Q8**, **BCI-Q12**, **BCI-Q15**, and **BCI-Q19** were ruled out by Petr Pudlak.

A.3 Some relatives of BCI

Restricted versions of **C** can be gotten by replacing one or more of its variables with distinct elementary implications.

$\mathbf{C}^{000}$. $(p \to (q \to r)) \to (q \to (p \to r))$
$\mathbf{C}^{100}$. $((p \to s) \to (q \to r)) \to (q \to ((p \to s) \to r))$
$\mathbf{C}^{010}$. $(p \to ((q \to t) \to r)) \to ((q \to t) \to (p \to r))$
$\mathbf{C}^{001}$. $(p \to (q \to (r \to u))) \to (q \to (p \to (r \to u)))$

$\mathbf{C}^{110}$. $((p \to s) \to ((q \to t) \to r)) \to ((q \to t) \to ((p \to s) \to r))$
$\mathbf{C}^{101}$. $((p \to s) \to (q \to (r \to u))) \to (q \to ((p \to s) \to (r \to u)))$
$\mathbf{C}^{011}$. $(p \to ((q \to t) \to (r \to u))) \to ((q \to t) \to (p \to (r \to u)))$
$\mathbf{C}^{111}$. $((p \to s) \to ((q \to t) \to (r \to u))) \to ((q \to t) \to ((p \to s) \to (r \to u)))$

Kashima and Kamide (1999), using Gentzen-type sequent calculi, initiated investigation of the variants of **BCI** that can be obtained by adding to **B** and **I** varying combinations of those eight formulas. Exactly five nonequivalent logics result: $\mathbf{BC}^{011}\mathbf{I}$, it's two proper extensions, $\mathbf{BC}^{010}\mathbf{I}$ and $\mathbf{BC}^{001}\mathbf{I}$, their common proper extension $\mathbf{BC}^{010}\mathbf{C}^{001}\mathbf{I}$, and its proper extension $\mathbf{BC}^{000}\mathbf{I}$, which is of course **BCI** itself. These four proper subsystems of **BCI** can be axiomatized with single axioms as well.

$\mathbf{BC}^{011}\mathbf{I}$.

Three-base: **B**, $\mathbf{C}^{011} = (p \to ((q \to t) \to (r \to u))) \to ((q \to t) \to (p \to (r \to u)))$, **I** [Kashima and Kamide (1999)]

Two-base: **B**, $((p \to p) \to (q \to r)) \to (q \to r)$

One-base: $(p \to q) \to (((r \to r) \to (q \to s)) \to (p \to s))$ [Ulrich (2006)]

This axiom, which results from the insertion of a copy of **I** into $\mathbf{B}'$at a critical point, is the shortest possible, and no others of its length will suffice [(Ulrich 2011).]

$\mathbf{BC}^{010}\mathbf{I}$.

Three-base: **B**, $\mathbf{C}^{010} = (p \to ((q \to t) \to r)) \to ((q \to t) \to (p \to r))$, **I** [Kashima and Kamide (1999)]

Two-base: **B**, **Specialized Assertion** $= ((p \to p) \to q) \to q$, **I** [Meredith]

One-bases: $(p \to q) \to (((((r \to r) \to s) \to s) \to (q \to t)) \to (p \to t))$ [Meredith]
$((p \to p) \to (q \to r)) \to ((r \to s) \to (((q \to s) \to t) \to t))$
$(((p \to q) \to (((r \to r) \to (q \to s)) \to (p \to s))) \to t) \to t$
$(p \to q) \to (((r \to r) \to (q \to s)) \to (((p \to s) \to t) \to t))$
$(p \to q) \to (((((r \to r) \to (q \to s)) \to (p \to s)) \to t) \to t)$

Anderson and Belnap (1975) report that the first of these single axioms, which results from the insertion of a copy of **Specialized Assertion** into $\mathbf{B}'$, was given by Meredith as a single axiom uniting the two. The others are in Ulrich (2011); the list is exhaustive, and these one-bases are shortest possible.

$\mathbf{BC}^{001}\mathbf{I}$.

Three-base: **B**, $\mathbf{C}^{001} = (p \to (q \to (r \to u))) \to (q \to (p \to (r \to u)))$, **I** [Kashima and Kamide (1999)]

Two-base: $\mathbf{B}'$, $p \to ((q \to r) \to ((p \to (r \to t)) \to (q \to t)))$

One-bases: $p \to ((q \to r) \to ((p \to ((s \to s) \to (r \to t))) \to (q \to t)))$
$p \to ((p \to (q \to r)) \to (((s \to s) \to (r \to t)) \to (q \to t)))$
$((p \to p) \to (q \to r)) \to (s \to ((s \to (r \to t)) \to (q \to t)))$

Ulrich (2011); no shorter theorems of $\mathbf{BC^{001}I}$ are single axioms for it, nor are any others of equal length.

$\mathbf{BC^{010}C^{001}I}$.

Four-base: **B**, $\mathbf{C^{010}} = (p \to ((q \to t) \to r)) \to ((q \to t) \to (p \to r))$, $\mathbf{C^{001}} = (p \to (q \to (r \to u))) \to (q \to (p \to (r \to u)))$, **I** [Kashima and Kamide (1999)]

Two-base: $\mathbf{B'}$, $p \to ((p \to ((q \to q) \to r)) \to r)$

One-bases: $(((p \to q) \to ((r \to (q \to s)) \to (p \to s))) \to ((t \to t) \to u)) \to (r \to u)$
$((p \to ((p \to (q \to r)) \to (s \to r))) \to ((t \to t) \to u))) \to ((s \to q) \to u)$
$(((p \to (q \to r)) \to ((s \to q) \to (s \to r))) \to ((t \to t) \to u)) \to (p \to u)$
$(((p \to (q \to r)) \to (s \to r)) \to ((t \to t) \to u)) \to ((s \to q) \to (p \to u))$
$((p \to p) \to (q \to r)) \to (s \to ((((s \to (r \to t)) \to (q \to t)) \to u) \to u))$
$(p \to q) \to ((((r \to (q \to s)) \to (p \to s)) \to ((t \to t) \to u)) \to (r \to u))$
$p \to ((((p \to (q \to r)) \to (s \to r)) \to ((t \to t) \to u)) \to ((s \to q) \to u))$
Ulrich (2011); no shorter single axioms for $\mathbf{BC^{001}C^{010}I}$ can be found, but the author assumes that there exist additional hitherto undiscovered single axioms of this same length.

$\mathbf{BCI'}$.

Three-base: **B**, **C**, $\mathbf{I'} = (p \to q) \to (p \to q)$

Two-base: $\mathbf{B'}$, $\mathbf{Pon} = p \to ((p \to q) \to q)$

One-base: $(p \to q) \to (((r \to ((r \to s) \to s) \to (q \to t)) \to (p \to t))$ [Ulrich (2005b)]

The author has been unable to locate references to this system in the literature.

$\mathbf{BCI^*}$/**Monothetic BCI**.

Three-base: **B**, **C**, $\mathbf{I^*} = (p \to p) \to (q \to q)$ [Bunder (1983)]

Two-bases: $\mathbf{BB'C\text{-}1} = (((p \to q) \to q) \to r) \to ((r \to s) \to (p \to s))$, $\mathbf{I^*} = (p \to p) \to (q \to q)$

$\mathbf{BB'C\text{-}2} = (p \to q) \to ((((r \to s) \to s) \to p) \to (r \to q))$, $\mathbf{I^*}$

One-bases: $(((p \to p) \to q) \to r) \to ((s \to (r \to t)) \to (q \to (s \to t)))$ [Ulrich (2007)]

$(p \to (q \to r)) \to ((((s \to s) \to t) \to q) \to (t \to (p \to r)))$ [ibid.]

$(p \to q) \to ((((r \to r) \to (s \to s)) \to (q \to t)) \to (p \to t))$ is a single axiom for $\mathbf{BC^{011}I^*}$, where, as above, $\mathbf{C^{011}} = (p \to ((q \to t) \to (r \to u))) \to ((q \to t) \to (p \to (r \to u)))$.

A.4 Entailment: $\mathbf{E}_{\to}$, the implicational fragment of E

Three-base: **B**, **Specialized Assertion** $= ((p \to p) \to q) \to q$, **W** [Anderson et al. (1960)]

Two-base: $\mathbf{BC^{010}I\text{-}1}$, $\mathbf{W} = (p \to (p \to q)) \to (p \to q)$ [Meredith; cf. Anderson and Belnap (1975). Of course, adding **W** to any of the other single axioms for $\mathbf{BC^{010}I}$ would do as well.]

One-base: Rezus (1982) gives instructions for constructing one, but it would be quite long. Perhaps the most outstanding current open problem involving constructing single axioms for implicational logics is that of finding one of reasonable length for $\mathbf{E}_{\to}$.

A.5 Relevant implication: $\mathbf{R}_{\to}$, the implicational fragment of R

Four-bases: **B′**, **Pon**, **I**, $\mathbf{W} = (p \to (p \to q)) \to (p \to q)$ [Moh (1950); called $\mathbf{R}_{\to 2}$ in Anderson and Belnap (1975).]
B, **C**, **I**, **W** [Church (1951); axiom set $\mathbf{R}_{\to 1}$ in Anderson and Belnap (1975, p. 88).]
B′, **C**, **I**, $\mathbf{S/Frege} = (p \to (q \to r)) \to ((p \to q) \to (p \to r))$ [Anderson and Belnap (1975, p. 20)]
B′, **C**, **I**, **W** [Anderson and Belnap's set $\mathbf{R}_{\to 3}$.]
B, **Pon**, **I**, **W** [Anderson and Belnap's set $\mathbf{R}_{\to 4}$.]
B′, **C**, **I**, $((p \to q) \to (r \to p)) \to ((p \to q) \to (r \to q))$

Three-bases: $(p \to (q \to r)) \to ((s \to p) \to (q \to (s \to r)))$, **I**, **W** [Rezus (1982)]
$\mathbf{BB'C\text{-}1} = (((p \to q) \to q) \to r) \to ((r \to s) \to (p \to s))$, **I**, **W**
$\mathbf{BB'C\text{-}2} = (p \to q) \to ((((r \to s) \to s) \to p) \to (r \to q))$, **I**, **W**

These are the shortest bases for $\mathbf{R}_{\to}$ of any kind that are known to the author.

Two-bases: **BB′C-1**, $(p \to (p \to q)) \to (((r \to r) \to p) \to q)$
BB′C-2, $(p \to (p \to q)) \to (((r \to r) \to p) \to q$

The commuted version of the shorter axiom, $((r \to r) \to p) \to ((p \to (p \to q)) \to q)$, can be used instead. And of course pairing **W** with any of the single axioms above for **BCI** will also provide a length-28 two-base.

One-bases: $((p \to (((q \to q) \to (p \to r)) \to r)) \to (((((s \to t) \to ((t \to u) \to (s \to u))) \to (((((v \to (v \to u)) \to (v \to u)) \to ((y \to z) \to (x \to z))) \to a)) \to a) \to b)) \to b) \to ((z \to (((c \to c) \to (z \to d)) \to d)) \to e))) \to e$ [Rezus (1982)]
$((((p \to q) \to (r \to p)) \to ((p \to q) \to (r \to q))) \to ((s \to s) \to (t \to (u \to v)))) \to ((w \to t) \to (u \to (w \to v)))$ [Ulrich (2012)]
$(((p \to q) \to (r \to p)) \to ((p \to q) \to (r \to q))) \to ((s \to s) \to (t \to u))) \to (t \to ((v \to (u \to w)) \to (v \to w)))$

The author conjectures that there exists a single axiom of length at most 29.

A.6 $\mathbf{RM}_{\to}$: The implicational fragment of R-Mingle

Four-base: **B′**, **Pon**, **W**, $((((p \to q) \to q) \to p) \to r) \to (((((q \to p) \to p) \to q) \to r) \to r)$ [Meyer and Parks (1972)]

Three-bases: **B′**, **Pon**, **RM1** $= ((p \to (((q \to p) \to r) \to q)) \to r) \to r$ [Ernst et al. (2001)]
B′, **Pon**, **RM2** $= ((((p \to q) \to r) \to (q \to p)) \to r) \to r$ [Ernst et al. (2001)]
B′, **Pon**, **RM3** $= ((((p \to q) \to r) \to (q \to p)) \to (s \to r)) \to (s \to r)$

Two-bases: Either **BB′ C-1** or **BB′C-2** together with one of **RM1**, **RM2**, or **RM3**. Either **BB′C-1** or **BB′C-2** with $(p \to (p \to q)) \to (((r \to r) \to p) \to q)$.

One-bases: $((((((p \to q) \to r) \to (q \to p)) \to (s \to r)) \to (s \to r)) \to ((t \to t) \to (u \to (v \to w)))) \to ((x \to u) \to (v \to (x \to w)))$ [Ulrich (2009)]
$((((((p \to q) \to r) \to (q \to p)) \to (s \to r)) \to (s \to r)) \to ((t \to t) \to (u \to v))) \to (u \to ((w \to (v \to x)) \to (w \to x)))$

Again, the author conjectures there exists a single axiom of length at most 29.

The very first base offered in the literature as a set of axioms for a three-valued logic in Sobociński (1952), which much later proved to be (Parks 1972) characteristic for $\mathbf{RM}_{\to}$, was that of Rose (1956). Not shown here, it consists of twenty-one axioms, some of them quite long.

A.7 BCK

Three-base: **B**, **C**, **K/Simp** $= p \to (q \to p)$ [Meredith and Prior (1963)]

Two-base: **B′**, $((p \to (q \to p)) \to r) \to r$

One-bases: **BCK-1**. $((p \to q) \to r) \to ((s \to (r \to t)) \to (q \to (s \to t)))$
BCK-2. $(p \to (q \to r)) \to (((p \to t) \to q) \to (t \to (s \to r)))$
BCK-3. $p \to ((q \to (r \to s)) \to (((t \to p) \to r) \to (q \to s)))$
BCK-4. $((p \to q) \to r) \to (q \to ((s \to (r \to t)) \to (s \to t)))$
BCK-5. $p \to (((q \to p) \to r) \to ((s \to (r \to t)) \to (s \to t)))$
BCK-6. $p \to (((q \to r) \to s) \to ((s \to (p \to t)) \to (r \to t)))$
BCK-7. $((p \to q) \to r) \to (s \to ((r \to (s \to t)) \to (q \to t)))$
BCK-8. $(p \to (q \to r)) \to (s \to (((t \to s) \to q) \to (p \to r)))$
BCK-9. $(p \to q) \to ((q \to ((r \to s) \to t)) \to ((s \to (p \to t)))$
BCK-10. $(p \to ((q \to r) \to s)) \to (r \to ((t \to p) \to (t \to s)))$
BCK-11. $p \to ((q \to ((r \to p) \to s)) \to ((t \to q) \to (t \to s)))$
BCK-12. $(p \to (q \to r)) \to (q \to (((s \to t) \to p) \to (t \to r)))$
BCK-13. $p \to ((q \to (p \to r)) \to (((s \to t) \to q) \to (t \to r)))$

BCK-1 is Meredith's (Meredith and Prior 1963). The others are new. John Halleck provided a proof for **BCK-9**. Proofs for **BCK-10** through **BCK-13** were supplied by Mark Stickel.

None of these axiom sets or single axioms for **BCK** are D-complete: in none of them is there a letter occurring more than twice, and by a result of Hindley (1993), each theorem derivable in any of them will have this feature as well. A D-complete single axiom for **BCK** is $((((((((p \to q) \to r) \to ((s \to (r \to t)) \to (q \to (s \to t)))) \to s) \to s) \to s) \to s)$ from Ulrich (2005a).

The author has completed a proof that no shorter single axioms for **BCK** exist, but there are twenty-seven 17-symbol theorems of **BCK** whose status remains unknown.

Numbering below is taken from a list of open questions for single **BCK** axioms that he has been circulating for several years.

BCK-Q1. $((p \to q) \to (r \to s)) \to ((t \to r) \to (t \to (q \to s)))$
BCK-Q2. $(p \to q) \to r) \to ((((r \to s) \to s) \to t) \to (q \to t))$
BCK-Q4. $(p \to (q \to r)) \to (((s \to t) \to p) \to (q \to (t \to r)))$
BCK-Q5. $(p \to q) \to (((r \to s) \to (q \to t)) \to (p \to (s \to t)))$
BCK-Q6. $((((p \to q) \to r) \to r) \to s) \to ((s \to t) \to (q \to t))$
BCK-Q9. $((p \to q) \to r) \to ((((r \to s) \to (q \to s)) \to t) \to t)$
BCK-Q10. $((p \to q) \to r) \to ((r \to (s \to t)) \to (s \to (q \to t)))$
BCK-Q11. $((p \to q) \to r) \to ((r \to s) \to (((q \to s) \to t) \to t))$
BCK-Q14. $(p \to q) \to (p \to (((r \to s) \to (q \to t)) \to (s \to t)))$
BCK-Q15. $((((p \to q) \to r) \to ((r \to s) \to (q \to s))) \to t) \to t$
BCK-Q16. $p \to ((p \to q) \to (((r \to s) \to (q \to t)) \to (s \to t)))$
BCK-Q17. $((p \to q) \to r) \to ((p \to (r \to s)) \to (q \to (p \to s)))$
BCK-Q18. $((p \to q) \to r) \to ((r \to (p \to s)) \to (p \to (q \to s)))$
BCK-Q19. $((p \to q) \to (r \to s)) \to ((p \to r) \to (p \to (q \to s)))$
BCK-Q20. $((p \to q) \to r) \to (p \to ((r \to (p \to s)) \to (q \to s)))$
BCK-Q21. $((p \to q) \to r) \to (q \to ((p \to (r \to s)) \to (p \to s)))$
BCK-Q22. $(p \to ((q \to r) \to s)) \to (r \to ((q \to p) \to (q \to s)))$
BCK-Q23. $(p \to (q \to r)) \to (((p \to s) \to q) \to (s \to (p \to r)))$
BCK-Q24. $(p \to (q \to r)) \to (q \to (((q \to s) \to p) \to (s \to r)))$
BCK-Q25. $(p \to (q \to r)) \to (s \to (((p \to s) \to q) \to (p \to r)))$
BCK-Q26. $(p \to q) \to (((p \to r) \to (q \to s)) \to (p \to (r \to s)))$
BCK-Q27. $(p \to q) \to (((q \to r) \to (p \to s)) \to (p \to (r \to s)))$
BCK-Q28. $(p \to q) \to ((q \to ((p \to r) \to s)) \to (r \to (p \to s)))$
BCK-Q30. $(p \to q) \to (p \to (((p \to r) \to (q \to s)) \to (r \to s)))$
BCK-Q31. $p \to (((p \to q) \to r) \to ((r \to (p \to s)) \to (q \to s)))$
BCK-Q32. $p \to (((q \to p) \to r) \to ((q \to (r \to s)) \to (q \to s)))$
BCK-Q33. $p \to ((p \to q) \to (((p \to r) \to (q \to s)) \to (r \to s)))$

Candidates 3, 7, 8, 12, and 13 formerly on this list were the ones Halleck and Stickel showed to be single axioms. The formula that originally appeared on the list as **BCK-Q29** was shown by Halleck not to be a single axiom for **BCK**.

A.8 $\mathbf{H}_{\to}$: The implicational fragment of Heyting's intuitionistic calculus

Three-base: **B**′, **K**, **W** [Hilbert's 1930 set, according to Prior (1962, p. 316).]

Two-bases: $((p \to q) \to r) \to ((q \to (r \to s)) \to (q \to s))$, **I**
$(p \to q) \to ((q \to (q \to r)) \to (p \to r))$, **K**

One-bases:

HI-1. $((p \to q) \to r) \to (s \to ((q \to (r \to t)) \to (q \to t)))$
[Meredith (1953)]
HI-2. $p \to ((q \to r) \to (((s \to q) \to (r \to t)) \to (q \to t)))$
[Meredith and Prior (1963)]
HI-3. $(p \to (q \to r)) \to (s \to (((t \to p) \to q) \to (p \to r)))$

HI-4. $p \to ((q \to (r \to s)) \to (((t \to q) \to r) \to (q \to s)))$
HI-5. $((p \to q) \to r) \to ((q \to (r \to s)) \to (t \to (q \to s)))$
HI-6. $(p \to (q \to r)) \to (((s \to p) \to q) \to (t \to (p \to r)))$
HI-7. $p \to (((q \to r) \to s) \to ((r \to (s \to t)) \to (r \to t)))$
HI-8. $(p \to q) \to (((r \to (s \to p)) \to (q \to t)) \to (p \to t))$
HI-9. $(p \to q) \to ((q \to ((r \to p) \to s)) \to (t \to (p \to s)))$
HI-10. $(p \to q) \to (r \to (((s \to p) \to (q \to t)) \to (p \to t)))$
HI-11. $p \to (((q \to r) \to s) \to ((s \to (s \to t)) \to (r \to t)))$
HI-12. $((p \to q) \to r) \to (s \to ((r \to (r \to t)) \to (q \to t)))$

HI-3 through **HI-7** are from Ulrich (1999). As noted there, they come from Meredith's **HI-1** by permuting its first three antecedents, that is, $(p \to q) \to r$, s and $q \to (r \to t)$, in all possible ways. **HI-8** through **HI-12** are new.

These are almost certainly shortest possible single axioms for $\mathbf{H}_{\to}$, but (cf. Ulrich (2001)) there are four shorter theorems of $\mathbf{H}_{\to}$ whose status is unknown, namely, **C1** $= ((p \to q) \to r) \to ((q \to (r \to s)) \to (q \to s))$, **C2** $= ((p \to q) \to r) \to ((r \to (r \to s)) \to (q \to s))$, **C3** $= (p \to (q \to r)) \to (((s \to p) \to q) \to (p \to r))$, and **C4** $= (p \to q) \to (((r \to p) \to (q \to s)) \to (p \to s))$.

Among the 17-symbol theorems of $\mathbf{H}_{\to}$, there remain thirty-one whose status is unknown. Notice that only the first nine have (as do all of the single axioms listed above) five distinct sentence letters occurring in them. The rest contain occurrences of just four distinct letters, and the author conjectures that none of them can serve as a single axiom for $\mathbf{H}_{\to}$. Once more, the numbers for the remaining candidates are from a list of such circulated by the author.

HI-Q1. $((p \to q) \to (r \to s)) \to (((t \to q) \to r) \to (q \to s))$
HI-Q2. $((p \to q) \to (r \to s)) \to ((q \to r) \to (t \to (q \to s)))$
HI-Q3. $((p \to q) \to r) \to (((s \to q) \to (r \to t)) \to (q \to t))$
HI-Q4. $((p \to q) \to r) \to (((s \to r) \to (r \to t)) \to (q \to t))$
HI-Q5. $((p \to q) \to r) \to ((r \to (r \to s)) \to (t \to (q \to s)))$
HI-Q6. $(p \to (q \to r)) \to (((s \to (t \to p)) \to q) \to (p \to r))$
HI-Q7. $(((p \to q) \to r) \to s) \to ((s \to (s \to t)) \to (r \to t))$
HI-Q8. $((p \to (q \to r)) \to (s \to t)) \to ((r \to s) \to (r \to t))$
HI-Q9. $((p \to q) \to (r \to s)) \to (t \to ((q \to r) \to (q \to s)))$
HI-Q10. $((p \to q) \to r) \to (((p \to q) \to (r \to s)) \to (q \to s))$
HI-Q11. $((p \to q) \to r) \to (((p \to r) \to (r \to s)) \to (q \to s))$
HI-Q12. $((p \to q) \to r) \to ((q \to (r \to (r \to s))) \to (q \to s))$
HI-Q15. $((p \to q) \to r) \to ((q \to (r \to s)) \to (p \to (q \to s)))$
HI-Q16. $((p \to q) \to r) \to ((r \to (r \to s)) \to (p \to (p \to s)))$
HI-Q17. $(((p \to q) \to r) \to s) \to ((s \to (s \to p)) \to (r \to p))$
HI-Q18. $(((p \to q) \to r) \to s) \to ((s \to (s \to q)) \to (r \to q))$
HI-Q19. $((p \to q) \to (r \to s)) \to ((q \to r) \to (p \to (q \to s)))$
HI-Q20. $((p \to q) \to (r \to s)) \to (p \to ((q \to r) \to (p \to s)))$
HI-Q21. $((p \to q) \to (r \to s)) \to (((p \to q) \to r) \to (q \to s))$
HI-Q22. $((p \to q) \to r) \to (p \to ((q \to (r \to s)) \to (q \to s)))$
HI-Q23. $((p \to q) \to r) \to (p \to ((r \to (r \to s)) \to (q \to s)))$
HI-Q24. $(p \to (q \to r)) \to (((s \to (s \to p)) \to q) \to (p \to r))$

HI-Q25. $(p \to (q \to r)) \to (((s \to p) \to (p \to q)) \to (p \to r))$
HI-Q26. $(p \to (q \to r)) \to (((s \to p) \to q) \to (s \to (p \to r)))$
HI-Q27. $(p \to (q \to r)) \to (s \to (((s \to p) \to q) \to (p \to r)))$
HI-Q28. $(p \to q) \to (((r \to (r \to p)) \to (q \to s)) \to (p \to s))$
HI-Q30. $(p \to q) \to (((r \to p) \to (q \to (p \to s))) \to (p \to s))$
HI-Q31. $(p \to q) \to (((r \to p) \to (q \to (q \to s))) \to (p \to s))$
HI-Q33. $(p \to q) \to ((q \to ((r \to p) \to s)) \to (r \to (p \to s)))$
HI-Q34. $(p \to q) \to (r \to (((r \to p) \to (q \to s)) \to (p \to s)))$
HI-Q35. $p \to (((p \to q) \to r) \to ((r \to (r \to s)) \to (q \to s)))$

Candidates **HI-Q13, HI-Q14, HI-Q29**, and **HI-Q32** from the list as originally circulated were shown by John Halleck to be incapable of being single axioms for $\mathbf{H}_{\to}$.

A.9 The implicational fragment of Dummett's LC

Four-bases: Add **Dummett** $= ((p \to q) \to r) \to (((q \to p) \to r) \to r)$ to any three-base for $\mathbf{H}_{\to}$ [(Bull 1962)].

Of course three- and two-bases result if **Dummett** is added to a smaller base for $\mathbf{H}_{\to}$.

One-base: $(((p \to (q \to p)) \to (((((r \to s) \to t) \to (((s \to r) \to t) \to r)) \to (u \to (u \to v))) \to (w \to v)) \to x)) \to ((w \to u) \to x))$

Shorter single axioms no doubt exist. The author has not, for example, experimented with inserting **Dummett** into various spots in any of the single axioms for $\mathbf{H}_{\to}$.

A.10 The implicational fragment of Łukasiewicz's infinite-valued sentential calculus $\mathbf{L}_{\aleph_0}$, et relata

Implicational $\mathbf{L}_{\aleph_0}$.

Four-base: **B**$'$, **K**, **Inversion** $= ((p \to q) \to q) \to ((q \to p) \to p)$, **Linearity** $= ((p \to q) \to (q \to p)) \to (q \to p)$ [Meyer (1966)]

One-bases: The methods of Rezus (1982) ensure the existence of single axioms for all logics whose theorems include the first two of these axioms, albeit typically those methods produce long axioms.

In the presence of **B** and **K**, **Inversion** and **Linearity** can be replaced with the single formula $((p \to q) \to q) \to (((p \to q) \to (q \to p)) \to p)$. As a result, we have (Ulrich 2008) the following 37-symbol single axiom for $\mathbf{L}_{\aleph_0 \to}$.
$((p \to (q \to p)) \to ((((((r \to s) \to s) \to (((r \to s) \to (s \to r)) \to r)) \to (t \to u))) \to (v \to u)) \to w) \to ((v \to t) \to w)$

Again, shorter single axioms surely exist.

BCK + Inversion.

Four-base: **B**, **C**, **K/Simp**, **Inversion**

One-base: $((p \to (q \to p)) \to ((((r \to s) \to s) \to ((s \to r) \to r)) \to (t \to u))) \to ((u \to v) \to (t \to v))$ [Of length 29, the author doubts that it is shortest possible.]

BCK + Linearity.

Four-base: **B**, **C**, **K/Simp**, **Linearity**

One-base: $((((p \to (q \to p)) \to r) \to r) \to (((s \to t) \to (((u \to t) \to (t \to u) \to (s \to u))) \to v)) \to (w \to v)$

37 symbols long, one would think shorter axioms surely exist.

A.11 Some systems of strict implication

C3, the strict-implicational fragment of S3.

Three-base: **B**′, **W**, **Weak Irrelevance** $= (p \to q) \to (r \to r)$ [Anderson and Belnap (1962)]

An equivalent base can be obtained by adding the axiom $\mathbf{I}^* = (p \to p) \to (q \to q)$, used for Monothetic **BCI** above, to any base for $\mathbf{E}_{\to}$; **C3** is thus "Monothetic $\mathbf{E}_{\to}$."

One-base: $(((p \to (p \to q)) \to ((r \to s) \to (p \to q))) \to (((t \to u) \to ((v \to w) \to (((x \to x) \to (w \to y))) \to (v \to y))) \to z)) \to z$ [Ulrich (2005a), where it is also shown that every finitely axiomatizable extension of **C3** can be axiomatized by a single axiom. Shorter axioms no doubt await discovery.]

C4, the strict-implicational fragment of S4.

Three-base: **Frege**, **Weak Irrelevance** $= (p \to q) \to (r \to (p \to q))$, **I** [Anderson and Belnap (1962)]

The strict-implicational fragments of all extensions of **S4** between it and **S4.3** are identical (Ulrich 1981).

Two-base: $(p \to (q \to r)) \to ((p \to q) \to (s \to (p \to r)))$, **Irrelevance** $= p \to (q \to q)$ [Ernst et al. (2002)]

The authors show that this 20-symbol two-base is a shortest such base for **C4**; they also found five additional two-bases of that same length, and have shown that no others exist.

One-base: $(p \to ((q \to (r \to r)) \to (p \to q)) \to ((s \to t) \to (u \to (p \to t)))$ [Ernst et al. (2002)]

The authors show that this 21-symbol axiom is not only a shortest possible axiom for **C4**, but that it is *the* shortest possible axiom for that system: no shorter formula nor even another of the same length will do. Note that **C4** is thus one more example (cf. **BCI** etc., above) of a system whose shortest possible two-bases are one symbol shorter than their shortest possible single axioms.

C5, the strict-implicational fragment of S5.

Three-bases: **B**′, $(((p \to q) \to r) \to (p \to q)) \to (p \to q)$, **Irrelevance** $= p \to (q \to q)$ [Meredith, in Lemmon et al. (1969).]

Two-bases: **Irrelevance**, **M2** $= (((p \to q) \to r) \to q) \to ((q \to s) \to (p \to s))$ [Meredith circa 1956; ibid.]

Irrelevance, $(((p \to q) \to r) \to s) \to ((q \to s) \to (p \to s))$ [Prior (1961)]
$(p \to q) \to (((q \to r) \to s) \to r) \to (p \to r)$, **I** [Ernst et al. (2002)]

One-bases: $(((p \to p) \to q) \to r)) \to (s \to t)) \to ((t \to q) \to (u \to (s \to q)))$
$(((p \to q) \to r) \to ((s \to s) \to q)) \to ((q \to t) \to (u \to (p \to t)))$
$(((p \to q) \to r) \to ((s \to s) \to q)) \to (t \to ((q \to u) \to (p \to u))$
$(((p \to q) \to r) \to ((s \to s) \to t)) \to (u \to ((q \to t) \to (p \to t)))$
$(((p \to p) \to q) \to r) \to (s \to q)) \to ((q \to t) \to ((u \to (s \to t)))$
$(((p \to p) \to q) \to r) \to (s \to q)) \to ((t \to s) \to (u \to (t \to q)))$
$((((p \to p) \to (q \to r)) \to s) \to r) \to ((r \to t) \to (u \to (q \to t)))$

The first single axiom above is Meredith's circa 1956 in Lemmon et al. (1969). The others are from Ernst et al. (2002). The authors of the latter show that these 21-symbol single axioms for **C5** are shortest possible, and that the list above is exhaustive.

Their two-base, however, is just 18 symbols in length and is the most severe example known to the author of a system being axiomatizable by a two-base shorter than its shortest possible single axioms. They also show it is *the* shortest possible base for **C5**: no other base of any kind can match its length.

References

Anderson, A. R., & Belnap, N. D. (1962). The pure calculus of entailment. *Journal of Symbolic Logic*, *27*, 19–52.

Anderson, A. R., & Belnap, N. D. (1975). *Entailment: The Logic of Relevance and Necessity* (Vol. I). Princeton, NJ: Princeton University Press.

Anderson, A. R., Belnap, N. D., & Wallace, J. R. (1960). Independent axiom-schemata for the pure theory of entailment. *Zeitschrift für mathematische Logik und Grundlagen der Mathematik*, *6*, 93–95.

Belnap, N. D. (1976). The two-property. *The Relevance Logic Newsletter*, *1*, 173–180.

Borokowski, I. (1970). *Selected Works*. Amsterdam: North-Holland.

Bull, R. A. (1962). The implicational fragment of Dummett's LC. *Journal of Symbolic Logic*, *17*, 189–194.

Bunder, M. W. (1983). The answer to a problem of Iséki on BCI-algebras. *Mathematics Seminar Notes, Kobe University*, *11*, 167–169.

Church, A. (1951). The weak theory of implication. In A. Menne, A. Wilhelmy, & H. Angsil (Eds.), *Kontrolliertes Denken* (pp. 22–37). Munich: Untersuchungen zum Logikkalkül und zur Logik der Einzelwissenschaften, Kommissions-Verlag Karl Alber.

Davis, J. W., Hockney, D. J., & Wilson, W. K. (1969). *Philosophical Logic*. Dordrecht: D. Reidel.

Dunn, J. M. (1970). Algebraic completeness results for R-Mingle and its extensions. *Journal of Symbolic Logic*, *35*, 1–13.

Ernst, Z., Fitelson, B., Harris, K., & Wos, L. (2001). A concise axiomatization of $RM_{\to}$. *Bulletin of the Section of Logic* (University of Łódź), *30*, 191–194.

Ernst, Z., Fitelson, B., Harris, K., & Wos, L. (2002). Shortest axiomatizations of implicational S4 and S5. *Notre Dame Journal of Formal Logic*, *43*, 169–179.

Girard, J.-Y. (1987). Linear logic. *Theoretical Computer Science*, *50*, 1–102.

Hindley, R. J. (1993). BCK and BCI logics, condensed detachment and the 2-property. *Notre Dame Journal of Formal Logic*, *34*, 231–250.

Kashima, R., & Kamide, N. (1999). Substructural implicational logics including the relevant logic E. *Studia Logica, 63*, 181–212.
Lemmon, E. J., Meredith, C. A., Meredith, D., Prior, A. N., & Thomas, I. (1969). Calculi of pure strict implication. In J. W. Davis, D. J. Hockney, & W. K. Wilson (Eds.), *Philosophical Logic* (pp. 215–250). Dordrecht: D. Reidel. Circulated as a mimeograph (by Canterbury University College, Christchurch, 1956/57) for several years.
Leśniewski, S. (1929). Grundzüge eines neuen Systems der Grundlagen der Mathematik. *Fundamenta Mathematicae, 14*, 1–81.
Łukasiewicz, J. (1948). The shortest axiom of the implicational calculus of propositions. *Proceedings of the Royal Irish Academy, Section A*, **52**(3), 25–33. Republished in Borokowski (1970).
Łukasiewicz, J., & Tarski, A. (1930). Untersuchungen über den Aussagenkalkül, *Comptes rendus des séances de la Société et des Lettres de Varsovie, Classe III, 23*, 49–50. English translation by J. H. Woodger in Tarski (1956) and in McCall, S. (1967), Polish logic 1920–1939, Clarendon Press, Oxford.
Meredith, C. A. (1953). A single axiom of positive logic. *Journal of Computing Systems, 1*, 169–170.
Meredith, C. A., & Prior, A. N. (1963). Notes on the axiomatics of the propositional calculus. *Notre Dame Journal of Formal Logic, 4*, 171–187.
Meyer, R. K. (1966). Pure denumerable Łukasiewicz implication. *Journal of Symbolic Logic, 31*, 575–580.
Meyer, R. K., & Parks, R. Z. (1972). Independent axioms for the implicational fragment of Sobociński's three-valued logic. *Zeitschrift für mathematische Logik und Grundlagen der Mathematik, 18*, 291–295.
Moh, S.-K. (1950). The deduction theorems and two new logical systems. *Methodos, 2*, 56–75.
Parks, R. Z. (1972). A note on R-Mingle and Sobociński's three-valued logic. *Notre Dame Journal of Formal Logic, 13*, 227–228.
Peirce, C. S. (1885). On the algebra of logic: A contribution to the philosophy of notation. *American Journal of Mathematics, 7*, 180–202.
Prior, A. N. (1956). Logicians at play; or Syll, Simp and Hilbert. *Australasian Journal of Philosophy, 34*, 182–192.
Prior, A. N. (1961). Some axiom-pairs for material and strict implication. *Zeitschrift für mathematische Logik und Grundlagen der Mathematik, 7*, 61–65.
Prior, A. N. (1962). *Formal Logic* (2nd ed.). New York, NY: Oxford University Press.
Rezus, A. (1982). On a theorem of Tarski. *Libertas mathematica, 2*, 63–97.
Rose, A. (1956). An alternative formalization of Sobociński's three-valued implicational propositional calculus. *Zeitschrift für mathematische Logik und Grundlagen der Mathematik, 2*, 166–172.
Sobociński, B. (1952). Axiomatization of a partial system of three-valued calculus of propositions. *Journal of Computing Systems, 1*, 23–55.
Tarski, A. (1956). Logic, Semantics, Metamathematics. Papers from 1923 to 1938. English translations by J. H. Woodger, Clarendon press, Oxford, UK.
Thomas, I. (1970). Final word on a shortest implicational axiom. *Notre Dame Journal of Formal Logic, 11*, 16.
Tursman, R. (1968). The shortest axioms of the implicational calculus. *Notre Dame Journal of Formal Logic, 9*, 351–358.
Ulrich, D. (1981). Strict implication in a sequence of extensions of S4. *Zeitschrift für mathematische Logik und Grundlagen der Mathematik, 27*, 201–212.
Ulrich, D. (1999). New axioms for positive implication. *Bulletin of the Section of Logic* (University of Łódź), *28*, 39–42.
Ulrich, D. (2001). A legacy recalled and a tradition continued. *Journal of Automated Reasoning, 27*, 97–122.
Ulrich, D. (2005a). D-complete axioms for the classical equivalential calculus. *Bulletin of the Section of Logic* (University of Łódź). *34*, 135–142.
Ulrich, D. (2005b). On the existence of a single axiom for implicational R-Mingle, (abstract). *Bulletin of Symbolic Logic, 11*, 459.

Ulrich, D. (2005c). Single axioms for the finitely axiomatizable extensions of strict-implicational S3 and of some of its weaker subsystems, (abstract). *Bulletin of Symbolic Logic*, *11*, 555.
Ulrich, D. (2006). Single axioms for a class of substructural logics contained in BCI, (abstract). *Bulletin of Symbolic Logic*, *12*, 166.
Ulrich, D. (2007). A short axiom pair and a shortest possible single axiom for monothetic BCI, (abstract). *Bulletin of Symbolic Logic*, *13*, 404.
Ulrich, D. (2008). A new three-base and a single axiom for pure denumerable Łukasiewicz implication, (abstract). *Bulletin of Symbolic Logic*, *14*, 416.
Ulrich, D. (2009). On two open questions concerning the implicational fragment of R-Mingle. *Bulletin of the Section of Logic* (University of Łódź). *38*, 1–4.
Ulrich, D. (2011). New results concerning single axioms for four subsystems of BCI, (abstract). *Bulletin of Symbolic Logic*, *17*, 153–154.
Ulrich, D. (2012). A single axiom for relevant implication. *Bulletin of the Section of Logic* (University of Łódź). *41*, 13–16.
Wajsberg, M. (1932). Über Axiomensysteme des Aussagenkalküls. *Monatshefte für Mathematik und Physik*, *39*, 119–156.
Wajsberg, M., (1937). Metalogische Beiträge. *Wiedomości matematyczne*, *43*, 131–168. English translation by Storrs McCall and Peter Woodruff in McCall, S., (1967). *Polish logic 1920–1939*. Oxford: Clarendon Press.
Wajsberg, M., (1939). Metalogische Beiträge II. *Wiedomości matematyczne*, *47*, 119–139. English translation by Storrs McCall in McCall, S. (1967). *Polish logic 1920–1939*. Oxford: Clarendon Press.
Wos, L., & Pieper, G. W. (2003). Automated Reasoning and the Discovery of Missing and Elegant Proofs, Paramus, NJ: Rinton Press.

LC and Its Pretabular Relatives

Larisa Maksimova

Abstract In 1970, J.M. Dunn published a paper on a logic called RM extending the logic R of relevance. He proved that the logic RM is pretabular, i.e., it has no finite characteristic matrix, but every proper extension of it has such a matrix. In 1971, J.M. Dunn and R.K. Meyer obtained a similar result for Dummett's logic LC. This is a brief survey of pretabular logics. We consider pretabularity and tabularity problems over the intuitionistic and minimal logics, and also in families of positive, modal and relevant logics.

Keywords Minimal logic · Modal logic · Pretabular logic · Superintuitionistic logic · Tabularity problem

1 Introduction

In this paper, we present a brief survey of pretabular logics. By a logic we mean any set of formulas closed under substitutions. A logic L is *tabular* if it can be characterized by a finite model, i.e., there is a finite model such that L is the set of all formulas valid in this model. A logic is *pretabular* if it is not tabular, but any of its extensions is tabular. These terms were introduced in Kuznetsov (1971).

In 1951, J. Scroggs proved that the modal logic S5 is pretabular and described all the extensions of S5 (Scroggs 1951). In 1970, J.M. Dunn published a paper (Dunn 1970) on a logic called RM extending the logic R of relevance. He proved that the logic RM is pretabular, i.e., it has no finite characteristic matrix, but every proper extension of it has such a matrix. In 1971, J.M. Dunn and R.K. Meyer obtained a similar result for Dummett's logic LC (Dunn and Meyer 1971).

The logic LC, introduced by Dummett (1959), is one of the superintuitionistic logics (SIL), i.e., extensions of the intuitionistic logic Int. The study of the family

L. Maksimova (✉)
Sobolev Institute of Mathematics, Novosibirsk, Russia
e-mail: lmaksi@math.nsc.ru
URL: http://www.math.nsc.ru/~larisa/

K. Bimbó (ed.), *J. Michael Dunn on Information Based Logics*,
Outstanding Contributions to Logic 8, DOI 10.1007/978-3-319-29300-4_5

of superintuitionistic logics started from a paper by Umezawa (1959) and was rather popular. Another two pretabular SILs were found by Hosoi and Ono (1970).

It turned out that there are exactly three pretabular SILs. It was stated in Maksimova (1971) that there are exactly three pretabular SILs having the finite model property (FMP). At the same conference, an abstract by Kuznetsov (1971) was published, which stated that every pretabular SIL has the finite model property.

In more detail, A.V.Kuznetsov had proved that every tabular SIL has finitely many immediate predecessors (with respect to inclusion), and all of them are tabular. In addition, every non-tabular SIL is contained in a pretabular one. At the conference, we explained our proofs to each other. It was stated in Maksimova (1971) that every non-tabular SIL with FMP is included in some of the three pretabular SILs, which were already known. The full proof of this statement was published in Maksimova (1972), where the duality theory between Heyting algebras and partially ordered frames was developed.

The description of all pretabular SILs made it possible to solve the tabularity problem over Int, the intuitionistic logic: there is an algorithm which, given a finite set Ax of additional axiom schemas, decides if the logic $\mathrm{Int} + Ax$ is tabular.

In this paper, we consider pretabularity and tabularity problems over the intuitionistic and minimal logics, and also in families of positive, modal and relevant logics.

In all families under consideration, any extension of a tabular logic is tabular, and every tabular logic is finitely axiomatizable. By Zorn's lemma, for these families we have the following.

Proposition 1 *Every non-tabular logic is contained in a pretabular one.*

2 Pretabular Superintuitionistic Logics and Their Extensions

In this section, we describe pretabular SILs in detail.

We define π_i as follows.

$$\pi_0 = p_0, \qquad \pi_{n+1} = p_{n+1} \vee (p_{n+1} \to \pi_n).$$

The following theorem was proved in Maksimova (1972) with the use of Kuznetsov's theorem on finite approximability of all pretabular SILs (Kuznetsov 1971).

Theorem 2.1 (Maksimova 1972)
There are exactly three pretabular superintuitionistic logics, namely,

1. $\mathrm{LC} = \mathrm{Int} + ((p \to q) \vee (q \to p))$,
2. $\mathrm{LP}_2 = \mathrm{Int} + \pi_2$,
3. $\mathrm{LQ}_3 = \mathrm{Int} + \pi_3 + (\neg p \vee \neg\neg p)$.

Algebraic completeness for LC and its extensions was proved by Dunn and Meyer (1971). Decidability of LC was proved by Dummett (1959). Two other pretabular SILs and their extensions were described by Hosoi and Ono (1970).

Every proper extension of LC is of the form LC + π_n, for some n. All the extensions of LP_2 and LQ_3 are also finitely axiomatizable.

Any SIL is complete under algebraic semantics. The logic LC is characterized by any infinite linearly ordered Heyting algebra, LP_2 by any Heyting algebra built from an infinite Boolean algebra by adding a new greatest element, and LQ_3 by any Heyting algebra built from an infinite Boolean algebra by adding new greatest and new least elements.

All the three pretabular logics are complete with respect to suitable classes of Kripke frames.

Recall that a *frame* is a pair $\mathbf{W} = (W, R)$, where W is a set and R is a binary relation on W. An *intuitionistic model* $\mathbf{M} = (W, R, \models)$ satisfies the following conditions.

(i) R is a partial ordering of W;
(ii) $(x \models p \text{ and } xRy) \Rightarrow y \models p$, for any variable p;
(iii) $x \models (A \to B) \iff \forall y\,(xRy \Rightarrow (y \models A \Rightarrow y \models B))$;
(iv) $x \models \neg A \iff \forall y\,(xRy \Rightarrow y \not\models A)$.

A formula A is (*intuitionistically*) *valid in* a partially ordered frame $\mathbf{W} = (W, R)$, if for any intuitionistic model $(W, R, \models)$, we have $x \models A$, for any $x \in W$.

The logic LC is characterized by the class of all intuitionistic models based on linearly ordered frames, LP_2 by partially ordered frames with no 3-element chain, and LQ_3 by partially ordered frames having a greatest element and no 4-element chain.

Define the frames $\mathbf{V}_n$, $\mathbf{U}_{n+1}$ and $\mathbf{Z}_n$ as follows.

1. $\mathbf{V}_n = (V_n, R)$, where $V_n = \{0, 1, \ldots, n\}$, and $xRy \iff (x = 0 \text{ or } x = y)$;
2. $\mathbf{U}_{n+1} = (U_{n+1}, R)$, where $U_{n+1} = \{0, 1, \ldots, n, n+1\}$, and $xRy \iff (x = 0$ or $y = n+1$ or $1 \leq x = y \leq n)$;
3. $\mathbf{Z}_n = (Z_n, R)$, where $Z_n = \{1, \ldots, n\}$, and $xRy \iff x \leq y$.

For each of the pretabular SILs, its proper extensions form a countable descending chain. Every consistent proper extension of LC is characterized by a frame $\mathbf{Z}_n$, for some n, every extension of LP_2 is characterized by $\mathbf{V}_n$, for some n, and every extension of LQ_3 by $\mathbf{Z}_1$ or by $\mathbf{U}_n$, for some n. The logic LC is complete under the class of all frames $\mathbf{Z}_n$, the logic LP_2 under the class of all $\mathbf{V}_n$ frames, and LQ_3 under the class of all $\mathbf{U}_n$ frames.

Let L be a finitely axiomatizable SIL. Its *recognition problem over* Int is defined as follows.

> For any finite system Ax of axiom schemas, decide if the logic Int + Ax coincides with L.

A finitely axiomatizable logic L is *recognizable over* Int if there is an algorithm for deciding its recognition problem. We have the following criterion for recognizability from Maksimova and Yun (2015).

Lemma 2.2 *A finitely axiomatizable* SIL, *L is recognizable over* Int *iff it is decidable and the inclusion problem* Int $+ Ax \supseteq L$ *is decidable.*

All the pretabular logics over Int are decidable. Applying Lemma 2.2 and Proposition 2.6 below we conclude that they are recognizable over Int. The complexity bounds are given in the next Theorem.

Theorem 2.3 (Maksimova and Voronkov 2003) *For every pretabular* SIL, *its decision problem is coNP-complete and its recognition problem over* Int *is DP-complete.*

Recall from Papadimitriou (1994) that X is in DP if $X = Y \cap Z$ where Y is in NP and Z in coNP.

Until now, we considered SIL defined by additional axiom schemas. Let us turn to the case where a logic is defined by additional axiom schemas and rules of inference. We consider only rules invariant under substitution.

A logic L is *strongly recognizable over* Int if there is an algorithm which, for every finite system Rul of axiom schemas and rules of inference, decides if the logic Int $+ Rul$ coincides with L.

It is proved in Maksimova (2000) that the logic LC, and also every SIL containing the formula π_n, for some n, is strongly recognizable over Int. It follows that

Theorem 2.4 *Every pretabular* SIL *is strongly recognizable over* Int.

This result essentially uses the theory of admissibility of inference rules developed by Rybakov (1997).

One can restrict himself to rules with one premise since several premises can be replaced by their conjunction.

A rule $A(p_1, \ldots, p_n)/B(p_1, \ldots, p_n)$ is said to be *admissible in* a logic L if for any formulas $\psi_1, \ldots, \psi_n$, the formula $B(\psi_1, \ldots, \psi_n)$ is valid in L whenever $A(\psi_1, \ldots, \psi_n)$ is valid in L.

In order to prove Theorem 2.4, we used the following sufficient condition.

Lemma 2.5 (Maksimova 2000) *Let L be a finitely axiomatizable* SIL. *If its admissibility problem is decidable and, moreover, there is an algorithm which, given a finite set Rul of axiom schemas and rules of inference, decides whether L is included in* Int $+ Rul$, *then L is strongly recognizable over* Int.

It is proved in Rybakov (1997) that the admissibility problem is decidable in all pretabular SILs. Consider the inclusion problem.

A rule A/B is *valid in* a partially ordered frame $\mathbf{W} = (W, R)$ if, for any intuitionistic model $(W, R, \models)$,

$$(\forall x \in W)\ x \models A \ \Rightarrow\ (\forall x \in W)\ x \models B,$$

and *refutable* in W otherwise. A system Rul of formulas and rules is *refutable* in W if some formula or rule of Rul is refutable in W.

In Maksimova (2000), we established the following.

Proposition 2.6 *Let* $L = \text{Int} + Rul$. *Then,*

1. $L \supseteq \text{LC}$ *iff* Rul *is refutable in both the frame* $\mathbf{V}_2$ *and the frame* $\mathbf{U}_3$*;*
2. $L \supseteq \text{LP}_2$ *iff* Rul *is refutable in* $\mathbf{Z}_3$*;*
3. $L \supseteq \text{LQ}_3$ *iff* Rul *is refutable in both the frame* $\mathbf{V}_2$ *and the frame* $\mathbf{Z}_4$.

Thus all the three pretabular SIL are strongly recognizable over Int.

3 Modal Logics Over S4

In this section, we consider only normal extensions of S4, i.e., logics closed under the necessitation rule $A/\Box A$.

As we already mentioned, Scroggs (1951) proved that the logic S5 is pretabular and he described all its extensions. Other pretabular logics over S4 were listed much later.

It is well known that the intuitionistic logic Int is embeddable into the modal S4 logic via the Gödel–Tarski translation T: a formula A is valid in Int iff $T(A)$ is valid in S4 (McKinsey and Tarski 1948). With every SIL L, one can associate the set of its modal companions. A modal logic M over S4 is a *modal companion of* L if $A \in L \iff T(A) \in M$, for any non-modal formula A. The greatest modal companion of Int is the Grzegorczyk logic $\text{Grz} = \text{S4} + (\Box(\Box(p \to \Box p) \to p) \to p)$, the greatest modal companion of LC is $\text{Grz.3} = \text{Grz} + (\Box(\Box p \to q) \vee \Box(\Box q \to p))$ (Blok 1976).

Inter-relations of superintuitionistic logics and extensions of S4 were developed in Maksimova and Rybakov (1974), which allowed us to describe all pretabular logics over S4 (Maksimova 1975).

Theorem 3.1 (Maksimova 1975)
There are exactly five pretabular logics over S4.

$\text{PM1} = \text{Grz.3} = \text{Grz} + (\Box(\Box p \to q) \vee \Box(\Box q \to p))$
$\text{PM2} = \text{Grz} + \sigma_2$, *where* $\sigma_2 = \Box p \vee \Box(\Box p \to \Box q \vee \Box\Diamond\neg q)$,
$\text{PM3} = \text{Grz} + (\Box r \vee \Box(\Box r \to \sigma_2)) + (\Box\Diamond p \leftrightarrow \Diamond\Box p)$,
$\text{PM4} = \text{S4} + \sigma_2 + (\Box\Diamond p \leftrightarrow \Diamond\Box p)$,
$\text{PM5} = \text{S5} = \text{S4} + (p \to \Box\Diamond p)$.

The same result was also announced by Meskhi and Esakia (1974) and published in Esakia and Meskhi (1977). Also it was independently proved by Rautenberg (1977).

The logics PM1–PM3 are the greatest modal companions of pretabular SILs. The logic $\text{PM1} = \text{Grz.3}$ is complete for the class of models based on the frames $\mathbf{Z}_n$, PM2 is characterized by all $\mathbf{V}_n$, and PM3 by $\mathbf{U}_n$. The logic PM4 is complete under the class of $\mathbf{Y}_n$, and $\text{PM5} = \text{S5}$ is characterized by $\mathbf{X}_n$, where $\mathbf{X}_n$ and $\mathbf{Y}_n$ are as follows.

$\mathbf{X}_n = (X_n, R)$, where $X_n = \{1, \ldots, n\}$ and xRy for all x, y;
$\mathbf{Y}_n = (Y_n, R)$, where $Y_n = \{1, \ldots, n\}$ and $xRy \iff (x \leq n-1 \text{ or } y = n)$.

For each of PM1–PM5 its proper extensions form a countable descending chain.

Further results are similar to those of the previous section. All the pretabular extensions of S4 are decidable and recognizable over S4.

Theorem 3.2 (Maksimova and Voronkov 2003)
For every pretabular extension of S4, *its decision problem is coNP-complete and its recognition problem over* S4 *is DP-complete.*

Theorem 3.3 (Maksimova 2000)
The logics PM1–PM5 *are strongly recognizable over* S4.

4 Tabularity Problem

The description of all pretabular logics over Int plays a central role in the solution of the tabularity problem over Int, which is the following problem.

For any finitely axiomatizable SIL L, determine if L is tabular.

We know that any non-tabular logic is contained in a pretabular one. So we have the following criterion.

Theorem 4.1 *A* SIL *is tabular iff it is contained in no pretabular* SIL.

As there are only finitely many pretabular SILs and each of them is decidable, the tabularity problem over Int is decidable. Its complexity is given in the next theorem.

Theorem 4.2 (Maksimova and Voronkov 2003)
The tabularity problem over Int *is NP-complete.*

Algorithms for automatically recognition of tabularity and pretabularity over Int are presented in Maksimova and Schreiner (2006).

The tabularity problem is also decidable (NP-complete) in the family of positive logics extending the positive fragment of Int (Verhozina 1978; Maksimova 2002). A similar statement holds for modal logics over S4.

Theorem 4.3 (Maksimova and Voronkov 2003)
The tabularity problem over S4 *is NP-complete.*

For the whole family of normal modal logics, the tabularity problem is undecidable (Chagrov 1989).

The tabularity problem is decidable over the provability logic GL (Chagrov 1989; Chagrov and Zakharyaschev 1997), and over D4 $=$ K4 $+ (\Box p \to \Diamond p)$ (Chagrov 1989). Note that the set of pretabular extensions of the provability logic GL is countable (Blok 1980). The tabularity problem is undecidable over K; also, it is undecidable over the class of all (not only normal) extensions of K4, and is open for the class of normal extensions of K4 (Chagrov 1994; Chagrov and Zakharyaschev 1997).

It is unknown if tabularity is strongly decidable over Int or S4. A property is said to be *strongly decidable over* Int if there is an algorithm which, for any finite set *Rul* of axiom schemas and rules of inference, decides if the logic Int + *Rul* has this property. For strong decidability of tabularity, it would be sufficient to establish the strong decidability of the inclusion in each of the pretabular logics. It is shown in Maksimova (2000) that some inclusions in pretabular logics are strongly decidable.

Theorem 4.4 (Maksimova 2000)

1. *The inclusion in* LC *is strongly decidable over* Int.
2. *The inclusions in the pretabular logics* PM1, PM4, PM5 *are strongly decidable over* S4.

5 Modal and Relevant Logics

In this paper, we have considered so far only classical normal modal logics, i.e., logics containing the modal logic K and closed under necessitation rule $A/\Box A$. By analogy with Proposition 1, we have that every non-tabular logic is contained in a pretabular one.

J. Scroggs, in 1951, proved that the modal logic S5 is pretabular and described all the extensions of S5 (Scroggs 1951). All the five pretabular logics over S4 are described in the Sect. 3. For each of these logics, its proper extensions form a countable descending chain. This does not hold for the logic K4.

The logic K4 has a continuum of pretabular extensions (Blok 1980). There are countably many pretabular logics over the provability logic GL (Blok 1980), and finitely many over D4 (Chagrov and Zakharyaschev 1997).

Blok (1980) proved that over K4, every immediate predecessor of a tabular logic is tabular. This implies the following.

Theorem 5.1 *Every pretabular logic over* K4 *has the finite model property.*

Let us turn to relevant logics. The first pretabular extension of the logic R of relevance was found by Dunn (1970). It was the logic RM (R-Mingle) characterized by a matrix based on the set of integers.

Swirydowicz (2008) has proved that the logic R has a continuum of pretabular extensions. He has constructed an uncountable set of pretabular extensions of the relevant logic R, where each logic of this set is generated by a variety of finite height.

6 Positive Logics and Extensions of Johansson's Minimal Logic

In this section, we consider positive logics extending the positive fragment Int^{+} of the intuitionistic logic and extensions of Johansson's minimal logic J. By analogy with Proposition 1, we have that every non-tabular logic is contained in a pretabular one.

Pretabular positive logics were described by Verhozina (1978). They are positive fragments of the pretabular SIL LC and LP_2. The positive fragment of LQ_3 is contained in LP_2^+.

Theorem 6.1 (Verhozina 1978)
There are exactly two pretabular logics over Int^+:

1. $LC^+ = \mathrm{Int}^+ + ((p \to q) \vee (q \to p))$;
2. $LP_2^+ = \mathrm{Int}^+ + \pi_2$.

Every proper extension of LC^+ is characterized by a frame $\mathbf{Z}_n$ for some n, and LC^+ itself by the class of all $\mathbf{Z}_n$. The logic LP_2^+ is characterized by the class of all $\mathbf{V}_n$, and every proper extension of it is characterized by a $\mathbf{V}_n$ for some n. Remember that we defined $\mathbf{V}_n$ and $\mathbf{Z}_n$ as follows.

$$\mathbf{V}_n = (V_n, R), \text{ where } V_n = \{0, 1, \ldots, n\}, \text{ and } xRy \iff (x = 0 \text{ or } x = y),$$
$$\mathbf{Z}_n = (Z_n, R), \text{ where } Z_n = \{1, \ldots, n\}, \text{ and } xRy \iff x \leq y.$$

Just as for SILs, we have the following result.

Theorem 6.2 (Maksimova 2002, 2003)

1. *The tabularity problem over* Int^+ *is NP-complete.*
2. *For every pretabular positive logic, its recognition problem over* Int^+ *is DP-complete.*

Now we turn to the J-logics, i.e., to extensions of Johansson's minimal logic J (Johansson 1937). This logic can be axiomatized by the same axiom schemas and rules as Int^+, but its language contains a propositional constant $\bot$, $\neg A = A \to \bot$. We have

$$\mathrm{Int} = \mathrm{J} + (\bot \to p).$$

J-logics were studied in Rautenberg (1979) and in Odintsov (2008). A J-logic is *negative* if it contains $\bot$. The least negative logic is $\mathrm{Neg} = \mathrm{J} + \bot$.

Any negative logic is—in some sense—equivalent to its positive fragment (Rautenberg 1979), because $\bot$ is equivalent in Neg to $(p \to p)$. As an immediate consequence of Theorem 6.1 we obtain the next proposition.

Proposition 6.3 *There are exactly two pretabular logics over* Neg:

1. $\mathrm{NC} = \mathrm{Neg} + ((p \to q) \vee (q \to p))$;
2. $\mathrm{NP}_2 = \mathrm{Neg} + \pi_2$.

Now we turn to the pretabular extensions of the logic J. In his book (Rautenberg 1979), p. 295, W. Rautenberg states that the logic J has exactly seven pretabular extensions. He gives this statement as an exercise, without any information about these logics. But the proof seems not to be too easy. We give an axiomatization of these logics and ideas of our proof.

Theorem 6.4 *There are exactly seven pretabular logics over* J. *They are*

- *three pretabular logics over* Int,
- *two pretabular logics over* Neg,
- $\mathrm{PJ6} = \mathrm{J} + \pi_2 + (\bot \to \pi_1) + (p \vee \neg p)$,
- $\mathrm{PJ7} = \mathrm{J} + \pi_2 + (\bot \to \pi_1) + \neg\neg(\bot \to p) + (\neg p \vee \neg\neg p)$.

First of all, it is easily seen that any non-tabular logic over J is contained in a pretabular one.

Further, Theorem 6.4 can be proved by the same method as in Maksimova (1972) with the use of the modified Kripke semantics introduced in Maksimova (2007). We develop a duality between J-algebras and modified frames similar to the duality theory for Heyting algebras introduced in Maksimova (1972); see also Gabbay and Maksimova (2005). It makes it possible to describe all pretabular logics together with proving the following.

Theorem 6.5 *Every pretabular logic over* J *has the finite model property.*

The full proof will be presented in a separate paper. Here we find a semantical description of pretabular logics using Segerberg's semantics (Segerberg 1968). A *model* is a quadruple $\mathbf{M} = (W, R, Q, \vDash)$, where $(W, R, \vDash)$ is an intuitionistic model for positive formulas, $Q \subseteq W$, and for any $x \in W$,

$$x \in Q \ \Rightarrow \ \forall y\,(xRy \Rightarrow y \in Q), \qquad x \vDash \bot \iff x \in Q.$$

We define a series of frames.

$\mathbf{V}_n^0 = (V_n, R, Q)$, where $V_n = \{0, 1, \dots, n\}$, $\quad xRy \iff (x = 0$ or $x = y)$, and $Q = V_n$,
$\mathbf{V}_n^1 = (V_n, R, Q)$, where $Q = V_n - \{0\}$,
$\mathbf{V}_n^2 = (V_n, R, Q)$, where $n > 0$, $\quad Q = V_n - \{0, 1\}$,
$\mathbf{V}_n^t = (V_n, R, Q)$, where $Q = \emptyset$;
$\mathbf{U}_{n+1}^t = (U_{n+1}, R, Q)$, where $U_{n+1} = \{0, 1, \dots, n, n+1\}$, $xRy \iff (x = 0$ or $y = n+1$ or $1 \le x = y \le n)$, and $Q = \emptyset$;
$\mathbf{Z}_n^0 = (Z_n, R, Q)$, where $Z_n = \{1, \dots, n\}$, $\quad xRy \iff x \le y$, and $Q = Z_n$,
$\mathbf{Z}_n^t = (Z_n, R, Q)$, where $Q = \emptyset$.

The SILs LC, LP_2 and LQ_3 are characterized by the classes of all $\mathbf{Z}_n^t$, $\mathbf{V}_n^t$ and $\mathbf{U}_{n+1}^t$, respectively. The negative logics NC and NP_2 are characterized by the classes of all $\mathbf{Z}_n^0$ and $\mathbf{V}_n^0$, respectively. The logic PJ6 is characterized by the class of all $\mathbf{V}_n^1$, and PJ7 by $\mathbf{V}_n^2$.

Recall that for each of the pretabular logics over Int or Neg, its proper extensions form a countable descending chain. This does not hold for PJ6 and PJ7. Every non-trivial proper extension of PJ6 is characterized by the frame $\mathbf{Z}_1^0$ or $\mathbf{V}_n^1$, for some n, or by the set $\{\mathbf{Z}_1^0, \mathbf{V}_0^1\}$. The logics of $\mathbf{Z}_1^0$ and $\mathbf{V}_0^1$ are incomparable.

Every non-trivial and proper extension of PJ7 is characterized by the frame $\mathbf{Z}_1^0$, or $\mathbf{Z}_1^t$, or $\mathbf{V}_n^2$ for some $n > 0$, or by the set $\{\mathbf{Z}_1^0, \mathbf{Z}_1^t\}$ or $\{\mathbf{Z}_1^0, \mathbf{V}_1^2\}$. The logics of $\mathbf{Z}_1^0$ and $\mathbf{Z}_1^t$ are incomparable, so are also the logics of $\mathbf{Z}_1^0$ and $\mathbf{V}_1^2$.

Using this description, by analogy with Theorem 4.2, one can prove the following.

Proposition 6.6 *The tabularity problem over* J *is NP-complete.*

We turn to the recognition problem. The logics Int and Neg are recognizable over J and their pretabular extensions are recognizable over them. Therefore all pretabular superintuitionistic and negative logics are recognizable over J. One can show that so are the logics PJ6 and PJ7. One can prove that, for each of the seven logics, its recognition problem over J is DP-complete, and the tabularity problem over J is NP-complete.

The strong recognition problems over J and Int^+ have not been investigated.

References

Blok, W. J. (1976). *Varieties of Interior Algebras*, PhD thesis, University of Amsterdam.

Blok, W. J. (1980). Pretabular varieties of modal algebras. *Studia Logica*, *39*(2–3), 101–124.

Chagrov, A. V. (1989). Nontabularity—pretabularity, antitabularity, coantitabularity. In *Algebraic and logical constructions* (pp. 105–111). Kalinin: Kalinin State University (Russian).

Chagrov, A. V. (1994). Undecidable properties of superintuitionistic logics. In S. V. Jablonskij (Ed.), *Mathematical problems of cybernetics* (Vol. 5, pp. 67–108). Moscow: Physmatlit (Russian).

Chagrov, A., & Zakharyaschev, M. (1997). *Modal logic*. Oxford: Clarendon Press.

Dummett, M. (1959). A propositional calculus with denumerable matrix. *Journal of Symbolic Logic*, *24*, 97–106.

Dunn, J. M. (1970). Algebraic completeness results for R-Mingle and its extensions. *Journal of Symbolic Logic*, *35*, 1–13.

Dunn, J. M., & Meyer, R. K. (1971). Algebraic completeness results for Dummett's LC and its extensions. *Zeitschrift für mathematische Logik und Grundlagen der Mathematik*, *17*, 225–230.

Esakia, L., & Meskhi, V. (1977). Five critical modal systems. *Theoria*, *43*(1), 52–60.

Gabbay, D. M., & Maksimova, L. (2005). *Interpolation and definability: Modal and intuitionistic logics*. Oxford: Clarendon Press.

Hosoi, T., & Ono, H. (1970). The intermediate logics of the second slice. *Journal of Faculty of Sciences, University of Tokyo, Sec IA*, *17*, 457–461.

Johansson, I. (1937). Der Minimalkalkül, ein reduzierter intuitionistischer Formalismus. *Compositio Mathematica*, *4*, 119–136.

Kuznetsov, A. V. (1971). Some properties of the lattice of varieties of pseudoboolean algebras. In *11th Soviet Algebraic Colloquium, Abstracts*, (pp. 255–256), Kishinev (Russian).

Maksimova, L. (2000). Strongly decidable properties of modal and intuitionistic calculi. *Logic Journal of the IGPL*, *8*(6), 797–819.

Maksimova, L. (2002). Complexity of interpolation and related properties in positive calculi. *Journal of Symbolic Logic*, *67*(1), 397–408.

Maksimova, L. (2003). Complexity of some problems in positive and related calculi. *Theoretical Computer Science*, *303*(1), 171–185.

Maksimova, L. L. (1971). Quasi-finite superintuitionistic logics. In *11th Soviet Algebraic Colloquium, Abstracts* (pp. 258–259), Kishinev (Russian).

Maksimova, L. L. (1972). Pretabular superintuitionistic logics. *Algebra and Logic*, *11*, 308–314.

Maksimova, L. L. (1975). Pretabular extensions of Lewis *S*4. *Algebra and Logic*, *14*, 16–33.

Maksimova, L. L. (2007). A method of proving interpolation in paraconsistent extensions of the minimal logic. *Algebra and Logic*, *46*(5), 341–353.

Maksimova, L. L., & Rybakov, V. V. (1974). Lattices of modal logics. *Algebra and Logic, 13*, 105–122.
Maksimova, L. L., & Schreiner, P. A. (2006). Algorithms for recognizing tabularity and pretabularity in extensions of the intuitionistic calculus. *Vestnik Novosibirskogo Gosudarstvennogo Universiteta. Seriya Matematika, Mekhanika, Informatika, 6*(2), 29–38.
Maksimova, L. L., & Yun, V. F. (2015). Recognizable logics. *Algebra and logic, 54*.
Maksimova, L., & Voronkov, A. (2003). Complexity of some problems in modal and intuitionistic calculi. In M. Baaz & J. A. Makowsky (Eds.), *Computer science logic 2003*, Springer. *Proceedings of 17th International Workshop, CSL 2003, 12th Annual Conference of the EACSL, and 8th Kurt Gödel Colloquium, KGC 2003*, Vienna, Austria, August 25–30, 2003.
McKinsey, J., & Tarski, A. (1948). Some theorems about the sentential calculi of Lewis and Heyting. *Journal of Symbolic Logic, 13*(1), 1–15.
Meskhi, V. Y., & Esakia, L. L. (1974). On five critical modal systems. In *Theory of logical derivation. Abstracts of reports of USSR symposium, I*, (pp. 76–79), Moscow.
Odintsov, S. (2008). *Constructive negations and paraconsistency, Trends in Logic* (Vol. 26). Dordrecht: Springer.
Papadimitriou, C. H. (1994). *Computational complexity*. Reading: Addison-Wesley.
Rautenberg, W. (1977). Der Verband der normalen verzweigten Modallogiken. *Mathematische Zeitschrift, 156*, 123–140.
Rautenberg, W. (1979). *Klassische und nichtklassische Aussagenlogik*. Braunschweig-Wiesbaden: Vieweg.
Rybakov, V. V. (1997). *Admissibility of logical inference rules*. Amsterdam: Elsevier.
Scroggs, S. J. (1951). Extensions of the Lewis system *S5*. *Journal of Symbolic Logic, 16*, 112–120.
Segerberg, K. (1968). Propositional logics related to Heyting's and Johansson's. *Theoria, 34*, 26–61.
Swirydowicz, K. (2008). There exists an uncountable set of pretabular extensions of the relevant logic R and each logic of this set is generated by a variety of finite height. *Journal of Symbolic Logic, 73*(4), 1249–1270.
Umezawa, T. (1959). On intermediate propositional logics. *Journal of Symbolic Logic, 24*, 20–36.
Verhozina, M. I. (1978). Intermediate positive logics. *Algorithmic problems of algebraic systems* (pp. 13–25). Irkutsk: Irkutsk State University.

The Story of γ

Alasdair Urquhart

Abstract This paper recounts the history and solution of the problem of admissibility of the rule γ in the context of **E** and other relevance logics.

Keywords Admissibility of γ · Algebraic semantics · Metavaluation · Relational semantics · Relevance logic

1 The Origins of the Problem

The problem of the admissibility of the rule γ in the system **E** of entailment may appear to be a purely technical problem; but as we shall see, it is an intriguing and challenging problem that makes contact with a remarkable variety of other, seemingly distant, problems of logical interest.

The problem arose from Anderson and Belnap's early investigations into Wilhelm Ackermann's system (Ackermann 1956) of "Strenge Implikation." Ackermann includes four rules of inference labeled α, β, γ and δ. The rule γ is as follows: From $\vdash \neg A \vee B$ and $\vdash A$ to infer $\vdash B$. This can be described as *modus ponens* for the material conditional.

Anderson and Belnap defined the system **E** by excising the third rule, γ. Their reasons for the excision are described by Anderson in his paper of 1962 as follows:

> Candor compels me to admit, again in the interests of historical accuracy, that one of the principal reasons for dropping this rule (which I shall hereafter refer to as "the disjunctive syllogism") was that in the presence of this primitive rule, almost none of the arguments in the papers of Belnap and myself, cited above, can be carried through; or so it seems. (Anderson 1963, p. 10)

A. Urquhart (✉)
Departments of Philosophy and Computer Science, University of Toronto, Toronto, Canada
e-mail: urquhart@cs.toronto.edu
URL: http://www.utoronto.academia.edu/AlasdairUrquhart

K. Bimbó (ed.), *J. Michael Dunn on Information Based Logics*,
Outstanding Contributions to Logic 8, DOI 10.1007/978-3-319-29300-4_6

Thus the initial reason for the removal of Ackermann's primitive rule was purely pragmatic—it was hard to prove results about the system in the presence of this rule. However, Anderson offers a second, more philosophically respectable reason:

> Moreover, since $(\overline{A} \vee B)\, A \rightarrow B$ is obviously not a valid entailment (it leads by distribution to $(\overline{A}A \vee BA) \rightarrow B$, thence by the valid entailment $\overline{A}A \rightarrow (\overline{A}A \vee BA)$, with the help of transitivity, to $\overline{A}A \rightarrow B$), dropping the disjunctive syllogism as a primitive rule lends a certain coherence to Ackermann's system which it otherwise lacks. (*ibid.*)

This leads Anderson to the first of the open problems listed in his paper of 1962:

> So the first problem I would like to pose is the following: is it true in E that whenever $\vdash A$ and $\vdash \overline{A} \vee B$, we also have $\vdash B$ (where $\vdash$ stands for provability in the system E, here and in what follows)?

He goes on to give an example where there are proofs in **E** of both A and $\neg A \vee B$; although there is in fact a proof of B, it is not related in any obvious way to the first two proofs. With characteristic humour, Anderson remarks that

> If you *tell* us (truly) that A and $\overline{A} \vee B$ are provable, then most likely Belnap or I can go off and find you a proof of B; but what principle is involved? (Belnap has suggested that the principle is "hard work and a clean life"; but this must be laid to his Protestant upbringing.)

This problem of the admissibility of γ remained open until the breakthrough of Meyer and Dunn described in Sect. 2. We give the solution for the system **E**, as this was the problem as originally posed. In the following sections, we discuss the problem in the context of the system **R**, the setting for most of the later work. However, it should be pointed out that these techniques are applicable to a much larger family of systems. Our concern here, though, is with the basic techniques; the reader can refer to the work of Meyer, Dunn and others to get an idea of the generality of these constructions.

The algebras corresponding to the logic **R** are the De Morgan monoids; their theory was first expounded in Dunn's thesis (Dunn 1966). De Morgan monoids play exactly the same role in the logic **R** as Boolean algebras in classical logic. For background in these algebras, the reader can consult (Anderson and Belnap 1975) or (Dunn and Restall 2002).

2 The First Solution by Meyer and Dunn

I have a vivid recollection of the arrival of the news at the Philosophy Department in Pittsburgh that Meyer and Dunn had solved the problem. Anderson and Belnap were understandably excited that the first of the open problems had finally been solved! (Both Meyer and Dunn had left Pittsburgh when I arrived there as a graduate student in 1967, but I was able to count them among my friends a few years later.)

The heart of the Meyer–Dunn solution of 1968 (Meyer and Dunn 1969) is an algebraic construction, sketched below. Before we describe this, though, we need some preliminary remarks and results. We take the system **E** to be formulated as in Anderson and Belnap (1975), with the primitive connectives $\wedge$, $\vee$, $\neg$ and $\rightarrow$. An **E**-theory is a collection of formulas of **E** containing all the axioms of **E** and closed under its rules (adjunction and *modus ponens*). If T is an **E**-theory, we write $\vdash_T A$ if $A \in T$. An **E**-theory is *prime* if whenever $\vdash_T A \vee B$, either $\vdash_T A$ or $\vdash_T B$. It is *consistent* if there is no formula A such that $\vdash_T A$ and $\vdash_T \neg A$, and *normal* if it is both prime and consistent.

Lemma 2.1 *For every nontheorem B of* **E**, *there is a prime* **E***-theory T such that not $\vdash_T B$.*

The proof of this lemma is by a standard inductive construction (Meyer and Dunn 1969, p. 462).

Let us see if we can use this lemma to prove γ admissible. So, assume that $\vdash_{\mathbf{E}} A$, $\vdash_{\mathbf{E}} \neg A \vee B$, but not $\vdash_{\mathbf{E}} B$. Then according to the lemma, there is a prime **E**-theory T so that $\vdash_T A, \vdash_T \neg A \vee B$ but not $\vdash_T B$. Since T is prime, either $\vdash_T \neg A$ or $\vdash_T B$. Now if we could be sure that the first disjunct is false, then a contradiction ensues immediately, proving γ. But there is no guarantee that $\nvdash_T \neg A$, since the logic **E** allows of the possibility of non-trivial inconsistent theories (one of the sources of paraconsistent logic!). This is where the real hard work begins; we have to improve Lemma 2.1 by replacing "prime" by "normal."

Meyer and Dunn's strategy proceeds by an algebraic construction. First, they show how to construct an **E**-*algebra* from an **E**-theory T; the elements of the algebra are the formulas of T, the *designated elements* of the algebra are the theorems of T. This is essentially the familiar Lindenbaum construction, where we can think of the **E**-algebra as a big multi-valued truth-table, designed to validate all the theorems of **E**.

An **E**-algebra is *prime* if whenever $a \vee b$ is designated, then one of a or b is designated. If we start from a prime **E**-theory, then the resulting **E**-algebra is prime. Thus Lemma 2.1 can be rephrased as follows: A formula is provable in **E** if and only if it is valid in all prime **E**-algebras (that is to say, it takes a designated value for any assignment of values in a prime **E**-algebra).

We now approach the heart of the construction. Meyer and Dunn start from an arbitrary prime **E**-algebra M; their aim is to "normalize" the algebra M so that the resulting normal **E**-algebra M^* can be used to invalidate any formula that is invalidated by M. The construction is fairly involved—rather than describe it in detail, I shall show how it works in a special case.

Our **E**-algebra M is a De Morgan monoid with four elements; the distributive lattice structure involving $\wedge$ and $\vee$ is represented in the left-hand diagram of Fig. 1. The reader will find other pictures of the construction in the paper by Meyer et al. (1974, p. 107). The operation of $\rightarrow$ in M is given in the accompanying table of the operations of M^*; the $\rightarrow$ of M is simply that of M^*, restricted to the elements of M.

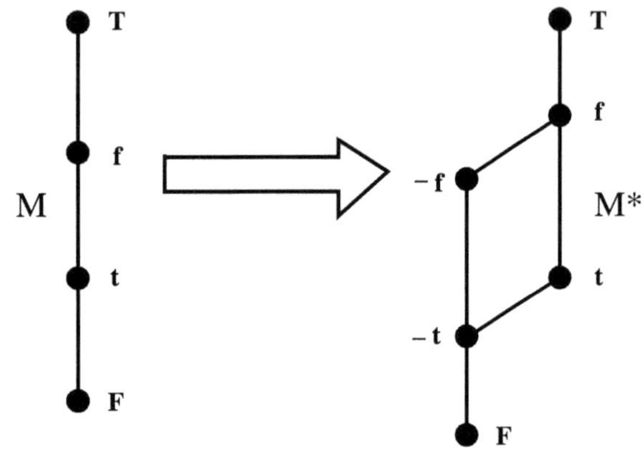

Fig. 1 The Meyer–Dunn construction

$\to$	T	f	t	F	$-f$	$-t$	$\neg$
T	T	F	F	F	F	F	F
f	T	t	F	F	$-t$	F	$-t$
t	T	f	t	F	$-f$	$-t$	$-f$
F	T	T	T	T	T	T	T
$-f$	T	t	F	F	t	F	t
$-t$	T	f	t	F	f	t	f

Negation in M is defined by the stipulation that T and F are each other's negations, as are t and f. The designated elements are those $\geq t$. Thus M is a prime algebra that validates all theorems of **R**, but is not normal, since t and $f = \neg t$ are both designated. It is these elements, dubbed *neuter* elements by Meyer and Dunn, that form the obstacle to normality; let us designate this set of neuter elements by N. Their solution to defining a normal algebra M^*, starting from M, is to *split* the elements in N. That is to say, for each such neuter element a in the original algebra M, we add a new element $-a$.

We shall not describe the somewhat complicated prescription that defines the operations on M^*; for our example, they are given in the accompanying table. The designated elements remain as before; clearly the new algebra is normal, thanks to the splitting procedure. Since the new algebra M^* invalidates any formula invalidated by M, the admissibility of γ follows.

3 The History of the Meyer–Dunn Proof

The splitting construction defined in the preceding section is subtle and complicated, and one may well wonder how Dunn and Meyer discovered it. In this section, we give an outline of the road that they followed. The path of discovery illustrates the methodological maxim that to solve a difficult problem, it helps to start from a simpler case.

The simpler context here is the logic **R-mingle**, which results from **R** by adding the axiom schema $A \to (A \to A)$. Certain matrices, the Sugihara matrices play an important role in the model theory of **R-mingle**. Following (Dunn 1970), we define a *Sugihara matrix* as a matrix defined on a chain with a bijective order-inverting mapping $a \mapsto \overline{a}$ defined on it; the operations are defined by $a \wedge b = \min(a, b)$, $a \vee b = \max(a, b)$ and $a \to b = \overline{a} \vee b$ if $a \leq b$, and otherwise $a \to b = \overline{a} \wedge b$. The designated elements are those satisfying the inequality $\overline{a} \leq a$.

The notion just defined is a generalization of the matrix from Sugihara (1955); this is the special case where the chain consists of all the non-zero integers. Certain finite matrices are particularly important in the model theory of **R-mingle**. The matrices S_n are the Sugihara matrices where the underlying chain consists of the non-zero integers from $-n$ to n; the matrices $S_n + 0$ are the Sugihara matrices defined on all the integers from $-n$ to n. Note that in the matrix $S_n + 0, \overline{0} = 0$, so that 0 is a neuter element, in the terminology of the preceding section.

Robert K. Meyer proved a completeness result (Meyer 1971b) for **R-mingle** with respect to the matrices S_n; specifically, he proved that if a formula contains at most n propositional variables, then it is a theorem of **R-mingle** if and only if it is valid in S_n. He remarked that since the matrices S_n are normal, this completeness result shows that γ holds for **R-mingle**.

The paper (Dunn 1970) gives a simpler proof of Meyer's completeness result by showing that if a formula is invalid in $S_n + 0$, then it is invalid in S_{n+1} by splitting the zero element. This is the first appearance of the splitting construction. Dunn remarks that

> Since Meyer had to do some hard, honest work to restrict his completeness result to the consistent Sugihara matrices S_n, it seems remarkable that we were able to get the same restriction by the simple expedient of "splitting" the "inconsistent" element 0 *via* our Theorem 4 into a "true half" and a "false half." Meyer has recently seen a way of generalizing this technique, and he and I use it in [7] [= Meyer and Dunn (1968)], [8] [= Meyer and Dunn (1969)] to show that γ is admissible for a number of relevant sentential calculi, including Anderson and Belnap's systems E and R. (Dunn 1970, pp. 8–9).

In an email of 23 March 2015 to me, Dunn provided some delightful reminiscences of the discovery of the first proof of the admissibility of γ.

> (1) Bob saw how his completeness theorem showed the admissibility of γ for **R-mingle**. The key thing here is that it is relative to the Sugihara matrices S_n, which do not contain 0. This is somewhat tricky since the natural completeness proof leads to Sugihara matrices which may contain 0, which is its own negation.
>
> (2) I saw a way to make this tricky step easier by the technique of "zero-splitting."
>
> (3) Meyer saw how to generalize this to the case where the matrix has more than one "inconsistent" element, where there is more than one element where both it and its negation are designated.
>
> I remember Bob calling early one evening when I was visiting at Yale and he was at Bryn Mawr, saying he had this insight and that he would drive up and be at our apartment in a few hours. He came and of course we stayed up late into the night pinning down the details of the construction of the expanded matrix containing the "splits" of all the inconsistent elements.

Bob never published anything but the abstract (Meyer 1971b) of his completeness of **R-mingle** until Sect. 29.3 of *Entailment I*.

It should be mentioned in conclusion that the logic **R-mingle**, a crucial testing ground for the splitting construction, was devised by Storrs McCall and J.M. Dunn when they were experimenting with the mingle rule of Ohnishi and Matsumoto in the context of a sequent calculus. A great deal of detailed information on this interesting system is to be found in Anderson and Belnap (1975).

4 The Second Solution by Routley and Meyer

The second solution was a spinoff of the relational semantics developed in the early 1970s by Routley and Meyer. We give a sketch of their solution for the case of **R** here. In spite of initial appearances, it is closely linked to the Meyer–Dunn solution described above, as we explain in the latter half of this section.

Routley and Meyer (1973) define a *relevant model structure*, or *r.m.s.* for short, as a quadruple $\langle 0, K, R, {}^* \rangle$ where K is a set, $0 \in K$, R is a ternary relation on K, and * is a one-place operation on K, satisfying the postulates:

1. $R0aa$,
2. $Raaa$,
3. $\exists x\,(Rabx \wedge Rxcd) \Rightarrow \exists y\,(Racy \wedge Rybd)$,
4. $(R0da \wedge Rabc) \Rightarrow Rdbc$,
5. $Rabc \Rightarrow Rac^*b^*$,
6. $a^{**} = a$,

for $a, b, c, d \in K$.

If we define $a \leq b$ as $R0ab$, then it is not hard to show that the relation $\leq$ is reflexive and transitive—in fact, we can assume in addition that it is a partial ordering, though this is not necessary for soundness. A subset S of K is *increasing* if it satisfies the condition: $(a \in S \wedge a \leq b) \Rightarrow b \in S$.

A *valuation* in an r.m.s. assigns an increasing subset $\Phi(P) \subseteq K$ to each propositional variable P. Given a valuation in an r.m.s., the forcing relation $\models$ for elements of K and formulas of **R** is defined by:

1. $a \models P \;\Leftrightarrow\; a \in \Phi(P)$,
2. $a \models A \wedge B \;\Leftrightarrow\; a \models A$ and $a \models B$,
3. $a \models A \vee B \;\Leftrightarrow\; a \models A$ or $a \models B$,
4. $a \models A \rightarrow B \;\Leftrightarrow\; \forall bc\,((b \models A \wedge Rabc) \Rightarrow c \models B)$,
5. $a \models \neg A \;\Leftrightarrow\; a^* \not\models A$.

A formula A is *valid* in an r.m.s. if $0 \models A$ for all valuations in the r.m.s. The main result of the 1973 paper (Routley and Meyer 1973) is a completeness proof for **R** relative to this semantics, proved by a canonical model construction.

Theorem 4.1 *A formula is a theorem of* **R** *if and only if it is valid in all relevant model structures.*

Our interest here, though, is in the simple proof of the admissibility of γ that the model theory permits. We define an r.m.s. to be *normal* if it satisfies the condition $0 = 0^*$. In a normal r.m.s., given a valuation, there can be no formula A so that $0 \models A$ and $0 \models \neg A$; thus the theory $\{ A : 0 \models A \}$ is normal. It follows that we can show γ admissible, provided we can strengthen Theorem 4.1 to completeness with respect to all normal relevant model structures.

The strategy of Routley and Meyer (1973, Sect. 8) is very similar to that of the first solution: starting from a non-normal model, we enlarge it to a normal model that continues to invalidate a given formula. If $\mathcal{M} = \langle 0, K, R, ^* \rangle$ is an r.m.s., then its *normalization* is $\mathcal{M}' = \langle 0', K', R', ^{*'} \rangle$, where $K' = K \cup \{ 0' \}$, $^{*'}$ is the extension of * defined by adding $0'^{*'} = 0'$, and R' is given by the definition:

1. $R'0'0'0'$,
2. $R'0'0'a \;\Leftrightarrow\; R00a$,
3. $(R'0'a0' \wedge R'a0'0') \;\Leftrightarrow\; R0a0^*$,
4. $R'ab0' \;\Leftrightarrow\; Rab0^*$,
5. $(R'0'ab \wedge R'a0'b) \;\Leftrightarrow\; R0ab$,
6. $R'abc \;\Leftrightarrow\; Rabc$,

where $a, b, c \in K$.

As an example of the construction, we show here an r.m.s. $\mathcal{M}' = \langle 0', K', R', ^{*'} \rangle$ in tabular form. The table is to be read as follows: given elements $a, b, c \in K'$, $Rabc$ holds if and only if c is in the set appearing in the a, b entry in the table (with the a elements on the left, the b elements along the top). We can also read the table as a multi-algebra, as explained in my paper (Urquhart 1996).

$\circ$	2	1	0	$0'$	*
2	$\{0, 0', 1, 2\}$	$\{0, 0', 1, 2\}$	$\{0, 0', 1, 2\}$	$\{0, 0', 1, 2\}$	0
1	$\{0, 0', 1, 2\}$	$\{0, 0', 1, 2\}$	$\{0, 1\}$	$\{0, 1\}$	1
0	$\{0, 0', 1, 2\}$	$\{0, 1\}$	$\{0\}$	$\{0\}$	2
$0'$	$\{0, 0', 1, 2\}$	$\{0, 1\}$	$\{0\}$	$\{0, 0'\}$	$0'$

The original r.m.s. from which $\mathcal{M}'$ was derived can be recovered from the table by the following process: first, delete the second-to-last column (headed by $0'$) and the bottom row of the table, second, delete $0'$ from all of the entries in the table.

It is fairly straightforward to verify that the normalized structure is in fact an r.m.s. Furthermore, if a formula A is invalidated by a valuation Φ in $\langle 0, K, R, ^* \rangle$, then Φ can be extended to a valuation Φ' on $\langle 0', K', R', ^{*'} \rangle$ by setting $\Phi'(0') = \Phi(0)$, so that A is invalidated there as well. This shows the admissibility of γ.

Although this proof looks quite different from the proof in Sect. 2, it is in essence the same proof, viewed through the lens of an algebraic duality theory.

Let $\mathcal{M} = \langle 0, K, R, ^* \rangle$ be an r.m.s. Then, starting from $\mathcal{M}$, we can define an algebra $\mathcal{A}(\mathcal{M})$ by the following prescription:

1. The universe of $\mathcal{A}(\mathcal{M})$ is the set of all increasing subsets of K,
2. $A \wedge B = A \cap B$,
3. $A \vee B = A \cup B$,
4. $A \circ B = \{ c\colon \exists ab\, (Rabc \wedge a \in A \wedge b \in B) \}$,
5. $e = \{ a\colon R00a \}$
6. $A \to B = \{ a\colon \forall bc\, ((Rabc \wedge b \in A) \Rightarrow c \in B) \}$,
7. $\neg A = K \setminus \{ a\colon a^* \in A \}$.

This definition is of course just an algebraic reformulation of the definition of the forcing relation in an r.m.s. given above. It is fairly easy to check that the resulting algebra is a De Morgan monoid—in fact, this is just the usual soundness argument for **R** relative to an r.m.s. The element e is the identity with respect to the monoid operation $\circ$, and serves to interpret the propositional constant t.

If $\mathcal{M} = \langle 0, K, R, ^* \rangle$ is an r.m.s., then starting from the De Morgan monoid $\mathcal{A}(\mathcal{M})$, we can define the De Morgan monoid $\mathcal{A}(\mathcal{M})^*$ following the construction of Sect. 2. We can also form the r.m.s. $\mathcal{M}' = \langle 0', K', R', ^{*'} \rangle$ following the Routley–Meyer method above. The following theorem shows that these are essentially the same construction.

Theorem 4.2 *Let $\mathcal{M}$ be an r.m.s. and $\mathcal{M}'$ the r.m.s. defined from it by the method of Routley and Meyer. Then the De Morgan monoids $\mathcal{A}(\mathcal{M})^*$ and $\mathcal{A}(\mathcal{M}')$ are isomorphic.*

Proof Define a mapping φ on $\mathcal{A}(\mathcal{M})^*$ to $\mathcal{A}(\mathcal{M}')$ as follows: If $A \in \mathcal{A}(\mathcal{M})$, then $\varphi(A) = A \cup \{0'\}$ if $0 \in A$; otherwise, $\varphi(A) = A$. Recall from Sect. 2 that the set of neuter elements in $\mathcal{A}(\mathcal{M})$ is denoted by N, while the set of added new elements in $\mathcal{A}(\mathcal{M})^*$ is written as $-N$. If $-A \in -N$, then $\varphi(-A) = A$. The proof that φ is an isomorphism from $\mathcal{A}(\mathcal{M})^*$ onto $\mathcal{A}(\mathcal{M}')$ is a rather lengthy case analysis and is omitted here. □

The r.m.s. described above in tabular form provides an illustration of Theorem 4.2. The reader is invited to check that if we compute the algebras $\mathcal{A}(\mathcal{M})$ and $\mathcal{A}(\mathcal{M}')$, then they are isomorphic to the algebras used as an example in Sect. 2.

The theory of De Morgan monoids admits an algebraic duality theory. A brief exposition of this theory is given in Urquhart (1996); for more extensive expositions, the reader is referred to the monographs Dunn and Hardegree (2001) and Bimbó and Dunn (2008). An upshot of this duality theory is that for every construction and theorem about De Morgan monoids, there are corresponding dual constructions and theorems about relevant model structures. Theorem 4.2 shows that the Meyer–Dunn construction and the later Routley–Meyer construction are each other's duals.

This last remark should be qualified—the earlier algebraic construction of Meyer and Dunn is somewhat more general, since it applies to arbitrary De Morgan monoids, whereas the algebras arising from model structures by the construction above are rather special, being *complete De Morgan monoids* (Routley and Meyer 1973, Sect. 12). However, this difference is not too significant in this context, since any De Morgan monoid can be embedded in a complete De Morgan monoid (Routley and Meyer 1973, Corollary 9.2).

5 Third Solution: Metavaluations

Our third and last proof of the admissibility of γ is the simplest and easiest of the three. It is due to Meyer (1976a), and has its origins in his work on metavaluations and metacompleteness (Meyer 1976b). Here, however, we shall present it as derived from the proof of the previous section.

Our goal, as before, is to prove the improved version of Lemma 2.1 where the adjective "prime" is replaced by "normal." As in Sect. 2, we start from a prime **R**-theory T that fails to contain a given formula B. Then by the basic model construction of Routley and Meyer (1973, Sect. 7), there is an r.m.s. $\mathcal{M} = \langle 0, K, R, {}^* \rangle$ and a valuation Φ in $\mathcal{M}$ so that $T = \{ A \colon 0 \models A \}$. We can now form the normalization $\mathcal{M}'$ as in Sect. 4, and extend the valuation Φ to a valuation Φ' on $\mathcal{M}'$, thus obtaining a normal subtheory of T.

The formulas that are forced by the new zero element $0'$ in $\mathcal{M}'$ are characterized in terms of T in the following lemma. We shall employ the notation $T \mid A$ as a synonym for $0' \models A$.

Lemma 5.1 *The relation $T \mid A$ is characterized by the following conditions.*

1. $T \mid P$ *if and only if* $T \vdash P$, *for P atomic;*
2. $T \mid A \wedge B$ *if and only if* $T \mid A$ *and* $T \mid B$;
3. $T \mid A \vee B$ *if and only if* $T \mid A$ *or* $T \mid B$;
4. $T \mid A \to B$ *if and only if* $T \vdash A \to B$ *and* ($T \nmid A$ *or* $T \mid B$);
5. $T \mid \neg A$ *if and only if* $T \vdash \neg A$ *and* $T \nmid A$.

Proof The first condition holds by definition, and the second and third are easy to verify.

For the fourth condition, if $0' \models A \to B$, then since $0' \leq' 0, 0 \models A \to B$, and since $R'0'0'0'$, either $0' \not\models A$, or $0' \models B$. For the converse, assume that $0 \models A \to B$, and either $0' \not\models A$, or $0' \models B$. In addition, assume that $R'0'xy$ and $x \models A$; we aim to show that $y \models B$. Four cases arise. If $x = y = 0'$, then $y \models B$, by the second condition. If $x = 0'$ and $y \in K$, then $R00y$; since $0' \leq' 0$, $0 \models A$, hence $y \models B$. If $x \in K$ and $y = 0'$, such that $R'0'x0'$, then $R0x0^*$, hence $0^* \models B$ and $y = 0' \models B$, since $0^* \leq' 0'$. Finally, if $x, y \in K$, then $R0xy$, so that $y \models B$.

For the fifth condition, assume first that $0' \models \neg A$. Then $0 \models \neg A$, since $0' \leq' 0$, and $0' \not\models A$, since $0'^* = 0'$. Conversely, if $0' \not\models A$, then $0' \models \neg A$, since $0'^* = 0'$. □

In Lemma 5.1, we have presented the characterization of the relation $T \mid A$ as a consequence of the Routley–Meyer construction of Sect. 4. However, we can just as well consider the relation as a primitive concept, defined by the conditions of the lemma. Viewed in this light, the relation corresponds exactly to the concept of a *canonical quasi-valuation* defined by Meyer (1976b, pp. 505–506).

Theorem 5.2 *Let T be a prime* **R***-theory, and $T \mid A$ be a relation satisfying the conditions of Lemma 5.1. Define $T|$ as $\{ A : T \mid A \}$. Then $T|$ is an* **R***-theory that is a normal subtheory of T.*

Proof First, we establish by induction on the complexity of the formula A that if $T \mid A$, then $T \vdash A$, and if $T \nmid A$, then $T \vdash \neg A$. (The second implication is only required for theorems of **R** that are not intuitionistically valid; the first alone is sufficient for theorems of intuitionistic logic.)

The proof that $T|$ is an **R**-theory is a straightforward exercise; for some details, the reader can consult Dunn and Restall's survey (Dunn and Restall 2002, Sect. 2.4). □

Theorem 5.2 provides us with our third proof of the admissibility of γ. We have presented it as derived from the proof of Sect. 4, but this is not in fact the way in which Meyer discovered the basic ideas. They arose from the philosophical ideas of coherence and metacompleteness, beginning in the early 1970s (Meyer 1971a, 1976b). The methodological problem that he was originally addressing in these papers was that of interpreting a logic in its own metatheory. It was only later that he realized that these technical developments could be used to provide a new proof of the admissibility of γ, as he reported in the addendum to his paper *Metacompleteness*, written in 1971, but only published in 1976 (Meyer 1976b, pp. 514–515).

Readers who are familiar with the metatheory of intuitionistic logic may have already recognized the relation $T \mid A$ characterized in Lemma 5.1—it is formally identical with the relation defined by Peter Aczel in his well-known 1968 paper (Aczel 1968), a relation now dubbed the "Aczel slash." Aczel employed it to prove the disjunction and existence properties for intuitionistic theories. He derived it from a model-theoretic proof of these properties that involves adding a new bottom element to a Kripke model for intuitionistic predicate calculus; our proof of Lemma 5.1 above mimics Aczel's proof.

Meyer was in fact aware of these connections with intuitionistic logic (Meyer, 1976b, p. 501); he specifically mentions the work of Harrop (1956, 1960), Rasiowa and Kleene (1962), that Saul Kripke brought to his attention. The Aczel slash is similar to "Kleene's slash" defined in Kleene (1962), but is not identical with it.

This brings to a close our survey of the various proofs of the admissibility of γ, and the evolution of ideas about the problem. Not only are the various solutions more closely related than appears at first sight, but they also make interesting contacts with neighbouring areas such as the metatheory of intuitionistic logic. In conclusion, we should mention that Dunn and Meyer (1989) have described a fourth solution due to Saul Kripke in 1978, using the ideas of the usual semantical proofs of cut elimination. This proof does not seem to have appeared in print.

6 Cut Elimination and an Open Problem

Kripke's proof, mentioned at the end of the last section, provides a hint that the problem of the admissibility of γ may have a close connection with classical proofs of cut elimination. Dunn and Meyer were the first to spell out this connection in their paper of 1989 (Dunn and Meyer 1989). They showed how the coherence techniques used to prove the admissibility of γ could be adapted directly to demonstrate cut elimination for a formulation of classical quantificational logic due to Schütte. This result is foreshadowed in the remark of Meyer et al. (1974, p. 120) that "the cut theorem ... is for classical theories simply γ in peculiar notation." Meyer went on in a paper of 1976 (Meyer 1976a) to extend these methods to higher-order logics.

I recall that as a graduate student in Pittsburgh, in conversation with Nuel Belnap, I raised an objection to the proof of Meyer and Dunn by saying that it was non-constructive. Nuel, in his usual sweetly reasonable manner, replied that it was nothing of the kind. If we have proofs of A and $\neg A \vee B$ in **E**, he remarked, then all we have to do is to enumerate all the proofs in the system; we know, thanks to the work of Meyer and Dunn, that we shall eventually hit on a proof of B!

This reply to my objection, is of course impeccable. Nevertheless, there is still a question that underlies my original dissatisfaction with the Meyer–Dunn proof. As Anderson observed in his original article on open problems, the proof of B in **E** or **R** may not seem to have any obvious relation to the proofs of $\neg A \vee B$ or A. We can perhaps gain some understanding of this situation if we look at the classical proofs of cut elimination.

The original constructions by Gentzen showing that the cut rule is unnecessary in classical logic are not simple transformations of the original proof that make a few local modifications. Rather, after the cut elimination procedure terminates, the entire proof has been reworked and reorganized in a radical manner. This fact is dramatized in the speedup phenomenon in classical quantification theory; elimination of the cut rule may result in a huge, and unavoidable, increase in the size of the proof. Define

$$2^{(0)} = 1; \qquad 2^{(n+1)} = 2^{2^{(n)}}.$$

Then we have the following speedup phenomenon for proofs with cut in classical logic over proofs without cut; the original speedup theorem was proved by Statman (1978).

Theorem 6.1 *There is a sequence of valid sentences of classical predicate logic* $\psi_1, \psi_2, \ldots$ *such that* ψ_n *has a proof of size* $p(n)$, $n = 1, 2, \ldots,$ *where* p *is a fixed polynomial, but there is no cut-free proof of* ψ_n *with less than* $2^{(n)}$ *proof lines for* $n = 1, 2, \ldots$.

Proof A very clear proof of this result can be found in Pudlák (1998). □

This result, together with the earlier observations of Dunn and Meyer about the relation between γ and cut elimination, suggests an interesting open problem.

Probelm 6.1 Can we prove a speedup theorem for the relevant quantificational logic **RQ** with γ, over **RQ** without γ, with respect to theorems of classical logic?

If we could prove a speedup comparable to that for classical logic, this would confirm the remarks I made earlier that the elimination of γ requires a radical reworking of proofs.

References

Ackermann, W. (1956). Begründung einer strengen Implikation. *Journal of Symbolic Logic*, *21*, 113–128.
Aczel, P. (1968). Saturated intuitionistic theories, Contributions to Mathematical Logic. *Proceedings of the Logic Colloquium*, Hannover 1966 (pp. 1–11), Vol. 50 of Studies in Logic and the Foundations of Mathematics, North Holland.
Anderson, A. R. (1963). Some open problems concerning the system E of entailment. *Proceedings of a Colloquium on Modal and Many-Valued Logics*, Helsinki. Retrieved August 23–26, 1962, number Fasc. 16. in Acta Philosophica Fennica, pp. 9–18.
Anderson, A. R., & Belnap, N. D. (1975). *Entailment: The logic of relevance and necessity* (Vol. I). Princeton, NJ: Princeton University Press.
Bimbó, K., & Dunn, J. M. (2008). *Generalized Galois Logics. Relational Semantics of Nonclassical Logical Calculi*, Vol. 188 of CSLI Lecture Notes. Stanford, CA: CSLI Publications.
Dunn, J. M. (1966). The Algebra of Intensional Logics, PhD thesis, University of Pittsburgh, Ann Arbor (UMI).
Dunn, J. M. (1970). Algebraic completeness results for R-Mingle and its extensions. *Journal of Symbolic Logic*, *35*, 1–13.
Dunn, J. M., & Hardegree, G. M. (2001). *Algebraic methods in philosophical logic*, Vol. 41 of Oxford Logic Guides. Oxford, UK: Oxford University Press.
Dunn, J. M., & Meyer, R. K. (1989). Gentzen's cut and Ackermann's gamma. In J. Norman & R. Sylvan (Eds.), *Directions in Relevant Logic* (pp. 229–240). Dordrecht: Kluwer.
Dunn, J. M., & Restall, G. (2002). Relevance logic. In D. Gabbay & F. Guenthner (Eds.), *Handbook of Philosophical Logic* (2nd ed., Vol. 6, pp. 1–128). Amsterdam: Kluwer.
Harrop, R. (1956). On disjunctions and existential statements in intuitionistic systems of logic. *Mathematische Annalen*, *132*, 342–361.
Harrop, R. (1960). Concerning formulas of the types $A \rightarrow B \vee C$, $A \rightarrow (\exists x)Bx$ in intuitionistic formal systems. *Journal of Symbolic Logic*, *25*, 27–32.
Kleene, S. C. (1962). Disjunction and existence under implication in elementary intuitionistic formalisms. *Journal of Symbolic Logic*, *27*, 11–17.
Meyer, R. K. (1971a). On coherence in modal logics. *Logique et Analyse*, *14*, 658–668.
Meyer, R. K. (1971b). R-mingle and relevant disjunction, (abstract). *Journal of Symbolic Logic*, *36*, 366.
Meyer, R. K. (1976a). Ackermann, Takeuti and Schnitt: γ for higher-order relevant logics. *Bulletin of the Section of Logic*, 138–144.
Meyer, R. K. (1976b). Metacompleteness. *Notre Dame Journal of Formal Logic*, *17*(4), 501–516.
Meyer, R. K., & Dunn, J. M. (1968). Entailment logics and material implication, (abstract). *Notices of the American Mathematical Society*, *15*, 1021–1022.
Meyer, R. K., & Dunn, J. M. (1969). E, R and γ. *Journal of Symbolic Logic*, *34*, 460–474.
Meyer, R. K., Dunn, J. M., & Leblanc, H. (1974). Completeness of relevant quantification theories. *Notre Dame Journal of Formal Logic*, *15*(1), 97–121.
Pudlák, P. (1998). The lengths of proofs. In S. R. Buss (Ed.), *Handbook of Proof Theory* (pp. 547–637). Amsterdam: North Holland.

Routley, R., & Meyer, R. K. (1973). The semantics of entailment. In H. Leblanc (Ed.), Truth, Syntax and Modality. Proceedings of the Temple University Conference on Alternative Semantics, (pp. 199–243). Amsterdam: North-Holland.

Statman, R. (1978). Bounds for proof-search and speed-up in the predicate calculus. *Annals of Mathematical Logic*, *15*, 225–287.

Sugihara, T. (1955). Strict implication free from implicational paradoxes. *Memoirs of the Faculty of Liberal Arts, Fukui University, Series*, *1*(4), 55–59.

Urquhart, A. (1996). Duality for algebras of relevant logics. *Studia Logica*, *56*, 263–276.

Manipulating Sources of Information: Towards an Interpretation of Linear Logic and Strong Relevance Logic

Edwin Mares

Abstract Relevance Logics are interpreted in terms of agents' comprehending and constructing sources of information. The rules governing these constructions are formulated in a natural deduction system. Two different sorts of interpretation are developed. On the productive interpretation, implications keep track of the number of times sources are to be applied to one another to produce a particular result. On the functional interpretation, only what is doable in principle (with whatever number of applications) is represented. The productive interpretation is used to understand the contraction-free logics, linear logic and RW. The functional approach is used to understand the logics LR and R.

Preface

Mike Dunn was my teacher. I studied logic with him when I was a graduate student and he supervised my PhD thesis. I cannot hope to list everything I learned from Mike in a short note such as this. I not only learned logical facts from him, but watching him prove theorems taught me how to think about logic and how to present it. He also taught me how to teach it, both to graduate students and to undergraduates. He taught me always to be sensitive both to the mathematical (and computational) aspects of logic and to its philosophical side. He was an excellent supervisor. With me and his other supervisees, he always seemed to know when to sit back and let us do our work and when to intervene. He always seemed to know when we needed help. Now, as a PhD supervisor myself, I now know how difficult these judgements are, and I admire Mike in this regard and wish I were similarly able. Mike was always able to approach problems, either of an abstract mathematical kind or a more practical interpersonal kind, with humour and intelligence. It is hard to say whether I admire Mike more for his superb contributions to research in logic or for his abilities as a teacher and communicator of logic.

E. Mares (✉)
Philosophy Programme, Victoria University of Wellington, Wellington, New Zealand
e-mail: edwin.mares@vuw.ac.nz
URL: http://www.victoria.ac.nz/hppi/about/staff/edwin-mares

K. Bimbó (ed.), *J. Michael Dunn on Information Based Logics*,
Outstanding Contributions to Logic 8, DOI 10.1007/978-3-319-29300-4_7

1 Introduction

One theme that is very clear in the work of J.M. Dunn is the idea that logic can be treated as a subtle epistemological tool (Dunn 1968, 1976, 2010). We use logic to reason about the information and misinformation that is presented to us. In thinking about logic, we can distinguish between the informational approach to logic from the approach that takes logic to represent certain structural characteristics of the world. This latter take on logic is the mainstream in philosophy and mathematics. Mathematicians use logic primarily to formulate theories, such as arithmetic and set theory. Philosophers such as Bertrand Russell and David Lewis use logic in which to formulate a metaphysics. W.V. Quine thinks that the main purpose of logic is to have a standard language in which to formulate theories and which we can use to evaluate those theories in terms of their ontological commitments. Dunn himself uses relevance logic as a tool to formulate a metaphysics in his theory of relevant predication (Dunn 1987, 1990a, b). There is no reason, however, why the informational approach and this structural/metaphysical approach cannot coexist. There is also no obvious reason why the logical systems used to formulate mathematical or metaphysical theories need to be the same as those that are used as tools with which to understand and manipulate information.

In this paper, I take up a version of the informational approach to logic in order to give an interpretation of relevance and linear logic. This interpretation is a descendant of Alasdair Urquhart's interpretation of his semi-lattice semantics (Urquhart 1972). Urquhart takes formulas to be true or false of "pieces of information." Pieces of information can be combined with one another to create new pieces of information. The combination operator is used to give a semantics for implication and properties imposed on this operator give us different systems of relevance logic. In the present paper, I develop Urquhart's ideas to give a semantics for four relevance and linear logics: the logic R of relevant implication, RW which is a strong contraction-free logic, LR which is R without the distribution of conjunction over disjunction, and the classical linear logic MALL (mulitiplicative-additive linear logic without exponentials).

The central ideas in this paper are that of an information source and the concept of the construction of information sources. An agent accepts some information sources as being veridical—as telling us the way the world really is. From those sources the agent constructs other sources. For example, like most other people in academic or office work, I am often sent emails about meetings that are to happen in the near future. Each of these emails is an information source and I accept them as veridical. If I wish to attend the meeting, I click a button that enters its date and time in my calendar, which is another information source constructed from a collection of such emails. I suggest that we construct sources of information all the time, and that studying such constructions is one fruitful way of understanding how people deal with and manipulate the information available to them.

One virtue, I suggest, of the present interpretation is that it extends to both to R and to the contraction-free logics RW and MALL and to the distribution-free systems LR and MALL. Contraction-free and distribution-free systems of relevance logic have received a lot of attention in the literature. They are decidable, which R is not, and MALL can be used as the basis of a naïve set theory and a naïve truth theory. While I am not sure that the interpretation I give here is appropriate for a logic of set theory (or any other mathematical theory), it is clear that contraction-free and distribution-free systems need more of a philosophical foundation than they currently have. Perhaps the present attempt will lead to a firmer philosophical foundation for these systems.

In other places, I have constructed an interpretation of the relevance logic R that has it represent the way in which we infer from the information available to us in the situations in which we find ourselves. I do not see these two interpretations as being in conflict, and I think that many different models of the way in which we reason are needed to capture all of the nuances of actual human behaviour and of the norms that we appeal to when reasoning or evaluating the reasoning of others.

The plan of the paper is as follows: In the first few sections of the paper, I engage in a philosophical discussion of informational semantics and introduce the notion of a source of information. I then develop a formal theory of information sources and a variation of Anderson and Belnap's natural deduction system for relevance logics that replaces relevance subscripts with terms that represent sources. I use this natural deduction system in Sects. 5 and 6 to distinguish between the functional and productive versions of the source interpretation of the logics. The functional interpretation corresponds to the two logics that contain the thesis of contraction $((A \rightarrow (A \rightarrow B)) \rightarrow (A \rightarrow B))$ and the productive interpretation is used to understand the two that do not contain it. I then look at three traditional semantical projects—proof theoretic semantics, and model theoretic semantics—and give a broad outline of how each can be given a reading in terms of information sources.

This paper is programmatic. It does not spell out the theory in great detail. Rather it gives a broad outline of a philosophical understanding of relevance logic.

2 Information and Truth

Most classical and non-classical logicians understand a logical system as a theory that is supposed to capture the "laws of truth." Frege, for example, famously said "the word "true" indicates the aim of logic as does "beautiful" that of aesthetics or "good" that of ethics" (Frege 1984). In their version of logical pluralism, Beall and Restall (2006) think of a logical system as capturing truth preservation over a class of "cases." Where logical systems differ from one another has to do with what sorts of cases they allow into their models. On Beall and Restall's view, relevance logic is committed to there being cases in which the principle of bivalence fails and cases in which the principle of consistency fails. Read this way, relevance logic, and other non-classical logics, are committed to a non-classical theory of truth.

If we think of information as something that is intermediary between our minds and the world, however, we can retain the classical notion of truth, for all of the extensional connectives (which I will specify later) and treat a logical system as a way of understanding and manipulating information. In other places (Mares 2009, 2010), I have understood logical consequence as information preservation, but here I take a more constructive approach. I discuss this new approach to information in Sects. 5 and 6. Right now I explore certain consequences of the separation of information and truth.

One of the problems with understanding the semantics for non-classical logics is that they sometimes lack *homomorphic* truth conditions for the connectives. For example, in the Routley–Meyer semantics for relevance logic, the clause for negation is

$$a \models \neg A \quad \text{iff} \quad a^* \not\models A.$$

Dunn (1993) explains that a^* is the index in the model that is maximally compatible with a. Two indices in a model are compatible if they do not say conflicting things. The star of an index is one that is maximally compatible with regard to that index.

If we see indices in models as worlds and the clause above as a truth condition, then there seems to be something deficient about the semantics. In classical semantics, the connectives are interpreted in a straightforward way in terms of the corresponding connectives of our metalanguage. To take the most straightforward example, the truth condition for conjunction in an indexed semantics for classical logic is usually given as

$$a \models A \wedge B \quad \text{iff} \quad a \models A \text{ and } a \models B.$$

Conjunction in the object language is understood straightforwardly in terms of conjunction in the metalanguage. Applying this idea to negation, it would seem that we obtain

$$a \models \neg A \quad \text{iff} \quad \text{not } a \models A.$$

When talking about truth, the truth of negations in the metalanguage should determine the truth of the corresponding negative statements in the object language.

When discussing information, however, the relationship between truth of metalinguistic statements and their corresponding object language statements is quite different. In other places, I have treated the indices of the Routley–Meyer semantics as situations that capture partial information about worlds. For example, I am currently sitting in my lounge. It is not the case that the information in this room allows me to discern where my dog Lola is. It does not tell me whether she is in a bedroom. But it does not contain the information that she is not in a bedroom. Whereas the truth of negative statements corresponds to the failure of the truth of closely connected positive statements, what negative information is in a situation does not always amount to the failure of the salient positive information in that situation. Dunn's analysis of negative information in terms of what information is incompatible with the information contained in a situation (or state, as Dunn calls them) is much more reasonable. I have the information that my coffee table is not green because I have the information it is brown all over, and so on. Thus, we can see that the move to understanding relevance logic in terms of information has the advantage of liberating us from the need to provide homomorphic truth conditions.

Although the informational approach to logic eschews truth preservation as the sole criterion of logical consequence, it must be that inferences judged valid on the informational approach do preserve truth. It seems a reasonable requirement of a system being classified as a *deductive* logic that truth preservation be a necessary condition for validity, even if it is not a sufficient condition.

A semantical interpretation of a logic is not only a justification for that logic, it also places constraints on the logic. Truth conditional semantics justifies the use of connectives by virtue of the truth conditions that they express. The fact that any complete formulation of classical logic is truth-functionally complete justifies the logic—it allows us to express whichever truth-functions we wish. But the need to provide a coherent truth function to correspond to any connective restricts the logic. Informational semantics justifies a logic in that it shows how it helps us to comprehend and manipulate information, but it also requires of the connectives of the logic that they each have an *informational role*. Their use in understanding and manipulating information must be made evident by their interpretation. In what follows, I develop an informational interpretation of relevance and linear logic to justify those logics and to provide an interpretation of their connectives.

3 Sources of Information

My interpretation requires a set of sources of information, an agent, and those sources that he or she takes to be veridical. As I said in the introduction above, sources of information are common sense objects such as news casts, signs, rumours, statements, an agent's perceptual input, memories, and so on. The theory, however, requires both actual and possible sources of information. It does not require that there be *impossible* sources of information. It does allow sources that carry contradictory information, but it is possible (and sometimes actual) that sources of information do have contradictory information.

Those sources that an agent thinks are veridical are the ones that he or she takes to be true. For example, right now I take my current visual perceptions of my living room to be accurate and they tell me that my dog is asleep on a rug. Thus, I believe that my dog is asleep on a rug.

When an agent takes a set of sources to be veridical, he or she is entitled to construct other sources from them. We often do this. We often apply sources of information to one another. Suppose that you are sitting watching television with someone else. The newsreader says that there is flooding in Ohakune. You don't know where Ohakune is so you ask the person sitting next to you. She tells you that Ohakune is in the central North Island of New Zealand at the foot of Mount Ruapehu. Now you apply the information that your friend has given you to what the newsreader has said to produce a new information source that tells you that there is flooding in the Central North Island at the foot of Mt. Ruapehu.

One difference between Urquhart's approach and my own is that I distinguish between the contents of a source of information and the source itself. For him, a piece of information is just a set of sentences:

> A piece of information is a set of basic sentences concerning a subject or subjects about which reasoning is being carried out. In physics the basic sentences might consist of statements of experimental results, in mathematics elementary facts about numbers, and so forth. (Urquhart 1972, p. 159)

The reason that I distinguish between a source and the set of statements that they make is that more than one source may tell us the same thing but we might have reasons to trust one source more than another that has the same content. I might trust the BBC News more than Fox News but they might both tell me the exact same information on a given occasion. Thus, if we wish to add an analysis of how agents assess risk with regard to information it will be useful to distinguish between what a source says and what it is. This idea is expressed very clearly by Dunn, who says:

> One source tells me *A*, another source tells me not *A*. What am I to believe? What am I to infer? How am I to act? Perhaps I trust the source that tells me *A* and I do not trust the source that tells me not *A*. Perhaps I trust the first source just a little bit more than the second, or perhaps a whole lot. And maybe there are many sources who tell me one thing, and just a few who say the opposite. But perhaps those few are regarded as experts. Etc. This is more nuanced than the simple 4-valued Logic and suggests strategies of "weighting." (Dunn 2010, p. 429)

I do not attempt to provide an analysis of how to weight sources and of the difficult task of determining how the weighting of sources affects degrees of rational belief in propositions on the part of agents. But it seems clear that in order to have an adequate theory of uncertainty and risk one needs to distinguish sharply between sources and their contents.[1]

Another difference between the current approach and Urquhart's is that I follow Slaney (1990) and Read (1988) in distinguishing between two ways in which we can construct a new source from two existing sources of information. (This distinction is an adaptation of the distinction between extensional and intensional context in Dunn's consecution calculus for R_+ in Anderson and Belnap (1975, Sect. 28.5).) The first sort of construction is *application*. This is the construction in which you take conditional information from one source and then use it to draw conclusions from the information in another source. Application was first described in this way by Fine (1974) and named "fusion." Suppose that you have accepted as veridical an automotive website that tells you that if there is blue smoke coming out of your car's exhaust pipe, then the car is burning oil. And your perception of your car tells you that blue smoke is coming out of your car's exhaust. Then you are entitled to construct a source that tells you that your car is burning oil.

---

[1]Perhaps the "opinion tetrahedron" that Dunn (2010) sets out can be extended to treat the full vocabulary of relevance logic, including the intensional connectives. At any rate, I do not know how to do that at this time. My only point here is to support a strong distinction between sources and their contents and the topic of uncertainty and risk helps to do that.

The second sort of construction merely collects together all the information in two sources. Let's call this sort of construction "collection." I get a phone call telling me that a chair we have ordered is ready and is in a shop across town and my calendar tells me that I have two hours free tomorrow afternoon. I put those two sources together to construct a single source that I will use to plan my activities tomorrow. When I do collect together all the information in these two sources what the sources tell me in conjunction, I know that I have two hours free and that I have a chair ready for me to retrieve.

Where α and β are sources of information, I use '$\alpha \circ \beta$' to denote the application of α and β and '$\alpha \sqcup \beta$' to denote the collecting together of α and β. Note that $\alpha \sqcup \beta$ and $\alpha \circ \beta$ may not denote unique entities. There may be more than one result of collecting together the information from two sources or applying one source to another. Consider, for example, my collecting the information from emails that tell me about my meetings for tomorrow. There is more than one way of constructing such a collection. I can click on a button in my email program and it can automatically put the meeting and time into my calendar. I can write the meetings and times down in my diary. These two methods might be importantly different. I might, for example, attribute a lower level of reliability to my writing in my diary than to the action of the email program. If I have left my computer glasses at home or am tired, I might write down the wrong time for a meeting. Thus, I suggest that source terms be taken to refer to a collection rather than to individual sources.[2]

To describe the construction of sources of information, I adopt a variant of Anderson and Belnap's natural deduction system for relevance logics (Anderson and Belnap 1975). I do not use their Fitch-style representation of natural deduction. Rather, I adopt a Prawitz/Martin-Löf style presentation that takes *judgments* to be the elements of an inference. A judgment is a string of symbols of the form $\alpha : A$, where α is a term that denotes sources of information and A is a logical formula. A judgement $\alpha : A$ is read as saying that α tells us A.

In this paper, I do not use β-reduction to reduce lambda terms. For example, I do not assume that $(\lambda x.(a \circ x)) \circ b$ reduces to $a \circ b$. If I were interested in constructing a logic to represent a sort of computer program (and a notion of computational efficiency), then reducing terms to normal forms would be extremely important. But I want to allow for the possibility that these sorts of terms actually represent different information sources.

3.1 Source Terms

A proof in the source interpretation is a tree of judgments. A judgment is an expression of the form $\alpha : A$, where α is a term that refers to a source of information—a *source term*—and A is a formula. I borrow this notation from Martin-Löf. The Martin-Löf

[2]See Sect. 12 for more discussion of the denotation of sources. The issue is similar to the relationship between relevance subscripts in the Anderson–Belnap natural deduction system and indices in the Routley–Meyer semantics. A set of numerals α picks out a collection of indices that are related to those denoted by the numerals in α.

type theory is used to derive the types of lambda terms, appealing to the Curry–Howard Isomorphism between proofs and computer programs. I use it for a very different purpose, to have terms for information sources that keep track of the way in which they are constructed.

I use $a_1, a_2, a_3, \ldots$ as source parameters, with $a, b, c, \ldots$ as metavariables ranging over source parameters. The set of source terms is the smallest set that satisfies the following clauses.

- If a is a source parameter, then a is a source term;
- if α and β are source terms, then $(\alpha \circ \beta)$ and $(\alpha \sqcup \beta)$ are source terms;
- if $\alpha(a)$ is a source term and $\alpha(x)$ results from the replacement of one or more occurrences of a in pure fusion contexts with x, then $(\lambda x.\, \alpha(x))$ is a source term.

An occurrence of a parameter a is in a pure fusion context in a term α if and only if it does not occur within the scope of an occurrence of $\sqcup$. For example, the expression $(\lambda x.\, (a \circ x)) \sqcup b$ is a source term, but $(\lambda x.\, ((a \circ x) \sqcup b))$ is not a source term.

Before I leave the topic of rules governing source construction, I wish to note that the link between relevance logics and the lambda calculus has a long history. Glen Helman's Sect. 71 in Anderson et al. (1992) shows that the pure implicational theorems of R are the same as the derivable type schemes in the λ-I calculus (the lambda calculus with the restriction that lambda abstracts can only bind variables that really occur in the term in their scope). Dunn and Bob Meyer—especially together with combinatorial logicians such as Mariangiola Dezani-Ciancaglini—developed a view of the relationship between relevance logic and combinatorial logic (which is very closely related to lambda calculus) relates fusion to application in much the same way as I am doing it here. Meyer came to refer to this relationship, and the correspondence between combinators and axiom schemes in relevance logic and their relationship to the postulates governing the ternary relation of the Routley–Meyer semantics as the "key to the universe" (see, e.g., Dunn and Meyer (1997); Dezani-Ciancaglini et al. (2002)). Baker-Finch (1992) uses this relationship between relevance logic and the λ-I calculus to give a formal analysis of strict computation and Neil Leslie and I (Leslie and Mares 2004) develop Baker Finch's ideas to produce a Martin-Löf style presentation of LR. Bunder (2003) shows relationships between the implication-conjunction fragment of various weak relevance logics and typed lambda calculus with intersection types. Katalin Bimbó has done work on the relationship between proof theory for substructural logics and combinatory logic and the lambda calculus (see, e.g., Bimbó (1999, 2004)).

4 Implication and Conjunction

The nature of the two sorts of construction immediately justify the following rules of inference.

$$\frac{\alpha : A \to B \qquad \beta : A}{\alpha \circ \beta : B} \qquad\qquad \frac{\alpha : A \qquad \beta : B}{\alpha \sqcup \beta : A \wedge B}$$

The rule on the left is the rule of implication elimination ($\rightarrow E$) and the rule on the right is conjunction introduction ($\wedge I$).

For the logics R and LR, we have the following implication introduction rule ($\rightarrow I$).

$$\frac{\begin{array}{c}[a: A] \\ \vdots \\ \beta(a): B\end{array}}{\lambda x.\beta(x): A \rightarrow B}$$

In order to accord with relevance logic, we need to constrain this rule in two ways. First, we need to restrict the use of the rule to cases in which a really occurs in $\beta(a)$. This is Anderson and Belnap's real use requirement. The real use requirement itself is not justified by the source interpretation. Rather, it seems to be a reasonable constraint that people do in fact place on their own construction of information sources. Second, we need to place a condition on the rule that allows the abstraction to take place on when *a does not* occur in the scope of any $\sqcup$. It needs to be in pure application contexts. This constraint makes sense because application is the operation on sources that is paired with implication. Collection has to do with conjunction. Conjunction is an extensional connective and implication is an intensional connective.

For the logics MALL and RW, the implication introduction rule is slightly different ($\rightarrow I^p$).

$$\frac{\begin{array}{c}[a: A] \\ \vdots \\ \beta(a): B\end{array}}{\lambda x.\beta(x): A \rightarrow_n B}$$

where $A \rightarrow_n B$ is defined inductively by: $A \rightarrow_1 B = A \rightarrow B$ and $A \rightarrow_{i+1} B = A \rightarrow (A \rightarrow_i B)$. In the rule, n is the number of occurrences of a that are replaced by x in β and $n \geq 1$. The restrictions for the rule $\rightarrow I$ apply to $\rightarrow I^p$ as well. The superscript 'p' stands for 'productive'. The reason I call MALL and RW productive is explained in Sect. 5.

The conjunction elimination rule is straightforward.

$$\frac{\alpha: A \wedge B}{\alpha: A} \qquad \frac{\alpha: A \wedge B}{\alpha: B}$$

The source meaning of conjunction is that $A \wedge B$ is in a source if and only if both A and B are in that source. This meaning justifies this conjunction elimination rule directly.

In order to make the present treatment of conjunction reasonable, I need to add the following rule.

$$\frac{\beta(\alpha \sqcup \alpha): A}{\beta(\alpha): A}$$

I need this rule to prove that $A \rightarrow (A \wedge A)$.

$$\dfrac{\dfrac{\dfrac{[a\colon A] \qquad [a\colon A]}{a \sqcup a\colon A \wedge A}}{a\colon A \wedge A}}{\lambda x.x\colon A \rightarrow (A \wedge A)}$$

The rule makes sense given the interpretation of $\sqcup$. $\alpha \sqcup \alpha$ contains exactly the same information as α so in combinations with other sources α and $\alpha \sqcup \alpha$ should be able to be interchanged producing sources with the same information.

I also add the intensional conjunction $\circ$ (called *fusion*), because it is the conjunction that is paired with implication and because it more directly represents fusion in source terms. The fusion introduction rule ($\circ I$) is straightforward.

$$\dfrac{\alpha\colon A \qquad \beta\colon B}{\alpha \circ \beta\colon A \circ B}$$

The fusion elimination rule, $\circ E$, displays the relationship between fusion and implication.

$$\dfrac{\alpha\colon A \circ B \qquad \beta\colon A \rightarrow (B \rightarrow C)}{\beta \circ \alpha\colon C}$$

5 The Functional Interpretation and the Productive Interpretation

The key idea in the source interpretation is that in manipulating sources of information, we *construct* new sources of information. I am watching on the news at this moment a story about a huge snowfall in Boston. I check an app on my tablet and find that there is forecast another snowfall tomorrow. I apply the information from the app to the news story and construct a source (the weather in Boston according to me) which says that the streets and footpaths in Boston will be completely clogged with snow for the next two days and that the subway will be slow, and may have to shut down.

The manner in which one constructs sources is captured by proofs. I discuss the status of the rules of proof at length in Sect. 6. For now, let us think of the rules of proof as norms that govern how we are permitted to construct sources. The meaning of the intensional connectives (implication and fusion) is given in terms of their relationship to these sorts of constructions. For example, the judgment $\alpha\colon A \rightarrow B$, is taken to mean that we can apply any source that tells us that A to obtain a source that tells us that B. But this reading is ambiguous. It is ambiguous in several ways, but the one that concerns me now is that it is ambiguous between telling us that this application is a single application and telling us that after one or more applications we are entitled to obtain a source that says that B. This ambiguity is allowed to stand in what I call the *functional* interpretation of the source semantics, but it is rejected by the *productive* interpretation.

According to the source semantics, a source can act as a function or as the argument of a function. As I have said, we apply sources to one another to construct other sources. In number theory and algebra, functions are often constructed so that the argument fits into more than one position in the function expression, as in x^x. If we think of a source in this way—as a function that may have more than one position that is to be filled with an argument—the implicative judgment above is read as saying that when we apply α to a source that tells us A *as many times as required* for us to construct a source of type B.

On the productive interpretation, on the other hand, the way in which a source is produced is tracked in a more fine grained manner. One level of fine-graining—the level I am treating in this paper—is one that keeps track of the number of times in which we have to apply a source that contains the information that A to produce one that contains B. In the next section, I explain why this fine-graining is of importance.

6 Idealization in Semantics

In a recent paper (Yap 2014), Audrey Yap argues that we should not view epistemic logics as normative theories, i.e., as theories about how we should reason. Rather, we should look at them as idealized descriptive theories. On her view, a system of epistemic logic should be viewed in the same way as we think of idealized scientific models. She thinks that various criticisms of epistemic logic can be met if we think of it as a way of describing what people, or agents in a computer program, do rather than a theory of what we should do. Idealizations in science, for example, neglect certain features of the actual world that make the application of mathematics or other conceptual structures too complicated. Similarly, simplifying assumptions, like assuming that agents have perfect memories or enough time (and interest) to do lengthy derivations are employed in constructing epistemic logics. Yap says:

> The previous examples have given a picture of some ongoing research programs in epistemic logic that showcase ways in which the field uses idealizing assumptions. While many of these projects do seek to describe the behavior of real agents, it is acknowledged that some simplifying assumptions will be required for the sake of tractability. The appropriateness of using formal models to study epistemological issues generally is a further issue, but, if this paper is right, then that appropriateness does not stand or fall with the presence of some idealizing assumptions. The question is whether, despite the inevitable idealization, the formal models can still give us insight into actual phenomena. (Yap 2014, p. 3365)

I suggest that we use Yap's view of logic as an idealization of human inferential thinking and behaviour in order to interpret the productive view of logic. I am ambivalent as to whether we should abandon the normative treatment of epistemic and relevance logic, but I do think that even if we take a normative view of logic we need to think of logic as an idealization of our actual inferential abilities. In what follows,

I consider the productive interpretation given both a descriptive and a normative view of logic taking into account the idealized relationship between logic and actual human inference.

The productive view differentiates between proofs according to the number of times in which sources are applied to one another. In human agents, these differences track differences in cognitive effort. If we differentiate between sources in terms of how much effort it takes an agent to produce, then we can represent in our theory a *decline in expectation* that the agent will produce sources that take more effort. I call this decline *expectation fade*.

On the functional view, the judgment

$$\alpha : A \rightarrow B$$

is read as

> If α is applied to any source that contains the information that A then it will produce a source that contains the information that B.

On the productive view, the same judgement is read as

> If one accepts α and any source $\beta : A$, then he or she is expected to create a source $\alpha \circ \beta : B$.

What is interesting about the productive view, is that we can interpret implications differently depending on how deeply nested they are within other implications. In a judgement $\gamma : A \rightarrow (B \rightarrow C)$, even if the agent accepts sources $\delta : A$ and $\sigma : B$, our expectation that he or she will produce a source $(\gamma \circ \delta) \circ \sigma : C$ is lower than our expectation that he or she will produce $\gamma \circ \delta : B \rightarrow C$. Thus, the productive reading understands the types of sources differently from the functional approach. One might object that the source terms themselves represent the number of applications of sources that is necessary to produce sources of particular types and so there would seem to be double counting of applications on the productive reading, so to speak. This objection would have weight if this particular natural deduction system were our only way (or even the preferred way) of representing the logics. What I am trying to do is to give an interpretation to the formulas of the logics and to everyday applications of logical reasoning, not just to the natural deduction system that I set out here. Natural language and other representations of the logic do not have source terms. Applied to these representations, the double counting disappears.

One way of understanding the productive view of logic is by contrasting it to the resource interpretation of linear logic. On the resource interpretation, the rule of contraction, which allows the inference of $\Gamma(A) \vdash B$ from $\Gamma(A; A) \vdash B$, is rejected because it treats premises in a sequent as representing resources and the conclusion as representing an action or a further resource that can be obtained. Just because an action can be obtained using two lots of some resource does not mean that it can be obtained using just one lot. On the productive interpretation, it is not resources but *effort* that is being tracked. The difference between a source that can be produced by one application of two sources is different in nature according to the productive interpretation from one that requires two applications of those sources to one another.

If we decide to adopt a normative reading of the logics, then an analogous decline that takes place with regard to sources that are more complicated to produce. According to the normative reading, an agent is committed to the veridicality of any source that is constructed from those sources that he or she has accepted by means of the rules of proof of the logical system. Norms are usually governed by the rule that what an agent is committed to is only what he or she can do—the principle of "ought implies can." An agent's actual abilities can be determined in various ways, taking into account or ignoring physical abilities, time constraints, opportunity costs, and so on. If we take into account all of these constraints, what an agent can do is quite limited. When we relax those constraints, we consider an agent to be able to do much more. The norms are more firmly binding that govern what an agent should do when we include more constraints. As we relax constraints, the norms in place are much weaker. Thus, we can see that there is a form of *commitment fade* that is analogous to expectation fade.

Marking out the differences between sources in terms of the number of implications involved in the production of those sources tracks these fades and indicates the level of expectation or commitment we should place on them for a given agent.

The functional interpretation idealizes away differences between the effort and time needed on behalf of an agent to do proofs of different lengths. This idealization is useful as well. If we are only interested in the commitments an agent has to non-implicational statements, then the productive approach imposes complications that may be completely irrelevant to the project at hand.

7 Disjunction

The introduction rules for disjunction ($\vee I$) are straightforward.

$$\frac{\alpha: A}{\alpha: A \vee B} \qquad \frac{\alpha: B}{\alpha: A \vee B}$$

Clearly, if a source tells us that A it also tells us that $A \vee B$.

The elimination rule is much more difficult to justify. The standard rule for relevance logic is ($\vee E$):

$$\frac{\alpha: A \vee B \qquad \begin{array}{c}[a: A] \\ \vdots \\ \beta(a): C\end{array} \qquad \begin{array}{c}[b: B] \\ \vdots \\ \beta(b): C\end{array}}{\beta(\alpha): C}$$

where a and b really occur in $\beta(a)$ and $\beta(b)$, respectively, and do not occur within the scope of a $\sqcup$. But this rule, together with the conjunction rules, does not entail the distribution of conjunction over disjunction. We can modify the disjunction elimination rule in the following way to yield distribution:

$$\frac{\alpha : A \vee B \qquad \alpha : D \qquad \begin{array}{c}[a : A \wedge D] \\ \vdots \\ \beta(a) : C\end{array} \qquad \begin{array}{c}[b : B \wedge D] \\ \vdots \\ \beta(b) : C\end{array}}{\beta(\alpha) : C}$$

where a and b are new and really occur in $\beta(a)$ and $\beta(b)$, respectively. This modified rule is called $(\vee E^d)$. The use of the metavariable D in $(\vee E^d)$ allows us to prove distribution.

$$\dfrac{\dfrac{\dfrac{[a : (A \vee B) \wedge C]}{a : A \vee B} \quad \dfrac{[a : (A \vee B) \wedge C]}{a : C} \quad \dfrac{[b : A \wedge C]}{b : (A \wedge C) \vee (B \wedge C)} \quad \dfrac{[c : B \wedge C]}{c : (A \wedge C) \vee (B \wedge C)}}{a : (A \wedge C) \vee (B \wedge C)}}{\lambda x . x : ((A \vee B) \wedge C) \rightarrow ((A \wedge C) \vee (B \wedge C))}$$

It is possible to derive a version of $(\vee E)$ from $(\vee E^d)$.

$$\dfrac{\alpha : A \vee B \quad \dfrac{\dfrac{\begin{array}{c}[a : A] \\ \vdots \\ \beta(a) : C\end{array}}{\lambda x . \beta(x) : A \rightarrow C} \quad \dfrac{[b : A \wedge (A \vee B)]}{b : A}}{\lambda x . \beta(x) \circ b : C} \quad \dfrac{\dfrac{\begin{array}{c}[c : B] \\ \vdots \\ \beta(c) : C\end{array}}{\lambda x . \beta(x) : B \rightarrow C} \quad \dfrac{[d : B \wedge (A \vee B)]}{d : B}}{\lambda x . \beta(x) \circ d : C}}{\lambda x . \beta(x) \circ \alpha : C}$$

The difference between this version of the rule and the original version concerns the source term in the conclusion. Differences of this sort between source terms are discussed in Sect. 9. As we shall see there, the differences between the versions of the rule have no serious consequences for the logics that they characterize.[3]

7.1 Distribution

Should a logic of information sources include the principle of the distribution? This is a difficult question to answer and I do not give a real answer to that question here. Rather I only show why the question is so difficult. Let us say that the truths of the actual world are closed under distribution. Even if this is the case, and even

[3]There are other ways of generating the distribution rule than this. Anderson and Belnap add a primitive rule. Ross Brady adopts a structural connective that corresponds in some sense to extensional disjunction. Dunn incorporates the mechanism that is found in his and Mints' sequent systems of having conjunctive hypotheses (Dunn and Restall 2002, Sect. 1.5). Dunn's proposal is particularly interesting and it might be illuminating to provide a source of information reading of it.

if the truths at any possible world are closed under disjunction, it need not be that distribution is a logical truth. As I explain in Sect. 9, a logical truth is a truth that we can derive *a priori* and one that an agent is entitled to use at any time in a derivation. We could justify distribution if we could find a good reason why it should be able to be brought into an inference at any time. But I cannot think of any such reason.

Omitting distribution, or $(\vee E^d)$, from our logic does have some good formal effects. As Girard (1998) shows, a naïve set theory can be constructed on the basis of Linear Logic. Moreover, Bob Meyer shows that LR is decidable (Thistlewaite et al. 1987), whereas, as Urquhart proved, R is undecidable. But it is neither clear that set theory should be understood in terms of information sources nor that an informational logic should be decidable. Hence the links between these facts about the formalism and the present interpretation have yet to be made.

On the other hand, Belnap (1993, p. 36) gives the following argument to show that from a semantic perspective it makes sense to include distribution in one's logic. He points out that the following three conditions force a semantics to include distribution.

1. The semantics evaluates sentences as being true or false at a point.
2. The semantics treats conjunction and disjunction extensionally.
3. The semantics takes logical consequence to be truth preservation at every point in a model.

If we have a two-valued semantics that gives an extensional treatment of conjunction and disjunction, then there is only one reasonable truth condition for disjunction and only one reasonable truth condition for conjunction, and these lead directly to points being closed under distribution. If logical consequence just is a theory of the closure of truths at points in models (as is stated in condition 3), then distribution is unavoidable.

The source interpretation, however, is not compatible with Belnap's first and second conditions. As we have said, an informational semantics talks about the information that we have (in some sense of 'have'). What is true or false often goes beyond the information that one has at hand. Moreover, disjunctive information should probably not be treated extensionally. We can have the information that a disjunction obtains without the information about which disjunct is true. I used to have two dogs. If I came home to find that their food bowl was empty, I would know that at least one of them ate the food, but not which of them had eaten it.

Belnap's argument does show that a certain sort of semantical approach is committed to distribution, but not that the source interpretation is.

8 Negation

In the natural deduction system, I use a falsum in order to formulate the negation rules. This is more a matter of choice than anything that is forced on me by the source interpretation. The falsum f, tells us that we are accepting or are committed to something impossible. The negation elimination rule for functional logics $(\neg E)$ can also be read as a falsum introduction rule.

$$\frac{\alpha : \neg A \quad \beta : A}{\alpha \circ \beta : f}$$

If an agent accepts two sources that have conflicting information, then she is committed to the falsum or, in other words, she is entitled to construct a source that tells us that the falsum is true.

The negation introduction rule $(\neg I)$ is quite standard.

$$\begin{array}{c} [a : A] \\ \vdots \\ \dfrac{\alpha(a) : f}{\lambda x . \alpha(x) : \neg A} \end{array}$$

where a really occurs in $\alpha(a)$. This rule indicates that negative statements in this system are really implications of a sort. For the productive interoperation, I restrict the rule to say that only one occurrence of a is to be replaced with x in $\alpha(a)$.

The two rules above give us only a sub-intuitionist negation. Missing from the negations of R, RW, LR and MALL are the double negation elimination rule, the intuitionistically illegitimate form of reductio, and excluded middle. These can be derived by adding the following form of Prawitz's classical reductio rule from (Red) (Prawitz 2006).

$$\begin{array}{c} [a : \neg A] \\ \vdots \\ \dfrac{\alpha(a) : f}{\lambda x . \alpha(x) : A} \end{array}$$

where a actually occurs in $\alpha(a)$. In the productive logics (MALL and RW), we need to place the constraint that exactly one occurrence of a is replaced by x to obtain $\lambda x . \alpha(x)$. The rule with this constraint is called (Red^p).

The inclusion of (Red) in the functional logics (R and LR) allows the proof of excluded middle.

$$\dfrac{\dfrac{[a : \neg(A \vee \neg A)] \qquad \dfrac{\dfrac{\dfrac{[a : \neg(A \vee \neg A)] \quad \dfrac{[b : A]}{b : A \vee \neg A}}{a \circ b : f}}{\lambda x . (a \circ x) : \neg A}}{\lambda x . (a \circ x) : A \vee \neg A}}{a \circ \lambda x . (a \circ x) : f}}{\lambda y . (y \circ \lambda x . (y \circ x)) : A \vee \neg A}$$

In the last step of the proof, in discharging the assumption $[a : \neg(A \vee \neg A)]$, I abstract on two occurrences of a in $a \circ \lambda x . (a \circ x)$. This is not allowed in productive proofs.

9 The Nature of Theorems

A theorem in the present systems is a formula A that can be proven in a judgment of the form $\alpha : A$, where α does not contain any source parameters. The source term shows that the judgement can be derived *a priori* in the sense that all it requires for its production are sources that can be constructed by anyone with the aid of the proof system.

In truth conditional semantics, a theorem of a logic is interpreted as a formula that is true in every context (as in Kripke semantics for normal modal logics) or as a formula that is true in all designated contexts (as in Kripke semantics for non-normal modal logics or Routley–Meyer semantics for relevance and substructural logics). In the information source interpretation of relevance logic, any agent is entitled to use a theorem source (one that does not contain any source parameters) in any context.

In informational semantics, the link between a theorem and always being true is partially broken. Theorems are always true, but not all statements that are true in all possible worlds need be theorems. Consider the law of excluded middle. It is provable in R and LR but not in MALL or RW. The latter two logics lack are productive and lack the contracted form of implication introduction. This, however, does not mean that an advocate of MALL or RW must reject the view that the law of excluded middle is a universal truth. It commits her only to holding that one is not always entitled to use instances of the law of excluded middle when constructing sources. The truth of the law of excluded middle is a semantic or perhaps metaphysical issue. One's entitlement to use it is an informational and epistemological issue (that I discuss further in Sect. 11).

10 The Logics

As I have said, the rules stated here are used to characterize four logics. All of the logics contain the rules $(\wedge I)$, $(\wedge E)$, $(\rightarrow E)$, $(\vee I)$, $(\circ I)$, $(\circ E)$, and $(\neg E)$. MALL also contains $(\rightarrow I^p)$, $(\vee E)$, $(\neg I^p)$ and (Red^p). RW is just like MALL except that $(\vee E)$ is replaced by $(\vee E^d)$. R contains all of the non-productive rules and $(\vee E^d)$. LR is just like R except that it has $(\vee E)$ instead of $(\vee E^d)$.

In this section, I present the Hilbert systems for the logics and briefly indicate how they are to be proven equivalent to their natural deduction formulations. To formulate Hilbert-style systems in a reasonably efficient manner, I begin with a smaller language, one that includes only implication, a falsum, and conjunction. The other connectives are defined as usual: $\neg A =_{df} A \rightarrow f$, $A \circ B =_{df} \neg(A \rightarrow \neg B)$, $A \leftrightarrow B =_{df} (A \rightarrow B) \wedge (B \rightarrow A)$, $A \vee B =_{df} \neg(\neg A \wedge \neg B)$, and $t =_{df} \neg f$.

In this section, in order to keep track of whether I am discussing the natural deduction system or the Hilbert system, I refer to the natural deduction system for a logic L as SL (for 'source system for L') and the Hilbert style system as HL.

I take the list of axioms for HMALL from Troelstra (1992, p. 67).

Axioms of HMALL

1. $A \to A$
2. $(A \to B) \to ((B \to C) \to (A \to C))$
3. $(A \to (B \to C)) \to (B \to (A \to C))$
4. $((A \to f) \to f) \to A$
5. $A \to (B \to (A \circ B))$
6. $((A \circ B) \to C) \to (A \to (B \to C))$
7. t
8. $t \to (A \to A)$
9. $(A \wedge B) \to A, \quad (A \wedge B) \to B$
10. $((A \to B) \wedge (A \to C)) \to (A \to (B \wedge C))$
11. $A \to (A \vee B), \quad B \to (A \vee B)$
12. $((A \to C) \wedge (B \to C)) \to ((A \vee B) \to C)$

Rules

$$\frac{\vdash A \to B \quad \vdash A}{\vdash B} \qquad \frac{\vdash A \quad \vdash B}{\vdash A \wedge B}$$

To obtain HRW, the distribution axiom $(A \wedge (B \vee C)) \to ((A \wedge B) \vee (A \wedge C))$ is added to HMALL. To obtain HLR, the contraction axiom $(A \to (A \to B)) \to (A \to B)$ is added to HMALL. To obtain HR, the contraction axiom is added to HRW or the distribution axiom is added to HLR.

The following is easily shown.

Lemma 10.1 *The following are provable in* MALL: *(i)* $(A \circ (B \circ C)) \leftrightarrow ((A \circ B) \circ C)$; *(ii)* $(A \circ B) \to (B \circ A)$.

The following theorem is proven merely by proving the axioms and the admissibility of the rules of each logic HL in the corresponding SL:

Theorem 10.2 *If* A *is provable in* HL, *then* $\alpha: A$ *is provable in* SL, *for some parameter-free source term* α, *for all* $L \in \{$MALL, LR, RW, R$\}$.

In order to prove the converse, I define a translation function between source terms and formulas in the context of proofs.

- $a^* = A$, where $a: A$ is an assumption in the proof;
- $x^* = t$ for free occurrences of x;
- $(\alpha \sqcap \beta)^* = \alpha^* \wedge \beta^*$;
- $(\alpha \circ \beta)^* = \alpha^* \circ \beta^*$;
- $(\lambda x.\alpha(x))^* = \alpha^*(t)$.

Now I prove that the source natural deduction system can only prove theorems of the Hilbert system for MALL.

Theorem 10.3 *If* $\alpha: A$ *is provable in* SMALL, *then* $\alpha^* \rightarrow A$ *is provable in* HMALL.

The proof is by induction on the length of proofs in SMALL. The method is fairly straightforward, and I only prove the implication cases to illustrate how the proof works.

First, I prove that implication elimination can be mimicked in HMALL. I show that from $\vdash \alpha^* \rightarrow (A \rightarrow B)$ and $\vdash \beta^* \rightarrow A$, we can derive $\vdash (\alpha^* \circ \beta^*) \rightarrow B$ in HMALL.

1. $\vdash \alpha^* \rightarrow (A \rightarrow B)$	assumption
2. $\vdash \beta^* \rightarrow A$	assumption
3. $\vdash A \rightarrow (\alpha^* \rightarrow B)$	1, axiom 3, Modus Ponens (MP)
4. $\vdash \beta^* \rightarrow (\alpha^* \rightarrow B)$	2, 3, axiom 2, MP
5. $\vdash \alpha^* \rightarrow (\beta^* \rightarrow B)$	4, axiom 3, MP
6. $\vdash (\alpha^* \circ \beta^*) \rightarrow B$	axiom 6, 5, axiom 2, MP

I turn now to implication introduction. Where $\alpha(c): B$ is derived in a subproof from $a: A$,

1. $\vdash (\alpha(a))^* \rightarrow B$	assumption
2. $\vdash \alpha^*(A) \rightarrow B$	1, def. $*$
3. $\vdash (\alpha^* \circ A) \rightarrow B$	2, Lemma 10.1, MP
4. $\vdash \alpha^* \rightarrow (A \rightarrow B)$	3, axiom 6, MP

Extending the proofs of Theorems 10.2 and 10.3 to the other three logics is quite easy. So I merely state them.

Theorem 10.4 *For* $L \in \{\text{MALL}, \text{LR}, \text{RW}, \text{R}\}$, *if* $\alpha: A$ *is provable in* SL, *then* $\alpha^* \rightarrow A$ *is provable in* HL.

Theorem 10.5 *For* $L \in \{\text{MALL}, \text{LR}, \text{RW}, \text{R}\}$, *for any formula* A, *if* A *is provable in* HL, *then* $\alpha: A$ *is provable in* SL, *for some parameter-free* α.

11 Information Sources and Proof Theoretic Semantics

The informational reading of the proof theory can be taken in two ways. First, it can be taken as a way of introducing another semantics of some sort for the proof theory. This semantics will be a collection of models of some sort (Kripke models, algebraic models, or whatever). The models, moreover, are meant both to characterize the logic mathematically and to capture the philosophical (here, informational) reading. On this approach, the meanings of the connectives are understood in terms of the relationship between the formulas of the language and the elements of the models. Second, the informational semantics may be seen as a way of understanding the proof theory directly and showing that the proof theory itself can be taken as a semantics. The meanings of the connectives on the proof theoretic approach are understood in terms of their roles in proofs. I explore this proof theoretic approach to the semantics first.

Proof theoretic semantics treats proofs as having a role in the norms of inference for a given linguistic or epistemic community. On the information source theory, these norms govern what an agent ought to infer and what he or she is permitted to infer. An agent's inferential obligations are called his or her *commitments* and what he or she is permitted to infer are his or her *entitlements*. As I have said, on the information source view, an agent accepts some sources as being veridical. Let's call the set of sources that some agent accepts, Γ.

The theory of information sources gives rise to a theory of entitlement and commitment. Agents are committed to collecting and applying the sources that they accept. For the moment, I set aside the topic of commitment fade and explain a simple theory. On this simple theory, there are some simple closure principles.

- If $\alpha \in \Gamma$, then the agent is committed to α;
- if the agent is committed to α and β then he or she is committed to some $\alpha \sqcup \beta$ and some $\alpha \circ \beta$;
- if the agent is committed to α, then he or she is committed to some $\lambda x.(\alpha \circ x)$ and some $\lambda x.(x \circ \alpha)$.

The last clause is supposed to commit agents to accepting those implications that are implicit in the sources that they accept or to which they are otherwise committed. For example, suppose that $\alpha : A$ and the agent accepts α. Here is a little proof to show that an agent who is committed to a source that contains A is also committed to a source that contains $(A \rightarrow B) \rightarrow B$:

$$\dfrac{\dfrac{[a: A \rightarrow B] \quad \alpha : A}{a \circ \alpha : B}}{\lambda x.(x \circ \alpha) : (A \rightarrow B) \rightarrow B}$$

By the closure principles, the agent is committed to $\lambda x.(x \circ \alpha)$ and hence is committed to a source that says that $(A \rightarrow B) \rightarrow B$.

If we do take risk and reliability into account, the notions of commitment and entitlement may have to change somewhat. Some sorts of constructions of sources are unreliable (writing down the contents of other sources when one is very tired or when one does not have his reading glasses, for example). Thus, there may need to be restriction on what sort of commitments agents have. The possibility of creating sources that are as reliable as the ones that one has already accepted may be limited, and so this could be a source of commitment fade. Similarly combining or applying sources of uneven levels of reliability may lead to a form of commitment fade. Thus, a final theory of commitment and entitlement requires some deep thinking about the effect of agents' judgments concerning the reliability of sources and the transmission of that reliability through constructions of new sources, but here are a few ideas along these lines.

A theory of commitment fade would seem most plausibly based on features of human inferential competence. In other words, commitments would fade as our abilities to make inferences also fade. If an agent is committed to a source $\alpha : B$ only after he or she makes a long chain of difficult inferences, then the level of her commitment should be quite low. This suggests a continuum of different levels of commitment between 0 and 1. Every time a collection or application is necessary, the level of commitment drops. How much it drops should depend on context and the agent.

In addition to commitments, there are entitlements. The agent's entitlements reach beyond his or her commitments.

- If the agent is (weakly) committed to α then he or she is entitled to α;
- if the agent is entitled to α and β then he or she is entitled to collect them together into some $\alpha \sqcup \beta$ and apply one to the other in some $\alpha \circ \beta$;
- if the agent is entitled to α then he or she is entitled to some $\lambda x.(\alpha \circ x)$ and some $\lambda x.(x \circ \alpha)$;
- if γ is a source term that contains no source parameters then the agent is entitled to γ.

The last condition tells us that the agent is entitled to any source that tells us just theorems of the logic. Thus, the agent's entitlements are closed under the implications of the logic and contain every theorem (and they are closed under conjunction).

Not every agent is committed to every theorem all the time. Yet theorems do have a universal nature. Every agent is entitled to appeal to any theorem at any time.

The justification of these norms and the theories of deduction might come from two different sources. One might take the theory to be a description of actual norms of reasoning in particular epistemic communities. It seems to me that the source interpretation of the logic does describe actual social practices, but a real justification of this sort requires serious empirical investigation. It also might be that the source interpretation fits well with some more general epistemological project.

What does not have to be justified is the fact that the source interpretation allows or leads one to accept deviant logical rules like Prior's tonk. The source semantics lives happily alongside a truth-conditional semantics (at least for the conjunction, disjunction and negation fragment of the language). Thus I can appeal to the fact that the logical rules are truth preserving in order to prove that they are safe. There is no need to appeal to anything like a notion of harmony to prove that the system does not lead us astray in this way.

12 Information Sources and Model-Theoretic Semantics

If one does not want to understand source meaning in purely proof-theoretic terms, then it seems possible to construct models that are based on collections of information sources. The best known and most widely discussed model theory for relevance logic is Routley and Meyer's semantics. Their semantics treats formulas as being true or false at points and relates these points to one another by accessibility relations that are used to formulate truth conditions for the intensional connectives of the language.

The Routley–Meyer semantics, in its traditional form, is unacceptable as a formalization of source semantics. A disjunction $A \vee B$ holds at a point x in a Routley–Meyer model if and only if at least one of A or B hold at x. Points in standard Routley–Meyer models, thus, cannot be taken to be sources. There are, however, variations of the Routley–Meyer semantics that may help in this regard.

Perhaps the best candidate for a source interpretation is Hiroakira Ono's semantics for substructural logics (Ono 1993). Ono includes an intersection operator, $\sqcap$, to treat disjunction. For any point in an Ono model, x, $x \models A \vee B$ if and only if there are y and z such that $y \models A$ and $z \models B$ and $y \sqcap z \leq x$. If we read the intersection operator as producing a source that contains whatever information is common to y and z, then this is an intuitive semantics for disjunction in the source context. Ono does not have the dual, $\sqcup$, to treat conjunction, but it would seem that we could add it to his semantics without difficulty. Together with Ono's clause for conjunction—$x \models A \wedge B$ if and only if $x \models A$ and $x \models B$—the logics characterized do not contain the distribution of conjunction over disjunction. There is a natural fit between this semantics and MALL and LR. But the condition for disjunction can easily be modified to obtain distribution. Merely set $x \models A \vee B$ if and only if there are $y \models A$ and $z \models B$ such that $x = y \sqcap z$.

Ono also employs the fusion operator to handle implication. Thus, there is a fairly good fit between the source interpretation and Ono's model theory. The problem with Ono's theory, however, is that it has no theory of negation. Whether an intuitive semantics of negation can be added to Ono's view I do not know, but it might be a worthwhile project to see whether it can be done.

Fine's semantics (Fine 1974; Anderson et al. 1992) does have a treatment of negation, and a non-standard treatment of disjunction. In Fine's theory, there is a distinguished set of points in each model that he calls "saturated." A saturated point is one at which the standard condition for disjunction holds. For other points x, a disjunction $A \vee B$ obtains if and only if for every saturated point that is greater than x either A or B obtains. It would seem possible to interpret Fine's semantics in terms of sources. A saturated point is one that accurately represents a part of a possible (or impossible) world called a "situation." This notion of a situation is due to Barwise and Perry (1983). (I have interpreted relevance logic in terms of situations elsewhere Mares (2004).) Negation is handled by the Routley-star operator, $*$. For all *saturated* points x, $x \models \neg A$ if and only if $x^* \not\models A$.[4]

Perhaps a better fit for the source interpretation than Kripke semantics is an algebraic semantics. We can treat the sources of information as an algebra. Collecting two sources, a and b, together (using $\sqcup$) is a lot like taking their algebraic meet, $a \wedge b$. Similarly, as in Ono's semantics, we can think of the intersection of a and b as a source that contains only the content that occurs in both a and b. We can repre-

[4]The semantics for linear logic created by Allwein and Dunn (1993) might also be a candidate for a source reading, but I do not have the room here to discuss its complexities.

sent this intersection as a join, $a \vee b$. There are well-known algebraic semantics for MALL—classical linear algebras and the theory of quantales. There is also a well-known algebraic semantics for R, namely, Dunn's theory of De Morgan monoids (Dunn 1966; Anderson and Belnap 1975, Sect. 28.2).

In De Morgan monoids and other algebraic structures used to interpret relevance logics, fusion obeys certain postulates, such as commutativity $x \circ y = y \circ x$ and associativity $x \circ (y \circ z) = (x \circ y) \circ z$. The proof theory captures something like commutativity by its use of lambda abstraction. Consider the following proof.

$$\dfrac{[c\colon A \circ B] \qquad \dfrac{\dfrac{\dfrac{[a\colon A] \quad [b\colon B]}{b \circ a\colon B \circ A}}{\lambda x.(b \circ x)\colon B \to (B \circ A)}}{\lambda y \lambda x.(y \circ x)\colon A \to (B \to (B \circ A))}}{\dfrac{\lambda y \lambda x.(y \circ x) \circ c\colon (B \circ A)}{\lambda z \lambda y \lambda x.(y \circ x) \circ z\colon (A \circ B) \to (B \circ A)}}$$

In this proof, there is no need to commute source terms. The effect of commutativity is captured by the order in which source variables are bound by lambda abstracts.

The fact that commutativity along with associativity and other other postulates of the algebra are not incorporated into the proof theory might cause some problems for a source interpretation of the algebra, but I think these problems can be avoided. As a base for an algebraic semantics, let us take a set S of sources and a congruence relation $\approx$, which means 'contains the same information as'. Then we can take S modulo $\approx$ as the carrier set of the algebra. The use of $\approx$ has another good consequence. As I said in Sect. 3, there may not be unique collections or fusions of sources. Taking the points of an algebra to be congruence classes of sources all of which contain the same information would seem to get around the difficulty of thinking of the meet, join and fusion of the algebra as operators. (This same tactic might be used with regard to a Kripke semantics for the same reason.)

13 Conclusion

In this paper I have set out a programme for the interpretation of four relevance logics. The interpretation is constructive in the sense that it is supposed to describe the way in which people understand and build sources of information. On the productive interpretation, relevant implication is understood as a device that keeps track of how many steps are taken in the construction of sources. On the functional interpretation, the number or steps is ignored—what is doable in principle (in whatever number of steps) is what is of interest. The productive interpretation is used to understand the contraction-free logics MALL and RW and the functional interpretation is used to understand LR and R.

Interpreting relevance logic in terms of the construction of sources of information removes relevance logic from the metaphysically extravagant realm of true contradictions and impossible worlds, and locates it as an epistemological tool.

Acknowledgments I presented a very early draft of this paper to the Pukeko Logic Group. I am grateful to Jeremy Seligman, Zach Weber, and Patrick Girard for their helpful comments. I am also grateful for the two anonymous referees' helpful comments.

References

Allwein, G., & Dunn, J. M. (1993). Kripke semantics for linear logic. *Journal of Symbolic Logic, 58*(2), 514–545.
Anderson, A. R., & Belnap, N. D. (1975). *Entailment: The logic of relevance and necessity* (Vol. I). Princeton, NJ: Princeton University Press.
Anderson, A. R., Belnap, N. D., & Dunn, J. M. (1992). *Entailment: The logic of relevance and necessity* (Vol. II). Princeton: Princeton University Press.
Baker-Finch, C., (1992). *Relevant logic and strictness analysis* (pp. 81–82). LaBRI, Bordeaux, Bigre: Workshop on Static Analysis.
Barwise, J., & Perry, J. (1983). *Situations and attitudes*. Cambridge, MA: MIT Press.
Beall, J., & Restall, G. (2006). *Logical pluralism*. Oxford: Oxford University Press.
Belnap, N. (1993). Life in the undistributed middle. In K. Došen & P. Schröder-Heister (Eds.), *Substructural Logics* (pp. 31–41). Oxford: Oxford University Press.
Bimbó, K. (1999). *Substructural Logics, Combinatory Logic, and* λ*-calculus*, PhD thesis, Indiana University, Bloomington.
Bimbó, K. (2004). Semantics for dual and symmetric combinatory calculi. *Journal of Philosophical Logic, 33*, 125–153.
Bunder, M. (2003). Intersection type systems and logics related to the Routley-Meyer system B^+. *Australasian Journal of Logic, 1*, 43–55.
Dezani-Ciancaglini, M., Meyer, R. K., & Motohama, Y. (2002). The semantics of entailment omega. *Notre Dame Journal of Formal Logic, 43*, 129–145.
Dunn, J. M. (1966). *The Algebra of Intensional Logics*, PhD thesis, University of Pittsburgh.
Dunn, J. M. (1968). Natural versus formal languages. Given at an American Philosophical Association meeting.
Dunn, J. M. (1976). Intuitive semantics for first-degree entailments and 'coupled trees'. *Philosophical Studies, 29*, 149–168.
Dunn, J. M. (1987). Relevant predication 1: The formal theory. *Journal of Philosophical Logic, 16*, 347–381.
Dunn, J. M. (1990a). Relevant predication 2: Intrinsic properties and internal relations. *Philosophical Studies, 60*, 177–206.
Dunn, J. M. (1990b). Relevant predication 3: Essential properties. In J. M. Dunn & A. Gupta (Eds.), *Truth or Consequences: Essays in Honour of Nuel Belnap* (pp. 77–95). Dordrecht: Kluwer.
Dunn, J. M. (1993). Star and perp: Two treatments of negation. *Philosophical Perspectives, 7*, 331–357. (Language and Logic, J. E. Tomberlin (ed.)).
Dunn, J. M. (2010). Inconsistent information: Too much of a good thing. *Journal of Philosophical Logic, 39*, 425–252.
Dunn, J. M., & Meyer, R. K. (1997). Combinators and structurally free logic. *Logic Journal of the IGPL, 5*, 505–537.
Dunn, J. M., & Restall, G. (2002). Relevance logic. In D. Gabbay & F. Guenthner (Eds.), *Handbook of Philosophical Logic* (2nd ed., Vol. 6, pp. 1–128). Amsterdam: Kluwer.
Fine, K. (1974). Models for entailment. *Journal of Philosophical Logic, 3*, 347–372.

Frege, G. (1984). Thoughts. In B. McGuinness (Ed.), *Collected Papers on Mathematics, Logic, and Philosophy* (pp. 351–372), Oxford: Blackwell. Originally published in 1918–1919.
Girard, J.-Y. (1998). Light linear logic. *Information and Computation, 143*, 175–204.
Leslie, N., & Mares, E. (2004). CHR: A constructive relevant natural deduction logic. *Electronic Notes on Theoretical Computer Science, 91*, 158–170.
Mares, E. (2004). *Relevant logic: A philosophical interpretation*. Cambridge: Cambridge University Press.
Mares, E. (2009). General information in relevant logic. *Synthese, 167*, 343–362.
Mares, E. (2010). The nature of information: A relevant approach. *Synthese, 175*(supplement 1), 111–132.
Ono, H. (1993). Semantics for substructural logics. In K. Došen & P. Schröder-Heister (Eds.), *Substructural Logics* (pp. 259–291). Oxford: Oxford University Press.
Prawitz, D. (2006). *Natural deduction: A proof-theoretic study*. New York: Dover.
Read, S. (1988). *Relevant logic: The philosophical interpretation of inference*. Oxford: Blackwell.
Slaney, J. (1990). A general logic. *Australasian Journal of Philosophy, 68*, 74–89.
Thistlewaite, P., McRobbie, M., & Meyer, R. K. (1987). *Automated theorem proving in non-classical logic*. London: Pitman.
Troelstra, A. S. (1992). *Lectures on linear logic*. Stanford: CSLI.
Urquhart, A. (1972). Semantics for relevance logics. *Journal of Symbolic Logic, 37*, 159–169.
Yap, A. (2014). Idealization, epistemic logic, and epistemology. *Synthese, 191*, 3351–3366.

Epistemic Relevance and Epistemic Actions

Sebastian Sequoiah-Grayson

Abstract An operational and informational semantics for the ternary relation *R* is explored as a framework for modeling informational relevance. We extend this framework into robustly epistemic terrain. We take a new perspective on the problem of logical omniscience, using informationalised operational semantics to model the properties of the epistemic actions that underpin the epistemic relevance of certain explicit epistemic states of an epistemic agent as that agent executes said actions.

Keywords Epistemic actions · Logical omniscience · Informativeness · Relevance logics · Structural rules

1 Introduction

The problem of logical omniscience is the problem faced by epistemic modeling given that basic epistemic logics assume that the epistemic agents are logically omniscient, but we are not. This is a hard problem. The scandal of deduction is the failure of philosophy to give a sensible account of how it could be that deductive reasoning can be informative for us given that such inferences deliver zero information. The scandal of deduction has a straightforward answer, and this answer illustrates a way in which the problem of logical omniscience might be overcome. The answer to the scandal is best illustrated via a walk-through on general (non-logical) omniscience, and the information that we get from our empirical environment.

We get information from our environment either distally via direct observation, or indirectly via announcements. Examples are familiar from the philosophical canon. Consider *grass is green*, *snow is white*, or *there are one hundred and one dormice in the room next door*. In order to get information from our environment, we need

S. Sequoiah-Grayson (✉)
Sydney Centre for the Foundations of Science, University of Sydney, Sydney, Australia
e-mail: sebastian.sequoiah.grayson@sydney.edu.au
URL: http://www.logic.tsd.net.au

K. Bimbó (ed.), *J. Michael Dunn on Information Based Logics*,
Outstanding Contributions to Logic 8, DOI 10.1007/978-3-319-29300-4_8

to perform certain *epistemic actions*. These are actions such as the aforementioned observations or announcements.[1] We need to perform epistemic actions of these types on account of our not being omniscient. If we were omniscient, then we would not need to perform such actions because we would have automatic, effortless access to all information in our environment simply by definition. This is just what it means to be omniscient. We may, with a little poetic license, think of omniscience as the *limit* of the epistemic action of observation. Omniscience is the epistemic state achieved when further observation could not add any more information to our information base.

Just as we are not omniscient, neither are we *logically omniscient*. In order to use the information that we get from our environment, we need to reason with it. Such reasoning is an epistemic action of a cognitive sort, insofar as it is an action of the mind. We need to carry out such epistemic actions in order to bring the information corresponding to both logical theorems, as well as the logical consequences of our environmentally acquired information, into our information base. Analogous with the point made about omniscience and observations above, if we were logically omniscient, then we would not need to carry out reasoning-style epistemic actions because we would have automatic, effortless access to all logical theorems as well as all logical consequences of the information gotten from our environment. This is just what it means to be logically omniscient. Again, with a little poetic license, we may think of logical omniscience as the limit of the epistemic action of deductive reasoning. Logical omniscience is the epistemic state achieved when further deduction could not add any more information to our information base.

To be sure, when we speak of an epistemic agent being omniscient in the general sense, we often take this to imply that the agent is logically omniscient also. Nonetheless, these two types of omniscience remain conceptually distinct. There is, to be sure again, a commonly recognized priority of sorts between the two omniscience types. It is not particularly useful to think about an omniscient, but non-logically-omniscient agent. Such an agent might not be able to do all that much with the information that it got from its environment, if that agent lacked suitable logical acumen.[2] Being omniscient entails being in possession of a great deal of information, hence some heavy duty logical acumen would be required to handle it.

When it comes to modeling epistemic actions of the observation sort, the agents being modeled are assumed often to be logically omniscient for just this reason (see van Ditmarsh et al. 2008). By abstracting away from the cognitive epistemic actions which underpin logical information handling, such frameworks—standard

[1]Announcements and observations may be run together as a single type of epistemic action if you assume that announcements are always truthful, always believed, and always non-noisy, van Ditmarsh et al. (2008).

[2]This point is similar to the one made by Frege in his letter to Jourdain. Frege entertains an agent who is able, in principle, to comprehend every atomic sentence, but does not have the ability to execute any semantic composition. Given language's essential productivity, such an ability is, according to Frege, of little general interest. I am indebted to an anonymous referee for bringing this to my attention.

dynamic epistemic logic (DEL) for example—may concentrate on the properties of the information-updates resulting from observation-type epistemic actions. In such frameworks, the epistemic agent is assumed to be logically omniscient, or ideally rational, and the cognitive epistemic actions executed by the agent get "black boxed."[3]

This essay is an attempt to say something philosophically substantial about the nature of the epistemic actions which underpin logical information handling—to shine some light inside the black box.[4]

2 Epistemic Relevance and Relevance Logics

Both information itself as well as epistemic actions may be *epistemically relevant*. Some information is epistemically relevant for an agent if it is relevant to the agent's epistemology, where by this we mean the agent's knowledge or beliefs. For example, if you need to know how many bottles of wine you might need for your dinner party, then the number of guests is epistemically relevant. Similarly, if you have the information that the terrorist cell will attack either the Sydney Harbour Bridge or the Sydney Opera House, then the information that the terrorist cell will not attack the Sydney Harbour Bridge is epistemically relevant to your counter-terrorist plans.

An epistemic action will be epistemically relevant for an agent if the execution of the action gets information for the agent such that this information is epistemically relevant in the manner described above. For example, the announcement from each of your dinner party's invitees that they are able to attend the party will be a collection of epistemically relevant epistemic actions. Similarly, an observation of the terrorist cell's moving their personnel away from the Sydney Harbour Bridge is epistemically relevant. Both the dinner party and terrorist cell examples assume that you are able to reason with, or integrate, or logically handle the information that you got from the announcement and observation actions. As we noted in the previous section, this handling of information in a logical manner is an epistemic action of an internal, cognitive sort. Logically handling or reasoning with information will be epistemically relevant for an agent if the execution of such reasoning gets information for the agent such that this information is epistemically relevant.

Both the nature of epistemic relevance and the nature of the cognitive epistemic actions which underpin deductive reasoning could do with clarification and elaboration. We can find both of these with some help from *relevance logics* (see Mares 2004 for a canonical introduction).

[3]Assuming logical omniscience for the epistemic agents in one's model makes perfect sense insofar as one wants to idealize away from variables.

[4]The motivation here is similar to that of (Duc 1997). The difference is that Duc has models for what the agent knows *after* she has executed some rule of inference or other, whereas here we will be modelling the properties of the epistemic actions which underpin the execution of such rules.

That relevance logics provide a logical framework for epistemic relevance and epistemic actions is at the very least not obvious. Such logics are neither thought of as particularly epistemic, nor as dynamic (and actions, epistemic or otherwise, are dynamic if anything is). To see how it is that we might be justified in thinking of relevance logics as being both, we will skip the detailed nomenclature of the logics' syntactic/proof theoretic properties and motivations, and go directly to the semantics.

In relevance logics, a *relevance frame* **F** is a pair $\langle S, R\rangle$ consisting of a set $x, y, z, \ldots \in S$ of points of evaluation, and a ternary relation R on this set. A *relevance model* **M** is a pair $\langle \mathbf{F}, \Vdash\rangle$ consisting of a relevance frame **F** and an evaluation relation $\Vdash$ which holds between the points of evaluation in S and formulas $\phi, \psi, \ldots$.

We may now state the evaluation conditions given by relevance logic for the conditional $\phi \rightarrow \psi$ as follows.

$$x \Vdash \phi \rightarrow \psi \quad \textit{iff} \quad \forall y, z : Rxyz, \text{ if } y \Vdash \phi, \text{ then } z \Vdash \psi. \tag{1}$$

Equation (1) is still slightly opaque. What are the points of evaluation $x, y, z, \ldots$, and what does R mean?

The points of evaluation work just like possible worlds, except that in the present case they may be both inconsistent and incomplete. It is common practice to speak of the points of evaluation as *information states*, since there is no obvious constraint on a body of information that it be complete or consistent. Making sense of such information states insofar as we want them to correspond to something in the real world is the task of Sect. 3.

How to make sense of $Rxyz$ is an infamous issue. We might understand $Rxyz$ as something like "*if you combine the things which are true at x with the things which are true at y then you get the things which are true at z*." This is a good start, but does $x \Vdash \phi$ mean that ϕ is true at x? It does not, not quite. Given that inconsistent propositions may hold at points, that is, given that we may have $x \Vdash \phi \wedge \neg\phi$, understanding $\Vdash$ as "true at" is a little too crude.

Instead, we may understand $x \Vdash \phi$ as "*x carries/stores the information that* ϕ." In this case, $Rxyz$ comes out as "*if you combine the information carried by/stored at x with the information which is carried by/stored at y then you get the information which is carried by/stored at z*." This is an improvement over a "true of" understanding, and it puts us in a position to use relevance frames (and their corresponding models) in order to understand both epistemic relevance and cognitive epistemic actions.[5]

[5]See Mares (1996) and Restall (1996) for the fine-grained details involved in "informationalising" the ternary relation R. See also Dunn and Hardegree (2001).

3 Epistemic Relevance and Epistemic Actions

Following Dunn (2015), we will take the partial order of information inclusion, $\sqsubseteq$, to indicate *information relevance*.[6] In this case, $x \sqsubseteq y$, "*the information at x is included in the information at y*," means that the information at x is *relevant to* information at y.

This does not seem to be too much of a stretch. If the information at x is included in the information at y, then the information at the former seems relevant to the information at the latter on account of the inclusion itself. The information at y takes the information at x to be relevant because the information at y just is an informational extension of the information at x.

The informational relevance indicated by $x \sqsubseteq y$ is *non-contextual* relevance insofar as the relevance of x to y does not depend on any further information (or further informational context, as we might say). Suppose that $x \Vdash A$ and $y \Vdash A, B$. In this case we might have it that $x \sqsubseteq y$.[7] However, suppose instead that $y \Vdash A$ and $z \Vdash B$. Is it the case that we might have it that $y \sqsubseteq z$? Not as things stand, which is to say not without some further informational context.

Such further informational context may be given as follows. Suppose that we have $x \Vdash A \rightarrow B$. In the context of x (and given that $y \Vdash A$ and $z \Vdash B$ as specified in the paragraph above), it is the case that $y \sqsubseteq z$. This is just to say that if we take the information in state x together with the information in state y, then these two information states, when taken together, carry information which is relevant to the information in state z. We may represent this taking together of, or combination of, two information states with a binary composition operation on information states, $\bullet$. Given that $x \Vdash A \rightarrow B$, $y \Vdash A$ and $z \Vdash B$, we have it that $x \bullet y \sqsubseteq z$. In other words, given the information carried by states x and y, their combination is relevant to the information carried by state z. Moreover, the very act of combining x and y is itself informationally relevant to z. This is because it is the operation of combining x and y which bring the information at both states together. Sans such an operation, the information at x and y are separate informational entities, neither of which, either considered independently or non-contextually, are informationally relevant to z.

We may now give a more thoroughgoing explanation of $Rxyz$. We may understand $Rxyz$ as $x \bullet y \sqsubseteq z$. In this case, our relevance frame becomes an *information frame* $\mathbf{I}$, which is a triple $\langle S, \sqsubseteq, \bullet \rangle$. Our relevance model becomes an information model $\mathbf{M}_I$. Given this much, (1) comes out as:

---

[6]The role of a partial or pre-order in the Routley–Meyer semantics for relevance logic is well-known and explored in some detail in (Bimbó and Dunn 2008, Chap. 2).

[7]This is not guaranteed, since there is no sensible requirement on an epistemic state that the state in question be itself epistemically relevant to another epistemic state that subsumes the information carried by the original state. For example, my knowing that *grass is green* at some point in time does not have to be an epistemically relevant episode to every future epistemic state or action involving the information that grass is green. That is, our epistemic states are not totally ordered.

$$x \Vdash \phi \rightarrow \psi \quad \textit{iff} \quad \forall y, z : x \bullet y \sqsubseteq z, \text{ if } y \Vdash \phi, \text{ then } z \Vdash \psi. \tag{2}$$

With the full informational relevance interpretation of the relevance semantic conditions for the conditional in hand, we are in a position to see how it is that relevance frames have a role to play with regard to understanding our target phenomena—epistemic relevance and cognitive epistemic actions.

We stated above that making sense of the information states insofar as we want them to correspond to something in the real world was the present task. By making such sense, we will be on our way to addressing the issue posed at the end of the previous paragraph. Here is the suggestion:

We may understand the information states $x, y, z, \ldots$ to be states of explicit knowledge/belief of an epistemic agent, in other words, as *explicit epistemic states*.

By understanding the information states to correspond to explicit epistemic states of an agent, we have a direct link between information relevance, on the one hand, and our target phenomena of epistemic relevance, on the other. Suppose again that $x \Vdash A$ and $y \Vdash A \wedge B$. Now the former states that some agent α knows/believes explicitly that A, with the latter now stating that α knows/believes explicitly that $A \wedge B$.[8] In this case, $x \sqsubseteq y$ states that the agent's explicit epistemic state x is non-contextually epistemically relevant to their explicit epistemic state y.[9] But it is with *contextual* epistemic relevance that things get interesting.

Suppose that α is in the states $x \Vdash A \rightarrow B$, and $y \Vdash A$. This alone is insufficient for α to be in the state z such that $z \Vdash B$. For α to be in a state z such that $z \Vdash B$, α needs to *combine* the information in her states x and y. This is just to say that having explicit knowledge/belief of/in premises is insufficient for explicit knowledge/belief of/in conclusions. In order for α to get to z, she has to *think about things in the right way*. To think about things in the right way just is to combine the information encoded by the premises in such a manner that the result of this combination will make the information encoded by the conclusion explicit to α. The act of combining explicit epistemic states is just that, an act, or action. And it is such epistemic actions that underpin logical information handling, or the *process* of deductive reasoning itself.

At the end of the introduction we said that we were working towards saying something philosophically substantial about the nature of these epistemic actions. Information frames allow us to now do so.

$x \bullet y$ is a representation of the very epistemic action that we are looking for. Given that $x \Vdash A \rightarrow B$ and $y \Vdash A$, then given that $z \Vdash B$, it will be the case that $x \bullet y \sqsubseteq z$. Given our understanding of information states as explicit epistemic states, and of $\sqsubseteq$ as epistemic relevance, and of $\bullet$ as the epistemic action of combining such states, $x \bullet y \sqsubseteq z$ says something significant. It says that α's being in the explicit epistemic

[8]Of course, we could write "α knows/believes explicitly that A" as $x \Vdash_{\alpha} A$ or some such, but typographical rigour has a tendency to get in the way of readability.

[9]As well might be the case, given that both x and y carry A.

states x and y, *and* the execution by α of the epistemic action of combining these states, are *both* epistemically relevant for α's being in the epistemic state z.[10]

This is exactly what we are after. Given that we have a sensible framework for representing the cognitive epistemic actions which underpin deductive reasoning, the task now is to use this framework to say something philosophically substantial about such actions. In particular, what properties might such epistemic actions possess which preserve epistemic relevance? In other words, what properties might an agent's cognitive epistemic actions possess such that the properties guarantee that the agent will arrive in the correct epistemic state?

4 Preserving Epistemic Relevance

The properties that an agent's cognitive epistemic actions will need to possess such that these properties guarantee that the agent will arrive in the correct epistemic state will vary. Their variance will depend upon the logical form of the information that is being handled by the epistemic action itself. We can capture the nature of these form-contingent action properties with *structural rules.*

A structural rule tells us what structural changes may be made to the body of information being processed, whilst preserving the given output of that same act of processing. Let's start with four basic structural rules, *Association*, *Commutation*, *Contraction* and *Weakening*. Where $\Longrightarrow$ is if-then in the metalanguage,

$$w \bullet (x \bullet y) \sqsubseteq z \Longleftrightarrow (w \bullet x) \bullet y \sqsubseteq z \qquad \text{(Association)}$$

$$x \bullet y \sqsubseteq z \Longrightarrow y \bullet x \sqsubseteq z \qquad \text{(Commutation)}$$

$$x \bullet x \sqsubseteq x \qquad \text{(Contraction)}$$

$$x \bullet y \sqsubseteq z \Longrightarrow x \sqsubseteq z \qquad \text{(Weakening)}$$

Association tells us that, given a sequence of information states, the order of pairwise composition within that sequence makes no difference to informational output.[11] Commutation tells us that given a pairwise composition of information states, the order of the information states in the pair being composed makes no difference to

[10] Although cognitive epistemic actions may, and often do, involve the combination more that two premises, the treatment of the two-premise case is privileged on several fronts. Firstly, it is the simplest possible case. Given this, any model of cognitive epistemic actions needs to be shown to handle such cases before being applied to more complex cases. Secondly, it seems to be at least plausible that the majority of deductive episodes do proceed via two-premise combinations. Witness the standard natural deduction rules and classical syllogisms as examples. A third reason is simply that the two-premise case is hard enough.

[11] Note that Association is given here in its readable, abbreviated form. The full form of Association is $\exists u((x \bullet y \sqsubseteq u) \wedge (w \bullet u \sqsubseteq z)) \Longleftrightarrow \exists t((w \bullet x \sqsubseteq t) \wedge (t \bullet y \sqsubseteq z))$. This makes sense if you think about it. In the abbreviated form above, we are merely cutting out explicit reference to the states u and t, which are the results of composing w and x on the one hand, and x and y on the other, respectively.

informational output. Contraction tells us that the composition of two information states that carry the same informational payload outputs no more information than that carried by one of the states. Weakening tells us that we can get the same informational output if we weaken the epistemically relevant information states.

Given that we are understanding the information states as explicit epistemic states, the composition operation as the epistemic action of combining such states, and the partial order of informational inclusion as epistemic relevance, then the epistemic action contexts in which the structural rules hold or fail become salient. They become salient because they specify the properties that said epistemic actions need to possess with regard to guaranteeing epistemic success.

We may begin by considering cases at the level of abstraction where the agent's epistemic states carry information of either atomic or conditional form. In this case, α's explicit epistemic state x may be such that either $x \Vdash p$, or $x \Vdash p \rightarrow q$. In other words, α knows/believes explicitly some information which may be of either two forms. In this case, we have three possible scenarios given α's epistemic action $x \bullet y$, or, with a bit of a push, three types of epistemic actions. Both epistemic states may carry atomic information, or one epistemic state may carry information of atomic form and the other of conditional form, or both epistemic states may carry information of conditional form. Following Dunn (2015), we will call the first scenario the *Data Combining* (DC) interpretation, the second scenario the *Program Applied to Data* (PD) interpretation, and the third scenario the *Program Combining* (PC) interpretation.[12] So we have things as follows (reading ":" as *such that*, and "$z \Vdash p, q$" as shorthand for "$z \Vdash p$ and $z \Vdash q$").

$$\text{If } x \Vdash p \text{ and } y \Vdash q, \text{ then } x \bullet y \sqsubseteq z : z \Vdash p, q \tag{DC}$$

$$\text{If } x \Vdash p \rightarrow q \text{ and } y \Vdash p, \text{ then } x \bullet y \sqsubseteq z : z \Vdash q \tag{PD}$$

$$\text{If } x \Vdash p \rightarrow q \text{ and } y \Vdash q \rightarrow r, \text{ then } x \bullet y \sqsubseteq z : z \Vdash p \rightarrow r \tag{PC}$$

The consequences for the structural rules given these three epistemic action scenarios are interesting insofar as we are using the structural rules to specify the epistemically salient properties of the epistemic actions themselves.

Let's start with Association.

$$w \bullet (x \bullet y) \sqsubseteq z \Longleftrightarrow (w \bullet x) \bullet y \sqsubseteq z \tag{Association}$$

Association fails for some epistemic actions in PD scenarios. Consider the following explicit epistemic states w, x, y, z of α.

[12]Dunn uses "data" to refer to static information p, q, etc., and "programs" to refer to dynamic information, or conditionals, $p \rightarrow q$, etc. As we will see in Sect. 5, agents may treat programs *as* data.

$$w \Vdash q \rightarrow r \qquad (3)$$
$$x \Vdash p$$
$$y \Vdash p \rightarrow q$$
$$z \Vdash q$$

In its left to right hand direction, Association fails for (3). This is just to say that although the epistemic action $w \bullet (x \bullet y)$ is epistemically relevant for α with respect to α's being in state z in a PD type epistemic action, that is to say, although we have it that $w \bullet (x \bullet y) \sqsubseteq z$, we do not have it that $(w \bullet x) \bullet y \sqsubseteq z$. Composing the information carried by w and x ($q \rightarrow r$ and p respectively) with PD type epistemic actions is an illegitimate epistemic action insofar as it will not get the agent anywhere, epistemically speaking. The result will not be anything which may be composed with the information carried by the α's state y ($p \rightarrow q$) such that is may be used to get α into state z. Via similar reasoning, we can see that Association will fail in its right to left hand direction where we have it that $w \Vdash p \rightarrow q$, $x \Vdash p$, $y \Vdash q \rightarrow r$ and $z \Vdash r$.[13]

However, there is no failure for Association for epistemic actions involving PC scenarios. Consider any three epistemic states w, x, and y, such that each state carries information of composable conditional form. In this case, any output state z such that the epistemic action is epistemically relevant for α with respect to z (i.e., $(w \bullet x) \bullet y \sqsubseteq z$) will be preserved under Association. Consider the following explicit epistemic states of α.

$$w \Vdash p \rightarrow q \qquad (4)$$
$$x \Vdash q \rightarrow r$$
$$y \Vdash r \rightarrow s$$
$$z \Vdash p \rightarrow s$$

Association holds for (4), as it will for any PC scenario where the information states carry information with composable conditional form.[14]

In contrast with Association, however, Commutation holds for epistemic actions consisting of PD scenarios, but fails for those consisting of PC scenarios.

$$x \bullet y \sqsubseteq z \Longrightarrow y \bullet x \sqsubseteq z \qquad \text{(Commutation)}$$

[13] There is a lot to say here about dynamic negation and negative information. One way to go is to say that there is a null object $\mathbf{0}$ such that $x \Vdash \mathbf{0}$ for no x. The way is then clear to define a dynamic negation $A^{\mathbf{0}}$ in terms of $A \rightarrow \mathbf{0}$, which will type information of the type that can never be combined with information of type A. Classical and other static negations rule out truth, whilst dynamic negations rule out certain operations or combinatorial procedures. See Dunn (1993, 1996) and Sequoiah-Grayson (2009).

[14] Since, as the category theory folks are fond of saying, "Arrows associate!".

Consider again the epistemic states specified by (3). A thoroughgoing application of Commutation to (3) would give us the following.

$$w \bullet (x \bullet y) \sqsubseteq z \Longrightarrow (y \bullet x) \bullet w \sqsubseteq z \tag{5}$$

Given the epistemic states of α specified by (3) and (5) holds (as is checked easily). In fact, Commutation will hold for any PD type collection of epistemic states whatsoever. This is because for any arbitrary pairwise composition of two pieces of information such that one piece is the input of the other piece, the composition will be order invariant. This is just a slick way of saying that for any two pieces of information such that one is of form A and the other is of form $A \rightarrow B$, the order of their composition is irrelevant insofar as deriving B is concerned, and similarly of course for the order of the epistemic states being composed by the relevant epistemic action.

However, Commutation fails for epistemic actions consisting of PC scenarios. Consider a simplified version of the scenario specified by (6).

$$\begin{aligned} x &\Vdash p \rightarrow q \\ y &\Vdash q \rightarrow r \\ z &\Vdash p \rightarrow r \end{aligned} \tag{6}$$

We have it that $x \bullet y \sqsubseteq z$. If Commutation held here, then we should have it that $y \bullet x \sqsubseteq z$, but this is not the case. This latter epistemic action is not epistemically relevant for α's being in the epistemic state z at all (since $(q \rightarrow r) \circ (p \rightarrow q)$ is the wrong order insofar as combining dynamic information is concerned).

The following related example emphasizes this point.

$$\begin{aligned} x &\Vdash p \rightarrow q \\ y &\Vdash q \rightarrow p \\ z &\Vdash p \rightarrow p \end{aligned} \tag{7}$$

With (7), we have it that $x \bullet y \sqsubseteq z$ also. But we do not have it that $y \bullet x \sqsubseteq z$. $y \bullet x$ results in a state $w \Vdash q \rightarrow q$, and $p \rightarrow p \neq q \rightarrow q$![15]

Consider Contraction.

$$x \bullet x \sqsubseteq x \tag{Contraction}$$

Contraction fails for epistemic actions of PC types in general, although it does hold for some special restricted cases. These cases are those where the antecedent and consequent of the relevant conditional encode the same information, as carried by the following explicit epistemic state.

[15] Although both formulas are classically (and non-classically in certain logics) equivalent, recall that our epistemic agent α is not logically omniscient.

$$x \Vdash p \rightarrow p \tag{8}$$

Contraction is preserved by epistemic actions that combine information of the type specified by (8), since the epistemic action in question is epistemically relevant to α's knowing explicitly that $p \rightarrow p$. Of course the epistemic action might well be *redundant*, but that is neither here nor there.

However, consider the following explicit epistemic state.

$$x \Vdash p \rightarrow q \tag{9}$$

Contraction fails for epistemic actions of the sort composed with epistemic states of the type specified by (9). Here, the situation is not that the epistemic action in question is redundant, but that it is *epistemically irrelevant*. The PC type epistemic action $(p \rightarrow q) \circ (p \rightarrow q)$ does *not* result in $p \rightarrow q$.

Contraction does not apply at all to PD type epistemic actions, on account of the epistemic states composing contracted epistemic actions carrying the same explicit informational payload (by definition), whilst the epistemic states composing PD type epistemic actions must be of different types (again by definition).

Consider Weakening.

$$x \bullet y \sqsubseteq z \Longrightarrow x \sqsubseteq z \tag{Weakening}$$

Weakening fails outrightly for both PD and PC scenarios. To see this, consider the following PD scenario.

$$\begin{aligned} &x \Vdash p \rightarrow q \\ &y \Vdash p \\ &z \Vdash q \end{aligned} \tag{10}$$

Given the explicit epistemic states specified by (10), we have it that $x \bullet y \sqsubseteq z$. α's epistemic action combining α's explicit knowledge/belief of $p \rightarrow q$ and p is, along with the relevant epistemic states themselves (x and y) epistemically relevant to α's knowing explicitly that q. In other words, it is epistemically relevant to α being in the epistemic state z. However, $x \sqsubseteq z$ states that α's being in the epistemic state x, that is, their explicit knowledge that $p \rightarrow q$, is *non-contextually* epistemically relevant to their being in the epistemic state z, that is, their explicit knowledge that q. This it most certainly is not.

Now consider again the PC type epistemic actions specified by (6).

$$\begin{aligned} &x \Vdash p \rightarrow q \\ &y \Vdash q \rightarrow r \\ &z \Vdash p \rightarrow r \end{aligned} \tag{6}$$

Reasoning directly analogous to that entertained with respect to (10) demonstrates that Weakening fails for PC scenarios such as (6) also.

The failure of Weakening and Contraction for epistemic action scenarios is not entirely surprising insofar as brute considerations with regard to informational resources are concerned. The curious behaviour of Association and Commutation in our epistemic context is however, rather surprising indeed. There is, we should hope, a great deal more to say here with respect to structural rules, epistemic relevance, and epistemic actions.

5 Treating Programs as Data

But what of DC scenarios? When α is reasoning from a state x such that $x \Vdash p \rightarrow q$ (or any other piece of conditional information), it does not have to be the case that α's epistemic action is *an attempt* to combine this state with one carrying input-information, $y \Vdash p$ for example. Neither must it be the case that α is attempting to merge the information carried by this state with another state carrying information in conditional form as with scenario (6). This is just to say that α does not always have to treat dynamic information dynamically, so to speak. Instead, α may treat a program *as* data of a complex, non-atomic sort.

This will be the situation with many of α's epistemic actions. Consider those actions underpinning the merging of $p \rightarrow q$ with $(p \rightarrow q) \rightarrow r$ for example. In the context of this epistemic action, the dynamic information $p \rightarrow q$ is being treated by α as static data, input into the dynamic $(p \rightarrow q) \rightarrow r$. DC type epistemic actions build on this idea. Suppose that α is in the explicit epistemic state $x \Vdash ((p \rightarrow q) \wedge r) \rightarrow s$. Suppose also that α enters into two distinct sequences of reasoning, one of which brings α to state y such that $y \Vdash p \rightarrow q$, and another of which brings α to a state z such that $z \Vdash r$. In this case, for α to get to state w such that $w \Vdash (p \rightarrow q) \wedge r$, α will need to combine her states y and z *in such a way* that $y \bullet z \sqsubseteq w$, such that $w \Vdash p \rightarrow q, r$.

Importantly however, the "way" in which α combines y and z will be a way that treats the information carried by y as *data* to be combined with the data carried by z. This ensures that the result of the epistemic action $y \bullet z$ is w such that $w \Vdash p \rightarrow q, r$, as opposed to some failed attempt to *input* the information carried by z to the information carried by y. In other words, α knows that the epistemic action that she is executing with $y \bullet z$ is a DC scenario and not a PD one. A PD type epistemic action will in this case not be epistemically relevant to w at all, hence we would not have it that $y \bullet z \sqsubseteq w$.

The exact status of Boolean connectives $\wedge$, $\vee$, is something of a delicate matter. Although an agent may be reasoning with complex bodies of information which *contain* Boolean connectives, it is unlikely that the agent's epistemic states inherit all of the properties of these connectives. Consider the following under our epistemic state interpretation.

$$x \Vdash p \wedge q \text{ iff } x \Vdash p \text{ and } x \Vdash q. \tag{11}$$

$$x \Vdash p \vee q \text{ iff } x \Vdash p \text{ or } x \Vdash q. \tag{12}$$

In its left to right direction, (11) is true straightforwardly. If α knows/believes explicitly that $p \wedge q$, then α knows/believes explicitly that p and knows/believes explicitly that q. The right to left hand direction is slightly trickier however. Equation (11) is true in its right to left hand direction, *given* the restricted case that it specifies. This is a consequence of it being the case that if α is in an explicit epistemic state x, which carries the information that p, and *that very same* explicit epistemic state x of α's carries the information that q, then x will carry $p \wedge q$. But this is not true of explicit epistemic states in general. It can be the case that α knows/believes explicitly that p, and that α knows/believes explicitly that q, without it being the case that α knows/believes explicitly that $p \wedge q$. Suppose that p and q are carried by explicit, but distinct epistemic states of α (x and y say). In this case there is no guarantee that α will have, or even so much as ever get to, some explicit epistemic state $z \Vdash p \wedge q$. For α to reach such a state z, α needs to execute a DC type epistemic action such that $x \bullet y \sqsubseteq z$.[16]

Equation (12) is even less well behaved in a robustly epistemic context than is (11). In its right to left direction, (12) is well behaved epistemically. In its left to right direction however, (12) fails for even the restricted case that it captures. It might well be true that α knows/believes explicitly that $p \vee q$, that is, α may be in state $x \Vdash p \vee q$, without it being the case that α knows/believes explicitly that p, or knows/believes explicitly that q. Consider an example from Dunn (2015), where you remember or believe that you left your keys either on the upstairs dresser, or on the basement workbench.[17] You could well be in the explicit epistemic state, without it being in that case that either of the disjuncts (considered independently) fall within the scope of that same epistemic state. Interestingly, there does not seem to be any obvious cognitive, or *a priori* executable epistemic action, DC type or otherwise, which would bring α to p or to q in this case. Rather, it would be an observation-type epistemic action.

This is to only touch on the issue of the DC type epistemic actions with regard to Boolean connectives. That there is more to say is obvious, but what to say is less so.

6 Conclusion

We have made a distinction between different types of omniscience, as well as different types of epistemic actions. Hopefully, a strong case has been made for a central role of such actions when it comes to *a priori* reasoning. Hopefully, a strong case has been made for the use of the structural rule architecture of relevance and related logics when it comes to modeling the properties of such actions for non-ideal, or non-logically omniscient agents also.

---

[16] For an investigation into the epistemic role of explicit conjunctions, especially with respect to the closure axiom and related modal-epistemic phenomena, see Sequoiah-Grayson (2013).

[17] Although Dunn's remarks are not framed in explicitly epistemic terms, all of his examples concerning disjunction are epistemic/doxastic in nature. This is presumably no mere coincidence!

There has been a slow, but reassuringly steady interest in the applicability of relevant and related substructural logics to epistemic phenomena. See for example Majer and Pelis (2009) and their followup paper Bilkova et al. (2010). Relatedly, Sedlar (2012) makes explicit connections between universal modal operators and the ternary relation R. Given the role that such modal operators have played in traditional epistemic logic, the future along this route is promising. Relatedly, Sedlar (2014) and Roy and Hjortland (2014) explore epistemicised substructural modal logics to explore various epistemic phenomena of the epistemic action and epistemic update variety. There is hopefully much more to come.[18]

References

Bilkova, M., Majer, O., Pelis, M., & Restall, G. (2010). Relevant Agents. *Advances in Modal Logic* (Vol. 8). London: College Publications.
Bimbó, K., & Dunn, J. M. (2008). *Generalized Galois logics. Relational semantics of nonclassical logical calculi, vol. 188 of CSLI Lecture Notes*. Stanford, CA: CSLI Publications.
Duc, H. N. (1997). Reasoning about rational, but not logically omniscient, agents. *Journal of Logic and Computation*, *7*(5), 633–648.
Dunn, J. M. (1993). Star and perp: Two treatments of negation. *Philosophical Perspectives, 7*, 331–357. (Language and Logic, J. E. Tomberlin (Ed.)).
Dunn, J. M. (1996). Generalised ortho negation. In H. Wansing (Ed.), *Negation: A notion in focus* (pp. 3–26). New York, NY: Walter de Gruyter.
Dunn, J. M. (2015). The relevance of relevance to relevance logic. In M. Banerjee & S. N. Krishna (Eds.), *Logic and its applications: 6th Indian conference, ICLA 2015, Mumbai, India, January 8–10, 2015, number 8923 in lecture notes in computer science* (pp. 11–29). Heidelberg: Springer.
Dunn, J. M., & Hardegree, G. M. (2001). *Algebraic methods in philosophical logic, vol. 41 of Oxford logic guides*. Oxford, UK: Oxford University Press.
Majer, O., & Pelis, M. (2009). Epistemic logic with relevant agents. In M. Pelis & V. Puncochar (Eds.) *The Logica Yearbook 2008* (pp. 123–135).
Mares, E. D. (1996). Relevant logic and the theory of information. *Synthese*, *109*, 345–360.
Mares, E. D. (2004). *Relevant logic: A philosophical interpretation.* Cambridge: Cambridge University Press.
Restall, G. (1996). In J. Seligman & D. Westerstahl (Eds.), *Logic, language, and computation* (Vol. 1, pp. 463–477)., Information flow and relevant logics Stanford: CSLI Publications.
Roy, O., & Hjortland, O. T. (2014). Dynamic consequences for soft information. *Journal of Logic and Computation*, 1–22.
Sedlar, I. (2012). Boxes are relevant. In M. Pelis & V. Puncochar (Eds.), *The Logica Yearbook 2011* (pp. 265–278). London: College Publications.
Sedlar, I. (2014). Epistemic extensions of modal distributive substructural logics. *Journal of Logic and Computation*, (online first).
Sequoiah-Grayson, S. (2009). Dynamic negation and negative information. *Review of Symbolic Logic*, *2*(1), 233–248.
Sequoiah-Grayson, S. (2013). Epistemic closure and commuting, nonassociating residuated structures. *Synthese*, *190*(1), 113–128.
van Ditmarsh, H., van der Hoek, W., & Kooi, B. (2008). *Dynamic epistemic logic*. Springer.

[18] I would like to thank two anonymous referees for their invaluable feedback. I would like to thank Katalin Bimbó also, for her tireless efforts and superhuman patience.

Comparing Contents with Information

Ross T. Brady

Abstract I first introduced my notion of logical content in 1988 and 1989. This was a broad concept providing the backbone for a style of algebraic semantics, called "content semantics", and covering a wide range of logics from the weak relevant logic BBQ right through to the classical predicate calculus. This concept was subsequently specialized in 1996, in such a way as to help conceptualize a particular logic DJ^d. This specialized concept was extended to quantifiers in 2006, and was modified, jointly with Meinander in 2013, to form the logic MCQ. In this paper, we contend that contents are best represented as analytic closures, with the appropriate entailments captured by this logic MC of meaning containment. On the other hand, the term "information" has been widely used in logical work, usually as a means of underpinning or understanding a semantics of a logic or logics. Floridi in his book of 2011, contends: "semantic information is well-formed, meaningful and truthful data". We pick up on this, by essentially adding the concept truth to that of contents to form information, appropriately chosen for its logical usage. We also divide information into two types: prime and non-prime information, and also determine their respective impacts on the proof theory and semantics of logical systems, with special interest in those of the relevant logics. We especially refer to the works of Carnap, Dunn and his former student, Mares.

Keywords Analytic closure · Contents · Information · Priming property · Veridicality thesis

1 Introduction

We propose to compare the two semantic concepts: contents and information, in special recognition of the work of Prof. Michael Dunn. I am most pleased to be asked to write such a contribution in honour of a logician I have always admired and with

---

R.T. Brady (✉)
La Trobe University, Melbourne, VIC 3086, Australia
e-mail: Ross.Brady@latrobe.edu.au
URL: http://www.latrobe.edu.au/humanities/about/staff/profile?uname=RTBrady

K. Bimbó (ed.), *J. Michael Dunn on Information Based Logics*,
Outstanding Contributions to Logic 8, DOI 10.1007/978-3-319-29300-4_9

whom I've had a close friendship and academic association over many years. His interest in the study of the concept of information stems from his lifetime's work in relevant logics and computation.[1] So, we will focus on the use of the concepts, contents and information, especially in the study of relevant logics, with some mention of computer databases.

Nevertheless, we will take as our starting point, Carnap's introduction of the two concepts from a classical perspective, before moving on to the two concepts separately, covering a wider range of logics. For contents, we focus on the author's work on logical contents within a content semantics, first introduced in Brady (1988, 1989), for relevant and other logics, and subsequently specialized in Brady (1996, 2006), in such a way as to help conceptualize a particular logic DJ^d (see Sect. 3 for its axiomatization). For information, though widely used as a means of underpinning or understanding a semantics of a logic or logics, we focus on the works of Dunn and his former student Mares, especially, in relation to the Routley–Meyer style of semantics of relevant logics, but also in relation to their natural deduction systems.

We then single out what appears to be the distinguishing differential features between contents and information. We will see that contents do not need to embody truth; they are just the static semantic content of a sentence or set of sentences. In comparison, we will argue that information does presuppose the truth of the sentence or sentences involved, this issue being discussed in Dunn (2013) with reference to Floridi (2011), the latter claiming that "semantic information is well-formed, meaningful and truthful data".

Further, information is very often studied in the context of informational inference where information is expected to be able to flow from one piece of information to another. In such a case, the truth is "assumed truth", assumed for the sake of a logical deduction formalizing such a flow of information. This will then relate information, not only to truth-theoretic semantics such as the Routley–Meyer semantics, but also to natural deduction systems, as can be seen in Mares (2010). However, there are two strands of information, depending on whether the Priming Property (if $A \vee B$ is included, then so is either A or B) is satisfied or not, a point made by Dunn (2015). Contents, on the other hand, relate closely to content semantics, that is an algebraic style of semantics, which in turn relates closely to Hilbert-style axiomatization.

So, we will finally distinguish contents, which are indifferent to truth and falsity, information with truth, whether satisfying the Priming Property or not, and information with assumed truth, for the purpose of capturing informational inference. We will see that information which includes truth is most likely to satisfy priming, whilst information including assumed truth may well not be prime. However, we stop short at taking the further step of quantifying information by talking about its amount in mathematical terms, as introduced in Carnap and Bar-Hillel (1964).

[1]I should have said 'relevance logic', using the American terminology, especially, in the light of Dunn's recent work "The relevance of relevance to relevance logic" (2015). However, I will stick to the current Australian terminology, in keeping with my earlier work in this area.

2 Carnap on Contents and Information

In Sect. 18 of his book (1942), Carnap introduces an L-state and an L-range thus:

> A possible state of affairs of all objects dealt with in a system S with respect to all properties and relations dealt with in S is called an *L-state* with respect to S. A sentence or sentential class designating an L-state is called a *state-description.* A given L-state leaves no question in S open; every sentence in S either admits or excludes that L-state. The class of the L-states admitted by A is called the *L-range* of A (LrA). Two postulates for L-ranges are laid down (P1 and 2). L-states are propositions.

[We replace his German letter by A, as this is more customary for a meta-linguistic variable over sentences.]

S is a semantical system (obviously classical), and an L-property is one that is logical, as opposed to factual. As we will see, L-ranges are introduced prior to L-contents as their definition is more immediate. The two postulates for L-ranges are:

P1. If A L-implies B (in S), then $\text{Lr}A \subseteq \text{Lr}B$.
P2. If $\text{Lr}A \subseteq \text{Lr}B$ (in S), then A L-implies B.

Just considering sentential logic with the truth-table semantics, an L-state is given by a row of a truth-table and an L-range of the sentence A is the class of rows of the truth-table with A true. So, given P1 and P2, $A \supset B$ is a tautology iff the set of rows of the truth-table that make A true will also make B true, since Carnap's "L-implies" is a material implication tautology.

Carnap then goes on to introduce an *L-content*, LcA of a sentence A, as a concept that satisfies the following postulates C1 and C2:

C1. If A L-implies B (in S), then $\text{Lc}B \subseteq \text{Lc}A$.
C2. If $\text{Lc}B \subseteq \text{Lc}A$ (in S), then A L-implies B.

Thus, the concept of an L-content is the dual of that of an L-range, where the content of the antecedent contains more than that of the consequent, rather than the converse as for ranges, assuming that A and B are not logically equivalent. In order to make this work for L-states, and hence, rows of a truth-table, Carnap, in Sect. 23 of (1942), goes on to properly define an *L-content* of A as the class of L-states that make A false. This L-content became, in the sentential logic, the class of rows of the truth-table making A false. Further, obviously, $A \supset B$ is a tautology iff the set of rows of the truth-table that make B false will also make A false, i.e., iff the set of rows of the truth-table that make A true will also make B true, since the logic is classical. One can also see from P1, P2, C1 and C2 that L-contents are conjunctive whilst L-ranges are disjunctive, in the sense that the following hold: $\text{Lc}A\&B \supseteq \text{Lc}A$, $\text{Lc}A\&B \supseteq \text{Lc}B$, $\text{Lr}A \vee B \supseteq \text{Lr}A$ and $\text{Lr}A \vee B \supseteq \text{Lr}B$. More inclusive contents are built up by conjunction of formulae, whilst more inclusive ranges are built up by disjunction.

Two key properties of L-contents that follow from the classicality of the logic are:

C3. Lc$B \subseteq$ Lc$A\&{\sim}A$, for every B.
C4. Lc$A \vee {\sim}A \subseteq$ LcB, for every B.

Here, C3 says that the content Lc$A\&{\sim}A$ is the universal set or, sententially, the set of all rows of the truth table, whilst C4 says that Lc$A \vee {\sim}A$ is the null set, or the set of no rows. That is, the content of a contradiction consists of every sentence, whilst the content of a tautology consists of no sentence.

Carnap and Bar-Hillel (1964), do introduce information and its amount but, as stated above, we will not be covering the amount. They first introduce information as satisfying a certain basic requirement: In(A) includes In(B) iff A L-implies B, where In(A) is the information carried by the sentence A. Thus, $A \supset B$ is a tautology iff In(B) $\subseteq$ In(A), which indeed takes the same shape as that for the properties, C1 and C2, for contents. They then introduce contents, defined as above, as the preferred of three given explications of information, all three of which satisfy this requirement. So, contents have a specific definition in terms of L-states, whilst information is a broader concept that just satisfies this basic requirement.

However, it is worth noting, for future reference, what the other two explications of information are. The first one, $\text{In}_1(A)$, is defined as the class of all sentences which are L-implied by A and not L-true. The second, $\text{In}_2(A)$, is the class of all sentences which are L-implied by A. Note that the first maintains some negativity which is characteristic of Carnap's definition of contents, whilst the second is the purely positive tautological implication, which is essentially strict implication when incorporating the necessity of the tautology. Note also that In_2 shows that the additional conjunct 'and not L-true' is redundant, this being because every A L-implies all L-truths.

3 Logical Contents Within a Content Semantics

An algebraic-style semantics, called content semantics, was introduced in Brady (1988) for the weak quantified relevant logics BBQ and BB^{d}Q (see below for BB and BB^{d}), and further in (1989) covering a wide range of quantified relevant logics and other logics extending BBQ and BB^{d}Q, right up to the classical predicate logic. As in semantics generally, the concepts are introduced in a structure, satisfying certain semantic postulates. Contents, in particular, would then be anything satisfying the postulates within the semantics and it is then left open to come to some understanding as to what they might be, especially in this case where a large range of logics are covered. We are able to show, however, that the correlate of Carnap and Bar-Hillel's property for information, viz. $I(A \to B) \in T$ iff $I(A) \leq I(B)$, holds generally for contents, where I is an interpretation within a model structure of the content semantics taking formulae A to their contents, T is the set of true contents and '$\leq$' satisfies enough properties for it to be construed as a containment, the '$\leq$' only being used due to standard algebraic practice. Indeed, the '$\leq$' would have been more appropriately symbolized as a '$\geq$' or a '$\supseteq$'.

The problem here is that the semantics is too broad to capture contents with an ideal level of specification; the logic needs to be more specialized. Indeed, Routley, on p. 935 in Sect. 11 of (1980), gives the following reasonable criterion for contents:

> A condition of adequacy on any account of logical content, or information, is that it leads to the results that A entails B iff (the meaning of) B is included in the meaning of A, and (the meaning of) B is included in the meaning of A iff the content of A includes the content of B.

This would require a logic of entailment which captures the concept of meaning containment, which, as the work of Brady (1996, 2006, 2013) (with Meinander) shows, can be well determined as a specific logic MC, after some tweaking in (2013) of the earlier candidate DJ^d of (1996, 2006). (The tweaking consisted of dropping the distribution axiom, but leaving it in rule-form via the use of the strengthened meta-rule MR1 below.) Stronger systems with an axiom such as contraction, $(A \rightarrow .A \rightarrow B) \rightarrow .A \rightarrow B$, would not be appropriate here, as the meaning of $A \rightarrow B$ being contained in that of A is hard to comprehend. This is even more so when it follows that both As contract to a single A, such contraction being more appropriate for truth-preservation than for MC.

We set out the logic MC as follows, focusing on the sentential logic to keep matters simple, and using the bracketing conventions of Anderson and Belnap (1975).

MC.

Primitives: $\sim, \&, \vee, \rightarrow$.

Axioms:

1. $A \rightarrow A$
2. $A\&B \rightarrow A$
3. $A\&B \rightarrow B$
4. $(A \rightarrow B)\&(A \rightarrow C) \rightarrow .A \rightarrow B\&C$
5. $A \rightarrow A \vee B$
6. $B \rightarrow A \vee B$
7. $(A \rightarrow C)\&(B \rightarrow C) \rightarrow .A \vee B \rightarrow C$
8. $\sim\sim A \rightarrow A$
9. $A \rightarrow \sim B \rightarrow .B \rightarrow \sim A$
10. $(A \rightarrow B)\&(B \rightarrow C) \rightarrow .A \rightarrow C$

Rules:

1. $A,\ A \rightarrow B \Rightarrow B$
2. $A,\ B \Rightarrow A\&B$
3. $A \rightarrow B,\ C \rightarrow D \Rightarrow B \rightarrow C \rightarrow .A \rightarrow D$

Meta-rule:

1. If $A,\ B \Rightarrow C$, then $D \vee A,\ D \vee B \Rightarrow D \vee C$.

We add in the logics DJ^d, BB and BB^d, mentioned earlier.

DJ^d = MC + the distribution axiom:

11. $A\&(B \vee C) \rightarrow (A\&B) \vee (A\&C)$

$\mathrm{BB} = \mathrm{DJ^d} - \mathrm{A4} - \mathrm{A7} - \mathrm{A9} - \mathrm{A10} - \mathrm{MR1} +$ the following 3 rules:

4. $A \to B,\ A \to C \ \Rightarrow\ A \to B\&C$
5. $A \to C,\ B \to C \ \Rightarrow\ A \vee B \to C$
6. $A \to {\sim}B \ \Rightarrow\ B \to {\sim}A$

$\mathrm{BB^d} = \mathrm{BB} + \mathrm{MR1}$.

The *content semantics for the logic MC* is set out as follows, as in Brady (2006), but taking into account the tweaking of the logic occurring in (2013). A *content model structure (c.m.s.)* consists of the following 5 concepts: $T, C, \cup, {}^*, c$, where C is a set of sets (called contents), $T \neq \emptyset, T \subseteq C$ (the non-empty set of all true contents), $\cup$ is a 2-place function on C (the closed union of contents), * is a 1-place function on C (the *-function on contents), and c is a 1-place function from containment sentences, $c_1 \supseteq c_2$ between contents c_1 and c_2 of C, to members of C, subject to the semantic postulates p1–p15, below.[2] The concepts $\cap$, $=$ and $\supseteq$, are taken from the background set theory, $\cap$ being a 2-place function on C (the intersection of contents), $=$ being a 2-place relation on C (identity), and $\supseteq$ being a 2-place relation on C (content containment).

The semantic *postulates* are:

p1. $c_1 \cup c_2 \supseteq c_1, \quad c_1 \cup c_2 \supseteq c_2$
p2. If $c_1 \supseteq c_2$ and $c_1 \supseteq c_3$, then $c_1 \supseteq c_2 \cup c_3$.
p3. $c_1 \supseteq c_1 \cap c_2, \quad c_2 \supseteq c_1 \cap c_2$
p4. If $c_1 \supseteq c_3$ and $c_2 \supseteq c_3$, then $c_1 \cap c_2 \supseteq c_3$.
p5. $c_1^{**} = c_1$
p6. If $c_1 \supseteq c_2$, then $c_2^* \supseteq c_1^*$.
p7. If $c_1 \supseteq c_2$ and $c_1 \in T$, then $c_2 \in T$.
p8. If $c_1 \in T$ and $c_2 \in T$, then $c_1 \cup c_2 \in T$.
p9. If $c_1 \cap c_2 \in T$, then $c_1 \in T$ or $c_2 \in T$.
p10. $c(c_1 \supseteq c_2) \cup c(c_2 \supseteq c_3) \ \supseteq\ c(c_1 \supseteq c_3)$
p11. $c(c_1 \supseteq c_2) \cup c(c_1 \supseteq c_3) \supseteq c(c_1 \supseteq c_2 \cup c_3)$
p12. $c(c_1 \supseteq c_3) \cup c(c_2 \supseteq c_3) \supseteq c(c_1 \cap c_2 \supseteq c_3)$
p13. $c(c_1 \supseteq c_2) \supseteq c(c_2^* \supseteq c_1^*)$
p14. $c(c_1 \supseteq c_2) \in T$ iff $c_1 \supseteq c_2$.
p15. If $c_1 \supseteq c_2$, then $c(c_3 \supseteq c_1) \supseteq c(c_3 \supseteq c_2)$ and $c(c_2 \supseteq c_3) \supseteq c(c_1 \supseteq c_3)$.

An *interpretation I on a c.m.s.* is an assignment, to each sentential variable, of an element of C. An interpretation I is extended to all formulae, inductively as follows:

(i) $I({\sim}A) = I(A)^*$
(ii) $I(A\&B) = I(A) \cup I(B)$
(iii) $I(A \vee B) = I(A) \cap I(B)$
(iv) $I(A \to B) = c(I(A) \supseteq I(B))$

[2]Note that the bar over the $\cup$, used in previous work to distinguish the closed union from the union, has been removed for convenience in this account. However, I do like the representation of closed union as $c(c_1 \cup c_2)$ in Mares (2004), where $\cup$ is the simple set-theoretic union.

A formula A is true under an interpretation I on a c.m.s. M iff $I(A) \in T$.
A formula A is valid in a c.m.s. M iff A is true under all interpretations I on M.
A formula A is valid in the content semantics iff A is valid in all c.m.s.

Soundness (if A is a theorem of MC then A is valid in the content semantics) follows readily and completeness (if A is valid in the content semantics then A is a theorem of MC) follows by the usual Lindenbaum method for algebraic-style semantics, but here there is a slight difference. In constructing the canonical models, instead of taking equivalence classes of formulae as the contents, we put the content $[A]$ of A as $\{C: A \to C \in T'\}$, where T' is constructed as a prime extension of the set of theorems which does not include a non-theorem B. This essentially means that these canonical contents are closed under entailment, i.e., they are *analytic closures* of the sentence (or sentences) involved, since the set T of theorems is already prime, due to the logic MC being metacomplete.[3] Since entailments here are understood as meaning containments, closure under entailment is closure under meaning containment and hence closure under the analysis of the meanings of words.[4] Whilst T' satisfies primeness to verify postulate p9, this extension of the set of theorems to T' is only needed when extending or applying the logic to a system which is not prime, e.g., a system that is not metacomplete. This answers the question of what contents are, in more specific terms than just the satisfaction of the correspondent, $I(A \to B) \in T$ iff $I(A) \supseteq I(B)$, of Carnap and Bar-Hillel's general property for information, which can be easily seen to hold, due to semantic postulate p14. Brady (2006) also introduces the dual concept of ranges.

To finish, we need to compare Carnap's definition of content with this definition. Before we do, we note that this definition turns out to be logically similar to Carnap and Bar-Hillel's second explication of information, $\text{In}_2(A)$, as the class of all sentences which are L-implied by A, or strictly implied by A. However, Routley in Sect. 11 of (1980) is critical of the use of strict implication here, as seen from Carnap's C3 and C4, which says that "some assertions, namely necessarily true ones, have no content, whereas others, the negations of necessary assertions, have total content." Routley contends that the transmission of necessary truths does send some information, e.g., as occurs in logic text-books. Routley further adds "despite the Carnap-Bar-Hillel strict theory of information, contradictions are commonly not so vastly informative." In particular, a contradiction $A\&{\sim}A$ should not inform us about a totally irrelevant sentence B, nor such a B inform us about $A \vee {\sim}A$. Floridi (2011), also makes a similar point in response to Routley's concern, which he calls the Bar-Hillel Carnap Paradox, but takes it further by giving the information of a contradiction a measure of 0, as with tautologies. The problem lies in the choice of logic as these problems disappear when the (classical) logic is replaced by a good entailment logic like MC, which is a weak relevant logic. Such a logic does not include $A \to B \vee {\sim}B$

[3]For the definition of metacompleteness and the proof of metacompleteness of MC, see Slaney (1984, 1987) and Sect. 4.2 of Brady (2006), noting that the dropping of the distribution axiom and the retaining of its rule-form to obtain MC from DJ^{d} has little effect on the proof.

[4]For a fuller description of analytic closure, see Sect. 1.5 of Brady (2006).

and $A\&{\sim}A \rightarrow B$ as theorems and so Carnap's C3 and C4 above are not derivable. Indeed, Routley (1980) lists the following desirable principles.

> Every formula has some content; thus every tautology has some content.
> No formula has total content; so no contradiction has total content.
> If A and B are disjoint formulae, then they have distinct [non-inclusive] contents. In particular, then, any two tautologies with distinct variables have distinct contents, and similarly [for] distinct contradictions.

These principles are appropriate for relevant logics, given the shape of the Routley–Meyer semantics where tautologies can be falsified in a world and contradictions can be true in a world. Indeed, Routley takes this further on p. 936 in Sect. 11 of (1980), where he defines the content $c(A)$ as $\{ a : A$ does not hold in $a \}$, where a is a world of a Routley–Meyer semantics, called there a "set-up."

So, we are left with comparing contents as analytic closures not only with Carnap's class of L-states that make A false, but now with Routley's definition using worlds of a Routley–Meyer semantics. However, the problem with Carnap's definition is that it still leads to C3 and C4 being derivable, which is subject to Routley's above concerns. And, this applies to all three of Carnap and Bar-Hillel's explications of information, as they all satisfy C3 and C4. As with Carnap's definition though, Routley's definition of the content of A as the set of worlds in which A is false is still oddly negative. One is still determining a content of a sentence by where it fails rather than where it holds. However, Routley's definition and analytic closure have the same key property for contents, for relevant logics like DJ^d which include the distribution axiom and the disjunctive meta-rule, as can be seen from the following four equivalences.[5]

(1) $\{ C\colon A \rightarrow C$ is provable$\} \supseteq \{ C\colon B \rightarrow C$ is provable$\}$. (using analytic closure)
(2) $A \rightarrow B$ is provable.
(3) $\{a\colon I(A,a) = T\} \subseteq \{a\colon I(B,a) = T\}$, for all interpretations I, for all Routley–Meyer reduced model structures.[6] (This requires soundness and completeness.)
(4) $\{a\colon I(A,a) = F\} \supseteq \{a\colon I(B,a) = F\}$, for all I, for all reduced model structures. (using Routley's definition)

So, analytic closure is the last content concept standing, but this definition does make sense, especially when the entailment logic is a logic of meaning containment, which is appropriate to the notion of entailment, and when contents are but meanings in a logical context. (See Routley's above condition of adequacy for contents.) Thus, we use the term 'logical content'. And, meanings can be understood in accordance with the 'meaning as use' approach, which translates into entailments in a deductive setting, as in our definition of contents.

---

[5]The following disjunctive meta-rule suffices here: If $A \Rightarrow B$ then $C \vee A \Rightarrow C \vee B$.

[6]Routley–Meyer reduced model structures satisfy: $I(A \rightarrow B, T) = T$ iff, for all worlds a, if $I(A,a) = T$, then $I(B,a) = T$, for all interpretations I. This can be seen from Lemma 4.2(3) on p. 302 of Routley et al. (1982). However, by Lemma 4.2(5) (ibid.), we can also work with unreduced modellings, though the reduced modellings better suit the sort of logics we are interested in using to capture the concept of meaning containment.

We now need to look into information by itself and, in passing, consider the adequacy of Carnap's account of it.

4 Dunn and Mares on Information

We will start by noting Floridi's definition of information in his book from (2011) setting out his philosophy of information, "Semantic information is well-formed, meaningful and truthful data," and considering Dunn's discussion of Floridi's book in (2013). Then, we move on to Mares' introduction of situated information, presented in his (2004, 2009), and with more depth in (2010). We conclude with Dunn's information states and his distinction between information with the Priming Property and information without it, discussed in his (2015).

Dunn seems to disagree with Floridi in his review essay (2013) saying "I feel that Floridi's 'veridicality thesis' for semantic information is merely a matter of 'semantics'." This would mean that it is all a matter of the interpretation of the words "true" and "information" as to whether semantic information has to be true. However, Dunn (2010) does not go so far as to consider information as contradictory saying, as per the title, that contradictory information is too much of a good thing. Here, Dunn regards information in itself as being a good thing.[7]

What I propose to do is to simplify the matter by focusing on a normative interpretation of information that is useful for logical purposes. So, we are not going down the path of considering the variety of interpretations of information used in natural language, interesting as that might be. What we are going to do is use the natural language readings to find features of information that are useful to us in our logical endeavours, just as has happened in the case of contents, which too has a variety of natural language readings. If the concept of information in natural language is somewhat vague, it is not of much use to logic; we need to focus on relatively sharp interpretations. The standout interpretation is that of truth, in conjunction with the meaning of an informative sentence, largely in agreement with Floridi, but without the semantic analysis. We can take the contents, discussed above, as representing such meanings, noting the preference for analytic closure as its interpretation. Truth is, of course, the other key semantic concept, i.e., other than meaning, and is used in every semantics of logical systems to define validity.

Further, information ought to be true, since otherwise it does not inform. Nevertheless, despite one's good intention, information can turn out to be false, leaving a co-operative informee neither impressed nor informed. (The word 'inform' has a slightly broader usage than 'information'.) Moreover, information sometimes has to be taken for granted, as in a computer database or as background information, for the purposes of a mechanized or human deduction.[8] Further, as such arguments are

---

[7]Dunn (2001), also provides an informative history of the usage of information in logic, down through the ages, providing a useful background for our endeavours.

[8]See Dunn (2008) for a detailed account of information in computer science.

generally considered in schematic rule-form (or through the use of variables in computer languages), it is the true substitutions into such forms that matter for determining validity, but there can also be false substitutions into premises, with the argument still being valid as a result of its form. So, though information should be true, false information could still be put as premises that are part of the rounding out of valid argumentation. In particular, in the case of informational inference, one takes the initial information for granted before determining further information that this may flow on to. However, we cannot go to the extreme of assuming information to be false. We compare all this with contents of a sentence or set of sentences, which are just logically focused meanings, without regard to truth or falsity, where there is no intention to inform built into such sentences.

Mares (2004) introduces situations as providing an understanding of set-ups (or worlds, as they are currently called) in the Routley–Meyer-style of semantics for relevant logics. The term 'situation' comes from Barwise and Perry's situation semantics, developed in their book (1983), and Mares uses it for partial information to distinguish it from the full information needed for Kripke-style worlds, which of course are negation-complete, as well as being simply consistent. Mares' situations are said to hold at worlds, which are Kripke-style possible worlds, but nevertheless situations can still be inconsistent, with negation being captured using Dunn's compatibility relation or by an intuitionistic definition $A \to f$, for some false constant f, in accordance with Mares (2010).[9] This is a departure from the standard Routley–Meyer semantics, which uses the Routley * with semantic postulates to capture the axiomatic negation properties. Section 2 of Mares (2009) provides further clarification:

> Situations carry information. ... A situation may accurately describe a possible world or may fail to do so. In accurately describing a world, a situation need not describe everything that is true of the world, but rather all the information carried by the situation must be true of that world. A situation which accurately describes a possible world is said to be a possible situation and a situation which does not accurately represent any possible world is said to be impossible.

Mares (2009) distinguishes truth conditions from information conditions in a semantics for relevant logics. Mares takes truth as classical propositional truth and truth conditions as the standard ones from classical logic. This contrasts with information conditions which apply to situations and follow the Routley–Meyer pattern, with some changes to negation, as above. Moreover, disjunction is also varied in (2010) along the lines of the natural deduction introduction and elimination rules. This allows disjunction not to be prime, i.e., a formula $A \vee B$ can hold in a situation, without either A or B holding. This is because $A \vee B$ can be a hypothesis, without A or B being in its subproof. (See Brady (2010) for what is called a free semantics

[9] Mares in Sect. 5.2 of (2004) presents Dunn's compatibility relation Cst, which says of situations s and t that they are compatible with one another. Goldblatt (1974) uses an incompatibility relation $\perp$, which is called orthogonality. He then gives the following truth condition for negation: $\sim A$ is true at a iff all outcomes which make A true are precluded by a. This compares with Dunn's truth condition: $\sim A$ is true at s iff $\forall x\,(Csx \supset \sim(A \text{ is true at } x))$, i.e., $\sim A$ is true iff every situation compatible with it fails to make A true.

for the logic LDW, which is set up using natural deduction in a similar way to this. Note that LDW is MC − A10 − MR1.)

This leads us to Dunn's discussion of the Priming Property in his (2015), which is included in a section on information states, which we first examine. He uses the term 'information states' for what Routley and Meyer initially call 'set-ups' in their semantics of relevant logics. Dunn uses the storage system of a computer to give a concrete representation of them, regarding an information state as a finite sequence of bits, either 1 or 0, this being similar to Carnap's state descriptions. He goes on to say that propositions can be thought of as sets of information states, each proposition P being true or false in a given information state a according to whether a is in P or not, i.e., what has previously been called a UCLA-proposition. Dunn goes on to say that the Routley–Meyer valuation-clause, $I(A \vee B, x) = T$ iff $I(A, x) = T$ or $I(B, x) = T$, for information states x, makes perfect sense in this setting. [I have used original symbolism here.]

Dunn goes on to point out that "For a more ordinary conception of an information state the left-to-right direction is problematic." He goes on to give an example, "Suppose I am about to throw a coin. I have the information that it will turn up heads or tails, but I do not have the information as to which." He suggests that we "also let the notation N (for neither) sometimes occur to indicate that the information state is not complete," i.e., in addition to the 1s and 0s above. He also says "there are also circumstances where we might have conflicting information," for which "we might use the notation B (for both)." The inclusion of N and B would bring information states into line with Dunn's 4-valued information states, as described in his (2008).

What we need to do next is to try to determine under what circumstances the Priming Property should hold or not. The cases where priming holds tend to be set within a semantical system, as above, and the key element of such a semantics is that disjunction is introduced inductively from one of its disjuncts, and that this is the only way disjunction is obtained. Hence, we have the Routley–Meyer valuation clause or such like, which contains the left-to-right Priming Property. The right-to-left is clear cut, regardless of the type of system. The key feature is the inductive determination of the connectives in setting up the semantics, that is, it is truth-functional, at least with respect to disjunction. However, one must avoid semantics such as the free semantics of Brady (2010) and the semantics in Sect. 10 of Mares (2010), both of which are based on natural deduction, that is, the meta-logic of the semantics itself is explicitly set up using natural deduction and this logic gets embedded into the valuation conditions, notably for disjunction.

This brings us to the cases where priming does not hold. The essence of this is natural reasoning or, as Dunn says, "a more ordinary conception of an information state," explained with true-to-life examples. Formally, we have considered natural deduction as an appropriate logical formalization into which to place Dunn's examples, but Hilbert-style deduction is also relevant. Indeed, any proof-theoretic system will do, except that metacomplete or intuitionist systems satisfy the Priming Property for theorems: If $A \vee B$ is a theorem, so is either A or B. This priming does not generally extend to subproofs of their natural deduction systems, as $A \vee B$ can be introduced as a hypothesis, without A or B being in its subproof. This is the case with Dunn's coin example, where the information that it will turn up heads or tails is regarded as a premise or hypothesis of the natural deduction system. So, basically, priming holds for standard semantical systems and fails for proof-theoretic systems, with the exceptions noted above.

5 Informational Concepts and Their Logical Usage

In conclusion, we briefly recap and compare the various concepts of contents and information, and then determine, for each of these concepts, where in logical study they best fit.

Carnap and Bar-Hillel's concept of information in (1964) is far too broad, in that it is any concept In satisfying the property, $\mathrm{In}(A)$ includes $\mathrm{In}(B)$ iff A L-implies B, for formulae A and B. It needs more detailed spelling out to narrow it down. Indeed, three content interpretations were proposed, all of which were discussed in Sect. 2, leading to the final preference for analytic closure in a logic of meaning containment.

As already presented, our preferred contents are placed within a content semantics, which is algebraic in style. Now, algebraic semantic postulates follow the Hilbert-style axioms and rules very closely, as can be seen in the case of MC and its content semantics. So, though true contents play the usual role in defining validity and the rules of MC preserve truth, contents of sentences, by themselves, are static and can be either true or false. Thus, they relate most closely with algebraic semantics and Hilbert-style axiomatization.

Information, on the other hand, is more complex. There are two distinctions that have to be drawn regarding truth and information, one from Mares' work and one from Dunn's work. Mares distinguishes truth conditions from information conditions, but we disregard the former which regards classical truth because of our interest in relevant logic. The latter gives rise to his situated information, which fits into a semantics for relevant logic, taking the role of formulae evaluated as true in a Routley–Meyer semantics, but modified to treat negation and possibly disjunction differently. This brings us to Dunn's distinction between prime information and non-prime information. Dunn gives examples of non-prime information, which best fits into formal natural deduction systems, as these can have disjunctive hypotheses, without a disjunct appearing in the subproof generated by the hypothesis. As argued above, non-prime information best fits proof-theoretic systems whilst prime information best fits truth-functional semantics. Whilst this is the case, any analysis of information can be uniformly carried out, firstly by examination of logical content then by determining its truth or assumed truth, but in the latter case one should bear in mind that priming may not hold. One final point is that MC, without distribution, does not have a truth-functional semantics, but with prime information for truth due to metacompleteness.

References

Anderson, A. R., & Belnap, N. D. (1975). *Entailment: The logic of relevance and necessity* (Vol. I). NJ: Princeton University Press.

Barwise, J., & Perry, J. (1983). *Situations and attitudes*. Cambridge: MIT Press.

Brady, R. T. (1988). A content semantics for quantified relevant logics—I. *Studia Logica, 47*, 111–127.

Brady, R. T. (1989). A content semantics for quantified relevant logics—II. *Studia Logica, 48*, 243–257.

Brady, R. T. (1996). Relevant implication and the case for a weaker logic. *Journal of Philosophical Logic, 25*, 151–183.

Brady, R. T. (2006). *Universal logic*. Stanford: CSLI Publications.

Brady, R. T. (2010). Free semantics. *Journal of Philosophical Logic, 39*, 511–529.

Brady, R. T., & Meinander, A. (2013). Distribution in the logic of meaning containment and in quantum mechanics. In K. Tanaka, F. Berto, E. Mares, & F. Paoli (Eds.), *Paraconsistency: logic and applications* (pp. 223–255). Dordrecht: Springer.

Carnap, R. (1942). *Introduction to semantics*. Cambridge: Harvard University Press.

Carnap, R., & Bar-Hillel, Y. (1964). An outline of a theory of semantic information. In Y. Bar-Hillel (Ed.) *Language and Information: Selected Essays on their Theory and Application* (221–274). MA: Addison-Wesley and The Jerusalem Academic Press. (Reprint of Technical Report No. 247, Research Laboratory of Electronics, MIT, 1952.).

Dunn, J. M. (2001). The concept of information and the development of modern logic. In W. Stelzner & M. Stöckler (Eds.), *Zwischen traditioneller und moderner Logik: nichtklassische Ansatze* (pp. 423–447). Paderborn: Mentis-Verlag.

Dunn, J. M. (2008). Information in computer science. In D. M. Gabbay, P. Thagard, J. Woods, P. Adriaans & J. van Benthem (Eds.) *Philosophy of Information, Vol. 8 of Handbook of the Philosophy of Science* (pp. 581–608). Amsterdam: Elsevier.

Dunn, J. M. (2010). Contradictory information: Too much of a good thing. *Journal of Philosophical Logic, 39*, 425–452.

Dunn, J. M. (2013). Guide to the Floridi keys: Essay review of Luciano Floridi's the philosophy of information. *MetaScience, 22*, 93–98.

Dunn, J. M. (2015). The relevance of relevance to relevance logic. In M. Banerjee & S. N. Krishna (Eds.), *Logic and its Applications: 6th Indian Conference, ICLA 2015, Mumbai, India, January 8–10, 2015, number 8923 in Lecture Notes in Computer Science* (pp. 11–29). Heidelberg: Springer.

Floridi, L. (2011). *The philosophy of information*. Oxford: Oxford University Press.

Goldblatt, R. (1974). Semantic analysis of orthologic. *Journal of Philosophical Logic, 3*, 19–35.

Mares, E. D. (2004). *Relevant logic: A philosophical interpretation*. Cambridge: Cambridge University Press.

Mares, E. D. (2009). General information in relevant logic. *Synthese, 167*, 343–362.

Mares, E. D. (2010). The nature of information: A relevant approach. *Synthese, 175*, 111–132.

Routley, R. (1980). *Ultralogic as universal?: Exploring Meinong's jungle and beyond*. Canberra: A.N.U.

Routley, R., Meyer, R. K., Plumwood, V., & Brady, R. T. (1982). *Relevant logics and their rivals* (Vol. I). Atascadero: Ridgeview Publishing Company.

Slaney, J. K. (1984). A metacompleteness theorem for contraction-free relevant logics. *Studia Logica, 43*, 159–168.

Slaney, J. K. (1987). Reduced models for relevant logics without WI. *Notre Dame Journal of Formal Logic, 28*, 395–407.

On Split Negation, Strong Negation, Information, Falsification, and Verification

Heinrich Wansing

Abstract This paper deals with some criticism that has been put forward against strong, constructive negation in comparison to a certain example of Galois connected negations. The general background to this discussion is the informational interpretation of substructural logics, and the key issue is whether there exists an asymmetry or not between positive and negative information and between verification and falsification. The present paper confirms the view that a symmetrical conception is adequate for *both* direct and indirect variants of verification and falsification.

Keywords Falsification · Information · Split negation · Strong negation · Verification

1 Introduction

In a series of papers (Dunn 1993, 1996, 1999; Dunn and Zhou 2005), J. Michael Dunn has presented a very influential in-depth investigation of negation as impossibility or "unnecessity." The basic setting Dunn starts with has room for what he calls *Galois connected negations* or, following Chrysafis Hartonas (Hartonas and Dunn 1993), *split negations*. A pair of negations $\sim_1$ and $\sim_2$ is a pair of Galois connected or split negations iff for all formulas A and B of the language under consideration it holds that $A \vdash \sim_1 B$ iff $B \vdash \sim_2 A$. Let us call this the split negation property. Dunn's work on negation in the algebraic framework of his gaggle theory (see Dunn 1991, 1993, 1995; Dunn and Hardegree 2001; Bimbó and Dunn 2008) and in the context of frame semantics continues earlier work on ortho-negation in quantum logic by Garrett Birkhoff and John von Neumann (1936) and Robert Goldblatt (1974, 1975), and work by Dimiter Vakarelov (1977, 1989), and Kosta Došen (1984, 1986, 1999), who treat negation as a modal operator defined with respect to a binary relation on a non-empty set of states, see also (Shramko 2005; Horn and Wansing 2015; Berto

H. Wansing (✉)
Department of Philosophy II, Ruhr-University Bochum, Bochum, Germany
e-mail: Heinrich.Wansing@rub.de
URL: http://www.ruhr-uni-bochum.de/philosophy/logic/

K. Bimbó (ed.), *J. Michael Dunn on Information Based Logics*,
Outstanding Contributions to Logic 8, DOI 10.1007/978-3-319-29300-4_10

2015). Negation as a modal operator of impossibility or unnecessity has given rise to interesting arrays of negation operations, starting form Dunn's "kite of negations" (Dunn 1993), and leading from Yaroslav Shramko's "lopsided kite" and his dual version of it (Shramko 2005), to the so-called "united kite of negations" (Dunn and Zhou 2005), a combination developed upon a suggestion by Shramko, see also (Horn and Wansing 2015; Onishi 2015). Among these negations there is intuitionistic negation, a negation usually understood as "implies falsity," and minimal negation, usually understood as "implies some designated atomic proposition."

Dunn is also one of the main contributors to an informational interpretation of substructural logics, and, together with Nuel Belnap, he developed what is now often called Belnap and Dunn's *useful four-valued logic*, also known as first-degree entailment logic, FDE, see (Belnap 1977a, b; Dunn 1976, 2000). The system FDE is the implication-free fragment of David Nelson's constructive, paraconsistent four-valued logic N4. Strong negation in the three-valued variant N3 of N4 has been proposed by Gurevich (1977) as a means to express direct falsification, and substructural subsystems of N4 have been advocated as logics of information processing for instance in Wansing (1993a, b), see also (Kamide and Wansing 2012, 2015). As a result of combining constructive implication with FDE, strong negation in Nelson's systems N3 and N4 does not satisfy contraposition as an inference rule. Strong negation in N3 and N4, therefore, is not captured by Dunn's perp semantics (cf. Sect. 2.1), but it is treated in Dunn's seminal paper on consequence relations in the context of a four-valued semantics (Dunn 2000), where he carefully explains that the contraposition rule is an admissible rule of FDE.

Sebastian Sequoiah-Grayson (2009) criticizes Gurevich and myself for arguing in favour of direct falsification and for having available a strong negation. The plea for strong negation is based on the assumption that it is desirable to have an information-based semantics in which positive and negative information are taken as independent, primitive notions that receive a symmetrical, or perhaps better analogical, analysis. Sequoiah-Grayson (2009, p. 233) claims "that a strong asymmetry between positive and negative information is in fact the case." I will scrutinize Sequoiah-Grayson's critique and explain why and where it goes astray. This discussion will then give me an opportunity to make some remarks on symmetry between verification and falsification in the setting of both the logic of generalized truth values and constructive logic.

2 Michael Dunn on Negation and Information

2.1 Negation

Central notions for this paper are the notions of negation and information. In (Gabbay and Wansing 1999) the question *What is negation?* was raised. Concerning that question, Michael Dunn (1999, p. 48) explains:

'What is truth?' asked Pontius Pilate. 'What is negation?' ask Dov Gabbay and Heinrich Wansing. Pilate never got an answer, and I do not answer Dov's and Heinrich's question either, unless a variety of answers can count as *the* answer. I instead show how many of the structural properties of negation can be obtained by various model-theoretic devices. The most powerful of these devices is perhaps the use of perp structures, since by various fine tunings of their properties one can get many of the well-known properties of negation.

This kind of fine-tuning is familiar from modal correspondence theory. A perp model is a structure $\mathcal{M} = (S, \perp, v)$, where S is a non-empty set of information states, $\perp$ ("perp") is a binary relation on S, and v is a valuation function. Dunn views $\perp$ as a relation of incompatibility or orthogonality between states. Negation as impossibility, $\sim$, is then semantically defined by postulating that $\sim A$ is true at a state $x \in S$ iff x is incompatible with all states $y \in S$ at which A is true: $\mathcal{M}, x \models \sim A$ iff $\forall y\,(\mathcal{M}, y \models A$ implies $x \perp y)$. Instead of an orthogonality relation one may use a relation of compatibility between states, denoted by C, and $\mathcal{M}, x \models \sim A$ is then defined by requiring that $\forall y\,(xCy$ implies $\mathcal{M}, y \not\models A)$. The negation $\sim$ is thus a universally quantifying "necessity" operator with respect to the relation C, whereas negation as unnecessity, $\neg$, is an existentially quantifying "possibility" operator with respect to C: $\mathcal{M}, x \models \neg A$ iff $\exists y\,(xCy$ and $\mathcal{M}, y \not\models A)$.

If the relation C is not required to be symmetric, then one may distinguish between two negation operations $\sim_1$ and $\sim_2$ that are defined as follows:

$\mathcal{M}, x \models \sim_1 A$ iff $\forall y\,(xCy$ implies $\mathcal{M}, y \not\models A)$;
$\mathcal{M}, x \models \sim_2 A$ iff $\forall y\,(yCx$ implies $\mathcal{M}, y \not\models A)$.

These negations satisfy the split negation property and the following interaction principles: $A \vdash \sim_1\sim_2 A$; $A \vdash \sim_2\sim_1 A$.

Every negation as impossibility, $\sim$, and every negation as unnecessity, $\neg$, is, in Dunn's terminology, *preminimal*, i.e., they satisfy the following contraposition rules:

$$A \vdash B \text{ implies } \sim B \vdash \sim A; \quad A \vdash B \text{ implies } \neg B \vdash \neg A.$$

Notions of negation stronger than preminimal negation are obtained semantically by imposing conditions on the compatibility relation C. Certain such conditions correspond with well-known and important negation principles in the sense that C satisfies the condition in question just in case the negation principle under consideration is validity preserving. In this way, a plurality of negation concepts emerges, see (Dunn 1993, 1996, 1999; Dunn and Zhou 2005; Restall 1999, 2000; Shramko 2005; Horn and Wansing 2015; Onishi 2015; Berto 2015).

However, Dunn does not completely leave it at that kind of pluralism with respect to negation and in Dunn (1999, p. 49) he writes:

But I must say that my own favourite is the 4-valued semantics. I am persuaded that '$\neg\varphi$ is true iff φ is false', and that '$\neg\varphi$ is false iff φ is true'. And now to paraphrase Pontius Pilate, we need to know more about 'What are truth and falsity?' It is of course the common view that they divide up the states into two exclusive kingdoms. But there are lots of reasons, motivated by applications, for thinking that this is too simple-minded.

From a constructive point of view, giving up the classical understanding of negation as Boolean complementation is only natural since classical negation permits, for example, non-constructive existence proofs. Moreover, reasoning with partial and contradictory information suggests the use of a paracomplete and paraconsistent logic, and Dunn's favoured four-valued semantics is indeed a natural candidate for modelling paracomplete and paraconsistent reasoning. Nelson's N4 combines the four-valued FDE with a constructive implication.

Intuitionistic negation fails to be a paraconsistent negation, at least with respect to provability. I will draw a distinction between direct falsification as expressed by strong negation and weak falsification as expressed by intuitionistic negation. I will, however, also distinguish between provability and a certain notion of dual provability. In that framework, intuitionistic negation is a paraconsistent negation with respect to dual provability, whereas another negation, co-negation, is a paraconsistent negation with respect to provability.

2.2 *Information*

The other 'what is'-question, i.e., *What is information?*, is addressed by Dunn (2001, 2008), see also (Dunn 2010). I cannot adequately discuss or even answer this fundamental question here, but I would like to emphasize that I agree with Dunn's general conception of information. In Dunn (2001, p. 423) he writes that "information is what is left of knowledge when one takes away belief, justification, and truth. ... Information is ... a kind of semantic content—the kind of thing that can be expressed by language," and in Dunn (2008, p. 582) he says that information is something like a Fregean thought. Although this characterization of information is quite broad, it excludes the factivity (truthfulness) of information and, therefore, it excludes so-called "semantic information" as considered by, for example, Luciano Floridi, see the surveys (Adriaans 2013; Floridi 2015).[1]

What is important for the present paper is that Dunn distinguishes between definitely positive and definitely negative information and that he admits both incomplete and inconsistent information states, i.e., states that provide neither positive nor negative information concerning a given proposition and states that provide both positive and negative information with respect to a given proposition. In his dissertation (Dunn 1966), Dunn introduced the notion of a "proposition surrogate" as a pair (A^+, A^-), where A^+, A^- are subsets of a universe of discourse U (a non-empty set of information states), and A^+ (A^-) is a set of topics the proposition gives definitely positive (negative) information about. Negation $\sim$ is interpreted as the exchange of positive

[1]The association with Fregean thoughts may prompt further discussion. In his famous paper on sense and reference (Frege 1892), Frege considers an example of a fictional sentence that has no truth value but nevertheless expresses a thought. In the later paper "Compound Thoughts" (Frege 1923), however, he characterizes a thought as "something which must be either true or false, *tertium non datur*."

and negative information: $\sim(A^+, A^-) := (A^-, A^+)$. Moreover, it is not required that $(A^+ \cap A^-) = \varnothing$ or $(A^+ \cup A^-) = U$.

Such an understanding of information and information states is suitable for an interpretation of the logical operations within a paraconsistent and paracomplete logic.

3 Split Negation and Information Models

Another aspect of information reflected in Dunn's work on relevance logic and gaggle theory is that information is a resource that comes in pieces. Logics of information processing are substructural logics; the contraction, the dilution, and the exchange of premises is not supported by applications in which inference is viewed as the processing of information pieces, see, for example, (Wansing 1993a, b; Restall 2000; Paoli 2002).

3.1 Split Negation

One way of obtaining a pair of split negations arises from a language, say $\mathcal{L}$, with a falsity constant, $\mathbf{0}$, two directional implications, $\rightarrow$ and $\leftarrow$, that are left and right residuals of a non-commutative (intensional, multiplicative) conjunction, $\otimes$, also known as fusion, if the right-searching and the left-searching negation of an $\mathcal{L}$-formula A are defined as "A implies falsity"[2]:

$$\neg^r A :\equiv \mathbf{0} \leftarrow A, \quad \neg^l A :\equiv A \rightarrow \mathbf{0}.$$

One should also keep in mind Dunn's (1991, p. 42) warning that "there is no consistency in the literature about which is the "left" residual and which is the "right". We follow Birkhoff." Calling $\rightarrow$ and $\leftarrow$ the left and right residual of multiplicative conjunction follows, e.g., (Paoli and Tsinakis 2012).

As a semantics for $\mathcal{L}$, Sequoiah-Grayson (2009) introduces information frames and models as follows:

Definition 3.1 An information frame is a triple $\langle S, \sqsubseteq, \bullet \rangle$, where $\langle S, \sqsubseteq \rangle$ is a partial order and $\bullet$ is a non-commutative binary operation on S. An information model is a pair $\langle \mathbf{F}, \Vdash \rangle$, where $\mathbf{F}$ is an information frame and $\Vdash$ is an evaluation relation between elements of S and $\mathcal{L}$-formulas that satisfies the following conditions for all formulas A, B, and every $x \in S$:

---

[2]Sequoiah-Grayson uses '$\neg$' instead of '$\neg^r$' and '$\sim$' instead of '$\neg^l$'. The superscripts have the advantage of reminding one of the directionality of the implication that is involved.

(1) For every $y, z \in S$, if $y \Vdash A$ and $y \sqsubseteq z$, then $z \Vdash A$ (persistency);
(2) $x \Vdash (A \otimes B)$ iff there exist $y, z \in S$ with $y \bullet z \sqsubseteq x$, $y \Vdash A$ and $z \Vdash B$;
(3) $x \Vdash (A \rightarrow B)$ iff for all $y, z \in S$, if $y \bullet x \sqsubseteq z$ and $y \Vdash A$, then $z \Vdash B$;
(4) $x \Vdash (B \leftarrow A)$ iff for all $y, z \in S$, if $x \bullet y \sqsubseteq z$ and $y \Vdash A$, then $z \Vdash B$;
(5) not $x \Vdash \mathbf{0}$.

Note that I have rephrased Sequoiah-Grayson's clauses (3) and (4) in such a way that $x \Vdash A \otimes (A \rightarrow B)$ implies $x \Vdash B$ and $x \Vdash (B \leftarrow A) \otimes A$ implies $x \Vdash B$. With these clauses we obtain the residuation principles that are stated in (Sequoiah-Grayson 2009, p. 337) as:

(6) $A \otimes B \vdash C$ iff $B \vdash A \rightarrow C$;
(7) $A \otimes B \vdash C$ iff $A \vdash C \leftarrow B$,

if we interpret $A \vdash B$ as coinciding with a semantical consequence relation $\models$ such that $A \models B$ holds iff for every information model $\langle S, \sqsubseteq, \bullet, \Vdash \rangle$ and every $x \in S$, $x \Vdash A$ implies $x \Vdash B$.[3]

In the rest of this subsection I will critically discuss the presentation of split negation in Sequoiah-Grayson (2009); this discussion is *not* a critique of split negation as such. As will become clear later, there is nothing wrong with split negation. It expresses a weak, indirect notion of falsification with respect to provability and can be supplemented by a weak, indirect notion of verification with respect to dual provability.

Sequoiah-Grayson (2009) does not give references for the above semantics but points out that clauses (2)–(4) amount to a particular, informational reading of the ternary relational frame semantics for $\otimes$, $\rightarrow$, and $\leftarrow$. The ternary frame semantics is due to Routley and Meyer (1972, 1973), and the informational reading is explained, for example, in Dunn's article on relevance logic and entailment in the *Handbook of Philosophical Logic*, (Dunn 1986), where one also can find more information on the semantics of relevance logic and its history. Sequoiah-Grayson suggests to understand $x \Vdash A$ following (Mares 2009) to mean that *state x carries the information that A* or that *x supports the information that A*.[4] He emphasizes (Sequoiah-Grayson

[3]We may assume that there is a typing error in the versions of (3) and (4) in Sequoiah-Grayson (2009). With these versions:

$(3)_{SG}$ $x \Vdash A \rightarrow B$ iff for all $y, z \in S$, if $x \bullet y \sqsubseteq z$ and $y \Vdash A$, then $z \Vdash B$
$(4)_{SG}$ $x \Vdash A \leftarrow B$ iff for all $y, z \in S$, if $y \bullet x \sqsubseteq z$ and $y \Vdash A$, then $z \Vdash B$

$x \Vdash (A \rightarrow B) \otimes A$ implies $x \Vdash B$ and $x \Vdash A \otimes (A \leftarrow B)$ implies $x \Vdash B$, and, as a result, we do not obtain (6) and (7).

[4]He writes (Sequoiah-Grayson 2009, p. 236):

> One might wish to understand 'supports' as 'makes true' if one holds to a *dialethic paraconsistentism* whereby at least some contradictions are taken to be true. However, we will sidestep this particular debate and stay with the interpretation of 'supports' that takes it to be the subtler relative of 'makes true' in a manner aligned with Mares' informational interpretation.

2009, p. 235) that he "want[s] to allow for the information at x being incomplete and/or inconsistent" and explains that "x may support A where A is 'p and not p'." But this 'not p' cannot be $\sim p$ or $\neg p$, i.e., $\neg^l p$ or $\neg^r p$ because, for example:

$$\begin{array}{ll} & x \Vdash p \otimes (p \rightarrow \mathbf{0}) \\ \text{iff} & \text{there are } y, z \text{ with } y \bullet z \sqsubseteq x, y \Vdash p \text{ and } z \Vdash p \rightarrow \mathbf{0} \\ \text{iff} & \text{there are } y, z \text{ with } y \bullet z \sqsubseteq x, y \Vdash p \text{ and} \\ & \qquad \text{for all } u, v, \text{ if } u \bullet z \sqsubseteq v \text{ and } u \Vdash p, \text{ then } v \Vdash \mathbf{0}. \end{array}$$

But $y \bullet z \sqsubseteq y \bullet z$, $y \Vdash p$ and *not* $y \bullet z \Vdash \mathbf{0}$.

Doubtlessly, it is a desideratum of any logic of information processing to adequately account for the processing of partial and contradictory information, in particular, to the effect that contradictory information does not allow one to derive any statement whatsoever. However, I do not see that Sequoiah-Grayson's semantics satisfies the latter desideratum. The split negations under consideration fail to be paraconsistent negations at least insofar as no state supports the information that $A \otimes \neg^l A$ and no state supports the information that $\neg^r A \otimes A$, so that if entailment is understood as preservation of support of information (or support of truth) in any state from any model, both $A \otimes \neg^l A$ and $\neg^r A \otimes A$ entail arbitrary formulas. Unfortunately, Sequoiah-Grayson offers no definition of $A \vdash B$ but merely explains that

> [i]n deductive information processing, we understand the premises as databases and the consequence relation '$\vdash$' as the information processing mechanism, a more brutally syntactic operation that the information carrying/supporting of $\Vdash$. In informational terms, we may read $A \vdash B$ as information of type B follows from information of type A, or the information in B follows from the information in A, and so forth. We can think of typing as encoding, in which case we might also read $A \vdash B$ as the information encoded by B follows from the information encoded by A. (Sequoiah-Grayson 2009, p. 237).

One thing to note here is that although $\vdash$ is explained only in informal, though informational terms and formally remains undefined, it seems that $\vdash$ is taken to be a relation between single formulas, so that expressions of the form $\vdash A$ with an empty antecedent appear to be excluded.[5]

Do we, under the suggested reading of $\vdash$, get $A \otimes \neg^l A \vdash B$ and $\neg^r A \otimes A \vdash B$ for any formulas A, B? Here is what Sequoiah-Grayson (2009, p. 238) says:

> [T]he suggestion is that we interpret $\sim A$ as the body of information that cannot be applied to bodies of information of type A, and that we interpret $\neg A$ as the body of information that cannot have bodies of information of type A applied to it. The interpretation is supported by the model theory; by the information states supporting $\sim A$, $\neg A$, and A. If x supports $\sim A$ and y supports A, x cannot be applied to y. Similarly, if x supports $\neg A$ and y supports A, then y cannot be applied to x. This is not because such an application will cause an explosion of information, but because it does not generate any information.

There are at least two things that are peculiar here. First, it is suggested to read $x \bullet y$ as "*x applied to y*" and it is emphasized that "'[a]pplying' is an order-sensitive

[5]Note that in this paper I do not pay close attention to the mention/use distinction when there is no risk of confusion.

notion." (Sequoiah-Grayson 2009, p. 236). Whereas it is quite natural to view the relation $\sqsubseteq$ as a relation of information containment or information development, in my opinion the reading of $\bullet$ in terms of application of bodies of information is unfortunate because "application" reminds one of functional application, and an operation that interprets non-commutative conjunction is a kind of composition different from the application of a function to its arguments, for example if we think of atomic bodies of information x and y. This peculiarity transmits to the understanding of the two negation operations. Sequoiah-Grayson argues that the particular form of split negation under consideration may be understood as the "prohibition" of information processing procedures. It is said that a state that supports $\sim A$ cannot be applied to a state that supports A, and that a state that supports A cannot be applied to a state supporting $\neg A$. We have the analogous asymmetry between $\sim A$ and $\neg A$ under the "bodies of information" reading of these formulas. But that's strange. Both negations are implications, and under a suitable formulas-as-types notion of construction, $\sim A$ and $\neg A$ both would be interpreted *on a par* as types of lambda-abstracted terms, terms that are *both applicable to* terms of type A, whereas $A \otimes B$ is the type of ordered pairs of terms of types A and B (cf. Wansing 1993a, b).

Secondly, it is said that a state y that supports A cannot be applied to a state x that supports $\neg A$ "not because such an application will cause an explosion of information, but because it does not generate any information." For the undefined derivability relation $\vdash$ this seems to imply that nothing can be derived from $A \otimes \neg^l A$ and nothing can be derived form $\neg^r A \otimes A$, and Sequoiah-Grayson (Sequoiah-Grayson 2009, pp. 239, 240) is explicitly talking about "get[ting] nothing, namely **0**." But then $\vdash$ cannot be reflexive because otherwise $A \otimes \neg^l A \vdash A \otimes \neg^l A$ and $\neg^r A \otimes A \vdash \neg^r A \otimes A$. If $\vdash$ is the syntactic counterpart of a relation that preserves support-of-information-that, however, then $\vdash$ *is* reflexive.

3.2 The Syntactic Characterization of Information Models

A considerable part of Sequoiah-Grayson (2009) consists of an elaboration of the above support-of-information-that conditions and their application in order to justify various inference patterns. To simplify the discussion, it is helpful to formally reconstruct what he does by characterizing the informational semantics syntactically. We also have to supplement the definition of a semantic consequence relation.

Depending on whether $\bullet$ is associative or not, information models are models for the associative or non-associative Lambek calculus, see Lambek (1958, 1961), extended by a falsity constant. Sequoiah-Grayson is indifferent with respect to whether $\bullet$ should be taken to be associative or not. Let us for definiteness agree on nonassociativity and briefly present the characterization.

We inductively define the set of all Gentzen terms. Every $\mathcal{L}$-formula is a Gentzen term. If Δ and Γ are Gentzen terms, then (Δ, Γ) is a Gentzen term as well. Nothing else is a Gentzen term. We use capital Greek letters Δ, Γ, Θ etc. to denote

Gentzen terms. A sequent is an expression of the form $\Delta \Rightarrow A$, so that sequents always have a non-void antecedent. The set $st(\Delta)$ of sub-terms of a Gentzen term Δ is inductively defined as one would expect: $st(A) = \{A\}$; $st((\Delta, \Gamma)) = st(\Delta) \cup st(\Gamma) \cup \{(\Delta, \Gamma)\}$. We write $\Delta[A]$ to highlight a certain occurrence of A as a sub-term in Δ. If $\Delta[A]$ appears in a premise of a sequent rule, then in the conclusion of that sequent rule $\Delta[\Gamma]$ is the result of replacing the highlighted occurrence of A in Δ by Γ.

Definition 3.2 The nonassociative Lambek calculus with falsity in the language $\mathcal{L}$, NL**0**, consists of the following rules:

$$\begin{array}{rl}
\text{(id)} & \vdash A \Rightarrow A \\
\text{(cut)} & \Gamma \Rightarrow A, \quad \Delta[A] \Rightarrow B \vdash \Delta[\Gamma] \Rightarrow B \\
(\rightarrow \Rightarrow) & \Delta \Rightarrow A, \quad \Gamma[B] \Rightarrow C \vdash \Gamma[(\Delta, (A \rightarrow B))] \Rightarrow C \\
(\Rightarrow \rightarrow) & (A, \Delta) \Rightarrow B \vdash \Delta \Rightarrow (A \rightarrow B) \\
(\leftarrow \Rightarrow) & \Delta \Rightarrow A, \quad \Gamma[B] \Rightarrow C \vdash \Gamma[((B \leftarrow A), \Delta)] \Rightarrow C \\
(\Rightarrow \leftarrow) & (\Delta, A) \Rightarrow B \vdash \Delta \Rightarrow (B \leftarrow A) \\
(\otimes \Rightarrow) & \Delta[(A_1, A_2)] \Rightarrow B \vdash \Delta[(A_1 \otimes A_2)] \Rightarrow B \\
(\Rightarrow \otimes) & \Delta \Rightarrow A, \quad \Gamma \Rightarrow B \vdash (\Delta, \Gamma) \Rightarrow (A \otimes B) \\
(\mathbf{0} \Rightarrow) & \vdash \Delta[\mathbf{0}] \Rightarrow A
\end{array}$$

The non-associative Lambek calculus, NL, is given by the rules (id)–($\Rightarrow \otimes$).

A sequent $\Delta \Rightarrow A$ is provable in NL**0** ($\vdash \Delta \Rightarrow A$) iff it is derivable in NL**0** from axiomatic sequents of the form $A \Rightarrow A$ or $\Delta[\mathbf{0}] \Rightarrow A$. We will thus understand Sequoiah-Grayson's $A \vdash B$ as $\vdash A \Rightarrow B$. NL has the cut-elimination property, and the additional rule ($\mathbf{0} \Rightarrow$) does not destroy that property. The rule (cut) is thus an admissible rule of NL**0**, i.e., a sequent is provable in NL**0** iff it is provable in NL**0** without (cut). For the semantical characterization we will make use of a very simple syntactical property of cut-free derivations in NL**0**.

Observation 3.3 *In* NL**0** *without* (cut), *every application of a right introduction rule can be permuted downwards over any application of a left introduction rule.*

Proof We consider the three right introduction rules.
($\Rightarrow \rightarrow$):

- Permutation over ($\rightarrow \Rightarrow$):

$$\dfrac{\dfrac{\vdots}{\Delta \Rightarrow A} \quad \dfrac{\dfrac{\vdots}{(A', \Gamma[B]) \Rightarrow B'}}{\Gamma[B] \Rightarrow (A' \rightarrow B')}\,(\Rightarrow \rightarrow)}{\Gamma[(\Delta, (A \rightarrow B))] \Rightarrow (A' \rightarrow B')}\,(\rightarrow \Rightarrow) \quad \rightsquigarrow \quad \dfrac{\dfrac{\dfrac{\vdots}{\Delta \Rightarrow A} \quad \dfrac{\vdots}{(A', \Gamma[B]) \Rightarrow B'}}{(A', \Gamma[(\Delta, (A \rightarrow B))]) \Rightarrow B'}\,(\rightarrow \Rightarrow)}{\Gamma[(\Delta, (A \rightarrow B))] \Rightarrow (A' \rightarrow B')}\,(\Rightarrow \rightarrow)$$

- Permutation over ($\leftarrow \Rightarrow$): Analogous to the previous subcase.

- Permutation over $(\otimes \Rightarrow)$:

$$\dfrac{\dfrac{\dfrac{\vdots}{(A', \Delta[(A_1, A_2)]) \Rightarrow B'}}{\Delta[(A_1, A_2)] \Rightarrow (A' \to B')}\,(\Rightarrow\to)}{\Delta[(A_1 \otimes A_2)] \Rightarrow (A' \to B')}\,(\otimes\Rightarrow) \quad \rightsquigarrow \quad \dfrac{\dfrac{\dfrac{\vdots}{(A', \Delta[(A_1, A_2)]) \Rightarrow B'}}{(A', \Delta[(A_1 \otimes A_2)]) \Rightarrow B'}\,(\otimes\Rightarrow)}{\Delta[(A_1 \otimes A_2)] \Rightarrow (A' \to B')}\,(\Rightarrow\to)$$

$(\Rightarrow \leftarrow)$: Analogous to the previous case.
$(\Rightarrow \otimes)$:

- Permutation over $(\leftarrow \Rightarrow)$:

$$\dfrac{\dfrac{\vdots}{\Theta \Rightarrow A} \quad \dfrac{\dfrac{\vdots}{\Delta \Rightarrow A'} \quad \dfrac{\vdots}{\Gamma[B] \Rightarrow B'}}{(\Delta, \Gamma[B]) \Rightarrow (A' \otimes B')}\,(\Rightarrow\otimes)}{(\Delta, \Gamma[((B \leftarrow A), \Theta)]) \Rightarrow (A' \otimes B')}\,(\leftarrow\Rightarrow) \quad \rightsquigarrow \quad \dfrac{\dfrac{\vdots}{\Delta \Rightarrow A'} \quad \dfrac{\dfrac{\vdots}{\Theta \Rightarrow A} \quad \dfrac{\vdots}{\Gamma[B] \Rightarrow B'}}{\Gamma[((B \leftarrow A), \Theta)] \Rightarrow B'}\,(\leftarrow\Rightarrow)}{(\Delta, \Gamma[((B \leftarrow A), \Theta)]) \Rightarrow (A' \otimes B')}\,(\Rightarrow\otimes)$$

- Permutation over $(\to \Rightarrow)$: Analogous to the previous subcase.
- Permutation over $(\otimes \Rightarrow)$:

$$\dfrac{\dfrac{\dfrac{\vdots}{\Delta[(A, B)] \Rightarrow A'} \quad \dfrac{\vdots}{\Gamma \Rightarrow B'}}{(\Delta[(A, B)], \Gamma) \Rightarrow (A' \otimes B')}\,(\Rightarrow\otimes)}{(\Delta[(A \otimes B)], \Gamma) \Rightarrow (A' \otimes B')}\,(\otimes\Rightarrow) \quad \rightsquigarrow \quad \dfrac{\dfrac{\dfrac{\vdots}{\Delta[(A, B)] \Rightarrow A'}}{\Delta[(A \otimes B)] \Rightarrow A'}\,(\otimes\Rightarrow) \quad \dfrac{\vdots}{\Gamma \Rightarrow B'}}{(\Delta[(A \otimes B)], \Gamma) \Rightarrow (A' \otimes B')}\,(\Rightarrow\otimes) \qquad \square$$

Let $\otimes\Delta$ be the result of replacing every occurrence of the comma in Δ by an occurrence of $\otimes$. Clearly, $\vdash \Delta \Rightarrow A$ iff $\vdash \otimes\Delta \Rightarrow A$. The left-to-right direction is obvious from rule $(\otimes \Rightarrow)$, for the right-to-left direction we use (cut):

$$\dfrac{\dfrac{A \Rightarrow A \quad B \Rightarrow B}{(A, B) \Rightarrow (A \otimes B)} \quad \dfrac{\vdots}{\Delta[(A \otimes B)] \Rightarrow C}}{\Delta[(A, B)] \Rightarrow C}\,(\text{cut})$$

A sequent $\Delta \Rightarrow A$ is valid in an information model $\mathcal{M} = \langle S, \sqsubseteq, \bullet, \Vdash \rangle$ ($\mathcal{M} \models \Delta \Rightarrow A$) iff for every $x \in S$, $x \Vdash \otimes\Delta$ implies $x \Vdash B$; $\Delta \Rightarrow A$ is valid ($\models \Delta \Rightarrow A$) iff it is valid in every information model. Soundness of **NL0** with respect to the class of all information models follows by induction on proofs in **NL0**. For completeness we may consider a characterizing model.[6]

Definition 3.4 The structure $\mathcal{M}^c = \langle G, \sqsubseteq^c, \bullet^c, \Vdash^c \rangle$ is defined as follows: G is the set of all Gentzen terms Δ such that $\Delta \Rightarrow A$ is not provable in **NL0** for at least one $\mathcal{L}$-formula A, $\sqsubseteq^c$ is the syntactic identity relation $\equiv$ between Gentzen terms, and $\bullet^c$ is the bracketing $(\cdot, \cdot)$ of Gentzen terms. The valuation relation $\Vdash^c$ is defined by stipulating that for every $\mathcal{L}$-formula A and Gentzen term Δ, $\Delta \Vdash^c A$ iff $\vdash \Delta \Rightarrow A$.

[6] An anonymous reviewer raised the question whether there are other ways of defining a semantics in order to obtain a formal reconstruction of the presentation in (Sequoiah-Grayson 2009). I have nothing else to offer than the preservation of support-of-information-that, which captures derivability in **NL0**.

Observation 3.5 $\mathcal{M}^c$ *is an information model.*

Proof The above persistency condition (1) is trivially satisfied. We consider conditions (2)–(5).
(2) By definition, $\Delta \Vdash^c (A \otimes B)$ iff $\vdash \Delta \Rightarrow (A \otimes B)$ iff (by Observation 3.3) there exist $\Gamma, \Theta \in G$ with $\Delta \equiv (\Gamma, \Theta)$, $\vdash \Gamma \Rightarrow A$ and $\vdash \Theta \Rightarrow B$ iff (by definition) there exist $\Gamma, \Theta \in G$ with $\Gamma \bullet^c \Theta \sqsubseteq^c \Delta$, $\Gamma \Vdash^c A$ and $\Theta \Vdash^c B$.
(3) By definition, $\Delta \Vdash^c (A \rightarrow B)$ iff $\vdash \Delta \Rightarrow (A \rightarrow B)$ iff (by Observation 3.3) $\vdash (A, \Delta) \Rightarrow B$ iff (by admissibility of (cut)) for all $\Gamma, \Theta \in G$, if $(\Gamma, \Delta) \equiv \Theta$ and $\vdash \Gamma \Rightarrow A$, then $\vdash \Theta \Rightarrow B$ iff (by definition) for all $\Gamma, \Theta \in G$, if $\Gamma \bullet^c \Delta \sqsubseteq^c \Theta$ and $\Gamma \Vdash^c A$, then $\Theta \Vdash^c B$.
(4) Analogous to (3).
(5) By the definition of G and admissibility of (cut), for no $\Delta \in G$ it holds that $\vdash \Delta \Rightarrow \mathbf{0}$. □

Observation 3.6 (Completeness) *If* $\models \Delta \Rightarrow A$, *then* $\vdash \Delta \Rightarrow A$.

Proof If it is not the case that $\vdash \Delta \Rightarrow A$, then in $\mathcal{M}^c$ it is not the case that $\Delta \Vdash^c A$. But in $\mathcal{M}^c$ it holds that $\Delta \Vdash^c A$ iff $\otimes\Delta \Vdash^c A$. Therefore, $\Delta \Rightarrow A$ is not valid. □

We may now use this characterization and consider the inference patterns that in (Sequoiah-Grayson 2009) are elaborated in terms of information models. The inference patterns (12)–(18) are easily provable, in particular the split negation property is easily verified:

$$\dfrac{\dfrac{\dfrac{\vdots}{A \Rightarrow B \rightarrow \mathbf{0}} \quad \dfrac{B \Rightarrow B \quad \mathbf{0} \Rightarrow \mathbf{0}}{(B, B \rightarrow \mathbf{0}) \Rightarrow \mathbf{0}}}{(B, A) \Rightarrow \mathbf{0}}\ (\text{cut})}{B \Rightarrow \mathbf{0} \leftarrow A} \qquad \dfrac{\dfrac{\dfrac{\vdots}{B \Rightarrow \mathbf{0} \leftarrow A} \quad \dfrac{A \Rightarrow A \quad \mathbf{0} \Rightarrow \mathbf{0}}{(\mathbf{0} \leftarrow A, A) \Rightarrow \mathbf{0}}}{(B, A) \Rightarrow \mathbf{0}}\ (\text{cut})}{A \Rightarrow B \rightarrow \mathbf{0}}$$

The pattern

(19) $A \Rightarrow \neg B \vdash B \Rightarrow {\sim}A$

however, of which Sequoiah-Grayson (2009, p. 240) says that it "makes procedural sense," fails to be (cut-free) provable, as can be seen from the failure of bottom-up proof search[7]:

$$\dfrac{\dfrac{\dfrac{B \Rightarrow A \quad (A, (\mathbf{0} \leftarrow B)) \Rightarrow \mathbf{0}}{(A, (B, (A \rightarrow (\mathbf{0} \leftarrow B)))) \Rightarrow \mathbf{0}}}{(B, (A \rightarrow (\mathbf{0} \leftarrow B))) \Rightarrow A \rightarrow \mathbf{0}}}{A \rightarrow (\mathbf{0} \leftarrow B) \Rightarrow B \rightarrow (A \rightarrow \mathbf{0})} \qquad \dfrac{\dfrac{B \Rightarrow A \quad \dfrac{(A, (\mathbf{0} \leftarrow B)) \Rightarrow \mathbf{0}}{\mathbf{0} \leftarrow B \Rightarrow A \rightarrow \mathbf{0}}}{(B, (A \rightarrow (\mathbf{0} \leftarrow B))) \Rightarrow A \rightarrow \mathbf{0}}}{A \rightarrow (\mathbf{0} \leftarrow B) \Rightarrow B \rightarrow (A \rightarrow \mathbf{0})}$$

Sequoiah-Grayson writes that (19) "prohibits the complex procedure"

[7] Assuming sequent rules that capture Sequoiah-Grayson's $(3)_{SG}$ and $(4)_{SG}$ instead of (3) and (4) would not help.

(20) $A \otimes (B \otimes (A \to \neg B))$.

By "prohibition" of $A \otimes (B \otimes (A \to \neg B))$ he means that $A \otimes (B \otimes (A \to \neg B)) \vdash \mathbf{0}$, which is, however, not the case[8]:

$$\dfrac{\dfrac{(A, (B, (A \to (\mathbf{0} \leftarrow B)))) \Rightarrow \mathbf{0}}{(A, (B \otimes (A \to (\mathbf{0} \leftarrow B)))) \Rightarrow \mathbf{0}}}{A \otimes (B \otimes (A \to \neg B)) \Rightarrow \mathbf{0}}$$

4 Symmetry and All that

I will now gradually come to a more constructive part of the present paper and return to information, verification, and falsification. A reply to Sequoiah-Grayson's criticism of strong negation will bring us to considering symmetry between verification and falsification. These considerations are independent from using a substructural logic with a non-commutative conjunction and a pair of directional implications since both intuitionistic logic and constructive logic with strong negation already have an informational interpretation in terms of Kripke models of a certain kind.

4.1 On an Alleged Asymmetry Between Positive and Negative Information

Negation understood as "implies falsity" is an important and interesting notion. It is not at all inadequate *as such*, however, as claimed in (Gurevich 1977) and (Wansing 1993a, b), it is inadequate to express negative information to the effect that a certain statement is definitely false. Moreover, if a state x supports the truth of the negation of A just in case for all states y and z, if y supports the truth of A and z extends the "fusion" of x and y ($y \bullet x$ or $x \bullet y$), then z supports the truth of $\mathbf{0}$, such a "negation as inconsistency" is a rather non-constructive notion of negation. David Nelson (1959) emphasized that intuitionistic negation, which is a negation as inconsistency in a non-substructural setting, is inadequate to represent the constructive meaning of negated formulas in intuitionistic arithmetic: "Under the recursive interpretation of a formal system for intuitionistic arithmetic, the provable implications of the form $A \supset 1 = 0$ receive a trivial interpretation."

[8] Analogous remarks apply to the inference pattern (21) and the "prohibited procedure" (22). Also the endorsed inference patterns

(23) $A \to \neg B \vdash {\sim}A \leftarrow B$, and
(24) $A \to {\sim}B \vdash \neg A \leftarrow B$

are not provable, so that they are not underpinned by the formulas numbered (25) and (26) in (Sequoiah-Grayson 2009).

Sequoiah-Grayson (2009, p. 244) presents the reasoning in favour of strong negation in (Gurevich 1977) and (Wansing 1993a, b) as follows:

> So far we have the following (translating from intuitionistic to ternary terminology):
>
> (a) Any *adequate* theory of information processing will allow for representing both positive and negative information.
>
> (b) The definition of split negation in a model **M** has the result that positive and negative information are treated asymmetrically; A may be verified "on the spot," while $\sim A$ and $\neg A$ may not.
>
> (c) Therefore, any theory of information processing based upon **M** will not be an adequate theory of information processing.

Sequoiah-Grayson notes that (c) does not follow from (a) and (b) because (a) does not say that representations of positive and negative information need to be in symmetry. According to him, the reasoning is incomplete and there are two candidates for a suitable missing premise:

> (a′) The representation of positive information must be in symmetry with the representation of negative information in order for a theory of information processing to be adequate.
>
> (a″) Bodies of either positive or negative information must be directly verifiable in order for a theory of information processing to be adequate.
>
> (a″) is the stronger claim, since if we have satisfied it then we have *ipso facto* satisfied (a′). We could, in principle at least, have a verification condition on A that was just as "off the spot" as are the present verification conditions on $\sim A$ and $\neg A$, in which case symmetry would be satisfied. Similarly, refuting (a′) refutes (a″), but not vice versa. Which of either (a′) and (a″) are the intended premise? Gurevich seems to be arguing for the weaker (a′) when he states that "[i]n many cases the falsehood of a simple scientific sentence can be ascertained as directly (or indirectly) as its truth" Gurevich (1977, p. 49) (my emphasis). Wansing sometimes seems to be arguing for (a″) when he states "Gurevich's remark amounts to the complaint that there is no possibility of direct falsification of [A] on the spot," Wansing (1993, p. 14). However, other comments such as "... the idea of taking negative information seriously and putting it on par with positive information leads Gurevich to intuitionistic logic with strong negation ...," Wansing (1993, p. 14), are much more in line with (a′). We take it then, that the more flexible (a′) is the missing premise. In this case, the full form of the argument is (a), (a′), (b), therefore (c). This argument is valid.

Sequoiah-Grayson then explains that he rejects (a′).

The first thing to ask is whether the above reconstruction is faithful to the sources and correct. Remarkably, the reasoning attributed to Gurevich and me is first reconstructed as enthymematic and is then supplemented by an additional premise that *is* traced back to (Gurevich 1977) and (Wansing 1993a, b). Gurevich takes up Andrzej Grzegorczyk's "philosophically plausible formal interpretation of intuitionistic logic" (Grzegorczyk 1964), in which Grzegorczyk holds that whereas atomic sentences are verified experimentally, compound sentences, including negated sentences "arise from reasoning." Gurevich (1977, p. 49) objects to this view by observing that often "the falsehood of a simple scientific sentence can be ascertained as directly (or indirectly) as its truth." In intuitionistic Kripke models, a state x may verify or fail to verify an atomic formula. Gurevich takes the verification of an atomic formula to be an evaluation as "true" and the failure of verification to be an evaluation as "uncertain." To overcome the limitations of intuitionistic Kripke models, he considers three-valued functions such that the value of an atomic formula at a state "can be equal to "false," "uncertain" or "true." That gives rise to a conservative extension of the intuitionistic logic which is nicer at least in one aspect: it is more symmetric, it satisfies very natural duality laws" (Gurevich 1977, p. 49). Although Gurevich emphasizes the symmetry of the resulting logic with strong negation, he does not explicitly endorse (a′), but it seems fair to read his comments on ascertaining the falsehood of simple scientific sentences "as directly (or indirectly) as its truth" as an endorsement of (a′). However, when we highlight that Gurevich wants to make room for *directly ascertaining falsehoods*, then this remark may also be seen as a plea for being able to model the direct falsification of atomic sentences besides their direct verification. I agree with Sequoiah-Grayson in that (a′) does not entail (a″) provided negative information is not associated with falsity but with the failure of verification in the sense of "implies falsity." The indirect falsification that comes with a proof that a statement implies falsity can indeed be supplemented by a notion of indirect verification. But then it is natural to require symmetry twice, namely in addition to being able to represent *both* direct verifications and direct falsifications (i.e., proofs and disproofs), also to represent *both* indirect verifications and indirect falsifications, see also (Wansing 2010) and Sect. 4.4.

Whether one, as Sequoiah-Grayson does, reads (a″) as entailing (a′) or not, there is some textual evidence that Gurevich (1977) and Wansing (1993a, b) endorse both (a′) and (a″), so that in any case their reasoning is reconstructed as valid. How does Sequoiah-Grayson then come to the conclusion that "a strong asymmetry between positive and negative information is in fact the case"? He offers three arguments.

First argument The first argument goes as follows (Sequoiah-Grayson 2009, p. 245):

> Gurevich's example of scientific sentences that have their falsehood directly ascertained is "The solution is acid" with regards to a litmus paper test. However, falsification can happen just as well as a result of positive information as negative information. In fact, contra Gurevich, the litmus paper example is an instance of just this.
>
> Suppose that we are testing for acid, and that the paper remains blue (blue litmus paper turns red in an acid, red litmus paper turns blue in a base). In this case, we have falsified "The solution is acid." But on the basis of what? The falsification proceeds via the positive information that the paper is still blue. The negative information concerning the falsification of "The solution is acid" is derivative upon the positive information concerning the blueness of the paper. Even in the restricted context of scientific reasoning, it is certainly not straightforward that negative information should be granted the status of an "informational primitive," on par with positive information. Adapting Grzegorczyk's point from the paragraph above, we see that the paper is blue, we do not see that it is not red. We do not "see that it is not red" any more than we "see that it is not a cat." We acquire the negative information *it is not the case that the solution is acid* on the basis of the positive information *the paper is blue* that is incompatible with the positive information *the solution is acid*. We ascertain that it is not red because we see that it is blue.

It is important to note here that direct falsification in the context of the conservative extension of intuitionistic logic considered by Gurevich, namely Nelson's constructive logic with strong negation N3, leads to contrary pairs. The adjectives 'red' and 'blue' give rise to contrary pairs of sentences, and so do, for example the pairs 'basic' versus 'acid', 'happy' versus 'unhappy', and 'polite' versus 'impolite'. Metaphysically speaking, a given person cannot *be* both happy and unhappy (at a given moment of time in one and the same respect), but she or he may well be neither happy nor unhappy. The situation is more complicated in a four-valued semantics.[9] The sentence 'Arthur is happy' may be told true and told false, so that a state x may support both the truth and the falsity of 'Arthur is happy', and if x supports the falsity of 'Arthur is happy' just in case x supports the truth of 'Arthur is unhappy', then x may support the truth of both 'Arthur is happy' and 'Arthur is unhappy'. Whereas prefixes such as 'un', 'im', and 'dis' suggest that a negation operation is used, pairs such as 'acid' versus 'basic' fail to suggest that the semantic opposition between 'acid' and 'basic' involves negation. What it certainly does not involve is classical negation (expressing falsity understood as absence of truth) or intuitionistic negation (expressing falsity as "implies falsity").

The symmetry argued for in (Gurevich 1977) and (Wansing 1993a, b) is the symmetry obtained by accounting for *both* direct verification and direct falsification as distinct types of reasoning procedures. One way of obtaining such a symmetry is to internalize support of falsity into the logical object language by means of a strong negation connective. The sentence 'Arthur is unhappy' is then analyzed as the strong negation of 'Arthur is happy', and semantically the strong negation $\sim A$ of A switches between support of truth, $\models^+$, and support of falsity, $\models^-$.[10] A state x supports the truth of $\sim A$ iff it supports the falsity of A and it supports the falsity of $\sim A$ iff it supports the truth of A: $x \models^+ \sim A$ iff $x \models^- A$; $x \models^- \sim A$ iff $x \models^+ A$. As a result, from the point of view of verification, the information that Arthur is unhappy is negative information; from the point of view of falsification, the information that Arthur is unhappy is positive information.

Sequoiah-Grayson tries to show that "a strong asymmetry between positive and negative information is in fact the case" by denying that there is a notion of direct falsification. According to him there is only direct verification based on positive information; "[t]he negative information concerning the falsification of "The solution is acid" is derivative upon the positive information concerning the blueness of the paper." Suppose that we are working with neutral litmus paper, so that by a change of colour we may test for both acids and bases. What is positive information from the point of view of verification is then negative information in the context of falsification (and vice versa). If we see that the violet litmus paper turns blue, then this observation verifies that the solution in question is basic *and* at the same time falsifies that the solution is acid (and it not merely shows that the solution is not acid, where "not" here and elsewhere in the metalanguage stands for classical negation). Moreover, the fact that the solution is basic is typically not ascertained by deriving a falsehood (an "absurdity") from the assumption that the solution is acid but by performing a litmus test. In order to show that at a given state a solution is basic, what we do

[9]It is complicated in the constructive setting of intuitionistic logic and N3 already, cf. (Wansing 2006).

[10]Note that the strong negation $\sim$ is different from Sequoiah-Grayson's $\sim$, i.e., $A \rightarrow \mathbf{0}$.

is precisely not to convince ourselves that in every future state the solution fails to be acid. One might object that observing the violet or red litmus paper turning blue falsifies the assumption that the solution is acid only against the background of a certain theory about acids and bases, call it $\mathcal{T}$, so that what the litmus paper turning blue really shows is that the assumption that the solution is not basic together with $\mathcal{T}$ implies a falsehood. But then we obtain a notion of verification that is in conflict with Sequoiah-Grayson's claim that there is a strong asymmetry between positive and negative information because the litmus paper turning blue would *verify* 'The solution is basic' also only indirectly, namely against the background of $\mathcal{T}$.

Sequoiah-Grayson considers the blue litmus paper as providing only positive information. Indeed, Nelson's N3 can be faithfully embedded into intuitionistic logic and Nelson's N4 can be faithfully embedded into positive intuitionistic logic by replacing every strongly negated atom $\sim p$ by a fresh sentence letter p', see (Gurevich 1977) and, for example, (Wansing 2001), (Kamide and Wansing 2012). But this does not mean that definitely negative information and direct falsification are not in fact represented in Nelson's logics. Since direct falsification in N3 and N4 is internalized into the object language by means of strong negation, reasoning from support of falsity to support of falsity, however, need not be defined as a separate entailment relation in addition to the conception of entailment as preservation of support of truth, cf. Sect. 4.4.

Second argument Another consideration in Sequoiah-Grayson (2009, p. 246) makes use of the assumption that "facts ground information," so that "an asymmetry between positive facts and negative facts will carry over into an asymmetry between positive and negative information." The assumption is certainly contentious. Sequoiah-Grayson emphasizes that it is more likely if information is taken to be veridical. We have embraced Dunn's conception of information according to which the propositional content of information may be untrue and therefore information is not always grounded on facts.

Third argument A third argument makes use of the notion of ruling out (Sequoiah-Grayson 2009, p. 246):

> We have a natural asymmetry with the very context of procedural interpretation of split negation that we are considering. In a system of procedural information processing, it is completely natural to interpret $\sim/\neg$ as the ruling out of a procedure. The procedure ruled out by $\sim/\neg A$ is just any procedure that involves combining $\sim/\neg A$ with A itself. In summary, insofar as general concerns regarding the indirectness of such a definition of negation in informational terms is concerned, it is worth considering the observation that an interpretation of negation in terms of ruling something out is about as direct as we could want. "Ruling out" is a direct notion. But to rule something out requires a "universal checking." This kind of *in*directness is required in order that something be truly ruled out. To put this another way, how can we rule something out without first considering all the possible cases?

The idea here seems to be that if $\sim/\neg A$ is true, then A is ruled out, and that ruling out is direct, although it involves a "universal checking." If it is assumed that verification based on positive information is direct and that split negation understood as ruling out is a direct notion, one may wonder how this consideration may then demonstrate that "a strong asymmetry between positive and negative information in fact is the case." In the case of the split negation of A the universal checking consists of checking whether it holds for all states y and z that if y supports the truth of A and z extends $y \bullet x$ or $x \bullet y$, then z supports the truth of $\mathbf{0}$. This *is* a kind of indirectness and it seems completely natural to accompany this indirect falsification of A by an analogous way of verification, namely *ruling in*. But this is then again leading to a symmetry between indirect ways of verification and falsification instead of an asymmetry, and that's what I will turn to in the next section.

4.2 Symmetry Between Verification and Falsification

So far I have argued, with Gurevich, that we may directly falsify elementary, atomic empirical statements. We may, for example, directly falsify that Arthur is polite by pointing to his impolite behaviour. This does not mean that direct falsification is the only kind of falsification. But if we recognize the dichotomy between direct verification and direct falsification, then it is natural to associate the former with definitely positive information that a certain proposition is true and the latter with definitely negative information that a certain proposition is false. This amounts to considering Belnap's (1977a, b) *told false* and *told true* values. In Nelson's logics with strong negation, definite falsity is expressed by means of strong negation, henceforth $\sim$, and in intuitionistic logic (and N3) indirect falsification is internalized into the logical object language by means of intuitionistic negation, henceforth $\neg$. Note that in Nelson's logics N3, N4, and in Odintsov's (2005, 2008) extension $N4^{\perp}$ of N4, there is not a total and perfect symmetry between the direct verification and direct falsification of all kinds of formulas. In these logics, atomic formulas are treated completely on a par with respect to verification and falsification. The verification and the falsification of atomic formulas is *static* insofar as only the state of evaluation is involved. The verification and falsification clauses for conjunctions and disjunctions are also static. However, whilst the verification conditions for constructive implication are *dynamic*, their falsification conditions are static:

$$x \models^{-} A \to B \;\text{ iff }\; (x \models^{+} A \text{ and } x \models^{-} B).$$

Additional symmetry with respect to dynamic versus static verification and falsification clauses for compound formulas is obtained by adopting a connexive understanding of negated implications, cf. (Wansing 2005, 2008, 2014; Kamide and Wansing 2011):

$$x \models^{-} A \to B \;\text{ iff }\; x \models^{+} (A \to \sim B).$$

In the remainder of this section and this paper, I will consider two other ways of separating reasoning about truth from reasoning about falsity.

4.3 Truth and Falsity Entailment

A separate treatment of truth and falsity with respect to entailment can, for example, be found in the theory of generalized truth values. The theory is rooted in Belnap and Dunn's semantics for FDE and, in particular, in Dunn's representation of the four truth values of the FDE semantics as the elements of the powerset $\mathcal{P}(\{T, F\})$ of the set of classical truth values $\{T, F\}$, see (Shramko and Wansing 2005, 2011). If one goes one step further and considers $\mathbf{16} = \mathcal{P}(\mathcal{P}(\{T, F\}))$, one can define an information ordering $\leq_i$, a truth ordering $\leq_t$, and a falsity ordering $\leq_f$ on the set of generalized truth values from **16** as follows:

Definition 4.1 For every x and y in **16**,

1. $x \leq_i y$ iff $x \subseteq y$;
2. $x \leq_t y$ iff $x^t \subseteq y^t$ and $y^{-t} \subseteq x^{-t}$, where $x^t := \{y \in x : T \in y\}$ and $x^{-t} := \{y \in x : T \notin y\}$;
3. $x \leq_f y$ iff $x^f \subseteq y^f$ and $y^{-f} \subseteq x^{-f}$, where $x^f := \{y \in x : F \in y\}$ and $x^{-f} := \{y \in x : F \notin y\}$.

The partial orders $\leq_t$ and $\leq_f$ may be seen as logical orderings. Clearly, meets and joins exist in **16** for $\leq_i$, $\leq_t$, and $\leq_f$. If we use $\sqcap$ and $\sqcup$ with the appropriate subscripts for these operations under the corresponding ordering relations, we obtain the trilattice $SIXTEEN_3 = \langle \mathbf{16}, \sqcap_i, \sqcup_i, \sqcap_t, \sqcup_t, \sqcap_f, \sqcup_f \rangle$. Moreover, one can define unary operations $-_t$, $-_f$, and $-_i$ which invert the respective lattice ordering, preserve the other orderings and satisfy $x = -_i -_i x$, $x = -_t -_t x$, $x = -_f -_f x$.

The languages $\mathcal{L}_t$, $\mathcal{L}_f$, and $\mathcal{L}_{tf}$ are defined in Backus–Naur form as follows:

$$\mathcal{L}_t \quad A ::= p \mid {\sim_t} A \mid (A \wedge_t A) \mid (A \vee_t A)$$
$$\mathcal{L}_f \quad A ::= p \mid {\sim_f} A \mid (A \wedge_f A) \mid (A \vee_f A)$$
$$\mathcal{L}_{tf} \quad A ::= p \mid {\sim_t} A \mid {\sim_f} A \mid (A \wedge_t A) \mid (A \wedge_f A) \mid (A \vee_t A) \mid (A \vee_f A),$$

where p is a propositional variable from some fixed infinite set.

In (Shramko and Wansing 2005) valuation functions v^{16} from the set $Prop$ of propositional variables into **16** are extended to the set of all $\mathcal{L}_{tf}$-formulas by requiring that for any A and $B \in \mathcal{L}_{tf}$:

1. $v^{16}(A \wedge_t B) = v^{16}(A) \sqcap_t v^{16}(B)$;
2. $v^{16}(A \vee_t B) = v^{16}(A) \sqcup_t v^{16}(B)$;
3. $v^{16}({\sim_t} A) = -_t v^{16}(A)$;
4. $v^{16}(A \wedge_f B) = v^{16}(A) \sqcup_f v^{16}(B)$;
5. $v^{16}(A \vee_f B) = v^{16}(A) \sqcap_f v^{16}(B)$;
6. $v^{16}({\sim_f} A) = -_f v^{16}(A)$

Definition 4.2 Relations of truth and falsity entailment between sentences $A, B \in \mathcal{L}_{tf}$ are defined as follows:

$$A \models_t^{16} B \text{ iff } \forall v^{16}\, (v^{16}(A) \leq_t v^{16}(B)); \qquad A \models_f^{16} B \text{ iff } \forall v^{16}\, (v^{16}(B) \leq_f v^{16}(A)).$$

The reason given in (Shramko and Wansing 2005) for defining falsity entailment as shown is that for defining entailment, the authors were interested in decreasing falsehood. We obtain a uniform definition of the two entailment relations if we denote by $\leq_f$ the relation inverse to the relation which is denoted by $\leq_f$ in (Shramko and Wansing 2005). As a consequence of this change, the operations $\sqcup_f$ and $\sqcap_f$ are interchanged. A **16**-valuation $v : Prop \longrightarrow \mathbf{16}$ can then be extended to the set of all $\mathcal{L}_{tf}$-formulas in a homomorphic way:

1. $v(A \wedge_t B) = v(A) \sqcap_t v(B)$;
2. $v(A \vee_t B) = v(A) \sqcup_t v(B)$;
3. $v(\sim_t A) = -_t v(A)$;
4. $v(A \wedge_f B) = v(A) \sqcap_f v(B)$;
5. $v(A \vee_f B) = v(A) \sqcup_f v(B)$;
6. $v(\sim_f A) = -_f v(A)$,

and the relations $\models_t$ and $\models_f$ are defined in a uniform way.

Definition 4.3 $A \models_t B$ iff $\forall v\, (v(A) \leq_t v(B))$; $A \models_f B$ iff $\forall v\, (v(A) \leq_f v(B))$.

In (Shramko and Wansing 2005), it is shown that the restrictions of the consequence relation $\models_t^{16}$ to the language $\mathcal{L}_t$ and of $\models_f^{16}$ to the language $\mathcal{L}_f$ both coincide with FDE-entailment. The problem of axiomatizing the consequence relations $\models_t^{16}$ and $\models_f^{16}$ in the full language $\mathcal{L}_{tf}$ remained open for a while. It has finally been solved in (Odintsov and Wansing 2015), where truth and falsity entailment are axiomatized by means of a first-degree bicalculus. Moreover, it is shown that the logic of $SIXTEEN_3$ in the propositional language $\mathcal{L}_{tf}$ is the logic of commutative distributive bilattices.

As far as symmetry versus asymmetry between verification and falsification is concerned, we may note that the truth and falsity orderings of $SIXTEEN_3$ are defined in a completely symmetrical way with regard to the presence and absence of the classical truth value T, respectively F. The two distinct relations of truth and falsity entailment treat truth and falsity as mutually independent dimensions of reasoning. As remarked in (Odintsov and Wansing 2015), we may think of $\models_t$, $\models_f$, and the analogously defined relation $\models_i$ as colouring reasoning in terms of truth, falsity, and information.

4.4 Ruling Out and Ruling In

In this section, it is shown that intuitionistic negation, expressing a notion of indirect falsification, can be symmetrically supplemented with a notion of indirect verification.

As we have seen, Sequoiah-Grayson suggests thinking of intuitionistic negation and its substructural variants as ways of representing the idea of ruling out. If A implies $\perp$ (in Sequoiah-Grayson's notation $\mathbf{0}$), then A is *ruled out*. In semantical terms, a state supports the truth of $\neg A$ iff every possible expansion of that state supports the truth of $\perp$. Since no state supports the truth of $\perp$, this amounts to requiring that a state x supports the truth of $\neg A$ iff no possible expansion of x supports the truth of A. If that gives one a notion of ruling out, what is *ruling in*? The idea is to have a negation, $-$, such that a state supports the falsity of $-A$ iff every possible expansion of that state supports the falsity of the constantly true zero-place connective $\top$. Since no state supports the falsity of $\top$, this amounts to requiring that a state x supports the falsity of $-A$ iff no possible expansion of x supports the falsity of A. In other words, A is ruled in at state x iff the assumption that A is false leads to the falsity of $\top$. In that way the state x makes allowance for A.

A logic that allows one to represent on the one hand ruling out and ruling in and on the other hand direct and indirect falsification is the bi-intuitionistic logic 2Int introduced in (Wansing 2013). The language $\mathcal{L}_{2\mathrm{Int}}$ of 2Int is defined in Backus–Naur form as follows:

$$A ::= p \mid \bot \mid \top \mid (A \wedge A) \mid (A \vee A) \mid (A \rightarrow A) \mid (A \prec A).$$

In 2int, the co-implication connective $A \prec B$ is in a sense dual to intuitionistic implication, it internalizes a relation of dual derivability into the logical object language. Dual derivability leads from counterassumptions (premises assumed to be false) to false conclusions. The co-negation $-A$ of A is defined as $\top \prec A$, and the intuitionistic negation $\neg A$ of A is defined as $A \rightarrow \bot$.

Definition 4.4 A model for 2Int is a structure $\mathcal{M} = \langle I, \leq, v^+, v^- \rangle$, where $\langle I, \leq \rangle$ is a pre-order and v^+, v^- are functions from the set of atomic formulas to subsets of the non-empty set of states I. For $x \in I$ the relations $\mathcal{M}, x \models^+ A$ ("x supports the truth of A in $\mathcal{M}$") and $\mathcal{M}, x \models^- A$ ("x supports the falsity of A in $\mathcal{M}$") are inductively defined as shown in Table 1. Moreover, support of truth and support of falsity are required to be persistent. For every atomic formula p, and all states x, x': if $x' \geq x$ and $\mathcal{M}, x \models^+ p$, then $\mathcal{M}, x' \models^+ p$ and if $x' \geq x$ and $\mathcal{M}, x \models^- p$, then $\mathcal{M}, x' \models^- p$.

Definition 4.5 An $\mathcal{L}_{2\mathrm{Int}}$-formula A is said to be valid in a model for 2Int $\mathcal{M} = \langle I, \leq, v^+, v^- \rangle$ iff for every $x \in I$, $\mathcal{M}, x \models^+ A$ (iff for every $x \in I$, $\mathcal{M}, x \models^- \neg A$); A is valid in 2Int ($\models_{2\mathrm{Int}} A$) iff A is valid in every model for 2Int.

An $\mathcal{L}_{2\mathrm{Int}}$-formula A is dually valid in a model for 2Int $\mathcal{M} = \langle I, \leq, v^+, v^- \rangle$ iff for every $x \in I$, $\mathcal{M}, x \models^- A$ (iff for every $x \in I$, $\mathcal{M}, x \models^+ -A$); A is dually valid in 2Int ($\models^d_{2\mathrm{Int}} A$) iff A is dually valid in every model for 2Int.

Definition 4.6 Let $\Delta \cup \{A\}$ be a set of $\mathcal{L}_{2\mathrm{Int}}$-formulas. Δ entails A ($\Delta \models A$) iff for every model for 2Int $\mathcal{M} = \langle I, \leq, v^+, v^- \rangle$ and every $x \in I$, it holds that if the truth of every element of Δ is supported by x, then the truth of A is supported by x.

Let $\Delta \cup \{A\}$ be a set of $\mathcal{L}_{2\mathrm{Int}}$-formulas. Δ dually entails A ($\Delta \models^d A$) iff for every model for 2Int $\mathcal{M} = \langle I, \leq, v^+, v^- \rangle$ and every $x \in I$, it holds that if the falsity of every element of Δ is supported by x, then the falsity of A is supported by x.

If truth and falsity do not divide up the totality of all states into two exclusive and exhaustive domains, then it is only natural to distinguish between support of truth, $\models^+$, and support of falsity, $\models^-$ and, moreover, to distinguish between truth preservation and falsity preservation, i.e., between the relations $\models$ and $\models^d$.

Table 1 Support of truth and support of falsity conditions for 2Int

$\mathcal{M}, x \models^+ p$ iff $x \in v^+(p)$

$\mathcal{M}, x \models^- p$ iff $x \in v^-(p)$

$\mathcal{M}, x \models^+ \top \qquad \mathcal{M}, x \not\models^- \top \qquad \mathcal{M}, x \not\models^+ \bot \qquad \mathcal{M}, x \models^- \bot$

$\mathcal{M}, x \models^+ (A \wedge B)$ iff $\mathcal{M}, x \models^+ A$ and $\mathcal{M}, x \models^+ B$

$\mathcal{M}, x \models^- (A \wedge B)$ iff $\mathcal{M}, x \models^- A$ or $\mathcal{M}, x \models^- B$

$\mathcal{M}, x \models^+ (A \vee B)$ iff $\mathcal{M}, x \models^+ A$ or $\mathcal{M}, x \models^+ B$

$\mathcal{M}, x \models^- (A \vee B)$ iff $\mathcal{M}, x \models^- A$ and $\mathcal{M}, x \models^- B$

$\mathcal{M}, x \models^+ (A \rightarrow B)$ iff for every $x' \geq x : \mathcal{M}, x' \not\models^+ A$ or $\mathcal{M}, x' \models^+ B$

$\mathcal{M}, x \models^- (A \rightarrow B)$ iff $\mathcal{M}, x \models^+ A$ and $\mathcal{M}, x \models^- B$

$\mathcal{M}, x \models^+ \neg A$ iff for every $x' \geq x : \mathcal{M}, x' \not\models^+ A$

$\mathcal{M}, x \models^- \neg A$ iff $\mathcal{M}, x \models^+ A$

$\mathcal{M}, x \models^+ -A$ iff $\mathcal{M}, x \models^- A$

$\mathcal{M}, x \models^- -A$ iff for every $x' \geq x : \mathcal{M}, x' \not\models^- A$

$\mathcal{M}, x \models^+ (A \prec B)$ iff $\mathcal{M}, x \models^+ A$ and $\mathcal{M}, x \models^- B$

$\mathcal{M}, x \models^- (A \prec B)$ iff for every $x' \geq x : \mathcal{M}, x' \not\models^- B$ or $\mathcal{M}, x' \models^- A$.

A sound and complete natural deduction proof system N2Int for 2Int is presented in (Wansing 2013). The system uses single-line rules for proofs and double-line rules for dual proofs. Derivations in N2Int combine proofs and dual proofs, so that a proof, in which the conclusion appears under a single line, may contain dual proofs as subderivations, and a dual proof, in which the conclusion appears under a double line, may contain proofs as subderivations. The conclusions of proofs and dual proofs depend on ordered pairs $(\Delta; \Gamma)$ of finite sets of premises, a set Δ of assumptions that are taken to be true, and a set Γ of "counterassumptions" that are taken to be false. Single square brackets [ ] are used to indicate that assumptions may be cancelled, and double-square brackets ⟦ ⟧ are used to indicate that counterassumptions may be cancelled. We write [A] instead of $[\overline{A}]$ and ⟦A⟧ instead of $[\![\overline{\overline{A}}]\!]$.

The proof rules for the connectives $\top$, $\bot$, $\wedge$, $\vee$, and $\rightarrow$ are basically those of intuitionistic logic; the rules for introducing (eliminating) the connectives of intuitionistic logic into (from) dual proofs are obtained by a dualization of their introduction and elimination rules for proofs. In 2Int the rules for introducing (eliminating) implications into (from) dual proofs are chosen in accordance with the usual understanding of the falsification conditions of implications, i.e., an implication $A \rightarrow B$ is false iff A is true and B is false. The rules for introducing (eliminating) co-implications into (from) proofs are such that the provability of $A \prec B$ amounts to the dual provability of $A \rightarrow B$. We will consider $\overline{A}$ as a proof of A from $(\{A\}; \varnothing)$ and $\overline{\overline{A}}$ as a dual proof of

Table 2 Introduction and elimination rules of N2Int w.r.t. proofs

$$\begin{array}{c}(\Delta;\,\Gamma)\\ \vdots\\ \overline{\bot}\\ \hline A\end{array}\ (\bot Ep)\qquad \begin{array}{cc}(\Delta;\,\Gamma) & (\Delta';\,\Gamma')\\ \vdots & \vdots\\ \overline{A} & \overline{B}\\ \hline \multicolumn{2}{c}{A\wedge B}\end{array}\ (\wedge Ip)\qquad \begin{array}{c}(\Delta;\,\Gamma)\\ \vdots\\ \overline{A\wedge B}\\ \hline A\end{array}\ (\wedge Ep)$$

$$\begin{array}{c}(\Delta;\,\Gamma)\\ \vdots\\ \overline{A\wedge B}\\ \hline B\end{array}\ (\wedge Ep)\qquad \begin{array}{c}(\Delta;\,\Gamma)\\ \vdots\\ \overline{A}\\ \hline A\vee B\end{array}\ (\vee Ip)\qquad \begin{array}{c}(\Delta;\,\Gamma)\\ \vdots\\ \overline{B}\\ \hline A\vee B\end{array}\ (\vee Ip)$$

$$\begin{array}{ccc}(\Delta;\,\Gamma) & ([A],\Delta';\,\Gamma') & ([B],\Delta'';\,\Gamma'')\\ \vdots & \vdots & \vdots\\ \overline{A\vee B} & \overline{C} & \overline{C}\\ \hline \multicolumn{3}{c}{C}\end{array}\ (\vee Ep)$$

$$\begin{array}{c}([A],\Delta;\,\Gamma)\\ \vdots\\ \overline{B}\\ \hline A\rightarrow B\end{array}\ (\rightarrow Ip)\qquad \begin{array}{cc}(\Delta;\,\Gamma) & (\Delta';\,\Gamma')\\ \vdots & \vdots\\ \overline{A} & \overline{A\rightarrow B}\\ \hline \multicolumn{2}{c}{B}\end{array}\ (\rightarrow Ep)$$

$$\begin{array}{cc}(\Delta;\,\Gamma) & (\Delta';\,\Gamma')\\ \vdots & \vdots\\ \overline{A} & \overline{\overline{B}}\\ \hline \multicolumn{2}{c}{A\prec B}\end{array}\ (\prec Ip)\qquad \begin{array}{c}(\Delta;\,\Gamma)\\ \vdots\\ \overline{A\prec B}\\ \hline A\end{array}\ (\prec Ep)\qquad \begin{array}{c}(\Delta;\,\Gamma)\\ \vdots\\ \overline{A\prec B}\\ \hline\hline B\end{array}\ (\prec Ep)$$

A from $(\varnothing;\, \{A\})$.[11] Moreover $\overline{\top}$ is a proof of $\top$ from $(\varnothing;\, \varnothing)$ and $\overline{\overline{\bot}}$ is a dual proof of $\bot$ from $(\varnothing;\, \varnothing)$. In addition to these stipulations, the system N2Int comprises the introduction and elimination rules listed in Tables 2 and 3.[12] We write $(\Delta;\, \Gamma) \vdash A$ if there is a proof of A from $(\Delta;\, \Gamma)$; and we write $(\Delta;\, \Gamma) \vdash^d A$ if there is a dual proof of A from $(\Delta;\, \Gamma)$. Moreover, we assume that if $(\Delta;\, \Gamma) \vdash A$, $\Delta \subseteq \Delta'$ and $\Gamma \subseteq \Gamma'$ for finite sets of $\mathcal{L}_{2\mathrm{Int}}$-formulas Δ' and Γ', then $(\Delta';\, \Gamma') \vdash A$. Similarly, we assume that if $(\Delta;\, \Gamma) \vdash^d A$, $\Delta \subseteq \Delta'$ and $\Gamma \subseteq \Gamma'$ for finite sets of $\mathcal{L}_{2\mathrm{Int}}$-formulas Δ' and Γ', then $(\Delta';\, \Gamma') \vdash^d A$.

How is ruling in related to ruling out? Note first that the intuitionistic negation $\neg$ allows one to switch from provability to dual provability, whereas the dual negation allows one to switch from dual provability to provability.

---

[11] Since in ordinary natural deduction a formula A is a proof of A from $\{A\}$, the cancellation of formulas amounts to the cancellation of proofs.

[12] In these tables, *Ep* stands for "elimination from proofs," *Ip* for "introduction into proofs," *Edp* for "elimination from dual proofs," and *Idp* for "introduction into dual proofs."

Table 3 Introduction and elimination rules of N2Int w.r.t. dual proofs

$$\dfrac{\begin{matrix}(\Delta;\Gamma)\\ \vdots\\ \overline{\overline{\top}}\end{matrix}}{\overline{\overline{A}}}\ (\top Edp) \qquad \dfrac{\begin{matrix}(\Delta;\Gamma)\\ \vdots\\ \overline{\overline{A}}\end{matrix} \quad \begin{matrix}(\Delta';\Gamma')\\ \vdots\\ \overline{\overline{B}}\end{matrix}}{A\vee B}\ (\vee Idp) \qquad \dfrac{\begin{matrix}(\Delta;\Gamma)\\ \vdots\\ \overline{\overline{A\vee B}}\end{matrix}}{A}\ (\vee Edp)$$

$$\dfrac{\begin{matrix}(\Delta;\Gamma)\\ \vdots\\ \overline{\overline{A\vee B}}\end{matrix}}{B}\ (\vee Edp) \qquad \dfrac{\begin{matrix}(\Delta;\Gamma)\\ \vdots\\ \overline{\overline{A}}\end{matrix}}{A\wedge B}\ (\wedge Idp) \qquad \dfrac{\begin{matrix}(\Delta;\Gamma)\\ \vdots\\ \overline{\overline{B}}\end{matrix}}{A\wedge B}\ (\wedge Idp)$$

$$\dfrac{\begin{matrix}(\Delta;\Gamma)\\ \vdots\\ \overline{\overline{A\wedge B}}\end{matrix} \quad \begin{matrix}(\Delta';\Gamma',[\![A]\!])\\ \vdots\\ \overline{\overline{C}}\end{matrix} \quad \begin{matrix}(\Delta'';\Gamma'',[\![B]\!])\\ \vdots\\ \overline{\overline{C}}\end{matrix}}{C}\ (\wedge Edp)$$

$$\dfrac{\begin{matrix}(\Delta;\Gamma,[\![A]\!])\\ \vdots\\ \overline{\overline{B}}\end{matrix}}{B \prec A}\ (\prec Idp) \qquad \dfrac{\begin{matrix}(\Delta';\Gamma')\\ \vdots\\ \overline{\overline{B \prec A}}\end{matrix} \quad \begin{matrix}(\Delta;\Gamma)\\ \vdots\\ \overline{A}\end{matrix}}{B}\ (\prec Edp)$$

$$\dfrac{\begin{matrix}(\Delta;\Gamma)\\ \vdots\\ \overline{A}\end{matrix} \quad \begin{matrix}(\Delta';\Gamma')\\ \vdots\\ \overline{\overline{B}}\end{matrix}}{A\to B}\ (\to Idp) \qquad \dfrac{\begin{matrix}(\Delta;\Gamma)\\ \vdots\\ \overline{\overline{A\to B}}\end{matrix}}{A}\ (\to Edp) \qquad \dfrac{\begin{matrix}(\Delta;\Gamma)\\ \vdots\\ \overline{\overline{A\to B}}\end{matrix}}{B}\ (\to Edp)$$

Observation 4.7 $(\Delta;\Gamma) \vdash A$ *iff* $(\Delta;\Gamma) \vdash^d \neg A$; $(\Delta;\Gamma) \vdash^d A$ *iff* $(\Delta;\Gamma) \vdash -A$.

Proof

$$\dfrac{\begin{matrix}(\Delta;\Gamma)\\ \vdots\\ \overline{A}\end{matrix} \quad \overline{\overline{\bot}}}{A\to\bot}\ (\to Idp) \qquad \dfrac{\begin{matrix}(\Delta;\Gamma)\\ \vdots\\ \overline{\overline{A\to\bot}}\end{matrix}}{A}\ (\to Edp) \qquad \dfrac{\overline{\top} \quad \begin{matrix}(\Delta;\Gamma)\\ \vdots\\ \overline{\overline{A}}\end{matrix}}{\top \prec A}\ (\prec Ip) \qquad \dfrac{\begin{matrix}(\Delta;\Gamma)\\ \vdots\\ \overline{\top \prec A}\end{matrix}}{A}\ (\prec Ep) \qquad \square$$

As a corollary to Observation 4.7 we may note that due to the presence of the two negations, the distinction between assumptions and counterassumptions could be dispensed with in 2Int.

Observation 4.8 *Let* $\neg\Theta := \{\neg A\colon A \in \Theta\}$ *and* $-\Theta := \{-A\colon A \in \Gamma\}$ *for a set of formulas* Θ*. Then* $(\Delta; \Gamma) \vdash A$ *iff* $(\Delta \cup -\Gamma; \varnothing) \vdash A$*, and* $(\Delta; \Gamma) \vdash^d A$ *iff* $(\varnothing; \Gamma \cup \neg\Delta) \vdash^d A$.

Observation 4.9 *In the system* N2Int *the negations* $\neg$ *and* $-$ *and the two notions of derivability* $\vdash$ *and* $\vdash^d$ *are related as follows:*

1. $(\Delta; \Gamma) \vdash \neg A$ *iff* $(\Delta; \Gamma) \vdash^d -\neg A$.
2. $(\Delta; \Gamma) \vdash^d \neg A$ *iff* $(\Delta; \Gamma) \vdash -\neg A$.
3. $(\Delta; \Gamma) \vdash -A$ *iff* $(\Delta; \Gamma) \vdash^d \neg - A$.
4. $(\Delta; \Gamma) \vdash^d -A$ *iff* $(\Delta; \Gamma) \vdash \neg - A$.

Proof 1. and 2. are just instantiations of the equivalences from Observation 4.7. As to 3. and 4. we have:

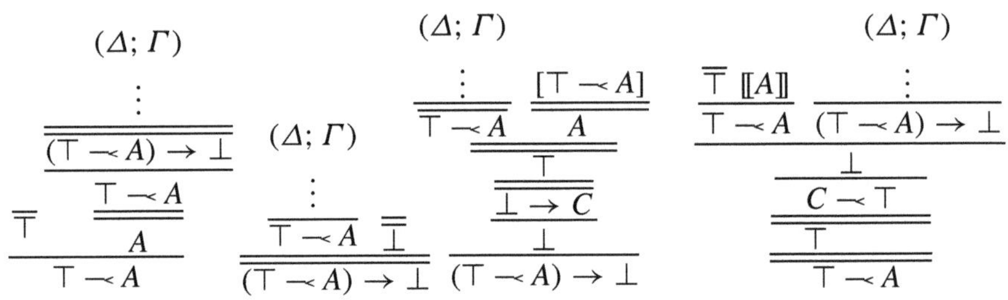

□

In (Wansing 2013), N2Int is shown to be weakly sound and complete with respect to 2Int: $\models_{2\mathrm{Int}} A$ iff $(\varnothing; \varnothing) \vdash A$ and $\models^d_{2\mathrm{Int}} A$ iff $(\varnothing; \varnothing) \vdash^d A$. The proof uses a faithful embedding of 2Int into intuitionistic logic with respect to entailment.[13] Strong soundness and completeness can easily be shown.

Observation 4.10 *Let* A *be an* $\mathcal{L}_{2Int}$*-formula and let* $\{A_1, \ldots, A_k\}$, $\{B_1, \ldots, B_m\}$ *be finite, possibly empty sets of* $\mathcal{L}_{2Int}$*-formulas. If both sets are empty, then let* $A \equiv (A_1 \to (A_2 \to (\ldots(-B_1 \to (\ldots(-B_m \to A)\ldots))\ldots)))$ *and* $A \equiv (((\ldots((\ldots(A \prec B_m)\ldots) \prec B_1)\ldots) \prec \neg A_2) \prec \neg A_1)$.

1. $(\{A_1, \ldots, A_k\};\ \{B_1, \ldots, B_m\}) \vdash A$ *iff* $\{A_1, \ldots, A_k, -B_1, \ldots, -B_m\} \models A$*;*
2. $(\{A_1, \ldots, A_k\};\ \{B_1, \ldots, B_m\}) \vdash^d A$ *iff* $\{\neg A_1, \ldots, \neg A_k, B_1, \ldots, B_m\} \models^d A$.

[13] Another translation presented in (Wansing 2013) gives one a faithful embedding of 2Int into dual intuitionistic logic with respect to dual entailment.

Proof

$$
\begin{aligned}
&(\{A_1, \ldots, A_k\}; \{B_1, \ldots, B_m\}) \vdash A \\
\text{iff } &(\{A_1, \ldots, A_k, -B_1, \ldots, -B_m\}; \varnothing) \vdash A \quad \text{by Obs. 4.8} \\
\text{iff } &(\varnothing; \varnothing) \vdash (A_1 \to (A_2 \to (\ldots(-B_1 \to (\ldots(-B_m \to A)\ldots))\ldots))) \\
\text{iff } &\models (A_1 \to (A_2 \to (\ldots(-B_1 \to (\ldots(-B_m \to A)\ldots))\ldots))) \\
\text{iff } &\{A_1, \ldots, A_k, -B_1, \ldots, -B_m\} \models A
\end{aligned}
$$

$$
\begin{aligned}
&(\{A_1, \ldots, A_k\}; \{B_1, \ldots, B_m\}) \vdash^d A \\
\text{iff } &(\varnothing; \{\neg A_1, \ldots, \neg A_k, B_1, \ldots, B_m\}) \vdash^d A \quad \text{by Obs. 4.8} \\
\text{iff } &(\varnothing; \varnothing) \vdash^d (((\ldots((\ldots(A \prec B_m)\ldots) \prec B_1)\ldots) \prec \neg A_2) \prec \neg A_1) \\
\text{iff } &\models^d (((\ldots((\ldots(A \prec B_m)\ldots) \prec B_1)\ldots) \prec \neg A_2) \prec \neg A_1) \\
\text{iff } &\{\neg A_1, \ldots, \neg A_k, B_1, \ldots, B_m\} \models^d A
\end{aligned}
$$

□

The language of Nelson's N4 contains the connectives $\wedge$, $\vee$, $\to$, and $\sim$. There is thus only one negation, namely the strong negation $\sim$. The relational semantics for N4 also comes with a distinction between support of truth and support of falsity.

Definition 4.11 A model for N4 is defined exactly as a model for 2Int, except that the relations $\mathcal{M}, x \models^+ A$ ("x supports the truth of A in $\mathcal{M}$") and $\mathcal{M}, x \models^- A$ ("x supports the falsity of A in $\mathcal{M}$") are inductively defined as shown in Table 4.

The proof theory of N4 and closely related systems is comprehensively dealt with in (Kamide and Wansing 2012), see also (Kamide and Wansing 2015). If we denote provability in N4 by $\vdash_{N4}$, we could define a notion of disprovability in N4, $\vdash^{dis}_{N4}$, by setting[14]

$$\vdash^{dis}_{N4} A \text{ iff } \vdash_{N4} \sim A.$$

We would then have $\vdash^{dis}_{N4} \sim A$ iff $\vdash_{N4} \sim\sim A$ iff $\vdash_{N4} A$, whereas in 2Int we have $\vdash^d \neg A$ iff $\vdash A$ and $\vdash^d A$ iff $\vdash -A$.

In Sect. 4 I argued that what is positive information in the context of verification may be negative information in the context of falsification, and vice versa. This idea is realized in different ways in the systems N4 and 2Int. In N4, the falsification of A (in the sense of support of falsity) *amounts to* the verification of $\sim A$ (in the sense of support of truth). If an information state x directly falsifies A, then x directly verifies $\sim A$, and if x directly verifies A, then x directly falsifies $\sim A$. In 2In, an information state x verifies (in the sense of support of truth) a formula A iff it falsifies (in the sense of support of falsity) its intuitionistic negation $\neg A$, whereas x falsifies A iff x verifies its co-negation $-A$.

[14]Cf. (Wansing 2010), where I consider extensions of Heyting–Brouwer logic (Rauszer 1980) by strong negation and refer to reductions to non-truth as dual proofs.

Table 4 Support of truth and support of falsity conditions for N4

$\mathcal{M}, x \models^+ p$ iff $x \in v^+(p)$
$\mathcal{M}, x \models^- p$ iff $x \in v^-(p)$
$\mathcal{M}, x \models^+ (A \wedge B)$ iff $\mathcal{M}, x \models^+ A$ and $\mathcal{M}, x \models^+ B$
$\mathcal{M}, x \models^- (A \wedge B)$ iff $\mathcal{M}, x \models^- A$ or $\mathcal{M}, x \models^- B$
$\mathcal{M}, x \models^+ (A \vee B)$ iff $\mathcal{M}, x \models^+ A$ or $\mathcal{M}, x \models^+ B$
$\mathcal{M}, x \models^- (A \vee B)$ iff $\mathcal{M}, x \models^- A$ and $\mathcal{M}, x \models^- B$
$\mathcal{M}, x \models^+ (A \rightarrow B)$ iff for every $x' \geq x : \mathcal{M}, x' \not\models^+ A$ or $\mathcal{M}, x' \models^+ B$
$\mathcal{M}, x \models^- (A \rightarrow B)$ iff $\mathcal{M}, x \models^+ A$ and $\mathcal{M}, x \models^- B$
$\mathcal{M}, x \models^+ {\sim}A$ iff $\mathcal{M}, x \models^- A$
$\mathcal{M}, x \models^- {\sim}A$ iff $\mathcal{M}, x \models^+ A$

5 Summary

We have seen that in accordance with Gurevich (1977) and Wansing (1993a, b), it makes sense to treat definitely positive and definitely negative information on an equal footing. In particular, in the system N4 a sentence letter p may be directly falsified (verified) by directly verifying (falsifying) its strong negation ${\sim}p$. In the system 2Int, p may be directly falsified by a direct verification of its co-negation $-p$, and p may be directly verified by a direct falsification of its intuitionistic negation $\neg p$. There is thus again a symmetry, though a more complex symmetry than in N4, between direct verification and direct falsification and between definitely positive and definitely negative information. Intuitionistic negation internalizes a notion of indirect falsification with respect to provability, $\vdash$. In the system 2Int we have both intuitionistic negation and co-negation, and the latter internalizes a notion of indirect verification with respect to dual provability, $\vdash^d$. If indirect falsification makes sense, and it does, indirect verification makes sense as well. Again we have symmetry instead of asymmetry.[15]

If we come back to the above central 'what is'-question, *What is negation?*, we may say that what strong negation, intuitionistic negation, and co-negation have in common is that they permit a passage from provability to its dual, or vice versa. Michael Dunn's favourite 4-valued semantics models strong negation, which expresses a symmetry between direct verification and direct falsification.

Acknowledgments I would like to thank Katalin Bimbó, Yaroslav Shramko, Stanislav Speranski and two anonymous referees for their very helpful comments.

[15] In (Kapsner 2014), a detailed analysis is presented of Michael Dummett's views on the interaction between verification and falsification in an inferentialist theory of meaning. In that context Kapsner argues for "the superiority of the Nelson account over the intuitionistic one," (Kapsner 2014, p. 198).

References

Adriaans, P. (2013). Information. In E. N. Zalta (Ed.), *The Stanford encyclopedia of philosophy*, fall 2013 ed. http://www.plato.stanford.edu/archives/fall2013/entries/information/.

Belnap, N. D. (1977a). How a computer should think. In G. Ryle (Ed.), *Contemporary aspects of philosophy* (pp. 30–56). Stocksfield: Oriel Press.

Belnap, N. D. (1977b). A useful four-valued logic. In J. M. Dunn & G. Epstein (Eds.), *Modern uses of multiple-valued logic* (pp. 8–37). Dordrecht: Reidel.

Berto, F. (2015). A modality called 'negation'. *Mind*, *124*, 761–793.

Bimbó, K. & Dunn, J. M. (2008). *Generalized Galois logics. Relational semantics of nonclassical logical calculi*, Volume 188 of CSLI Lecture Notes. Stanford, CA: CSLI Publications.

Birkhoff, G., & von Neumann, J. (1936). The logic of quantum mechanics. *Annals of Mathematics*, *37*, 823–843.

Došen, K. (1984). Negative modal operators in intuitionistic logic. *Publications de l'Institut Mathématique. Nouvelle Série*, *35*(49), 3–14. See also at www.emis.de/journals/PIMB/049/n049p003.pdf.

Došen, K. (1986). Negation as a modal operator. *Reports on Mathematical Logic*, *20*, 15–27.

Došen, K. (1999). Negation in the light of modal logic. In D. Gabbay & H. Wansing (Eds.), *What is negation?* (pp. 77–86). Dordrecht: Kluwer.

Dunn, J. M. (1966). The algebra of intensional logics, PhD thesis. Ann Arbor (UMI): University of Pittsburgh.

Dunn, J. M. (1976). Intuitive semantics for first-degree entailments and 'coupled trees'. *Philosophical Studies*, *29*, 149–168.

Dunn, J. M. (1986). Relevance logic and entailment. In D. Gabbay & F. Guenthner (Eds.), *Handbook of philosophical logic* (1st ed., Vol. 3, pp. 117–224). Dordrecht: D. Reidel.

Dunn, J. M. (1991). Gaggle theory: An abstraction of Galois connections and residuation with applications to negation, implication, and various logical operators. In van Eijck, J. (Ed.), *Logics in AI: European Workshop JELIA'90, Volume 478 of Lecture Notes in Computer Science* (pp. 31–51). Berlin: Springer.

Dunn, J. M. (1993). Star and perp: Two treatments of negation. In Tomberlin, J. (Ed.), *Philosophical perspectives, language and logic*, (Vol. 7, pp. 331–357)

Dunn, J. M. (1995). Gaggle theory applied to intuitionistic, modal and relevance logic. In: I. Max & W. Stelzner (Eds.), *Logik und Mathematik. Frege-Kolloquim Jena* (pp. 335–368). Berlin: W. de Gruyter.

Dunn, J. M. (1996). Generalised ortho negation. In H. Wansing (Ed.), *Negation: A notion in focus* (pp. 3–26). Berlin: W. de Gruyter.

Dunn, J. M. (1999). A comparative study of various model-theoretic treatments of negation: A history of formal negation. In D. M. Gabbay & H. Wansing (Eds.), *What is negation?* (pp. 23–51). Dordrecht: Kluwer.

Dunn, J. M. (2000). Partiality and its dual. *Studia Logica*, *65*, 5–40.

Dunn, J. M. (2001). The concept of information and the development of modern logic. In W. Stelzner & M. Stöckler (Eds.), *Zwischen traditioneller und moderner Logik: Nichtklassische Ansätze* (pp. 423–447). Paderborn: Mentis-Verlag.

Dunn, J. M. (2008). Information in computer science. In: P. Adriaans & J. van Benthem (Eds.), *Philosophy of information. Handbook of the philosophy of science, (D. M. Gabbay, P. Thagard, J. Woods (Eds.))* (Vol.8, pp. 581–608). Amsterdam: Elsevier.

Dunn, J. M. (2010). Contradictory information: too much of a good thing. *Journal of Philosophical Logic*, *39*, 425–452.

Dunn, J. M. & Hardegree, G. M. (2001). *Algebraic methods in philosophical logic*, Vol. 41 of Oxford Logic Guides. Oxford, UK: Oxford University Press.

Dunn, J. M., & Zhou, C. (2005). Negation in the context of gaggle theory. *Studia Logica*, *80*, 235–264.

Floridi, L. (2015). Semantic conceptions of information. In E. N. Zalta (Ed.), *The Stanford encyclopedia of philosophy* (spring 2015 ed.). http://www.plato.stanford.edu/archives/spr2015/entries/information-semantic/.

Frege, G. (1892). Über Sinn und Bedeutung. *Zeitschrift für Philosophie und Philosophische Kritik NF, 100*, 25–50.

Frege, G. (1923). Gedankengefüge. *Beiträge zur Philosophie des Deutschen Idealismus, 3*, 36–51.

Gabbay, D., & Wansing, H. (Eds.). (1999). *What is negation?*. Dordrecht: Kluwer.

Goldblatt, R. (1974). Semantic analysis of orthologic. *Journal of Philosophical Logic, 3*, 19–35.

Goldblatt, R. (1975). The Stone space of an ortholattice. *Bulletin of the London Mathematical Society, 7*, 45–48.

Grzegorczyk, A. (1964). A philosophically plausible formal interpretation of intuitionistic logic. *Indagationes Mathematicae, 26*, 596–601.

Gurevich, Y. (1977). Intuitionistic logic with strong negation. *Studia Logica, 36*, 49–59.

Hartonas, C. & Dunn, J. M. (1993). Duality theorems for partial orders, semilattices, Galois connections and lattices, Technical Report Preprint No. IULG-93-26, Indiana University Logic Preprint Series.

Horn, L. R. & Wansing, H. (2015). Negation. In: E. N. Zalta (Ed.), *The Stanford encyclopedia of philosophy* (spring 2015 ed.). http://www.plato.stanford.edu/archives/spr2015/entries/negation/.

Kamide, N. & Wansing, H. (2011). Connexive modal logic based on positive S4. In J.-Y. Béziau & M. Coniglio (Eds.), *Logic without Frontiers. Festschrift for Walter Alexandre Carnielli on the occasion of his 60th birthday* (pp. 389–409). London: College Publications.

Kamide, N., & Wansing, H. (2012). Proof theory of Nelson's paraconsistent logic: A uniform perspective. *Theoretical Computer Science, 415*, 1–38.

Kamide, N., & Wansing, H. (2015). *Proof theory of N4-related paraconsistent logics*. London: College Publications.

Kapsner, A. (2014). *Logics and falsifications, A new perspective on constructivist semantics*. Dordrecht: Springer.

Lambek, J. (1958). The mathematics of sentence structure. *American Mathematical Monthly, 65*, 154–169.

Lambek, J. (1961). On the calculus of syntactic types. In R. Jacobson (Ed.), *Structure of language and its mathematical aspects* (pp. 166–178). RI: American Mathematical Society, Providence.

Mares, E. D. (2009). General information in relevant logic. *Synthese, 167*, 343–362.

Nelson, D. (1959). Negation and separation of concepts in constructive systems. In A. Heyting (Ed.), *Constructivity in mathematics* (pp. 208–225). Amsterdam: North-Holland.

Odintsov, S. (2008). *Constructive negations and paraconsistency*, Vol. 26. of Trends in Logic. Dordrecht: Springer.

Odintsov, S. P. (2005). The class of extensions of Nelson's paraconsistent logic. *Studia Logica, 80*, 291–320.

Odintsov, S. P., & Wansing, H. (2015). The logic of generalized truth values and the logic of bilattices. *Studia Logica, 103*, 91–112.

Onishi, T. (2015). Substructural negations. *Australasian Journal of Logic, 12*, 177–203.

Paoli, F. (2002). *Substructural logics: A primer*. Dordrecht: Kluwer.

Paoli, F., & Tsinakis, C. (2012). On Birkhoff's common abstraction problem. *Studia Logica, 100*, 1079–1105.

Rauszer, C. (1980). *An algebraic and Kripke-style approach to a certain extension of intuitionistic logic, Dissertationes Mathematicae 167*. Polish Academy of Sciences, Warsaw: Institute of Mathematics.

Restall, G. (1999). Negation in relevant logics (How I stopped worrying and learned to love the Routley star). In D. Gabbay & H. Wansing (Eds.), *What is negation?* (pp. 53–76). Dordrecht: Kluwer.

Restall, G. (2000). *An introduction to substructural logics*. London: Routledge.

Routley, R. & Meyer, R. K. (1972). The semantics of entailment, II-III. *Journal of Philosophical Logic, 1*, 53–73 and 192–208.

Routley, R. & Meyer, R. K. (1973). The semantics of entailment. In: H. Leblanc (Ed.), *Truth, syntax and modality, Proceedings of the Temple University Conference on Alternative Semantics* (pp. 199–243). North-Holland, Amsterdam.

Sequoiah-Grayson, S. (2009). Dynamic negation and negative information. *Review of Symbolic Logic*, *2*(1), 233–248.

Shramko, Y. (2005). Dual intuitionistic logic and a variety of negations: The logic of scientific research. *Studia Logica*, *80*, 347–367.

Shramko, Y., & Wansing, H. (2005). Some useful 16-valued logics: How a computer network should think. *Journal of Philosophical Logic*, *34*, 121–153.

Shramko, Y., & Wansing, H. (2011). *Truth and falsehood: An inquiry into generalized logical values*. Dordrecht: Springer.

Vakarelov, D. (1977). Theory of negation in certain logical systems: Algebraic and semantic approach, PhD thesis, University of Warsaw.

Vakarelov, D. (1989). Consistency, completeness and negations. In G. Priest, R. Routley, & J. Norman (Eds.), *Paraconsistent logic: Essays on the inconsistent* (pp. 328–368). Munich: Philosophia Verlag.

Wansing, H. (1993a). Informational interpretation of substructural propositional logics. *Journal of Logic, Language and Information*, *2*, 285–308.

Wansing, H. (1993b). *The logic of information structures*. Vol. 681 of Lecture Notes in AI. Berlin: Springer.

Wansing, H. (2001). Negation. In L. Goble (Ed.), *The Blackwell guide to philosophical logic* (pp. 415–436). Oxford: Blackwell.

Wansing, H. (2005). Connexive modal logic. In R. Schmidt, I. Pratt-Hartmann, M. Reynolds, & H. Wansing (Eds.), *Advances in modal logic* (Vol. 5, pp. 367–383). London: College Publications.

Wansing, H. (2006). Contradiction and contrariety. Priest on negation. In J. Malinowski & A. Pietruszczak (Eds.), *Essays in logic and ontology* (pp. 81–93). Amsterdam: Rodopi.

Wansing, H. (2008). Constructive negation, implication, and co-implication. *Journal of Non-Classical Logics*, *18*, 341–364.

Wansing, H. (2010). Proofs, disproofs, and their duals. In V. Goranko, L. Beklemishev, & V. Shehtman (Eds.), *Advances in modal logic* (Vol. 8, pp. 483–505). London: College Publications.

Wansing, H. (2013). Falsification, natural deduction and bi-intuitionistic logic. *Journal of Logic and Computation*. Published as "first online" on July 17th, 2013. doi:10.1093/logcom/ext035.

Wansing, H. (2014). Connexive logic. In: E. N. Zalta (Ed.), *The Stanford encyclopedia of philosophy*, fall 2014 ed. http://www.plato.stanford.edu/archives/fall2014/entries/logic-connexive/.

Truth, Falsehood, Information and Beyond: The American Plan Generalized

Yaroslav Shramko

Abstract This paper highlights the importance of a strategy for semantic analysis initiated by J. Michael Dunn, known in the literature as the "American Plan." The key insight of the plan relies on allowing under-determined and over-determined logical valuations, which prove to be essential for a logical analysis of information structures. The main directions in the development of this fundamental idea are explained, and an implementation of the possible generalization thereof is briefly reviewed, culminating in the notion of a multi-consequence logic.

Keywords American plan · Bilattice · Multi-consequence logic · Multilattice · Over-determined valuations · Trilattice · Under-determined valuations

1 Introducing Semantics on the American Plan: Four Versions

1.1 Preliminaries

Among the many important achievements of J. Michael Dunn in various areas of modern non-classical logic is his fundamental contribution, essential for the general development of the entire field, namely, introducing and thoroughly elaborating a revolutionary strategy for semantic analysis, whereby a sentence can *rationally* be considered to be not only just true or just false (Fregean *das Wahre* and *das Falsche*), but also *neither* true nor false, as well as simultaneously *both* true and false. The strategy implies abandoning some core principles of classical logic, including the principles of *bi-valence* (every sentence is true or false), and *unique-valence* (no sentence can be true and false simultaneously), cf. (Dunn 2000, p. 5). Since these principles seem to be rather restrictive for certain purposes of scientific inquiry, under- and over-determined logical valuations are allowed.

Y. Shramko (✉)
Department of Philosophy, Kryvyi Rih State Pedagogical University, Kryvyi Rih, Ukraine
e-mail: shramko@rocketmail.com
URL: http://kdpu.edu.ua/shramko/yse.htm

K. Bimbó (ed.), *J. Michael Dunn on Information Based Logics*,
Outstanding Contributions to Logic 8, DOI 10.1007/978-3-319-29300-4_11

This idea was initially proposed and developed in Dunn (1966) doctoral thesis, presented and discussed thereafter in a number of conference talks, seminars and abstracts such as (Dunn 1967, 1971), and finally exposed in detail in his seminal paper (Dunn 1976), see also a comprehensive discussion (and generalization) of the subject in (Dunn 1999, 2000).[1] At first, Dunn's idea of expanding logical valuations emerged in the context of investigations into relevant logic and the problem of entailment, conceived as an effective way of eliminating the so-called "paradoxes of classical entailment and material implication." But, as time passed, it became clear that it could find fruitful applications in many other fields, especially in a logical analysis of *information structures*, see, e.g., the comprehensive study by Wansing (1993).

Robert K. Meyer in (1978) wittily dubbed this strategy the "American Plan" to contrast the approach of *Americans* Dunn and Nuel Belnap with the semantic analysis of the *Australians* Richard Routley (later Sylvan) and himself.[2] The burden of the plan can be roughly defined as set out below.

Let $\mathcal{L}$ be some language (a set of sentences), and $\mathbf{2} = \{ F, T \}$ be the set of classical truth-values. A standard classical valuation v^2 (a 2-valuation) is then a *function* from $\mathcal{L}$ to $\mathbf{2}$. Since this is a total function, every sentence is allocated *one and only one* element from $\mathbf{2}$, i.e., either true or false. Now, to allow the non-standard, to wit, under-determined and over-determined valuations mentioned above, the classical truth-value function must be replaced with some other valuation procedure. Such a procedure can be constructed in several ways, thereby forming particular *versions* of a concrete implementation of the American Plan.

In this section *four* such versions are delineated, all of which are due to Dunn, except for the last one, which was elaborated by Belnap (although initiated also by Dunn). These versions are all formally equivalent and inter-definable, differing mainly in their philosophical background and informal motivations. Section 2 shows how Dunn's ideas can be grasped algebraically using a suitable notion of a bilattice, and presents an adequate logical formalism of this structure. In Sect. 3 a method for generalizing the American Plan through the notion of a trilattice and corresponding bi-consequence system is described. Section 4 completes the generalization, culminating in the notion of a multilattice which finds its deductive representation in a multi-consequence logical system. Section 5 presents another system for reasoning with logical multilattices.

In summary, the contributions of this paper are as follows: (1) a brief survey of the work by Dunn (and others) on four-valued semantics in accordance with the American Plan; (2) an exposition of some key results concerning a possible

[1] It should be pointed out that classical principles of bi-valence and unique-valence were occasionally criticized long before Dunn's work, see e.g., (Łukasiewicz 1920, 1993). However, it was in fact Dunn, who not only challenged particular principles, but also initiated a ground-breaking research program (paradigm) on semantic analysis, in which abandoning certain classical principles turned out to be not a starting point, but rather the effect of more general philosophical considerations.

[2] Meyer and Routley (Sylvan) are inseparably associated with Australia owing to their long and fruitful service at the Australian National University, even though Meyer originally came from the United States and Routley from New Zealand.

generalization thereof, as developed and summarized most notably in (Shramko and Wansing 2011); and (3) a method for implementing a *complete generalization* of the American Plan.

1.2 Version 1: Aboutness Valuation

In (Dunn 1966, pp. 121–132), driven by the idea of content containment for statements of relevant entailment, Dunn proposed replacing the classical truth-value function by a so-called "aboutness valuation"—a function that ascribes to each propositional variable p a pair of sets (X_1, X_2). These sets are subsets of some more general "set of topics" X called "the universe of discourse." X_1 is conceived as the set of topics about which the proposition gives definite *positive* information, while X_2 is the set of topics about which the proposition gives definite *negative* information, cf. (Dunn 1986, p. 191). The pair (X_1, X_2) is called a "proposition surrogate," since "such a pair when assigned to a formula gives a partial representation of the meaning of the formula" (Dunn 1966, p. 126).

The aboutness valuation is extendable to all sentences in our language by the following natural definition[3] of logical connectives:

Definition 1.1

$$
\begin{aligned}
(X_1, X_2) \wedge (Y_1, Y_2) &= (X_1 \cup Y_1, X_2 \cap Y_2);\\
(X_1, X_2) \vee (Y_1, Y_2) &= (X_1 \cap Y_1, X_2 \cup Y_2);\\
{\sim}(X_1, X_2) &= (X_2, X_1).
\end{aligned}
$$

Note, that proposition surrogates need be neither disjoint $(X_1 \cap X_2 = \varnothing)$ nor exhaustive $(X_1 \cup X_2 = X)$, and thus, topics can contradict each other, or—on some issue—we can have no topic at all. If the universe of discourse consist of a single topic x, then aboutness valuation gives to every sentence one of the following four assignments: $(\varnothing, \{x\})$, $(\{x\}, \varnothing)$, $(\{\varnothing\}, \{\varnothing\})$, and $(\{x\}, \{x\})$. In effect, the first two assignments correspond to classical falsity and truth, whereas the next two pairs represent under-determined and over-determined valuations respectively.

It is worth noting that in his thesis Dunn formulated the semantics without explicit reference to the terminology of truth-values, employing instead such euphemisms as "topics," "definite information about," and "aboutness valuation" among others.[4] There were both philosophical and social-psychological reasons for this. Philosophically a strong distinction should be made between ontological and epistemological

---

[3]This definition shows that X_1 represents the "falsity domain" of the corresponding propositional surrogate, while X_2 is its "truth domain." If one were to interpret the pair (X_1, X_2) as a "possible world semantics," then X_1 and X_2 would be the sets of "worlds" in which the corresponding proposition are false and true respectively.

[4]On truth-values and their importance for logic and philosophy see, e.g., (Shramko and Wansing 2014).

(informational) interpretations of truth-values. Without such a distinction, not many in the logical community in the 1960s, and even nowadays, were and are ready to talk about sentences as being both true and false, or neither true nor false. As Dunn remarked in (1976, Footnote 10), in his dissertation he "lacked the philosophical nerve to embrace this as a serious way of talking." However, he immediately admitted use of these phrases "in conversation."

1.3 Version 2: Valuation as a Relation

In (Dunn 1976) the American Plan is presented in full detail as the "intuitive semantics" for a so-called "first-degree relevant entailment," see also in (Anderson et al. 1992, Sect. 50). Dunn develops here a certain generalization of a possible-world semantics. Instead of the notion of a "possible world" Dunn employs the notion of a "situation," stressing in particular that "there are plenty of situations where we suppose, assert, believe, etc., contradictory sentences to be true" (Dunn 1976, p. 157). Hence, situations acquire an explicitly epistemic (informational) characterization, and as such they may well be inconsistent and/or incomplete.

Within the framework of a possible world semantics a proposition can be realized as a function from a set of possible worlds to the set of truth-values, see (Dunn 1976, p. 154). This is just another representation of a classical truth-value function with respect to possible worlds. In contrast, replacing possible worlds with (abstract epistemic) situations allows one to construe of a proposition "as relational but not necessarily functional in character." In other words, instead of the classical truth-value function Dunn introduces a valuation, which is a "three-placed relation φ relating sentences, situations, and truth-values." For the sake of further simplification "we can forget situations and deal just with two-placed relations simply relating sentences to truth-values" (Dunn 1976, p. 155). Thus, we obtain a valuation v that relates sentences in our language to elements from **2**.

The main idea of the American Plan can be implemented by generalizing the notion of valuation, defining it as a binary *relation* between sets $\mathcal{L}$ and **2**, which need not be total and functional in all cases. Given an atomic proposition p, this valuation either assigns to it one of the two classical truth-values (in this case the valuation behaves exactly like the classical truth-value function), and we then state that p is T or p is F, or fails to relate any of the values (partial function), and we say that p is neither T nor F, or assigns to it both values simultaneously (non-functional relation), and we say that p is both T and F.

Valuation v, being defined for atomic propositions, can be extended to compound sentences in a routine way, with the specificity that truth conditions need to be introduced alongside the falsity conditions. Then, falsity as an autonomous, independent notion is no longer equal to non-truth:

Definition 1.2 For any A and B,

$A \wedge B$ is T iff A is T and B is T, $A \wedge B$ is F iff A is F or B is F;
$A \vee B$ is T iff A is T or B is T, $A \vee B$ is F iff A is F and B is F;
$\sim A$ is T iff A is F, $\sim A$ is F iff A is T.

1.4 Version 3: Generalized Truth-Value Function

Another way of generalizing the classical truth-value function with the same effect as abandoning the principles of bi-valence and unique-valence with respect to classical truth and falsity, but by maintaining the functional character of the valuation procedure is also given in (Dunn 1976). For a given valuation relation v defined as above, Dunn defines $v^*(A)$ as "the image of A under v (i.e., the set of truth-values to which A is related by v)" (Dunn 1976, p. 156, notation adjusted). That is, v^* is a *function* from $\mathcal{L}$ to the *power-set* of **2**.

In (Shramko et al. 2001, p. 762) this kind of valuation is called *multivaluation*, while in (Shramko and Wansing 2005, p. 122) a truth-value function conceived in this way is called a *generalized truth-value function*. A generalized truth-value function defined on **2**, and applied to some proposition, produces exactly four possible assignments: $\varnothing$, $\{F\}$, $\{T\}$, $\{F, T\}$. These assignments are analogous to those of the aboutness valuation given above.

1.5 Version 4: Generalized Truth-Values

The idea of a generalized truth-value function not only implies a far-reaching generalization of the notion of a classical truth-value function, but also leads to an important generalization of the notion of the truth-value itself, obtained by hypostatizing the assignments of the generalized truth-value function given above.

Belnap (1977a, b) developed the generalization suggested by Dunn, and implemented it in the form of a "useful four-valued logic" for "how a computer should think," by devising a highly heuristic "computerized" interpretation of the situations considered in (Dunn 1976). Indeed, computers often have to work with incomplete or inconsistent information. Nevertheless, it is desirable that even when dealing with this kind of information a computer remains functional and more or less reliable, "without letting minor inconsistencies in its data lead to terrible consequences" (Dunn 1986, p. 193).

Think of the truth-value of some sentence as the information that "is told" to a computer about the sentence. Then, alongside the standard ("normal" or classical) situations when a computer is told that the sentence is true or false, we have to take into account situations in which the computer does not receive any information about the sentence, or is told that the sentence is both true and false, i.e., it receives (possibly

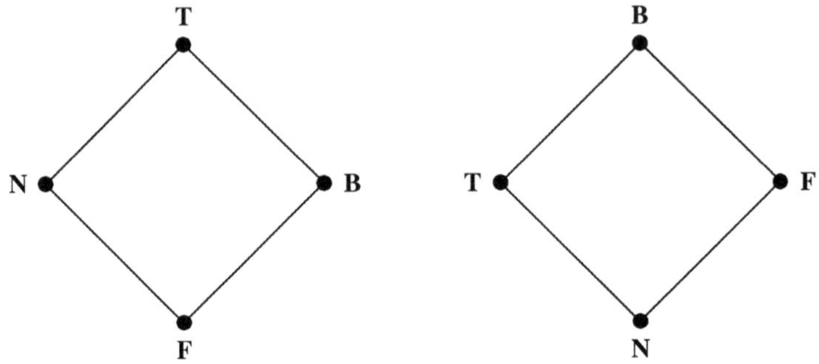

Fig. 1 Logical lattice L4 and approximation lattice A4

from different sources or implicitly) inconsistent information about the sentence. In this way we obtain the set of four "told values" $\mathbf{4} = \mathcal{P}(\mathbf{2})$, which correspond to the set of above assignments of a generalized truth-value function:

$\mathbf{N} = \{\ \}$—none ("told neither falsity nor truth");
$\mathbf{F} = \{ F \}$—plain falsehood ("told only falsity");
$\mathbf{T} = \{ T \}$—plain truth ("told only truth");
$\mathbf{B} = \{ F, T \}$—both falsehood and truth ("told both falsity and truth").

In (Shramko et al. 2001, p. 763) these "told values" are called "generalized truth-values." In other words, the application of a generalized truth-value function to some basic set of (initial) truth-values produces generalized truth-values, each of which is a subset of the basic set (including, of course, the empty set).

Belnap shows how the elements of **4** can be organized into two distinct lattices, a "logical lattice" L4, and an "approximation lattice" A4, as presented in Fig. 1. L4 is "logical" because the ordering on it is in effect a *logical order* with the usual truth-functional conjunction and disjunction as meet and join respectively, and negation as the operation that inverts this order. Moreover, the relation of logical entailment can be defined through a conformity with the logical order. The ordering of A4 can be naturally explicated as "approximates the information in." The idea of this lattice can be traced back to Dana Scott, see, e.g., (Scott 1973), who considers various examples of an approximation order. Belnap remarks that **N** is at the bottom of A4 because it gives no information at all, whereas **B** is at the top because it gives too much (inconsistent) information.

2 Truth-Value Bilattices and First-Degree Consequence System

Ginsberg (1986, 1988) noticed the possibility of uniting a logical lattice and an approximation lattice into one algebraic structure, which he called the *bilattice*. Bilattices have found fruitful applications in algebraic logic, logic programming, theory of deductive databases, and some other areas; they were studied by many authors, see, e.g., (Arieli and Avron 1996; Bou and Rivieccio 2011; Fitting 2006) and references therein.

Consider definitions of some important notions suitable for further generalization.

Definition 2.1

1. A *bilattice* is a structure $\mathcal{B}_2 = (S, \sqsubseteq_1, \sqsubseteq_2)$ in which S is a non-empty set, and $\sqsubseteq_1$ and $\sqsubseteq_2$ are partial orderings each giving S the structure of a lattice, determining thus for each of the two lattices the corresponding operations of meet and join denoted by $\sqcap_1$, $\sqcup_1$ and $\sqcap_2$, $\sqcup_2$.
2. A bilattice is called *complete* iff all meets and joins exist, with respect to both orderings.
3. A bilattice is called *interlaced* iff each of the operations $\sqcap_1, \sqcup_1, \sqcap_2, \sqcup_2$ is monotone with respect to both orderings.
4. A bilattice is called *distributive* iff all the twelve distributive laws hold: $x \circ (y \bullet z) = (x \circ y) \bullet (x \circ z)$, where $\circ, \bullet \in \{\sqcap_1, \sqcup_1, \sqcap_2, \sqcup_2\}$, $\circ \neq \bullet$.

A bilattice may be equipped with inversion operations, the main feature of which is to invert one of the bilattice orderings leaving the other unchanged:

Definition 2.2 Let $(S, \sqsubseteq_1, \sqsubseteq_2)$ be a bilattice. Then a unary operation $-_1$ is called 1-inversion iff it satisfies the following conditions for any $x, y \in S$:

$$\begin{array}{rrcl} \text{(anti)} & x \sqsubseteq_1 y & \Rightarrow & -_1 y \sqsubseteq_1 -_1 x; \\ \text{(iso)} & x \sqsubseteq_2 y & \Rightarrow & -_1 x \sqsubseteq_2 -_1 y; \\ \text{(per2)} & -_1 -_1 x & = & x. \end{array}$$

Operation of 2-inversion is defined analogously.

That is, 1-inversion (2-inversion) is an operation of period two (involution), antitone with respect to $\sqsubseteq_1$ ($\sqsubseteq_2$), and isotone with respect to $\sqsubseteq_2$ ($\sqsubseteq_1$). If operation $-_1$ ($-_2$) exists for a given bilattice, then it is called a bilattice with 1-inversion (2-inversion).[5]

If the carrier set of a bilattice is a set of truth-values, then we obtain a *truth-value bilattice*. It turns out that the set of Belnapian generalized truth-values constitutes the smallest non-trivial truth-value bilattice $\langle \mathbf{4}, \sqsubseteq_i, \sqsubseteq_t \rangle$ known in the literature as $FOUR_2$, and presented in Fig. 2. It is equipped with an *information* order and a *truth* order. These orderings represent an increase in information and truth respectively, i.e., $x \sqsubseteq_i y$ means that y is "at least as informative" as x, and $x \sqsubseteq_t y$ means that y is "at least as true" as x. $FOUR_2$ is complete, interlaced and distributive. The inversion operations exist for both orderings, t-inversion is usually called *negation*, and i-inversion is known as *conflation*, see (Fitting 2006).

[5]By Definitions 2.1 and 2.2, I try in a way to systematize and regularize the bilattice-terminology which is sometimes incoordinate [or "not uniform," see (Mobasher et al. 2000, p. 111)] in works by different authors. Moreover, these definitions are formulated in such a way as to enable further generalizations when it comes to trilattices and multilattices, see subsequent sections.

Fig. 2 Bilattice $FOUR_2$

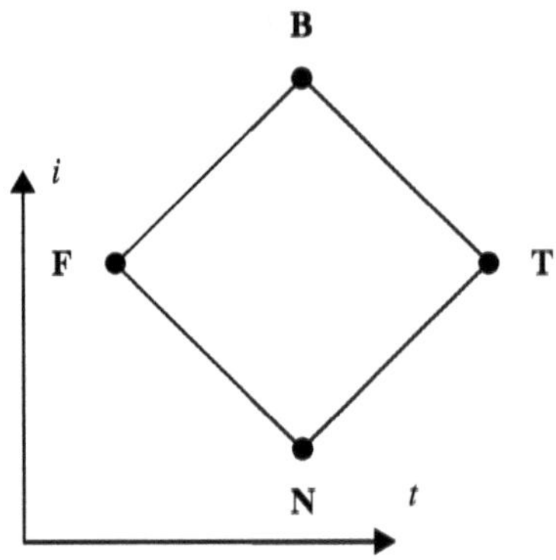

Truth ordering can also be viewed as a "logical order," because it determines the central logical notions: of logical connectives and the relation of entailment. Let *Prop* be a set of propositional variables, and let $p \in Prop$. Consider language $\mathcal{L}$ defined as follows:

$$\mathcal{L}: \quad A ::= p \mid {\sim}A \mid A \wedge A \mid A \vee A.$$

Let a valuation v^4 (4-valuation) be defined as a map from *Prop* into **4**. We have then the following definition of the truth conditions for propositional connectives:

Definition 2.3 For any A and B from $\mathcal{L}$,

1. $v^4(A \wedge B) = v^4(A) \sqcap_t v^4(B)$;
2. $v^4(A \vee B) = v^4(A) \sqcup_t v^4(B)$;
3. $v^4({\sim}A) = -_t v^4(A)$.

Entailment relation between any $A, B \in \mathcal{L}$ can be defined as follows.

Definition 2.4 $A \models^4 B$ iff $\forall v^4\ (v^4(A) \sqsubseteq_t v^4(B))$.

This relation is axiomatized by a system of "tautological entailments" from (Anderson and Belnap 1975, Sect. 15.2) called also *First Degree Entailment*. It is a so-called (single premiss–single conclusion) *consequence system*, the expressions of which are of the form $A \vdash B$ to be read as "A has B as a consequence," see, e.g., (Dunn 1995, p. 302). This system is often denoted as **FDE**, and consists of the following axiom schemata and rules of inference[6]:

System **FDE**:

$a1.\quad A \wedge B \vdash A$
$a2.\quad A \wedge B \vdash B$

[6]Note again that in the first degree entailment systems, a consequence is standardly considered to be a relation between (single) formulas, with a usual generalization in mind to a relation between sets of formulas (so that $\{A_1, \ldots, A_m\} \vdash \{B_1, \ldots, B_n\}$ can be represented by $A_1 \wedge \cdots \wedge A_m \vdash B_1 \vee \cdots \vee B_n$).

$a3.\ A \vdash A \vee B$
$a4.\ B \vdash A \vee B$
$a5.\ A \wedge (B \vee C) \vdash (A \wedge B) \vee C$
$a6.\ A \vdash {\sim\sim}A$
$a7.\ {\sim\sim}A \vdash A$
$r1.\ A \vdash B,\ B \vdash C\ /\ A \vdash C$
$r2.\ A \vdash B,\ A \vdash C\ /\ A \vdash B \wedge C$
$r3.\ B \vdash A,\ C \vdash A\ /\ B \vee C \vdash A$
$r4.\ A \vdash B\ /\ {\sim}B \vdash {\sim}A.$

3 One Step Further: From Bilattices to Trilattices and a Bi-consequence System

In (Shramko and Wansing 2005) a "powerset formation procedure" was continued as applied to the set of Belnap's truth-values **4**. In this way we obtain the next member among the sets of generalized truth-values, $\mathbf{16} = \mathcal{P}(\mathbf{4})$:

1. $\mathbf{N} = \varnothing$	9. $\mathbf{FT} = \{\{F\}, \{T\}\}$
2. $N = \{\varnothing\}$	10. $\mathbf{FB} = \{\{F\}, \{F, T\}\}$
3. $F = \{\{F\}\}$	11. $\mathbf{TB} = \{\{T\}, \{F, T\}\}$
4. $T = \{\{T\}\}$	12. $\mathbf{NFT} = \{\varnothing, \{F\}, \{T\}\}$
5. $B = \{\{F, T\}\}$	13. $\mathbf{NFB} = \{\varnothing, \{F\}, \{F, T\}\}$
6. $\mathbf{NF} = \{\varnothing, \{F\}\}$	14. $\mathbf{NTB} = \{\varnothing, \{T\}, \{F, T\}\}$
7. $\mathbf{NT} = \{\varnothing, \{T\}\}$	15. $\mathbf{FTB} = \{\{F\}, \{T\}, \{F, T\}\}$
8. $\mathbf{NB} = \{\varnothing, \{F, T\}\}$	16. $\mathbf{A} = \{\varnothing, \{T\}, \{F\}, \{F, T\}\}$.

The transition from **4** to **16** can be motivated and justified very naturally by a transition from single computers to *computer networks*, see (Shramko and Wansing 2005) for details. It also turns out that an adequate algebraic framework for **16** requires a transition from bilattices to *trilattices*.

Definition 3.1

1. A *trilattice* is a structure $\mathcal{T}_3 = (S, \sqsubseteq_1, \sqsubseteq_2, \sqsubseteq_3)$ in which S is a non-empty set, and $\sqsubseteq_1, \sqsubseteq_2, \sqsubseteq_3$ are partial orderings each giving S the structure of a lattice, determining thus for each of the three lattices the corresponding pairs of meet and join operations denoted by $\langle \sqcap_1, \sqcup_1 \rangle$, $\langle \sqcap_2, \sqcup_2 \rangle$, $\langle \sqcap_3, \sqcup_3 \rangle$.
2. A trilattice is called *complete* iff all meets and joins exist, with respect to all three orderings.
3. A trilattice is called *interlaced* iff each of the operations $\sqcap_1, \sqcup_1, \sqcap_2, \sqcup_2, \sqcap_3, \sqcup_3$ is monotone with respect to all three orderings.
4. A trilattice is called *distributive* iff all 30 distributive laws hold:
$x \circ (y \bullet z) = (x \circ y) \bullet (x \circ z)$, where $\circ, \bullet \in \{\sqcap_1, \sqcup_1, \sqcap_2, \sqcup_2, \sqcap_3, \sqcup_3\}$, $\circ \neq \bullet$.

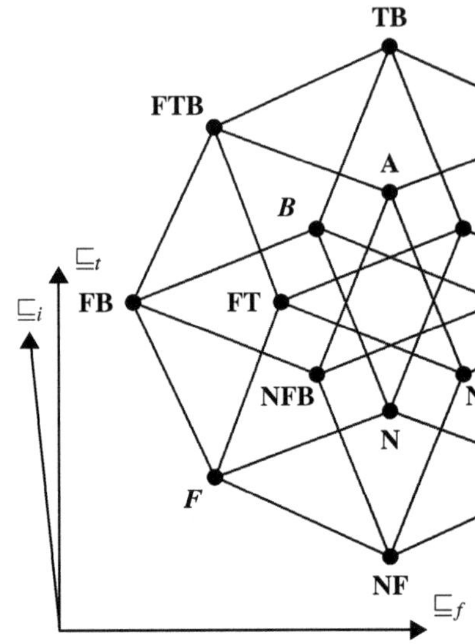

Fig. 3 Trilattice *SIXTEEN*$_3$

As to inversion operations, we have the following natural extension of Definition 2.2:

Definition 3.2 Let $(S, \sqsubseteq_1, \sqsubseteq_2, \sqsubseteq_3)$ be a trilattice. Then a unary operation $-_1$ is called 1-inversion iff it satisfies the following conditions for any $x, y \in S$:

$$\begin{array}{rl}
\text{(anti)} & x \sqsubseteq_1 y \Rightarrow -_1 y \sqsubseteq_1 -_1 x; \\
\text{(iso)} & x \sqsubseteq_2 y \Rightarrow -_1 x \sqsubseteq_2 -_1 y; \\
\text{(iso)} & x \sqsubseteq_3 y \Rightarrow -_1 x \sqsubseteq_3 -_1 y; \\
\text{(per2)} & -_1 -_1 x = x.
\end{array}$$

Operations of 2-inversion and 3-inversion are defined analogously.

The notion of a trilattice was introduced in (Shramko et al. 2001) in the context of a generalized truth-value space of *constructive logic*, with three partial orderings that represented respectively an increase in information, truth and *constructivity*.

In its turn, the elements of **16** above also constitute a trilattice *SIXTEEN*$_3$ as presented on Fig. 3. As explained in (Shramko and Wansing 2005), the truth order in bilattice *FOUR*$_2$ is in fact a *truth-and-falsity order*, since an increase in truth means here a simultaneous decrease in falsity. In contrast to this *SIXTEEN*$_3$ allows to define a pure falsity order side by side with pure truth order as totally independent of each other (for formal definitions see (Shramko and Wansing 2005, p. 128) with the necessary modifications in definition of the (non-)falsity ordering).[7]

[7]Following a suggestion by Odintsov in (2009) I reverse here the falsity ordering as compared to its definition in (Shramko and Wansing 2005). As Odintsov remarks, such a reversion allows us to define logical connectives and entailment relations for the "truth-language" and "falsity-language" in a homomorphic and uniform way.

Thus, whereas in *FOUR*$_2$ truth and falsity orders are merged into one logical order, *SIXTEEN*$_3$ effectively discriminates between truth and falsity, and an increase in truth does not necessarily mean here a decrease in falsehood (and *vice versa*). Hence, within *SIXTEEN*$_3$ we have in fact *two distinct logical orders*: a truth order $\sqsubseteq_t$ representing an *increase in truth*, and a non-falsity order $\sqsubseteq_f$ representing a *decrease in falsity*.

As noted in (Shramko and Wansing 2005, p. 134), operations $\sqcap_t$, $\sqcup_t$, and $-_t$ in *SIXTEEN*$_3$ are not the only algebraic operations that naturally correspond to logical conjunction, disjunction, and negation; $\sqcap_f$, $\sqcup_f$, and $-_f$ may play this role as well. And taking into account the fact that $x \sqcap_t y \neq x \sqcap_f y$, $x \sqcup_t y \neq x \sqcup_f y$ and $-_t x \neq -_f x$, we can state that both logical orders bring into existence "parallel" and, in fact, *distinct* logical connectives.

Consider languages $\mathcal{L}_t$, $\mathcal{L}_f$, and $\mathcal{L}_{tf}$ defined as follows:

$$\begin{aligned}
\mathcal{L}_t &: \ A ::= p \mid {\sim_t} A \mid A \wedge_t A \mid A \vee_t A;\\
\mathcal{L}_f &: \ A ::= p \mid {\sim_f} A \mid A \wedge_f A \mid A \vee_f A;\\
\mathcal{L}_{tf} &: \ A ::= p \mid {\sim_t} A \mid {\sim_f} A \mid A \wedge_t A \mid A \vee_t A \mid A \wedge_f A \mid A \vee_f A.
\end{aligned}$$

Then a valuation function v^{16} (a 16-valuation) can be defined as a map from *Prop* into **16** extended to all formulas of $\mathcal{L}_{tf}$ as follows:

Definition 3.3 For any A and B from $\mathcal{L}_{tf}$:

1. $v^{16}(A \wedge_t B) = v^{16}(A) \sqcap_t v^{16}(B)$;
2. $v^{16}(A \vee_t B) = v^{16}(A) \sqcup_t v^{16}(B)$;
3. $v^{16}({\sim_t} A) = -_t v^{16}(A)$;
4. $v^{16}(A \wedge_f B) = v^{16}(A) \sqcap_f v^{16}(B)$;
5. $v^{16}(A \vee_f B) = v^{16}(A) \sqcup_f v^{16}(B)$;
6. $v^{16}({\sim_f} A) = -_f v^{16}(A)$.

Thus, *SIXTEEN*$_3$ allows a nontrivial coexistence of pairs of different (although analogous) logical connectives without collapsing them into each other. As observed in (Shramko and Wansing 2005, p. 135), it might be helpful to think of $\wedge_t$, $\vee_t$, $\sim_t$ in terms of the *presence of truth* and to treat $\wedge_f$, $\vee_f$, $\sim_f$ as essentially highlighting the *absence of falsity*.

Each logical order determines now independent entailment relation between any sentences $A, B \in \mathcal{L}_{tf}$:

Definition 3.4 $A \models_t^{16} B$ iff $\forall v^{16}\, (v^{16}(A) \sqsubseteq_t v^{16}(B))$.

Definition 3.5 $A \models_f^{16} B$ iff $\forall v^{16}\, (v^{16}(A) \sqsubseteq_f v^{16}(B))$.

Certain important fragments of these logics were investigated in (Shramko and Wansing 2005). In particular, it was shown that the logics generated *separately* by the algebraic operations under the truth order and under the non-falsity order in *SIXTEEN*$_3$ coincide with the logic of *FOUR*$_2$, namely it remains First Degree Entailment.[8]

[8] In (Shramko and Wansing 2006) this result was extended to the infinite case, showing that Belnap's strategy of generalizing the set $\mathbf{2} = \{T, F\}$ of classical truth-values not only is coherent but

The corresponding systems can be denoted by $\mathbf{FDE}_t^t$ for the pure logic of truth order, and $\mathbf{FDE}_f^f$ for the pure logic of non-falsity order. The superscript indicates the type of language used, and the subscript explicates the kind of consequence. The system $\mathbf{FDE}_t^t$ is thus a pair $(\mathcal{L}_t, \vdash_t)$, where $\vdash_t$ is a binary relation (consequence) on the language $\mathcal{L}_t$ satisfying the axiom schemes and rules of inference for **FDE** above with the suitable subscripts, and analogously for $\mathbf{FDE}_f^f$.

Taking into account that trilattice *SIXTEEN*$_3$ comes with *two* natural definitions of (non-equivalent) entailment relations reflecting increase of truth and decrease of falsity, we put forward in (Shramko and Wansing 2005) an idea of a natural unified logic of *SIXTEEN*$_3$ as a *bi-consequence system* comprising two kinds of entailment relations. This idea finds its expression in the the following definition:

Definition 3.6 The bi-consequence logic $(\mathcal{L}_{tf}, \models_t^{16}, \models_f^{16})$ is the set of all valid statements $A \models_x^{16} B$, where $A, B \in \mathcal{L}_{tf}$, and $x = t$ or $x = f$, (cf. Definitions 3.4 and 3.5, respectively).

In accordance with this definition, one obtains an idea of a bi-consequence system $\mathbf{FDE}_{tf}^{tf} = (\mathcal{L}_{tf}, \vdash_t, \vdash_f)$. A peaceful co-existence of two entailment and two deductibility relations in one and the same logic is useful, because—as we have seen—it may well make a difference whether we move along the truth order or the non-falsity order.

In (Shramko and Wansing 2005) we axiomatized some important fragments of $(\mathcal{L}_{tf}, \models_t^{16}, \models_f^{16})$, but the problem of finding a complete formulation of the whole $\mathbf{FDE}_{tf}^{tf}$ remained there open. This problem was solved by Odintsov and Wansing in (2015), where they constructed a so-called *bi-calculus*, which is exactly the bi-consequence system $\mathbf{FDE}_{tf}^{tf}$ as defined in (Shramko and Wansing 2005, p. 144).[9] It is determined by the following axiom schemata and rules of inference:

FDE axioms for $\vdash_t$:	**FDE** axioms for $\vdash_f$:
$a1_t.\ A \wedge_t B \vdash_t A$	$a1_f.\ A \wedge_f B \vdash_f A$
$a2_t.\ A \wedge_t B \vdash_t B$	$a2_f.\ A \wedge_f B \vdash_f B$
$a3_t.\ A \vdash_t A \vee_t B$	$a3_f.\ A \vdash_f A \vee_f B$
$a4_t.\ B \vdash_t A \vee_t B$	$a4_f.\ B \vdash_f A \vee_f B$
$a5_t.\ A \vdash_t {\sim_t}{\sim_t} A$	$a5_f.\ A \vdash_f {\sim_f}{\sim_f} A$
$a6_t.\ {\sim_t}{\sim_t} A \vdash_t A$	$a6_f.\ {\sim_f}{\sim_f} A \vdash_f A$

Distributivity axioms:
$a7_t.\ A \circ (B \bullet C) \vdash_t (A \circ B) \bullet C$, where $\circ, \bullet \in \{\wedge_t, \vee_t, \wedge_f, \vee_f\}, \circ \neq \bullet$

(Footnote 8 continued)
stabilizes. At any stage, no matter how far it goes, the logic of the truth (non-falsity) order is again First Degree Entailment.

[9]Odintsov and Wansing denote their system **BiCalc**, but I prefer to retain the original label as more instructive. I also slightly modify the formulation from (Odintsov and Wansing 2015) to minimize the set of axioms and rules, and to visualize its further generalization.

Commutativity axiom:
$a8_t.\ \sim_f\sim_t A \vdash_t \sim_t\sim_f A$

FDE rules for $\vdash_t$:
$r1_t.\ A \vdash_t B,\ B \vdash_t C\ /\ A \vdash_t C$
$r2_t.\ A \vdash_t B,\ A \vdash_t C\ /\ A \vdash_t B \wedge_t C$
$r3_t.\ B \vdash_t A,\ C \vdash_t A\ /\ B \vee_t C \vdash_t A$
$r4_t.\ A \vdash_t B\ /\ \sim_t B \vdash_t \sim_t A$

FDE rules for $\vdash_f$:
$r1_f.\ A \vdash_f B,\ B \vdash_f C\ /\ A \vdash_f C$
$r2_f.\ A \vdash_f B,\ A \vdash_f C\ /\ A \vdash_f B \wedge_f C$
$r3_f.\ B \vdash_f A,\ C \vdash_f A\ /\ B \vee_f C \vdash_f A$
$r4_f.\ A \vdash_f B\ /\ \sim_f B \vdash_f \sim_f A$

Monotonicity rules:
$r5_t.\ A \vdash_t B\ /\ \sim_f A \vdash_t \sim_f B$
$r5_f.\ A \vdash_f B\ /\ \sim_t A \vdash_f \sim_t B$

Interconnection rules:
$r6_{tf}.\ A \vdash_t B,\ B \vdash_t A\ /\ A \vdash_f B$
$r6_{ft}.\ A \vdash_f B,\ B \vdash_f A\ /\ A \vdash_t B.$

Note, that the converse of $a8_t$ is provable in the system. Namely, by $a5_f$, $a6_f$ and using $r6_{ft}$ we have $A \vdash_t \sim_f\sim_f A$. From this by $r4_t$ we get $\sim_t\sim_f\sim_f A \vdash_t \sim_t A$, and then by $a8_t$ $\sim_f\sim_t\sim_f A \vdash_t \sim_t A$. Using $r5_t$ we obtain $\sim_f\sim_f\sim_t\sim_f A \vdash_t \sim_f\sim_t A$, and finally, $\sim_t\sim_f A \vdash_t \sim_f\sim_t A$.

$\mathbf{FDE}^{tf}_{tf}$ adequately axiomatizes the bi-consequence logic $(\mathcal{L}_{tf}, \models^{16}_t, \models^{16}_f)$, as the following soundness and completeness theorem states, see (Odintsov and Wansing 2015, Theorem 4.1):

Theorem 3.7 *For any $A, B \in \mathcal{L}_{tf}$:*

1. $A \models^{16}_t B$ *iff* $A \vdash_t B$;
2. $A \models^{16}_f B$ *iff* $A \vdash_f B$.

It is also possible to extract two important mono-consequence subsystems:

(1) $\mathbf{FDE}^{tf}_t = (\mathcal{L}_{tf}, \vdash_t)$, obtained from $\mathbf{FDE}^{tf}_{tf}$ by omitting FDE axiom schemata and rules for $\vdash_f$, as well as monotonicity rule $r5_f$, and interconnection rules $r6_{tf}$, $r6_{ft}$;
(2) $\mathbf{FDE}^{tf}_f = (\mathcal{L}_{tf}, \vdash_f)$, obtained from $\mathbf{FDE}^{tf}_{tf}$ by omitting FDE axiom schemata and rules for $\vdash_t$, as well as monotonicity rule $r5_t$, and interconnection rules $r6_{tf}$, $r6_{ft}$, and also changing axiom schemata $a7_t$ and $a8_t$ to their counterparts $a7_f$ and $a8_f$.

4 Generalization Completed: Truth-Value Multilattices and a Multi-consequence System

Bilattices and trilattices present particular cases of more general algebraic structures called in (Shramko and Wansing 2005, p. 126) *multilattices*. The idea of a multilattice can be systematized and generalized in the following definitions:

Definition 4.1

1. An n-dimensional *multilattice* (or just n-lattice) is a structure $\mathcal{M}_n = (S, \sqsubseteq_1, \ldots, \sqsubseteq_n)$ in which S is a non-empty set, and $\sqsubseteq_1, \ldots, \sqsubseteq_n$ are partial orderings each giving S the structure of a lattice, determining thus for each of n lattices the corresponding pairs of meet and join operations denoted by $\langle \sqcap_1, \sqcup_1 \rangle, \ldots, \langle \sqcap_n, \sqcup_n \rangle$.
2. An n-lattice is called *complete* iff all meets and joins exist, with respect to all n orderings.
3. An n-lattice is called *interlaced* iff each of the operations $\sqcap_1, \sqcup_1, \ldots, \sqcap_n, \sqcup_n$ is monotone with respect to all n orderings.
4. An n-lattice is called *distributive* iff all $2(2n^2 - n)$ distributive laws hold: $x \circ (y \bullet z) = (x \circ y) \bullet (x \circ z)$, where $\circ, \bullet \in \{\sqcap_1, \sqcup_1, \ldots, \sqcap_n, \sqcup_n\}$, $\circ \neq \bullet$.

One can omit an explicit indication of a dimension if it is inessential or clear from a context, and speak simply of a multilattice. The inversion operations on multilattices can now be defined in full generality, cf. (Shramko and Wansing 2006, p. 411).

Definition 4.2 Let $\mathcal{M}_n = (S, \sqsubseteq_1, \ldots, \sqsubseteq_n)$ be a multilattice. Then for any $j \leq n$ an unary operation $-_j$ on S is said to be a (pure) j-inversion iff for any $k \leq n$, $k \neq j$ the following conditions are satisfied:

$$\begin{array}{ll} \text{(anti)} & x \sqsubseteq_j y \Rightarrow -_j y \sqsubseteq_j -_j x; \\ \text{(iso)} & x \sqsubseteq_k y \Rightarrow -_j x \sqsubseteq_k -_j y; \\ \text{(per2)} & -_j -_j x = x. \end{array}$$

Thus, in a multilattice-framework any inversion defined relative to some partial order is an involution operation antitone with respect to this particular order and isotone with respect to all the remaining orderings.[10]

Multilattices present a natural algebraic framework for dealing with generalized truth-values, so that if the set S in a multilattice $\mathcal{M}_n$ consists of (generalized) truth-values, we can speak of a *generalized truth-value multilattice*.

We can define most abstractly the notions of a generalized truth-value and a generalized truth-value function, according to the method initially developed by Dunn and Belnap as applied to classical truth-values, cf. (Zaitsev and Shramko 2013, p. 1300):

Definition 4.3 Let X be a (basic) set of initial truth-values, and $\mathcal{P}(X)$ be the power-set of X. Then the elements of $\mathcal{P}(X)$ are called *generalized truth-values* defined on the basis of X.

Definition 4.4 Let X be a (basic) set of initial truth-values, $\mathcal{P}(X)$ be the set of generalized truth-values defined on the basis of X, and $\mathcal{L}$ be a given language. Then a *generalized truth-value function* (defined on the basis of X) is a function from the set of sentences of $\mathcal{L}$ to $\mathcal{P}(X)$.

[10]Moreover, for certain orderings it could be useful to consider combined inversion operations, so that, e.g., 23-inversion would invert simultaneously both $\sqsubseteq_2$ and $\sqsubseteq_3$, leaving the other partial orders untouched, but I skip this subject here, cf. (Shramko and Wansing 2006, p. 411).

By Definition 4.3, generalized truth-values are constructed on the basis of some set of primitive truth-values of an underlying "low-level logic" (or several such logics with their truth-values "lumped together"). Definition 4.4 explains, in what sense a sentence can simultaneously "possess" several truth-values from an underlying logic, or fail to possess some (maybe all) of them.[11]

The carrier set of a generalized truth-value multilattice constitutes a natural domain for a generalized truth-value function. At the same time, this domain is structured by specific logical orders for determining sets of logical connectives and possible entailment relations in agreement with the acceptable patterns of logical reasoning. Should we, in advance, rule out the possibility of the simultaneous coexistence of several such patterns? The construction of the bi-consequence system $\mathbf{FDE}_{tf}^{tf}$ above, which explicitly deals with two equal consequence relations, clearly demonstrates the impropriety of such a restriction.

Moreover, Arieli and Avron (1996) explicated $FOUR_2$ as a *logical bilattice*—an algebraic basis for a proof system, involving operations not only from a truth order, but also from an information order. As observed in (Shramko and Wansing 2005, p. 146), $SIXTEEN_3$ can also be construed as a logical bilattice if we discard, say, the information order, and consider just the structure $(\mathbf{16}, \sqsubseteq_t, \sqsubseteq_f)$ with two natural logical orders—for truth and non-falsity. In fact, Arieli and Avron conclusively show that the information order can play a logical role as well, being incorporated in the structure of a logical bilattice. But what can prevent us from considering logical trilattices, tetralattices, or most generally, *logical multilattices*? In the next section a method for generalizing the approach by Arieli and Avron towards an overall theory of logical multilattices is briefly outlined.

For now, it is important to stress that by a logical reasoning we can be interested not only in informational content, truth content, or falsity content, but also in some other possible characterizations of the given truth-values, such as constructivity, cf. (Shramko et al. 2001), (un)certainty, cf. (Zaitsev 2009), modality, cf. (Rescher 1965), or other kinds of "adverbial qualifications," cf. (MacIntosh 1991), by which truth-values can naturally be ordered. This motivates the possibility of a *multi-consequence logic* that comprises several entailment relations based on the partial orderings of a given truth-value multilattice viewed as logical orders.[12]

Consider language $\mathcal{L}_n$ defined as follows:

$$\mathcal{L}_n: \quad A ::= p \mid \sim_1 A \mid \ldots \mid \sim_n A \mid A \wedge_1 A \mid \ldots \mid A \wedge_n A \mid A \vee_1 A \mid \ldots \mid A \vee_n A.$$

[11] Precisely in the sense in which a generalized truth value of a "higher degree" may contain several truth values of a "lower degree" or fail to contain some (maybe all) of them.

[12] Interestingly, Arieli and Avron also admit the possibility that "more than one consequence relation is relevant," thereby enabling "the use of corresponding implication connectives," that "allow us also to express higher-order connections among those relations" (Arieli and Avron 1996, p. 44).

Let $\mathcal{M}_n = (S, \sqsubseteq_1, \ldots, \sqsubseteq_n)$ be a generalized distributive truth-value n-lattice, with pairs of meet and join operations $\langle \sqcap_1, \sqcup_1 \rangle, \ldots, \langle \sqcap_n, \sqcup_n \rangle$, and operations of j-inversions defined for every $\sqsubseteq_j$ $(j \leq n)$. Define valuation (generalized truth-value function) v^s as a map from *Prop* into S. Then, for any $j \leq n$, we have:

Definition 4.5 For any $A, B \in \mathcal{L}_n$,

1. $v^s(A \wedge_j B) = v^s(A) \sqcap_j v^s(B)$;
2. $v^s(A \vee_j B) = v^s(A) \sqcup_j v^s(B)$;
3. $v^s(\sim_j A) = -_j v^s(A)$.

Consider the set of all generalized distributive truth-value n-lattices. We can define for every $j \leq n$ the entailment relation between any $A, B \in \mathcal{L}_n$:

Definition 4.6 $A \models_j B$ iff $\forall \mathcal{M}_n$ $\forall v^s$ defined on $\mathcal{M}_n : v^s(A) \sqsubseteq_j v^s(B)$.

Now, for any n we can define a general notion of a multi-consequence logic with respect to the set of generalized distributive truth-value multilattices $\mathcal{M}_n$:

Definition 4.7 A multi-consequence logic $(\mathcal{L}_n, \models_1, \ldots, \models_n)$ is the set of all valid statements $A \models_j B$, where $A, B \in \mathcal{L}_n$, and $j \leq n$.

To grasp the multi-consequence logic deductively, I formulate a multi-consequence system $\mathbf{FDE}^n_n = (\mathcal{L}_n, \vdash_1, \ldots, \vdash_n)$, which is a straightforward generalization of $\mathbf{FDE}^{tf}_{tf}$. In axiom schemata and rules of inference below $j, k \leq n$ and $j \neq k$.

System $\mathbf{FDE}^n_n$:

$a1_j$. $A \wedge_j B \vdash_j A$
$a2_j$. $A \wedge_j B \vdash_j B$
$a3_j$. $A \vdash_j A \vee_j B$
$a4_j$. $B \vdash_j A \vee_j B$
$a5_j$. $A \vdash_j \sim_j\sim_j A$
$a6_j$. $\sim_j\sim_j A \vdash_j A$
$a7_j$. $A \circ (B \bullet C) \vdash_j (A \circ B) \bullet C$, where $\circ, \bullet \in \{\wedge_1, \vee_1, \ldots, \wedge_n, \vee_n\}$, $\circ \neq \bullet$
$a8_j$. $\sim_k\sim_j A \vdash_j \sim_j\sim_k A$
$r1_j$. $A \vdash_j B,\ B \vdash_j C\ /\ A \vdash_j C$
$r2_j$. $A \vdash_j B,\ A \vdash_j C\ /\ A \vdash_j B \wedge_j C$
$r3_j$. $B \vdash_j A,\ C \vdash_j A\ /\ B \vee_j C \vdash_j A$
$r4_j$. $A \vdash_j B\ /\ \sim_j B \vdash_j \sim_j A$
$r5_j$. $A \vdash_j B\ /\ \sim_k A \vdash_j \sim_k B$
$r6_{jk}$. $A \vdash_j B, B \vdash_j A\ /\ A \vdash_k B$.

Soundness and completeness can be proved by a generalization of the corresponding theorem from (Odintsov and Wansing 2015) (see Theorem 3.7).

Theorem 4.8 *For any* $A, B \in \mathcal{L}_n$, *for any* $j \leq n$: $A \models_j B$ *iff* $A \vdash_j B$.

Clearly, for any n the set of generalized distributive truth-value n-lattices forms a variety. An interesting problem consist in finding for any n a natural multilattice $\mathcal{M}_n = (S, \sqsubseteq_1, \ldots, \sqsubseteq_n)$ which generates the given variety. Then entailment relation $\models_j$ for the given variety could be identified with the relation $\models^s_j$ defined with respect to this very multilattice only: $A \models^s_j B$ iff $\forall v^s\ (v^s(A) \sqsubseteq_j v^s(B))$.

5 On Reasoning with Logical Multilattices

Arieli and Avron developed in (1996) a theory of logical bilattices and formulated the corresponding proof systems for reasoning in accordance with this theory. Having a bilattice, they introduce the notions of *prime bifilter* and *ultrabifilter* each determined by *both* bilattice orderings, and then define logical bilattice (ultralogical bilattice) as a pair $(\mathcal{B}, \mathcal{F})$, where $\mathcal{B}$ is a bilattice (bilattice with conflation) and $\mathcal{F}$ is a prime bifilter (ultrabifilter) on $\mathcal{B}$. They use then logical bilattices "for defining logics in a way which is completely analogous to the way Boolean algebras and prime filters are used in classical logic" (Arieli and Avron 1996, pp. 30–31).

The logics for (ultra)logical bilattices are mono-consequence systems which involve operations with respect to both bilattice-orderings. The basic system for capturing the bilattice meets, joins and inversions is presented in (Arieli and Avron 1996, p. 37–40) in a form of a Gentzen-type sequent calculus GBL. In what follows some central notions of a theory of (ultra)logical multillatices in spirit of Arieli and Avron are briefly sketched, and the corresponding logical system for reasoning with (ultra)logical multilattices is formulated.

Definition 5.1 Let $\mathcal{M}_n = (S, \sqsubseteq_1, \ldots, \sqsubseteq_n)$ be an n-lattice, with pairs of meet and join operations $\langle \sqcap_1, \sqcup_1 \rangle, \ldots, \langle \sqcap_n, \sqcup_n \rangle$. An *n-filter* (multifilter) on $\mathcal{M}_n$ is a nonempty proper subset $\mathcal{F}_n \subset S$, such that for every $j \leq n$:

$$x \sqcap_j y \in \mathcal{F}_n \;\Leftrightarrow\; x \in \mathcal{F}_n \text{ and } y \in \mathcal{F}_n.$$

A multifilter $\mathcal{F}_n$ is said to be *prime* iff it satisfies for every $j \leq n$:

$$x \sqcup_j y \in \mathcal{F}_n \;\Leftrightarrow\; x \in \mathcal{F}_n \text{ or } y \in \mathcal{F}.$$

A pair $(\mathcal{M}_n, \mathcal{F}_n)$ is called a *logical n-lattice* (logical multilattice) iff $\mathcal{M}_n$ is a multilattice, and $\mathcal{F}_n$ is a prime multifilter on $\mathcal{M}_n$.

Moreover, since we are interested in n-lattices with inversions defined for every $j \leq n$, we need stronger notions of an *ultramultifilter* and *ultralogical mulitlattice*.

Definition 5.2 Let $\mathcal{M}_n = (S, \sqsubseteq_1, \ldots, \sqsubseteq_n)$ be an n-lattice, with j-inversions defined with respect to every $\sqsubseteq_j$ $(j \leq n)$. Then $\mathcal{F}_n$ is an *n-ultrafilter* (ultramultifilter) on $\mathcal{M}_n$ if and only if it is a prime multifilter, such that for every $j, k \leq n$, $j \neq k$: $x \in$

$\mathcal{F}_n \Leftrightarrow -_j -_k x \notin \mathcal{F}_n$. A pair $(\mathcal{M}_n, \mathcal{F}_n)$ is called an *ultralogical n-lattice* (ultralogical multilattice) iff $\mathcal{M}_n$ is a multilattice, and $\mathcal{F}_n$ is an ultramultifilter on $\mathcal{M}_n$.

Consider language $\mathcal{L}_n$ defined as above. We can define entailment relations between any sets of formulas from $\mathcal{L}_n$ as follows:

Definition 5.3 Let $(\mathcal{M}_n, \mathcal{F}_n)$ be an (ultra)logical n-lattice, and let a valuation function v^s map. Atomic formulas from $\mathcal{L}_n$ to elements of S in $\mathcal{M}_n$ and be extended to compound formulas by Definition 4.5. Let Γ, Δ be finite sets of formulas from $\mathcal{L}_n$. Then:

1. $\Gamma \models_{ML(\mathcal{M}_n,\mathcal{F})} \Delta$ iff for every valuation v^s, such that $v^s(A) \in \mathcal{F}_n$ for each $A \in \Gamma$, there exists some $B \in \Delta$ with $v^s(B) \in \mathcal{F}_n$.
2. $\Gamma \models_{ML_n} \Delta$ iff for every multilattice $(\mathcal{M}_n, \mathcal{F}_n)$, $\Gamma \models_{ML(\mathcal{M}_n,\mathcal{F}_n)} \Delta$.

Theorem 5.4 *Let $A_1, \ldots, A_l$, $B_1, \ldots, B_m$ be finite sets of formulas from $\mathcal{L}_n$. Then, for every $j \leq n$: $A_1, \ldots, A_l \models_{ML_n} B_1, \ldots, B_m$ iff $A_1 \wedge_j \ldots \wedge_j A_l \models_j B_1 \vee_j \ldots \vee_j B_m$.*

Proof Consider an arbitrary $j \leq n$. Let for every (ultra)logical multilattice $(\mathcal{M}_n, \mathcal{F}_n)$, for every valuation v^s, if $v^s(A_1), \ldots, v^s(A_l) \in \mathcal{F}_n$, then $v^s(B_i) \in \mathcal{F}_n$ for some $i \leq m$. By generalizing to multilattices claim (∗) and case (2) in the proof of Lemma 4.3 from (Shramko and Wansing 2005), and reformulating them in a multifilters-terminology, we get for any v^s: $v^s(A_1 \wedge_j \ldots \wedge_j A_l) \sqsubseteq_j v^s(B_i)$, and hence $v^s(A_1 \wedge_j \ldots \wedge_j A_l) \sqsubseteq_j v^s(B_1 \vee_j \ldots \vee_j B_m)$. For the converse, consider an arbitrary n-lattice; assume that for any v^s, $v^s(A_1 \wedge_j \ldots \wedge_j A_l) \sqsubseteq_j v^s(B_1 \vee_j \ldots \vee_j B_m)$. Let $v^s(A_1), \ldots, v^s(A_l) \in \mathcal{F}_n$. Then $v^s(A_1 \wedge_j \ldots \wedge_j A_l) \in \mathcal{F}_n$. By a standard property of lattice filters, we get $v^s(B_1 \vee_j \ldots \vee_j B_m) \in \mathcal{F}_n$, and hence, $v^s(B_i) \in \mathcal{F}_n$ for some $i \leq m$.

A Gentzen-type sequent calculus for reasoning with ultralogical multilattices of an arbitrary dimension n, analogous to the system for (ultra)logical bilattices from (Arieli and Avron 1996), can be formulated as follows:

System **GML**$_n$:

Axiom: $\Gamma, A \rightarrow A, \Delta$

Rules:

Exchange, Contraction, and the following logical rules (for any $j, k \leq n$; $j \neq k$):

$$(\wedge_j\rightarrow)\ \frac{\Gamma, A, B \rightarrow \Delta}{\Gamma, A \wedge_j B \rightarrow \Delta} \qquad \frac{\Gamma \rightarrow \Delta, A \quad \Gamma \rightarrow \Delta, B}{\Gamma \rightarrow \Delta, A \wedge_j B}\ (\rightarrow\wedge_j)$$

$$(\vee_j\rightarrow)\ \frac{\Gamma, A \rightarrow \Delta \quad \Gamma, B \rightarrow \Delta}{\Gamma, A \vee_j B \rightarrow \Delta} \qquad \frac{\Gamma \rightarrow \Delta, A, B}{\Gamma \rightarrow \Delta, A \vee_j B}\ (\rightarrow\vee_j)$$

$$(\sim_j\wedge_j\rightarrow)\ \frac{\Gamma, \sim_j A \rightarrow \Delta \quad \Gamma, \sim_j B \rightarrow \Delta}{\Gamma, \sim_j(A \wedge_j B) \rightarrow \Delta} \qquad \frac{\Gamma \rightarrow \Delta, \sim_j A, \sim_j B}{\Gamma \rightarrow \Delta, \sim_j(A \wedge_j B)}\ (\rightarrow\sim_j\wedge_j)$$

$$(\sim_j\vee_j\to)\ \frac{\Gamma, \sim_j A, \sim_j B \to \Delta}{\Gamma, \sim_j(A \vee_j B) \to \Delta} \qquad \frac{\Gamma \to \Delta, \sim_j A \quad \Gamma \to \Delta, \sim_j B}{\Gamma \to \Delta, \sim_j(A \vee_j B)}\ (\to\sim_j\vee_j)$$

$$(\sim_j\sim_j\to)\ \frac{\Gamma, A \to \Delta}{\Gamma, \sim_j\sim_j A \to \Delta} \qquad \frac{\Gamma \to \Delta, A}{\Gamma \to \Delta, \sim_j\sim_j A}\ (\to\sim_j\sim_j)$$

$$(\sim_k\wedge_j\to)\ \frac{\Gamma, \sim_k A, \sim_k B \to \Delta}{\Gamma, \sim_k(A \wedge_j B) \to \Delta} \qquad \frac{\Gamma \to \Delta, \sim_k A \quad \Gamma \to \Delta, \sim_k B}{\Gamma \to \Delta, \sim_k(A \wedge_j B)}\ (\to\sim_k\wedge_j)$$

$$(\sim_k\vee_j\to)\ \frac{\Gamma, \sim_k A \to \Delta \quad \Gamma, \sim_k B \to \Delta}{\Gamma, \sim_k(A \vee_j B) \to \Delta} \qquad \frac{\Gamma \to \Delta, \sim_k A, \sim_k B}{\Gamma \to \Delta, \sim_k(A \vee_j B)}\ (\to\sim_k\vee_j)$$

$$(\sim_k\sim_j\to)\ \frac{\Gamma \to \Delta, A}{\Gamma, \sim_k\sim_j A \to \Delta} \qquad \frac{\Gamma, A \to \Delta}{\Gamma \to \Delta, \sim_k\sim_j A}\ (\to\sim_k\sim_j)$$

We have an analogue of Theorem 3.7 from (Arieli and Avron 1996) (soundness, completeness and cut elimination), which can be proved along similar lines (with $\models_{ML_n}$ defined with respect to ultralogical multilattices):

Theorem 5.5

(1) $\Gamma \models_{ML_n} \Delta$ *iff* $\Gamma \vdash_{GML_n} \Delta$;
(2) *If* $\Gamma_1 \vdash_{GML_n} \Delta_1, A$ *and* $\Gamma_2, A \vdash_{GML_n} \Delta_2$, *then* $\Gamma_1, \Gamma_2 \vdash_{GML_n} \Delta_1, \Delta_2$.

A subsystem of $\mathbf{GML}_n$ for reasoning with pure logical multilattices deals only with one inversion operator $\sim_j$ taken for an arbitrary (fixed) $j \leq n$. It is obtained from $\mathbf{GML}_n$ by omitting the rules $(\sim_k\sim_j\to)$ and $(\to\sim_k\sim_j)$, and changing $\sim_k$ to $\sim_j$, $\wedge_j$ to $\wedge_k$, as well as $\vee_j$ to $\vee_k$ in the rules $(\sim_k\wedge_j \to)$–$(\to\sim_k \vee_j)$.

6 Concluding Remark: Dunn Faces Suszko

Introducing multi-consequence logics (and the corresponding multi-consequence systems) means a significant generalization of the very notion of a logical system. $\mathbf{FDE}_n^n$ is a particular (single premiss–single conclusion) case of what in (Wansing and Shramko 2008b, p. 422) more abstractly was called a *Tarskian k-dimensional logic* (Tarskian k-logic). The latter was defined there as a $k+1$-tuple $\Lambda = (\mathcal{L}, \vdash_1, \ldots, \vdash_k)$ such that (i) $\mathcal{L}$ is a language in a denumerable set of sentence letters and a finite non-empty set $\mathcal{C}$ of finitary connectives, (ii) for every $i \leq k$, $\vdash_i \subseteq \mathcal{P}(\mathcal{L}) \times \mathcal{L}$, and (iii) every relation $\vdash_i$ satisfies the standard Tarskian conditions for a consequence relation, known as *Reflexivity*, *Monotonicity*, and *Cut*, see e.g., (Wansing and Shramko 2008b, p. 408).[13]

[13]For a single premiss–single conclusion case these conditions can be reformulated as follows: $B \wedge A \vdash A$; $B \vdash A \Rightarrow B \wedge C \vdash A$; $B \vdash A$ and $C \wedge A \vdash D \Rightarrow B \wedge C \vdash D$, which obviously hold for every $\vdash_j$ of $\mathbf{FDE}_n^n$.

The next notion is the one of an n-valued k-dimensional matrix (k-matrix), defined in (Wansing and Shramko 2008b, p. 422) as a structure $\mathfrak{M} = \langle \mathcal{V}, \mathcal{D}_1, \ldots, \mathcal{D}_k, \{ f_c : c \in \mathcal{C} \} \rangle$, where $\mathcal{C}$ is a finite non-empty set of finitary connectives in $\mathcal{L}$, $\mathcal{V}$ is a non-empty set (of values) of cardinality n $(2 \leq n)$, $2 \leq k$, every $\mathcal{D}_i$ $(1 \leq i \leq k)$ is a non-empty proper subset of $\mathcal{V}$, the sets $\mathcal{D}_i$ are pairwise distinct, and every f_c is a function on $\mathcal{V}$ with the same arity as c. The sets $\mathcal{D}_i$ are called *distinguished* sets. A function from $\mathcal{L}$ to $\mathcal{V}$ is called a valuation in $\mathfrak{M}$, and a pair $\mathcal{M} = \langle \mathfrak{M}, v \rangle$ is called an n-valued k-model based on $\mathfrak{M}$.

Every $\mathcal{D}_i$ can be used for defining its own consequence relation in a standard way, as a relation that ensures preservation of the elements from $\mathcal{D}_i$ in a course of reasoning.[14] Then n-valued k-models based on $\mathfrak{M}$ can be used for semantic characterization of the corresponding Tarskian k-logics.

The notion of a Tarskian k-logic was used in (Wansing and Shramko 2008b) as a case against a much-debated *thesis* advanced by Roman Suszko to the effect that "every logic is (logically) two-valued" (Suszko 1977, p. 378), or "there are but two logical values, true and false" (Caleiro et al. 2005, p. 169). Suszko justifies his thesis by a special *reduction* procedure that enables a characterization of every Tarskian consequence relation by a bivalent semantics. Yet, if a given logic necessarily contains *more than one* distinct consequence relations, then it is impossible to do only with two logical values (one logical ordering), and "logically n-valued logics" naturally arise, see (Wansing and Shramko 2008b, p. 422).

In this way the strategy of semantic analysis according to Dunn and Belnap's American Plan and its generalizations provides a reliable shield against Suszko's bold attack on the very idea of logical many-valuedeness.

Acknowledgments My work on this paper is a part of Marie Curie project PIRSES-GA-2012-318986 *Generalizing Truth Functionality* within the Seventh Framework Programme for Research funded by EU. I am grateful to Heinrich Wansing, Sergei Odintsov and Norihiro Kamide for helpful comments.

References

Anderson, A. R., & Belnap, N. D. (1975). *Entailment: The logic of relevance and necessity* (Vol. I). NJ: Princeton University Press.

Anderson, A. R., Belnap, N. D., & Dunn, J. M. (1992). *Entailment: The logic of relevance and necessity* (Vol. II). Princeton: Princeton University Press.

Arieli, O., & Avron, A. (1996). Reasoning with logical bilattices. *Journal of Logic, Language and Information*, *5*, 25–63.

Belnap, N. D. (1977a). How a computer should think. In G. Ryle (Ed.), *Contemporary aspects of philosophy* (pp. 30–56). Stocksfield: Oriel Press.

Belnap, N. D. (1977b). A useful four-valued logic. In J. M. Dunn & G. Epstein (Eds.), *Modern uses of multiple-valued logic* (pp. 8–37). Dordrecht: Reidel.

[14]For a comparison of defining consequence (entailment) relations through designated truth values and trough logical orderings, see e.g., (Wansing and Shramko 2008a).

Bou, F., & Rivieccio, U. (2011). The logic of distributive bilattices. *Logic Journal of the IGPL, 19*, 183–216.

Caleiro, C., Carnielli, W., Coniglio, M., & Marcos, J. (2005). Two's company: "The humbug of many logical values". In J.-Y. Beziau (Ed.), *Logica universalis* (pp. 169–189). Basel: Birkhäuser.

Dunn, J. M. (1966). The algebra of intensional logics, PhD thesis, University of Pittsburgh, Ann Arbor (UMI).

Dunn, J. M. (1967). The effective equivalence of certain propositions about De Morgan lattices, (abstract). *Journal of Symbolic Logic, 32*, 433–434.

Dunn, J. M. (1971). An intuitive semantics for first degree relevant implications, (abstract). *Journal of Symbolic Logic, 36*, 362–363.

Dunn, J. M. (1976). Intuitive semantics for first-degree entailments and 'coupled trees'. *Philosophical Studies, 29*, 149–168.

Dunn, J. M. (1986). Relevance logic and entailment. In D. Gabbay & F. Guenthner (Eds.), *Handbook of philosophical logic,* (1st ed., Vol. 3, pp. 117–224). Dordrecht: D. Reidel.

Dunn, J. M. (1995). Positive modal logic. *Studia Logica, 55*, 301–317.

Dunn, J. M. (1999). A comparative study of various model-theoretic treatments of negation: A history of formal negation. In D. M. Gabbay & H. Wansing (Eds.), *What is negation?* (pp. 23–51). Dordrecht: Kluwer.

Dunn, J. M. (2000). Partiality and its dual. *Studia Logica, 65*, 5–40.

Fitting, M. (2006). Bilattices are nice things. In T. Bolander, V. Hendricks, & S. Pedersen (Eds.), *Self-reference* (pp. 53–77). Stanford: CSLI Publications.

Ginsberg, M. (1986). Multi-valued logics, *Proceedings of AAAI-86, Fifth National Conference on Artificial Intellegence* (pp. 243–247). Los Altos: Morgan Kaufman Publishers.

Ginsberg, M. (1988). Multivalued logics: A uniform approach to reasoning in AI. *Computer Intelligence, 4*, 256–316.

Łukasiewicz, J. (1920). O logice trójwartościowej. *Ruch Filozoficny, 5*, 170–171. (English translation as "On three-valued logic". In: Lukasiewicz (1970), pp. 87–88.).

Łukasiewicz, J. (1970). *Selected works*. Amsterdam and Warsaw: North-Holland and PWN.

Łukasiewicz, J. (1993). *Über den Satz des Widerspruch bei Aristoteles*. New York: Georg Olms Verlag.

MacIntosh, J. J. (1991). Adverbially qualified truth values. *Pacific Philosophical Quarterly, 72*, 131–142.

Meyer, R. K. (1978). Why I am not a relevantist, Technical report, Australian National University, Logic Group, Research School of the Social Sciences, Canberra.

Mobasher, B., Pigozzi, D., Slutzki, G., & Voutsadakis, G. (2000). A duality theory for bilattices. *Algebra Universalis, 43*, 109–125.

Odintsov, S. (2009). On axiomatizing Shramko-Wansing's logic. *Studia Logica, 93*, 407–428.

Odintsov, S. P., & Wansing, H. (2015). The logic of generalized truth values and the logic of bilattices. *Studia Logica, 103*, 91–112.

Rescher, N. (1965). An intuitive interpretation of systems of four-valued logic. *Notre Dame Journal of Formal Logic, 6*, 154–156.

Scott, D. (1973). Models for various type-free calculi. In P. Suppes, E. Nagel, & A. Tarski (Eds.), *Logic, methodology and philosophy of science* (Vol. 4, pp. 157–187). Amsterdam: North-Holland.

Shramko, Y., Dunn, J. M., & Takenaka, T. (2001). The trilattice of constructive truth-values. *Journal of Logic and Computation, 11*, 761–788.

Shramko, Y., & Wansing, H. (2005). Some useful 16-valued logics: How a computer network should think. *Journal of Philosophical Logic, 34*, 121–153.

Shramko, Y., & Wansing, H. (2006). Hyper-contradictions, generalized truth-values and logics of truth and falsehood. *Journal of Logic, Language and Information, 15*, 403–424.

Shramko, Y., & Wansing, H. (2011). *Truth and falsehood.*, An inquiry into generalized logical values Dordrecht: Springer.

Shramko, Y., & Wansing, H. (2014). Truth values. In E. N. Zalta (Ed.) *The Stanford Encyclopedia of Philosophy, summer 2014 edn.* http://plato.stanford.edu/archives/sum2014/entries/truth-values/.

Suszko, R. (1977). The Fregean axiom and Polish mathematical logic in the 1920's. *Studia Logica, 36*, 373–380.
Wansing, H. (1993). *The logic of information structures, number 681 in lecture notes in AI*. Berlin: Springer.
Wansing, H., & Shramko, Y. (2008a). Harmonious many-valued propositional logics and the logic of computer networks. In C. Dégremont, L. Keiff, & H. Rückert (Eds.), *Dialogues, logics and other strange things* (pp. 491–516)., Essays in honour of Shahid Rahman London: College Publications.
Wansing, H., & Shramko, Y. (2008b). Suszko's thesis, inferential many-valuedness, and the notion of a logical system. *Studia Logica, 88*, 405–429.
Zaitsev, D. (2009). A few more useful 8-valued logics for reasoning with tetralattice $EIGHT_4$. *Studia Logica, 92*, 265–280.
Zaitsev, D., & Shramko, Y. (2013). Bi-facial truth: A case for generalized truth-values. *Studia Logica, 101*, 1299–1318.

Logical Foundations of Evidential Reasoning with Contradictory Information

Chunlai Zhou

Abstract Inconsistent or contradictory information is quite common in modern information technology such as the Web or unstructured databases. In this paper, we employ two levels of epistemic logics to provide logical foundations for evidential reasoning with this kind of information. The first-level logic is the well-known Belnap–Dunn four-valued logic. This logic provides a formalism for reasoning about both incomplete and contradictory information. In addition to the two standard Boolean truth values T and F, there are two new values: N and B. They are used to designate incomplete and contradictory information, respectively. The four-valued logic is *externally epistemic* in the sense that the truth values are intended to reflect what the agents may have been informed about and are passed over to the agents from the external environment. By using the semantics for this logic, we enrich Carnap's universe for consistent information by replacing standard possible worlds with states, set-ups or situations where a proposition may be both true and false. We shall call such a universe a Belnap–Dunn universe. The second-level logic is epistemic logic $S5$. When the information is uncertain and imprecise, it usually fails to provide probability values for *every* subset of the Belnap–Dunn universe. Probabilities are defined only on those subsets which are *known* with certainty. We employ epistemic logic $S5$ to distinguish those known subsets and to characterize the notion that such known part of the information improves our knowledge by reducing the scope of possible valid states. $S5$ is *internally* epistemic in the sense that the knowledge is determined by the agents. Probabilistic reasoning with the combination of the four-valued logic and epistemic logic $S5$ is nothing but evidential reasoning over bilattices or de Morgan lattices.

Keywords Belnap–Dunn four-valued logic · de Morgan lattices · Dempster–Shafer theory · Epistemic logic · First degree entailment

---

C. Zhou (✉)
Department of Computer Science and Technology, Renmin University,
Beijing, China
e-mail: czhou@ruc.edu.cn
URL: http://info.ruc.edu.cn/en/faculty_detail.php?id=38

K. Bimbó (ed.), *J. Michael Dunn on Information Based Logics*,
Outstanding Contributions to Logic 8, DOI 10.1007/978-3-319-29300-4_12

1 Introduction

Dealing with uncertainty is a fundamental issue for Artificial Intelligence (Halpern 2005). Numerous approaches have been proposed, including Dempster–Shafer theory of belief functions (also called Dempster–Shafer theory of evidence). Ever since the pioneering works by Dempster (1967) and Shafer (1976), belief functions were brought into a practically usable form by Smets and Kennes (1994) and have become a standard tool in Artificial Intelligence for knowledge representation and decision-making.

Dempster–Shafer belief functions on a finite frame of discernment S are defined on the power set of S, which is a *Boolean algebra*. They have an attractive mathematical theory and many intuitively appealing properties. Belief functions satisfy the three axioms which generalize the Kolmogorov axioms for probability functions. Interestingly enough, they can also be characterized in terms of *mass functions* m. Intuitively, for a subset event A, $m(A)$ measures the belief that an agent commits *exactly* to A, not the total belief that an agent commits to A. Shafer (1976) showed that a belief in A is the sum of the masses assigned to all the subsets of A. This characterization of belief functions through mass functions is simply an example of the well-known Inclusion-Exclusion principle in Enumerative Combinatorics (Stanley 1997) and hence has a strong combinatorial flavor. In this theory, mass functions are recognized as Möbius transforms of belief functions.

The Dempster–Shafer theory of belief functions is closely related to other approaches dealing with uncertainty. It includes the Bayesian theory (Savage 1972) as a special case. The first three rules of the Bayesian theory are simply those three axioms for probability functions. It is shown in Shafer (1976) that a belief function on S is *Bayesian* (also a probability function) if and only if its corresponding mass function assigns *positive* weights only to singletons. So a Bayesian belief function μ is more like a point function than a set function in its level of complexity in the sense that μ is determined by its values at *singletons* rather by its values at all events (its values at other non-singletons are 0).

The first investigation of mathematical properties of belief functions on more general lattices was initiated by Barthélemy (2000) with the combinatorial theory on lattices by Rota (1964), which was motivated by possible applications of belief functions for *non-standard* representation of knowledge. Grabisch (2009) continued along this direction and showed that such properties as Dempster's rule of combination and Smets's canonical decomposition (Smets 1995) in the case of Boolean algebras can be transposed in *general lattice setting*. This generalized theory has been applied to many objects in real world problems that may not form a Boolean algebra. An optimal balance between utility and elegance of a theory of belief functions is achieved for *distributive lattices*. We have developed a general theory for belief functions on distributive lattices (Zhou 2012, 2013). Not only does our approach for distributive lattices yield a mathematical theory as appealing as Dempster–Shafer theory, but also its applications extend to many *non-classical* formalisms of structures in Artificial Intelligence. After establishing the mathematical theory for belief functions on distributive lattices, we used this more general theory to provide a framework for reasoning about belief functions in a deductive approach on non-classical formalisms which assume a distributive lattice.

The integration of belief functions and non-classical formalisms is intended to master two sources of ignorance. While belief functions take care of the limitation of the information that the agents have *at their disposal*, non-classical formalisms usually take care of the imprecision, uncertainty or inconsistency in the knowledge-base due to imperfect data. As an illustration of this deductive approach, we dealt with belief functions on two particular classes of distributive lattices: bilattices and de Morgan lattices, which are actually mathematical objects in reasoning under incomplete and inconsistent information. A well-known simple *non-Boolean* epistemic logic the first-degree-entailment fragment $\mathbf{R}_{fde}$ of relevance logic $\mathbf{R}$ (Anderson and Belnap 1975) provides a complete deductive system for this type of non-classical information, which is used to deal with the famous *logical-omniscience problem* in the foundations of Knowledge Representation (Fagin et al. 1995; Levesque 1984), and used for reasoning in the presence of inconsistency in knowledge base systems (Lin 1996). A sound and complete axiomatization is provided for the integration of belief functions and the non-classical logic $\mathbf{R}_{fde}$, and finally the complexity of the satisfiability problem of a belief formula with respect to the class of the corresponding Dempster–Shafer structures is shown to be *NP*-complete. For the detailed proofs, one may consult (Zhou 2012, 2013).

The present research work is motivated by the need to improve the understanding of issues in the analysis and interpretation of evidence. In the context of this paper, the term evidence is used to describe information, which is usually imprecise, uncertain or *inconsistent* (contradictory). Here we employ the Belnap–Dunn four-valued logic (Belnap 1977; Dunn 1976a) and epistemic logic to provide logical foundations for evidential reasoning with contradictory (or inconsistent) information. Our approach is strongly based on Carnap's methodology (Carnap 1962) for the development of the logical foundations of probability theory and Ruspini's ideas (Ruspini 1987) of the logical foundations for evidential reasoning. In his formulation, Carnap developed a universe of possible worlds that encompasses all possible states of a real-world system. Information about that system, if precise and certain, identifies its actual state. If imprecise but certain, this information identifies a subset of possible system states. Such kind of subsets are called *truth sets*. If uncertain, then the information induces a probability distribution over system states which is defined on *all* subsets of the universe.

Ruspini (1987) noted that Carnap's characterization does not distinguish degrees of precision when the information is uncertain and Carnap's logical approach, while enabling a clearer understanding of the relations between logical and probabilistic concepts, suffers from a major handicap: it assumes that observations of the real world always determine unambiguously probability values for *every* subset in the universe. But uncertain information generates a probability function for all subsets of the universe only if it is precise. When information is imprecise, this probability function is defined on some subsets of possible states, which is not discussed in Carnap's methodology. This type of information, which provides some knowledge about the underlying probability distributions but not all values of the distributions, is quite common in practical applications and motivated Dempster's original formulation of evidential reasoning (Dempster 1967). In order to distinguish the degrees of precision, Ruspini employed epistemic logics—a form of modal logics developed to deal with problems of representation and manipulation of the states of knowledge of rational agents—to generalize Carnap's space of possible worlds, or universe. This generalization is obtained by considering the combination of representations of both the state of the possible world and the knowledge of rational agents, which is called the *epistemic universe*. Uncertain evidence is represented as a conventional probability function on the sigma-algebra of epistemic sets in the epistemic universe. Epistemic sets are interpreted as truth sets which are *known* with certainty and probability function is defined only on those sets that the agents know for sure. Such a probability function defined on epistemic sets can be regarded as a kind of constraint on possible probability functions defined on *all* subsets of the universe and hence is equivalent to a belief function in the Dempster–Shafer theory. So the combination of epistemic logic and probability theory provides an analysis of logical foundations of evidential reasoning.

However, Ruspini failed to consider the case that sometimes information is inconsistent. Both Carnap and Ruspini considered only consistent information. But inconsistent or contradictory information is quite common in modern information technology such as the Web and unstructured databases. In contrast to the conventional databases, unstructured databases allow for negative as well as positive information. In other words, they use the Open-world Assumption, which is different from the Closed-World Assumption in structured databases where negation is interpreted as absence. Belnap–Dunn four-valued logic (Belnap 1977; Dunn 1976a) is a well-known machinery to provide a deductive reasoning for this kind of information. In this logic, each proposition is assigned one of the four possible *epistemic values*: T, F, B and N. The meaning of such epistemic truth values highly differs from the meaning of standard Boolean truth values since they are not intrinsic to propositions but are intended to reflect what an agent may have been *informed* about (regarding these propositions). Thus, interpreting a proposition ϕ as F (resp., T) does not mean that ϕ is false (resp., true) but that the agent under consideration has some reasons to consider that ϕ is false (resp., true) or is told that ϕ is "false" (resp., "true"). The agent may have some reasons to consider that ϕ is false and other reasons to consider that ϕ is true, and the epistemic truth value B reflects this situation. Similarly, the agent may have no reasons to consider ϕ as true and no reasons to consider it as false; in this situation, ϕ is given the epistemic truth value N. So B reflects a situation of inconsistency and N of ignorance. (See also Dunn 2010.) By using the semantics for this logic, we enrich Carnap's universe by incorporating inconsistent information and by replacing standard possible worlds with worlds, set-ups or situations where a proposition may be both true and false. Such a universe is called a *Belnap–Dunn universe*. The epistemic nature of the Belnap–Dunn universe is *external* in the sense that the truth values are passed over to the agent from the external environment.

But deductive reasoning is a limited form of reasoning. In a world of certainty, the deductive information is capable of identifying only a subset of possibilities and is unable to distinguish degrees of uncertainty. In contrast, a weighting calculus, probabilistic reasoning is more applicable in real life. It is desirable to integrate the four-valued logic with probabilistic reasoning. Just as in Ruspini's analysis (Ruspini 1987), probability functions are not necessarily defined over all subsets of the Belnap–Dunn universe. So we follow Ruspini's idea to employ the framework of epistemic logic *S5* to distinguish those defined subsets, which are *known* with certainty. *S5* is *internally* epistemic in the sense that the knowledge is determined by the agent himself. So the epistemic nature of *S5* at this level is different from that of the Belnap–Dunn universe. Probabilistic reasoning over such (internally) epistemic structure on the Belnap–Dunn universe provides logical foundations of evidential reasoning with contradictory information.

The rest of the paper is organized as follows. In Sect. 2, we provide a background on lattice theory and belief functions on (distributive) lattices. Section 3 is the main part of the paper. There we employ both four-valued logic and epistemic logic *S5* to present the semantical framework for belief functions on both distributive bilattices and de Morgan lattices. In Sect. 4, we provide a sound and complete deductive system for reasoning about belief functions for the first-degree-entailment fragment of **R** and show that the complexity of the satisfiability problem of belief formulas with respect to the class of the corresponding Dempster–Shafer structures is *NP*-complete. And in the final section, we discuss some related work. The Appendix provides the duality theorem for finite de Morgan lattices which is parallel to Birkhoff's representation theorem for finite distributive lattices and is hence of independent interest.

2 Belief Functions on Distributive Lattices

We will first recall some basic definitions about lattices. Next Dempster–Shafer theory of belief functions on Boolean algebras will be generalized to this more general setting. All posets and lattices occurring in this paper are supposed to be finite. All lattice-theoretical notation and terminology in this paper follows Stanley (1997).

2.1 Lattices

Let P be a poset. A subset I of P is called an *order ideal* (*co-cone*, or *semi-ideal* or *down-set* or *decreasing subset*) if, for any $x, y \in L$, $x \in I$ provided that $x \leq y$ and $y \in I$. I is called a *principal order ideal* if $I = \{y \in L : y \leq x\}$ for some $x \in I$. Otherwise, it is called a *non-principal order ideal*. Dually, a subset F of L is called a *dual order ideal* (*cone* or *up-set* or *increasing subset* or *filter*) if, for any $x, y \in L$, $y \in F$ provided that $x \leq y$ and $x \in F$. F is called a *prime filter* if it satisfies the following additional condition: for any $a, b \in F$, $a \in F$ or $b \in F$ whenever $a \vee b \in F$. A strict partial ordering $<$ is defined from $\leq$ as $x < y$ if $x \leq y$ and $x \neq y$.

A lattice L is *distributive* if $(x \vee y) \wedge z = (x \wedge z) \vee (y \wedge z)$ holds for all $x, y, z \in L$. For any $x \in L$, we say that x has a *complement* in L if there exists $x' \in L$ such that $x \wedge x' = \bot$ and $x \vee x' = \top$. L is said to be *complemented* if every element has a complement. *Boolean lattices (algebras)* are distributive and complemented lattices. In a Boolean lattice, every element has a unique complement. According to the famous Stone representation theorem, every Boolean lattice is isomorphic to a subalgebra of the *concrete* Boolean lattice $\langle 2^S, \subseteq \rangle$ for some set S. For the lattice $\langle 2^S, \subseteq \rangle$, we have $\vee = \cup$, $\wedge = \cap$, $\top = S$ and $\bot = \emptyset$.

A *de Morgan lattice* D is a bounded distributive lattice $\langle D, \vee, \wedge, \top, \bot \rangle$ with an *involution* $\neg$ which satisfies the following *de Morgan's laws*:

$$\neg(x \wedge y) = \neg x \vee \neg y \text{ and } \neg\neg x = x, \quad \text{for all } x, y \in D.$$

It follows immediately that $\neg(x \vee y) = \neg x \wedge \neg y$, $\neg\top = \bot$ and $\neg\bot = \top$. So $\neg$ is a *dual automorphism*. Note that in a de Morgan lattice, the following laws may not hold:

$$\neg x \vee x = \top \text{ and } x \wedge \neg x = \bot.$$

De Morgan lattices are important for the study of the mathematical aspects of fuzzy logic (Zadeh 1988). The standard fuzzy algebra $F = \langle [0, 1], \max(x, y), \min(x, y), 0, 1, 1 - x \rangle$ is an example of a de Morgan lattice. Moreover, de Morgan monoids are an algebraic semantics for a non-classical formalism, namely, for the relevance logic $\mathbf{R}^t$ (Dunn 1986).

2.2 Belief Functions

There are two *equivalent* approaches to belief functions on lattices: one is mass functions and the other totally monotone capacities. The equivalence is characterized by *Möbius functions*. Let $(L, \leq)$ be a poset having a bottom element $\bot$ and a top one $\top$ and $\mathbb{R}$ be the real field. Without further notice, every function in this paper is meant to be a real-valued map. The *Möbius function* $\mu: L^2 \to \mathbb{R}$ of L is defined recursively by

$$\mu(x, y) = \begin{cases} 1 & \text{if } x = y, \\ -\sum_{x \leq t < y} \mu(x, t) & \text{if } x < y, \\ 0 & \text{otherwise.} \end{cases}$$

Note that μ solely depends on L.

Proposition 2.1 (Möbius inversion formula, Proposition 3.7.1 in Stanley 1997) *Let P be a poset. Let f and g be two functions. Then*

$$g(t) = \sum_{s \leq t} f(s) \quad \textit{for all } t \in P \tag{1}$$

if and only if

$$f(t) = \sum_{s \leq t} g(s)\mu(s, t) \quad \textit{for all } t \in L, \tag{2}$$

where μ *is the Möbius function of* P.

The function g in the above proposition is called the *Möbius transform* of f.

Definition 2.2 Given a lattice $\langle L, \leq \rangle$, a function f on L is called a *capacity* if it satisfies the following three conditions:

1. $f(\bot) = 0$;
2. $f(\top) = 1$;
3. $x \leq y$ implies $f(x) \leq f(y)$.

A function $bel\colon L \to [0, 1]$ is called a *belief function* if $bel(\top) = 1$, $bel(\bot) = 0$ and its Möbius transform m is non-negative. m is also called the *mass function* or *mass assignment* of f. For each element $a \in L$, the quantity $m(a)$ is intended to measure the belief that one commits *exactly* to a, not the total belief that one commits to a. To obtain the measure of the total belief committed to a, one must add to $m(a)$ the quantities $m(b)$ for all elements that are strictly smaller than a:

$$bel(a) = \sum_{b \leq a} m(a).$$

An element $a \in L$ is called a *focal element* of L if $m(a) > 0$.

Note that any belief function is a monotonic function by non-negativity of m, and hence a capacity.

Example 2.1 In the above definition, if L is a Boolean algebra, then bel on L is defined in the same way as in the Dempster–Shafer theory (Shafer 1976). Let Ω be a finite space. In this case, a function $m\colon 2^{\Omega} \to [0, 1]$ is a mass allocation function if $m(\emptyset) = 0$ and $\sum_{A \subseteq \Omega} m(A) = 1$. A belief function on Ω is a function $bel\colon 2^{\Omega} \to [0, 1]$ generated by a mass allocation function as follows. For $A \subseteq \Omega$,

$$bel(A) := \sum_{B \subseteq A} m(B).$$

Note that $bel(\emptyset) = 0$ and $bel(\Omega) = 1$. The Möbius function $\mu\colon 2^{\Omega} \times 2^{\Omega} \to [0, 1]$ is

$$\mu(A, B) = \begin{cases} (-1)^{|B \setminus A|} & \text{if } A \subseteq B, \\ 0 & \text{otherwise.} \end{cases}$$

m is the Möbius transform of bel and is expressed as the following formula:

$$m(A) = \sum_{B \subseteq A} (-1)^{|A \setminus B|} bel(B).$$

Belief functions on Boolean algebras are a generalization of probability functions in the sense that a belief function is a probability function iff its focal elements are singletons. Now consider a simple example how to assign beliefs in terms of mass functions. A murder has been committed. There are three suspects $\Omega = \{\text{John, Mary, Peter}\}$. A witness saw the murderer going away in the darkness and he can only assert that it was a man. However, we know that the witness is drunk 20 % of the time. When he is drunk, he cannot distinguish among the three suspects. So a mass of 0.2 is assigned to the whole Ω. When he is sober 80 % of the time, he asserts that the murderer is a man. But he does not know which man. So a mass 0.8 is assigned to the subset $\{\text{John, Peter}\}$ of the two men. This piece of evidence can be represented by the following mass function:

$$m(\{\text{John, Peter}\}) = 0.8, \text{ and } m(\Omega) = 0.2.$$

Definition 2.3 Let K denote the set $\{1, 2, \ldots, k\}$. Given a lattice $\langle L, \leq \rangle$, a function f on L is called a *k-monotone* whenever for each $(x_1, \ldots, x_k) \in L^k$, the k-product of lattice L, we have

$$f(\bigvee_{1 \leq i \leq k} x_i) \geq \sum_{J \subseteq K, J \neq \emptyset} (-1)^{|J|+1} f(\bigwedge_{j \in J} x_j) \tag{3}$$

A capacity is *totally monotone* if it is k-monotone for every $k \geq 2$. A k-monotone function f is called a *k-valuation* if the above inequality degenerates into the following equality:

$$f(\bigvee_{1 \leq i \leq k} x_i) = \sum_{J \subseteq K, J \neq \emptyset} (-1)^{|J|+1} f(\bigwedge_{j \in J} x_j) \tag{4}$$

It is an ∞-*valuation* if it is a k-valuation for each integer k. f is called a *probability function* if it is both a capacity and an ∞-valuation.

Lemma 2.4 *If L is distributive, then the following equality is sufficient for an ∞-valuation f:*

$$f(a \wedge b) + f(a \vee b) = f(a) + f(b), \quad \textit{for all } a, b \in L. \tag{5}$$

The following proposition in Barthélemy (2000) tells us that every belief function is totally monotone.

Proposition 2.5 *Let $f : L \to [0, 1]$ be a capacity and m be its Möbius transform. If f is a belief function, then it is totally monotone.*

Shafer proved that the converse is also true for any belief function on *Boolean algebras* (Theorem 2.1. in Shafer 1976). We have shown (Zhou 2013, 2012) that it actually holds *generally* for any lattice, which answers an open question raised in Grabisch (2009).

Theorem 2.6 *Let L be a lattice and* $f: L \to [0, 1]$ *be a capacity on L and m be its Möbius transform. The following two statements are equivalent:*

- *m is non-negative;*
- *f is totally monotone.*

Given a poset P, $J(P)$ $(F(P))$ denotes the lattice of order ideals (filters) of P with the ordinary union and intersection (on subsets of P). So $J(P)$ $(F(P))$ is distributive. Conversely, from Birkhoff's fundamental theorem for finite distributive lattices (Stanley 1997), we know that for any finite distributive lattice L, there is a unique (up to isomorphism) finite poset P for which $L \cong J(P)$ $(F(P))$. Usually P is chosen to be the poset of join-irreducibles (meet-irreducibles) in L. The following two propositions provide formulas for Möbius functions and Möbius transforms in distributive lattices.

Proposition 2.7 (Example 3.9.6 in Stanley 1997)
The Möbius function of the distributive lattice $L = J(P)$*, where P is a poset, is definable as follows. For any* $I, I' \in J(P)$*,*

$$\mu(I, I') = \begin{cases} (-1)^{|I' \setminus I|} & \text{if } [I, I'] \text{ is a Boolean algebra,} \\ 0 & \text{otherwise.} \end{cases}$$

where $[I, I']$ *denotes the interval* $\{K \in J(P) : I \subseteq K \subseteq I'\}$.

From this proposition, we immediately obtain a nice formula for Möbius transforms.

Theorem 2.8 *Let* $L = J(P)$ *be a distributive lattice for some poset P. Suppose* $Bel: L \to [0, 1]$ *is the belief function given by the mass assignment* $m: L \to [0, 1]$*. Then, for all* $A \in J(P)$*,*

$$m(A) = \sum_{\substack{[B,A] \text{ is a} \\ \text{Boolean algebra}}} (-1)^{|A \setminus B|} Bel(B).$$

Definition 2.9 Given a distributive lattice $L = \langle L, \leq \rangle$, a belief function $Bel: L \to [0, 1]$ is called *Bayesian* if,

$$Bel(a \vee b) + Bel(a \wedge b) = Bel(a) + Bel(b) \quad \text{whenever } a, b \in L. \tag{6}$$

From Lemma 2.4, for any distributive lattice $L = \langle L, \leq \rangle$, a belief function $Bel: L \to [0, 1]$ is Bayesian if and only if it is a probability function. The following proposition generalizes the natural analogy in lattice theory that join-irreducibles to distributive lattices are the same as singletons to Boolean algebras.

Proposition 2.10 *A belief function Bel on a distributive lattice D is Bayesian (or a probability function) iff all its focal elements are join-irreducibles.*

Most important properties of belief functions on Boolean algebras can be transposed naturally to the more general setting of distributive lattices (Zhou 2013).

3 Belief Functions on Non-classical Formalisms

The integration of belief functions and non-classical formalisms is intended to master two sources of ignorance. Non-classical formalisms usually take care of the incompleteness or inconsistency in the knowledge-base due to imperfect data while belief functions take care of the limitation of the information that the agents have *at their disposal*. In artificial intelligence especially in Knowledge Representation, non-classical formalisms play an important role in handling *imperfect* information in different forms. Most of these non-classical formalisms assume a mathematical setting of distributive lattices. (Quantum logic is probably one of the very few important exceptions (Birkhoff and von Neumann 1936; Ying 2010), with linear logic being another one (Girard 1987).) Each of these formalisms was intended for reasoning about some specific form of information.

For example, Kleene's three-valued logic has been used to take into account "undefined" as a third truth value, which is useful to model the situation in computer science when a computation does not return any result (Fitting 1994). Paraconsistent logics have been used to deal with contradictory knowledge bases (Arieli et al. 2011; Priest 1979), and relevance logic is used to deal with the famous logical omniscience problem in the foundation of knowledge representation, and used for reasoning in the presence of inconsistency in knowledge base systems (Levesque 1984; Lin 1996; Fagin et al. 1990).

Belief functions in the Dempster–Shafer theory are defined on Boolean algebras (Shafer 1976). One essential difference of non-classical formalisms from the Boolean setting is their specific treatments of negation. Negation is closely related to the treatment of *bipolarity* in information (Dubois and Prade 2008), which means that there is an intrinsic positive and negative affect in dealing with information. In the classical Dempster–Shafer theory, negation is assumed to be *Boolean*, i.e., every element has a complement (for any element a, there is an element a' such that $a \vee a'$ is the top element and $a \wedge a'$ is the bottom), which is used to represent *complete* information. A distributive lattice with a Boolean negation is a Boolean algebra and any Boolean algebra can be represented as a power set with the usual set operations (Example 2.1). In Kleene's three-valued logic, there are three truth values: true, false and undetermined. Logically, the treatment of negation considers some formulas to be neither true nor false (undetermined) but forbids any formula to be both true and false. In other words, positive and negative sides don't exhaust all possibilities. So this logic is used to represent incompleteness in information. There is an *implicit intuitionistic* negation in any finite distributive lattice. Since any finite distributive lattice $D = J(P)$ for some poset P is also a Heyting algebra, $\max\{I' \in J(P) : I' \cap I = \emptyset\}$ exists for any $I \in J(P)$, and is defined to be the negation of I (denoted by $\sim I$), which is in $J(P)$. One may also reason about belief functions in this case by replacing the classical proposition logic by intuitionistic propositional logic. (See the discussion in Sect. 5).

Kleene's three-valued logic finds a natural generalization in the Belnap–Dunn four-valued logic (Belnap 1977; Dunn 1976a), which can be naturally extended to distributive bilattices. A distributive bilattice is a distributive lattice with a second ordering which interacts with the original one in a certain way. In addition to their applications in logic programming (Fitting 1994), distributive bilattices are also used to represent the inconsistency in knowledge base systems. In these structures, another form of negation called de Morgan negation is employed. The most important aspect of de Morgan negations is their intrinsic ability to model inconsistency in knowledge base systems. In this paper, we will consider this type of negation and integrate belief functions with de Morgan lattices, which are distributive lattices with de Morgan negations. We will provide an axiomatization of reasoning about belief functions over such non-classical structures and discuss the computational complexity of different problems in this setting. More importantly, this approach to reasoning about belief functions on de Morgan lattices also provides a framework to reason about belief functions on other non-classical structures. For a comprehensive algebraic treatment of other negations in non-classical formalisms, one may refer to Dunn (1999) and Dunn and Zhou (2005).

A well-known slogan in algebraic methods for non-classical logics (Sect. 18 of Anderson and Belnap 1975; Dunn 1966) tells us that the algebraist and the logician are *dual* to each other in the sense that algebra and logic are dual to each other. Given a non-classical logic $\mathbb{L}$, the Lindenbaum algebra $\mathbb{A}_{\mathbb{L}}$ of this logic is in the class of algebras which characterize the logic but also the possible-world-like structure derived from the Lindenbaum algebra $\mathbb{A}_{\mathbb{L}}$ through the Duality theorem for this class of algebras is the *canonical structure* for this logic. On the other hand, given any possible-world-like structure for the logic $\mathbb{L}$, the interpretations of the formulas is an algebra that characterizes $\mathbb{L}$. So, in the following section, before we reason about belief functions on specific classes of algebras, we elaborate on this kind of duality between logics and their corresponding algebraic structures (and possible-world-like semantics) and won't distinguish belief functions on algebras and for their logics.

In addition to this duality, algebra and logic are dual to each other in the sense that (order) ideals to algebra are the same as filters to logic. If we replace all posets $\langle P, \leq\rangle$ with its dual $\langle P, \leq^{\partial}\rangle$ (where $x \leq^{\partial} y$ iff $y \leq x$) and all $\leq$-order-ideals by $\leq^{\partial}$-filters, the dual forms of all propositions there remain valid. So, in order to apply the algebraic propositions to non-classical logics, we have to keep this duality in mind.

3.1 Reasoning About Bilattices

Bilattices are algebras with two separate lattice structures. Ginsberg (1988) suggested using bilattices as the underlying framework for various AI inference systems including those based on default logics, truth maintenance systems, probabilistic logics, and others. These ideas were later pursued in the context of logic programming semantics (Fitting 1991). Moreover, bilattices and their extensions have been used in the literature to model a variety of reasoning mechanisms about uncertainty in the presence of incomplete or contradictory information (Kifer and Lozinskii 1992; Arieli et al. 2011). Also they have been employed to represent bipolar information (Konieczny 2008). In the following, we first present a well-known algebraic result about the representation of bilattices. We employ the Belnap–Dunn four-valued logic to *decouple* the interpretation of each formula ϕ into the set of states where ϕ is "true" and that of states where it is "false" (Dunn 1976a). In this way, the interpretations of all formulas form a distributive bilattice with two partial orderings: the *truth ordering* and the *knowledge ordering*.

One may refer to Mobasher et al. (2000), Fitting (1994) for the technical details about the duality theorem of bilattices which is presented below. For similar duality results, one may also consult (Jung and Rivieccio 2012) and (Jung and Rivieccio 2013). Recall that all lattices are assumed to be finite.

Definition 3.1 A *bilattice* is an algebra $\mathbf{B} = \langle B, \wedge_1, \vee_1, \bot_1, \top_1, \wedge_2, \vee_2, \bot_2, \top_2\rangle$ such that $\mathbf{B}_1 = \langle B, \wedge_1, \vee_1, \bot_1, \top_1\rangle$ and $\mathbf{B}_2 = \langle B, \wedge_2, \vee_2, \bot_2, \top_2\rangle$ are lattices. By a *negation* on $\mathbf{B}$ we mean a unary operation $\neg$ on B satisfying the conditions:

1. $\neg\neg x = x$;
2. $\neg(x \vee_1 y) = \neg x \wedge_1 \neg y, \quad \neg(x \wedge_1 y) = \neg x \vee_1 \neg y$;
3. $\neg(x \vee_2 y) = \neg x \vee_2 \neg y, \quad \neg(x \wedge_2 y) = \neg x \wedge_2 \neg y$.

B is called *distributive* if, for every $\Diamond, \Box \in \{\wedge_1, \vee_1, \wedge_2, \vee_2\}$ and for all $x, y, z \in B$, $x \Diamond (y \Box z) = (x \Diamond y) \Box (x \Diamond z)$.

The lattice ordering corresponding to the lattice $\mathbf{B}_1$ will be denoted by $\leq_1$ and the lattice ordering corresponding to $\mathbf{B}_2$ by $\leq_2$; often the bilattice $\mathbf{B}$ is written in the form $\langle B, \leq_1, \leq_2\rangle$. Alternatively, $\leq_1$ and $\leq_2$ are often denoted by $\leq_t$ and $\leq_k$, respectively, reflecting the fact that they represent the "truth" and "knowledge" orderings, which will become clear in the following definition of *four-valued model*.

Definition 3.2 Let $\mathbf{L} = \langle L, \wedge, \vee, \bot, \top\rangle$ and $\mathbf{L}' = \langle L', \wedge', \vee', \bot', \top'\rangle$ be lattices. Define $\mathcal{B}(\mathbf{L}, \mathbf{L}') = \langle L \times L', \sqcap_1, \sqcup_1, \bot_1, \top_1, \sqcap_2, \sqcup_2, \bot_2, \top_2\rangle$ as follows. For all $(x, x'), (y, y') \in L \times L'$,

- $(x, x') \sqcap_1 (y, y') = (x \wedge y, x' \vee' y'), \quad (x, x') \sqcup_1 (y, y') = (x \vee y, x' \wedge' y')$;
- $(x, x') \sqcap_2 (y, y') = (x \wedge y, x' \wedge' y'), \quad (x, x') \sqcup_2 (y, y') = (x \vee y, x' \vee' y')$;
- $\bot_1 = (\bot, \top'), \quad \top_1 = (\top, \bot'), \quad \bot_2 = (\bot, \bot'), \quad \top_2 = (\top, \top')$.

$\mathcal{B}(\mathbf{L}, \mathbf{L}')$ is called the *product bilattice associated with* L and L'. If $L = L'$, then we define

$$\sim(x, x') = (x', x).$$

$\mathcal{B}(\mathbf{L}, \mathbf{L})$ is called the *square bilattice with negation associated with* $\mathbf{L}$. Usually we write $\mathcal{B}(L)$ for $\mathcal{B}(\mathbf{L}, \mathbf{L})$.

Theorem 3.3 (Corollary 9 in Mobasher et al. 2000)
$\mathbf{B}$ is a distributive bilattice with negation if and only if there exists a distributive lattice L such that $\mathbf{B} \cong \mathcal{B}(L)$.

Let $\mathbf{B} = \langle B, \leq_1, \leq_2 \rangle$ be a distributive bilattice. An element $x \in B$ is called *positive* if, for every $y \in B$, $x \leq_1 y$ implies $x \leq_2 y$. It is called *negative* if, for every $y \in B$, $y \leq_1 x$ implies $x \leq_2 y$. Intuitively, an element x is positive (negative) if it should increase in the knowledge ordering whenever it increases (decreases) in the truth order. Denote by $POS(B)$ and $NEG(B)$ the set of positive and negative elements [they are called t-grounded and f-grounded in Ginsberg (1988)], respectively, of B. An element $x \in B$ is called *positive (resp., negative)* $\leq_2$*-join-irreducible* if it is positive (resp., negative) and join-irreducible with respect to $\leq_2$-ordering. We denote by $\Im_2^+(B)$ (resp., $\Im_2^-(B)$) the set of non-bottom positive (resp., negative) $\leq_2$-join-irreducible elements of B. Moreover, $\Im_2(B)$ denotes the set of all $\leq_2$-join-irreducible elements of B. If a bilattice $\mathbf{B}$ is distributive and is represented as a square bilattice, then it is easy to recognize those positive and negative elements, as shown in the following proposition.

Proposition 3.4 (Corollary 8 in Mobasher et al. 2000)
Let L be a distributive lattice and let 0 be the least element. An element x of $\mathcal{B}(L)$ is positive iff $x = (y, 0)$ for some $y \in L$, and is negative iff $x = (0, y)$ for some $y \in L$.

Let $P = \langle P, \preceq \rangle$ and $Q = \langle Q, \sqsubseteq \rangle$ be two disjoint partially ordered sets. Define the *lift* of P, denoted by $P_\perp = \langle P \cup \{0\}, \leq \rangle$, where $0 \notin P$ and $x \leq y$ in $P_\perp$ iff $x = 0$ or $x \preceq y$ in P. Define the *disjoint union* $P \uplus Q = \langle P \cup Q, \leq \rangle$ to be the partially ordered set with $x \leq y$ iff either $x, y \in P$ and $x \preceq y$ or $x, y \in Q$ and $x \sqsubseteq y$.

Given two partially ordered sets P and Q, define the *separated sum* of P and Q, denoted $P \oplus_\perp Q$, to be the poset $P \oplus_\perp Q = (P \uplus Q)_\perp$.

Theorem 3.5 (Corollary 23 in Mobasher et al. 2000)
Let B be a distributive bilattice with negation. Then $\Im_2(B) \cong \Im_2^+(B) \oplus_\perp \Im_2^-(B)$ and $\Im_2^+(B) \cong \Im_2^-(B)$. Conversely, for any finite poset P, there is a finite distributive bilattice $\mathbf{B}$ such that $\Im_2(B) \cong P \oplus_\perp P$. So there is a one-to-one correspondence between finite distributive bilattices and finite posets.

Let $J_2(\Im_2(B))$ denote $\leq_2$-order ideals in $\Im_2(B)$. It is easy to check that it is the same as the Cartesian product of the set of order ideals in P with itself. In other words, $J_2(\Im_2(B)) \approx J(P) \times J(P)$. For any two elements (I_1, I_2), $(I_1', I_2') \in J_2(\Im_2(B))$,

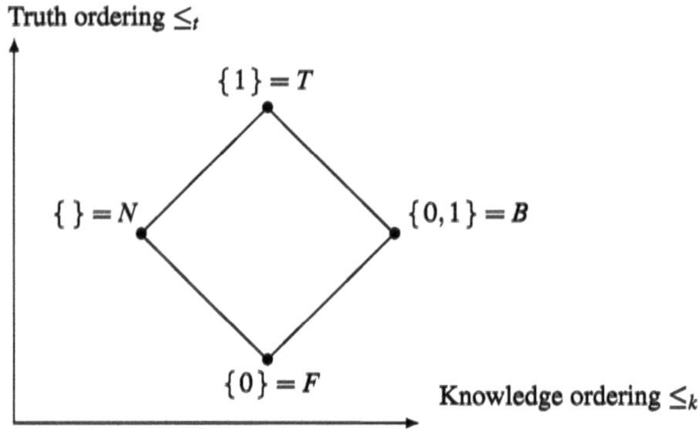

Fig. 1 Lattice *FOUR* interpreted by subsets of $\{0, 1\}$

- $(I_1, I_2) \leq_1 (I'_1, I'_2)$ if $I_1 \subseteq I_2$ and $I'_1 \supseteq I'_2$;
- $(I_1, I_2) \leq_2 (I'_1, I'_2)$ if $I_1 \subseteq I_2$ and $I'_1 \subseteq I'_2$

FOUR, the structure that corresponds to the Belnap–Dunn four-valued logic (Belnap 1977; Dunn 1976a), is the minimal bilattice, exactly as the structure $\mathbf{2} = \{\mathit{true}, \mathit{false}\}$ or $\{0, 1\}$ that is based on the classical two valued logic is the minimal Boolean algebra. It plays an important role in bilattice-based multi-valued logics. The Hasse diagram of the lattice *FOUR* is illustrated in Fig. 1.

Following Dunn (1976a; 2000), we interpret the lattice *FOUR* in terms of the power set of $\{0, 1\}$. The meaning of the capital letters attached to the elements of *FOUR* in the above figure is obvious from this type of interpretation. For example, B informally means "both" and can be translated as both "true" (1) and "false" (0). The truth ordering $\leq_t$ can be formalized as follows: for any two elements $x, y \in \{T, F, B, N\}$,

$$x \leq_t y \text{ if both } 1 \in x \text{ implies } 1 \in y, \text{ and } 0 \in y \text{ implies } 0 \in x.$$

The lattice *FOUR* is the tuple $\langle\{T, F, N, B\}, \vee, \wedge, \sim\rangle$ where $\wedge$ and $\vee$ are the lattice operations associated with the above truth ordering. Also we define $\sim$ as an order inverting operation that leaves N and B as fixed points, i.e., $\sim N = N$, $\sim B = B$, $\sim T = F$, $\sim F = T$. The meaning of this negation $\sim$ will be clear from the following semantical meaning of formulas.

It is interesting to note that there is a natural *knowledge* ordering implicit in the above lattice *FOUR*. The knowledge ordering $\leq_k$ is defined as follows: for any two elements $x, y \in \{T, F, N, B\}$,

$$x \leq_k y \text{ if both } 1 \in x \text{ implies } 1 \in y, \text{ and } 0 \in x \text{ implies } 0 \in y.$$

So the lattice operations $\sqcap$ and $\sqcup$ associated with $\leq_k$ are simply the usual set operations $\cap$ and $\cup$. It is easy to check that $\langle\{T, F, N, F\}, \vee, \wedge, \sqcap, \sqcup, \sim\rangle$ is a distributive bilattice. It is actually the smallest distributive bilattice. $\mathbf{2}$ denotes the sublattice $\langle\{T, F\}, \wedge, \vee\rangle$ of the reduct $\langle\{T, F, N, B\}, \wedge, \vee\rangle$.

The interpretations of standard propositional formulas in *FOUR* form exactly a distributive bilattice. Moreover, it was shown in Arieli and Avron (1998) that all the natural bilattice-valued logics that we had introduced for various purposes can be characterized using only the four basic "epistemic truth values." The meaning of such epistemic truth values highly differs from the meaning of standard Boolean truth values since they are not intrinsic to propositions but are intended to reflect what an agent may have been informed about (regarding these propositions). Thus, interpreting a proposition ϕ as 0 (resp., 1) does not mean that ϕ is false (resp., true) but that the agent under consideration has some reasons to consider that ϕ is false (resp., true) or is told that ϕ is "false" (resp., "true"). The agent may have some reasons to consider that ϕ is false and other reasons to consider that ϕ is true, and the epistemic truth value B reflects this situation. Similarly, the agent may have no reasons to consider ϕ as true and no reasons to consider it as false; in this situation, ϕ is given the epistemic truth value N (or $\emptyset$). So B reflects a situation of inconsistency and N of ignorance. The epistemic nature in the interpretations into *FOUR* agrees well with Shafer's emphasis of the epistemic nature of the set of possibilities on the frame of discernment in his theory (Shafer 1976, p. 36), but under incomplete or inconsistent information.

In order to reason about belief functions for the four-valued logic, we expand Φ_0 in the last section to Φ by adding the connective negation $\sim$. In other words, a formula in Φ is formed by the following syntax:

$$\phi := p \mid {\sim}\phi \mid \phi_1 \wedge \phi_2 \mid \phi_1 \vee \phi_2,$$

where p is a propositional letter.

A *valuation* v into the lattice *FOUR* is a function from the set P of propositional letters into *FOUR*. It is easy to see that v can be extended to the set of formulas naturally as follows:

- $v({\sim}\phi) = {\sim}v(\phi)$;
- $v(\phi \wedge \psi) = v(\phi) \wedge v(\psi)$;
- $v(\phi \vee \psi) = v(\phi) \vee v(\psi)$.

Following Belnap (1977), we simply say "ϕ is at least true" if $1 \in v(\phi)$; "ϕ is at least false" if $0 \in v(\phi)$. It follows immediately that

- $1 \in v(\phi)$ iff $0 \in v({\sim}\phi)$, $0 \in v(\phi)$ iff $1 \in v({\sim}\phi)$;
- $1 \in v(\phi \vee \psi)$ iff $1 \in v(\phi)$ or $1 \in v(\psi)$,
  $0 \in v(\phi \vee \psi)$ iff $0 \in v(\phi)$ and $0 \in v(\psi)$;
- $1 \in v(\phi \wedge \psi)$ iff $1 \in v(\phi)$ and $1 \in v(\psi)$,
  $0 \in v(\phi \wedge \psi)$ iff $0 \in v(\phi)$ or $0 \in v(\psi)$.

In order to introduce the bilattice into our setting, we simulate the Kripke semantics for intuitionistic logic (Kripke 1965; van Dalen 2004) to define a semantics for four-valued logic (Definition 3.7). In this semantics, a *four-valued model* is a tuple $S = \langle S, v \rangle$, where

- S is a non-empty set of states, which are like possible worlds except that they are not required to be either consistent or complete;
- v is a valuation which is a function at each state such that $v(s)(p) \subseteq \{1, 0\}$ for all propositional letters p. In other words, $v(s)(p) \in \{T, N, B, F\}$.

Two *support relations* between states and formulas are defined inductively as follows:

- $S, s \models_T p$ if $1 \in v(s)(p)$,
  $S, s \models_F p$ if $0 \in v(s)(p)$;
- $S, s \models_T \phi_1 \wedge \phi_2$ if $S, s \models_T \phi_1$ and $S, s \models_T \phi_2$;
  $S, s \models_F \phi_1 \wedge \phi_2$ if $S, s \models_F \phi_1$ or $S, s \models_F \phi_2$;
- $S, s \models_T \phi_1 \vee \phi_2$ if $S, s \models_T \phi_1$ or $S, s \models_T \phi_2$;
  $S, s \models_F \phi_1 \vee \phi_2$ if $S, s \models_F \phi_1$ and $S, s \models_F \phi_2$;
- $S, s \models_T \sim\phi$ if $S, s \models_F \phi$;
  $S, s \models_F \sim\phi$ if $S, s \models_T \phi$;

Note that, for any $s \in S$ and for any formula ϕ, $1 \in v(s)(\phi)$ iff $S, s \models_T \phi$ and $0 \in v(s)(\phi)$ iff $S, s \models_F \phi$.

We may group all the states according to the following equivalence relation:

$$s_1 \approx_\Phi s_2 \text{ if, for all formulas } \phi \in \Phi,\ S, s_1 \models_X \phi \Leftrightarrow S, s_2 \models_X \phi \text{ for } X \in \{T, F\}.$$

$[s]_\Phi$ denotes the equivalence class including s. A partial relation $\leq_\Phi$ on $S_\Phi = \{[s]_\Phi : s \in S\}$ can be defined as follows:

$$[s_1]_\Phi \leq_\Phi [s_2]_\Phi \text{ if, for any formula } \phi,$$
$$S, s_1 \models_T \phi \text{ implies } S, s_2 \models_T \phi \text{ and } S, s_1 \models_F \phi \text{ implies } S, s_2 \models_F \phi.$$

Moreover, a corresponding valuation v_Φ is defined on S_Φ as follows. For any $s \in S$ and propositional letter p,

$$v_\Phi([s]_\Phi)(p) = v(s)(p).$$

A satisfaction relation between states in S_Φ and formulas can be defined inductively as usual.

Proposition 3.6 *$\langle S, v\rangle$ and $\langle S_\Phi, \leq_\Phi, v_\Phi\rangle$ are equivalent in the sense that, for any $s \in S$ and any $\phi \in \Phi$,*

- $S, s \models_T \phi$ *if and only if* $S_\Phi, [s]_\Phi \models_T \phi$;
- $S, s \models_F \phi$ *if and only if* $S_\Phi, [s]_\Phi \models_F \phi$.

Proof We prove this by induction on the complexity of ϕ. Here we only show the non-trivial case when $\phi = \sim\phi'$.

$$S_\Phi, [s]_\Phi \models_T \sim\phi' \Leftrightarrow S_\Phi, [s]_\Phi \models_F \phi' \Leftrightarrow S, s \models_F \phi' \Leftrightarrow S, s \models_T \sim\phi'$$

Note that the second equivalence is based on induction hypothesis and the other two on semantical clause for $\sim$. □

Let $[\![\phi]\!]_T$ denote the set of all states where ϕ is "told true" $\{[s]_\Phi \in S_\Phi : S_\Phi, [s]_\Phi \models_T \phi\}$ and $[\![\phi]\!]_F = \{[s]_\Phi \in S_\Phi : S_\Phi, [s]_\Phi \models_F \phi\}$, for all formulas ϕ. It is easy to check that both of them are filters in $(S_\Phi, \leq_\Phi)$. Denote

$$B_\Phi = \{([\![\phi]\!]_T, [\![\phi]\!]_F) : \phi \in \Phi\}.$$

We can define two partial orders on B_Φ.

- $([\![\phi_1]\!]_T, [\![\phi_1]\!]_F) \leq^1_\Phi ([\![\phi_2]\!]_T, [\![\phi_2]\!]_F)$ if $[\![\phi_1]\!]_T \subseteq [\![\phi_2]\!]_T$ and $[\![\phi_1]\!]_F \supseteq [\![\phi_2]\!]_F$;
- $([\![\phi_1]\!]_T, [\![\phi_1]\!]_F) \leq^2_\Phi ([\![\phi_2]\!]_T, [\![\phi_2]\!]_F)$ if $[\![\phi_1]\!]_T \subseteq [\![\phi_2]\!]_T$ and $[\![\phi_1]\!]_F \subseteq [\![\phi_2]\!]_F$;

Remark 3.1 The four-valued semantics is equivalent to Dunn's semantics of proposition surrogates from Dunn (1966). Moreover, the four valued logic is employed to decouple the bipolar information in the semantics. And the interpretation of each formula ϕ is decomposed into two parts: the part for the epistemic truth and the other part for the epistemic falsity, as explained in Dunn (1976a, 1986). For the partial ordering $\leq^1_\Phi$, the agent has more reasons to consider ϕ_2 as true than ϕ_1 and more reasons to consider ϕ_1 as false than ϕ_2. In other words, ϕ_2 is *considered* at least as true as and at most as false as ϕ_1. The agent is more confident in considering that *overall* ϕ_2 is at least as true as ϕ_1. This is the reason why $\leq^1_\Phi$ is also called the truth ordering. For the other ordering $\leq^2_\Phi$, the agent has both more reasons to consider ϕ_2 as true than ϕ_1 and more reasons to consider ϕ_2 as false than ϕ_1. So the agent is more *informative* (in reasons) about ϕ_2 than about ϕ_1. This is the reason why $\leq^2_\Phi$ is called the information or knowledge ordering.

It is easy to check that the associated structure $\mathcal{B}_\Phi := \langle B_\Phi, \leq^1_\Phi, \leq^2_\Phi \rangle$ is a distributive bilattice with the following negation:

$$\sim_\Phi([\![\phi]\!]_T, [\![\phi]\!]_F) = ([\![\phi]\!]_F, [\![\phi]\!]_T) = ([\![\sim\phi]\!]_T, [\![\sim\phi]\!]_F).$$

According to a dual form of Theorem 3.5, we have that

$$\mathsf{M}_2(\mathcal{B}_\Phi) \cong S_\Phi \oplus_\perp S_\Phi,$$

where $\mathsf{M}_2(\mathcal{B}_\Phi)$ is the set of $\leq^2_\Phi$-meet-irreducibles in $\mathcal{B}_\Phi$ and S_Φ is the state space with the partial ordering $\leq_\Phi$.

So each four-valued model can be regarded as a poset with a valuation into *FOUR*. In order to simulate Kripke semantics for intuitionistic logic, we choose to define belief structures in a more abstract form.

Definition 3.7 A *Belnap–Dunn structure* is a tuple $S = \langle S, \leq, v\rangle$ where

- $\langle S, \leq\rangle$ is a poset;
- v is a valuation into *FOUR* on the set of propositional letters satisfying the following *persistency* condition: for any $s_1, s_2 \in S$ and propositional letter p,

$$\text{if } s_1 \leq s_2, \text{ then } \begin{array}{l} 1 \in v(s_1)(p) \text{ implies } 1 \in v(s_2)(p), \quad \text{and} \\ 0 \in v(s_1)(p) \text{ implies } 0 \in v(s_2)(p). \end{array}$$

It is a *Boolean structure* if $\leq$ is the identity relation and v is a valuation into **2**.

Two support relations $\models_T$ and $\models_F$ between states and formulas can be defined exactly as in the above four-valued-structures. Actually the persistency condition for the valuation is satisfied by all formulas, as shown by the following lemma.

Lemma 3.8 *Let $\langle S, \leq, v\rangle$ be a Belnap–Dunn structure. If $s_1, s_2 \in S$ and $s_1 \leq s_2$, then, for any formula ϕ,*

1. $S, s_1 \models_T \phi$ *implies* $S, s_2 \models_T \phi$*;*
2. $S, s_1 \models_F \phi$ *implies* $S, s_2 \models_F \phi$*.*

Proof Let $\langle S, \leq, v\rangle$ be a Belnap–Dunn structure and $s_1, s_2 \in S$ and $s_1 \leq s_2$. We prove by induction on the complexity of ϕ. Here we only prove the case that $\phi = {\sim}\phi'$. The proof of the other cases is straightforward.

We reason as follows:

$$\begin{aligned} S, s_1 \models_T \phi &\Rightarrow S, s_1 \models_F \phi' \\ &\Rightarrow S, s_2 \models_F \phi' \quad \text{(Induction hypothesis)} \\ &\Rightarrow S, s_2 \models_T {\sim}\phi' \end{aligned}$$

and

$$\begin{aligned} S, s_1 \models_F \phi &\Rightarrow S, s_1 \models_T \phi' \\ &\Rightarrow S, s_1 \models_T \phi' \quad \text{(Induction hypothesis)} \\ &\Rightarrow S, s_1 \models_F {\sim}\phi' \end{aligned}$$

□

Remark 3.2 The persistency condition in a Belnap–Dunn structure is quite similar to that in Kripke's semantics for intuitionistic logic (Kripke 1965; van Dalen 2004) and in Dunn's semantics for **R**-mingle (Dunn 1976b). Each state s can be seen as a pair of knowledge bases. The set of formulas that are at least true at s is the knowledge base of true facts and the set of formulas which are at least false at s constitutes the knowledge base for the false facts. This pair of knowledge evolves in the course of time. Both the knowledge base of true facts and that of false facts expand at every later stage. They are considered to be independent of each other and both take the *open-world assumption*, which also explains the incompleteness in the information in another way. So the essential difference from that in intuitionistic logic is that the persistency condition here concerns not only the knowledge of true facts but also that of false facts.

Definition 3.9 A formula ψ is a *logical consequence* of a formula ϕ (ϕ *logically implies* ψ) with respect to the class $\mathcal{B}$ of Belnap–Dunn structures (denoted $\phi \models^{\mathcal{B}} \psi$) if, for any Belnap–Dunn structure $S = \langle S, \leq, v\rangle$ and any $s \in S$, $S, s \models_T \phi$ implies $S, s \models_T \psi$ and $S, s \models_F \psi$ implies $S, s \models_F \phi$.

Now we investigate a deductive system for this logical implication with respect to the class of Belnap–Dunn structures. The following is the deductive system $\mathbf{R}_{fde}$ which is the well-known first-degree-entailment fragment of the relevance logic **R** (Anderson and Belnap 1975; Dunn 1986). Without further notice, $\vdash$ denotes $\vdash_{\mathbf{R}_{fde}}$ and $\phi \dashv\vdash \psi$ is short for both $\phi \vdash \psi$ and $\psi \vdash \phi$.

Axioms:

$\phi \vdash \phi$	Self-implication
$\phi \wedge \psi \vdash \phi, \quad \phi \wedge \psi \vdash \psi$	($\wedge$-elimination)
$\phi \vdash \phi \vee \psi, \quad \psi \vdash \phi \vee \psi$	($\vee$-introduction)
$\phi \wedge (\psi \vee \gamma) \vdash (\phi \vee \psi) \wedge (\psi \vee \gamma)$	(Distribution)
$\phi \dashv\vdash {\sim}{\sim}\phi$	(Double Negation)
${\sim}(\phi \wedge \psi) \dashv\vdash {\sim}\phi \vee {\sim}\psi \quad {\sim}(\phi \vee \psi) \dashv\vdash {\sim}\phi \wedge {\sim}\psi$	(de Morgan laws)

Rules:

- From $\phi \vdash \psi$ and $\psi \vdash \gamma$, infer $\phi \vdash \gamma$. (Transitivity)
- From $\phi \vdash \psi$ and $\phi \vdash \gamma$, infer $\phi \vdash \psi \wedge \gamma$. ($\wedge$-introduction)
- From $\phi \vdash \gamma$ and $\psi \vdash \gamma$, infer $\phi \vee \psi \vdash \gamma$. ($\vee$-elimination)
- From $\phi \vdash \psi$, infer ${\sim}\psi \vdash {\sim}\phi$. (Contraposition)

Actually the logical implication relation in the class of Belnap–Dunn structures coincides with the above consequence relation $\vdash_{\mathbf{R}_{fde}}$.

Theorem 3.10 (Theorem 7 in Dunn 2000)
For any two formulas ϕ and ψ in Φ,

$$\phi \vdash \psi \quad \textit{iff} \quad \phi \models^{\mathcal{B}} \psi.$$

3.2 *Reasoning About de Morgan Lattices*

De Morgan lattices are important for the study of the mathematical aspects of fuzzy logic (Zadeh 1988). The standard fuzzy algebra $F = \langle [0, 1], \max(x, y), \min(x, y), 0, 1, 1 - x\rangle$ is an example of a de Morgan lattice. Moreover, de Morgan monoids are an algebraic semantics for the relevance logic $\mathbf{R}^t$, as was shown by Dunn in his (1966) and in Anderson and Belnap (1975, Sect. 28.2). In this part, we investigate belief functions on de Morgan lattices which covers those for fuzzy events (Zadeh 1979; Smets 1981; Yen 1990). It is interesting to note that the first degree entailment $\mathbf{R}_{fde}$ also provides a calculus for reasoning about de Morgan lattices. In the following, we will give a presentation of the semantics in terms of de Morgan lattices which is in parallel to that for bilattices. For simplicity, we will not repeat those proofs which are similar to those in last part about bilattices but simply present the main ideas. Note that the $\wedge_1$-$\vee_1$-$\neg$-reduct of a distributive bilattice $\mathbf{B} = \langle B, \wedge_1, \vee_1, \wedge_1', \vee_2', \neg\rangle$ is a de Morgan lattice.

We need the following duality theorem for finite de Morgan lattices which is based on Białynicki-Birula and Rasiowa (1957), Dunn (1986), Urquhart (1979) and Priestley (1970).

Theorem 3.11 *Any finite de Morgan lattice D can be represented as the lattice $J(P_D)$ of order ideals in the sub-poset P_D of join-irreducibles with an order-reversing involution g. There is a one-to-one correspondence between de Morgan lattices and posets with order-reversing involutions.*

The interested reader may find a detailed proof of this theorem in the Appendix, which is of independent interest. An $\mathbf{R}_{fde}$*-structure* (Routley and Routley 1972; Dunn 1986, 1966) is a tuple $S = \langle S, *, v\rangle$, where

- S is a non-empty set of states, which are like possible worlds except that they are not required to be either consistent or complete;
- $*$ is an involution on S and is usually called Routley star;
- v is a valuation which is a function at each state such that $v(s)(p) \in \{\mathit{true}, \mathit{false}\}$ for all propositional letters p.

A satisfaction relation between states and formulas is defined inductively as:

- $S, s \models p \;$ if $\; v(s)(p) = \mathit{true}$;
- $S, s \models \phi_1 \wedge \phi_2 \;$ if $\; S, s \models \phi_1$ and $S, s \models \phi_2$;
- $S, s \models \phi_1 \vee \phi_2 \;$ if $\; S, s \models \phi_1$ or $S, s \models \phi_2$;
- $S, s \models {\sim}\phi \;$ if $\; S, s^* \not\models \phi$.

The truth value of the negation of a formula ϕ at a state s is determined by that of ϕ at its adjunct state s^*. We may group all the states according to the following equivalence relation:

$$s_1 \approx_\Phi s_2 \text{ if, for all formulas } \phi \in \Phi,\; S, s_1 \models \phi \Leftrightarrow S, s_2 \models \phi.$$

$[s]_\Phi$ denotes the equivalence class including s. A partial relation $\leq_\Phi$ on $S_\Phi = \{[s]_\Phi : s \in S\}$ can be defined as follows:

$$[s_1]_\Phi \leq_\Phi [s_2]_\Phi \text{ if, for any formula } \phi,\; S, s_1 \models \phi \text{ implies } S, s_2 \models \phi.$$

Further we define the unary operation g_Φ:

$$g_\Phi([s]_\Phi) = [s^*]_\Phi \text{ for each } s \in S.$$

It is easy to check that g_Φ is well-defined and is an order-reversing involution on S_Φ. Moreover, a corresponding valuation v_Φ is defined on S_Φ as follows: for any $s \in S$ and propositional letter p,

$$v_\Phi([s]_\Phi)(p) = \mathit{true} \text{ iff } v(s)(p) = \mathit{true}.$$

A satisfaction relation between states in S_Φ and formulas can be defined inductively as usual.

Proposition 3.12 *$\langle S, *, v\rangle$ and $\langle S_\Phi, \leq_\Phi, g_\Phi, v_\Phi\rangle$ are equivalent in the sense that, for any $s \in S$ and any $\phi \in \Phi$,*

$$S, s \models \phi \textit{ if and only if } S_\Phi, [s]_\Phi \models \phi.$$

So each $\mathbf{R}_{fde}$-structure can be regarded as a poset with an order-reversing involution. In order to reason about belief functions on de Morgan lattices, we choose to define structures for $\mathbf{R}_{fde}$ by simulating Kripke semantics for intuitionistic logic.

Definition 3.13 A *Routley structure* is a tuple $S = \langle S, \leq, g, v\rangle$, where

- $\langle S, \leq, g\rangle$ is a poset with an order-reversing involution g;
- v is a valuation on the set of propositional letters satisfying the following *persistency* condition: for any $s_1, s_2 \in S$ and propositional letter p,

$$\text{if } s_1 \leq s_2 \text{ and } v(s_1)(p) = \mathit{true}, \text{ then } v(s_2)(p) = \mathit{true}.$$

S is a Boolean structure when $\leq$ is the identity relation and g is the identity function.

From the persistency condition, we may immediately derive a "reverse" persistency condition as follows: for any $s_1, s_2 \in S$ and propositional letter p,

$$\text{if } s_1 \leq s_2 \text{ and } v(g(s_2))(p) = \mathit{true}, \text{ then } v(g(s_1))(p) = \mathit{true}.$$

A satisfaction relation between states and formulas can be defined exactly as in the above $\mathbf{R}_{fde}$-structure. Actually the persistency condition for the valuation is satisfied by all formulas, as shown in the following proposition.

Lemma 3.14 *Let $\langle S, \leq, g, v\rangle$ be a Routley structure. If $s_1, s_2 \in S$ and $s_1 \leq s_2$, then for any formula ϕ,*

1. *$S, s_1 \models \phi$ implies $S, s_2 \models \phi$;*
2. *$S, g(s_2) \models \phi$ implies $S, g(s_1) \models \phi$.*

Remark 3.3 The persistency condition in a Routley structure is quite similar to that in the Kripke semantics for intuitionistic logic (van Dalen 2004). Each pair $(s, g(s))$ of a state s and its adjunct can be seen as a pair of knowledge bases. s is the knowledge base *consisting of* true facts and $g(s)$ is the knowledge base *for* the false facts. Here we take the *closed-world assumption* (Reiter 1978) for $g(s)$ in the sense that, if a proposition is not implied in $g(s)$, then the negation of this proposition is implied at s. This pair of knowledge evolves in the course of time. The knowledge base of true facts expands at every later stage while the knowledge base for false facts decreases. So the essential difference from those in Belnap–Dunn structures is that the knowledge of true facts and that of false facts are dual to each other, rather than independent of each other as in Belnap–Dunn structures.

If the second (reverse) persistency condition is replaced by a new *adjunct but independent* valuation v^* which is defined as $v^*(s) = v(g(s))$, then we may also define a Routley structure as a poset $\langle S, \leq, g \rangle$ with an order-reversing involution g and two "independent" but adjunct-to-each-other valuations v and v^* satisfying the following two persistency conditions: for all propositional letters p,

- if $s_1 \leq s_2$, then $v(s_1)(p) = true$ implies $v(s_2)(p) = true$;
- if $s_1 \leq s_2$, then $v^*(s_2)(p) = true$ implies $v^*(s_1)(p) = true$.

The new valuation v^* is for false facts. So the semantics with this new type of Routley structures is the same as the above except that for the negated formulas:

$$(S, v), s \models {\sim}\phi \quad \text{if} \quad (S, v^*), s \not\models \phi.$$

The adjunct valuation is used to decouple the semantics to interpret negation. Since there is a straightforward interpretation between this semantics and the above one, they are equivalent. So each possibility in the epistemic frame for belief functions on Routley structures consists of a *pair* of valuations. It would be interesting to compare this pair with the valuation for the four-valued logic.

The following proposition tells us that the notion of satisfiability in both Routley semantics and Belnap–Dunn semantics are equivalent (Proposition 9.1 in Fagin et al. 1995).

Proposition 3.15 *For each Routley structure $M = \langle S, g, v \rangle$ and state s in M, there is a Belnap–Dunn structure $M' = \langle S', v' \rangle$ and state $s' \in S'$ such that for each formula ϕ,*

$$M, s \models \phi \quad \textit{iff} \quad M', s' \models_T \phi \tag{7}$$
$$M, s \models {\sim}\phi \quad \textit{iff} \quad M', s' \models_F \phi \tag{8}$$

Conversely, for any Belnap–Dunn structure $M' = \langle S', v' \rangle$ and state $s' \in S'$, there is a Routley structure $M = \langle S, g, v \rangle$ and state s in M such that the above equivalences (7) and (8) hold.

Proof For each Routley structure $M = \langle S, g, v \rangle$ and state s in M, we define a Belnap–Dunn structure as follows:

- $S' = S$;
- $1 \in v'(s)(p)$ iff $v(s)(p) = true$;
- $0 \in v'(s)(p)$ iff $v(s^*)(p) = false$.

It is easy to check that the following equivalences hold: for all $\phi \in \Phi$,

$$M, s \models \phi \text{ iff } M', s \models_T \phi \tag{9}$$

$$M, s \models {\sim}\phi \text{ iff } M', s \models_F \phi \tag{10}$$

Conversely, for any Belnap–Dunn structure $M' = \langle S', v' \rangle$, we define a Routley structure as follows:

- $S = S' \cup S'^{*}$ where $S'^{*} = \{ s^{*} : s \in S' \}$ and for each $s \in S'$, s^{*} is a new state;
- For each $s \in S$,
  - $v(s)(p) = \mathit{true}$ iff $1 \in v'(s)(p)$;
  - $v(s^{*})(p) = \mathit{false}$ iff $0 \in v'(s)(p)$.

It is easy to check that the equivalences (7) and (8) hold. □

Remark 3.4 Define an equivalence relation $\asymp$ on Φ,

$$\phi_1 \asymp \phi_2 \text{ iff } \phi_1 \vdash \phi_2 \text{ and } \phi_2 \vdash \phi_1.$$

Let $\Phi/_{\asymp}$ denote the set of $\asymp$-equivalence classes $[\phi]_{\asymp}$. Now we define the operations on this set as follows:

- $[\phi_1]_{\asymp} \wedge_{\asymp} [\phi_2]_{\asymp} = [\phi_1 \wedge \phi_2]_{\asymp}$;
- $[\phi_1]_{\asymp} \vee_{\asymp} [\phi_2]_{\asymp} = [\phi_1 \vee \phi_2]_{\asymp}$;
- ${\sim}_{\asymp}[\phi]_{\asymp} = [{\sim}\phi]_{\asymp}$.

It is easy to check that these operations are well-defined. Such a defined algebra $\langle \Phi/_{\asymp}, \wedge_{\asymp}, \vee_{\asymp}, {\sim}_{\asymp} \rangle$ is the Lindenbaum algebra on $\Phi/_{\asymp}$ and is actually a de Morgan lattice.

With respect to Φ, we define the *canonical* Routley structure as follows.

- S_Φ is the set of all prime filters in $\Phi/_{\asymp}$;
- $g_\Phi : S_\Phi \to S_\Phi$ is defined: for each $F \in S_\Phi$, $g_\Phi(F) = \Phi/_{\asymp} \setminus {\sim}_{\asymp}F$ where ${\sim}_{\asymp}F = \{ {\sim}_{\asymp}[\phi]_{\asymp} : [\phi]_{\asymp} \in F \}$, which is a prime filer;
- $v_\Phi(F)(p) = \mathit{true}$ if $[p]_{\asymp} \in F$ for any $F \in S_\Phi$.

It is easy to check that $\langle S_\Phi, \leq_\Phi, g_\Phi \rangle$ is a poset with the order-reversing involution g_Φ where $\leq_\Phi$ is the subset relation. If Φ is finite, according to the dual form of the representation theorem for finite de Morgan lattices, the Lindenbaum algebra $\langle \Phi/_{\asymp}, \wedge_{\asymp}, \vee_{\asymp}, {\sim}_{\asymp} \rangle$ is isomorphic to the concrete lattice of filters in the poset $\langle S_\Phi, \leq_\Phi, g_\Phi \rangle$ which underlies the canonical Routley structure.

Definition 3.16 A formula ϕ *logically implies* a formula ψ with respect to the class of Routley structures (denoted as $\phi \models^R \psi$) if, for any Routley structure $S = \langle S, g, v \rangle$, $S, s \models \phi$ implies $S, s \models \psi$.

The difference of this definition from that for Belnap–Dunn structures (Definition 3.9) is that we don't need to consider the "negative side." Actually the logical implication relation in the class of Routley structures coincides with the above consequence relation $\vdash_{\mathbf{R}_{fde}}$.

Theorem 3.17 (Theorem 7 in Dunn 2000)
For any two formulas ϕ and ψ in Φ,

$$\phi \vdash \psi \quad \textit{iff} \quad \phi \models^R \psi.$$

Theorem 3.18 *The complexity of deciding logical implication with respect to the class of Routley structures (Belnap–Dunn structures) is co-NP-complete.*

Proof The proof can be easily adapted from that of the similar problem in Sect. 8 in Fagin et al. (1995) (see also Urquhart 1990; Levesque 1984). □

3.3 Reasoning About Epistemic Structures

In order to accommodate internal epistemic considerations, we expand the syntax of the four-valued logic to include the $S5$-operator K and denote the expanded syntax as Φ^K.

Definition 3.19 An *epistemic Belnap–Dunn structure* is a tuple $S = \langle (S, \leq, v), K^+, K^- \rangle$, where

- $(S, \leq, v)$ is a Belnap–Dunn structure;
- K^+ and K^- are two equivalence relations on S such that $(\leq \circ K^+) \subseteq (K^+ \circ \leq)$ and $(\leq \circ K^-) \supseteq (K^- \circ \leq)$ where $\circ$ is the composition of relations.

In addition to those defined for the Belnap–Dunn structure, the two support relations between states and formulas of the form $K\phi$ are defined as follows:

- $S, s \models_T K\phi$ if $S, t \models_T \phi$ for all t such that sK^+t;
- $S, s \models_F K\phi$ if $S, t \models_F \phi$ for some t such that sK^-t.

Lemma 3.20 *Let $S = \langle (S, \leq, v), K^+, K^- \rangle$ be an epistemic Belnap–Dunn structure. If $s_1, s_2 \in S$ and $s_1 \leq s_2$, then, for any formula $\phi \in \Phi^K$,*

1. $S, s_1 \models_T \phi$ *implies* $S, s_2 \models_T \phi$;
2. $S, s_1 \models_F \phi$ *implies* $S, s_2 \models_F \phi$.

Proof We only need to prove the case when $\phi = K\phi'$. We reason as follows. Assume that $S, s_1 \models_T K\phi'$. That is to say, $S, t \models_T \phi'$ for all t such that sK^+t. We need to show that, for any t_2 such that $s_2K^+t_2$, $S, t_2 \models_T \phi'$. It is easy to see that $(s_1, t_2) \in (\leq \circ K^+)$. Since S satisfies the constraint that $(\leq \circ K^+) \subseteq (K^+ \circ \leq)$, $(s_1, t_2) \in (K^+ \circ \leq)$. In other words, there is a world $s_1' \in S$ such that $s_1K^+s_1'$ and $s_1' \leq t_2$. According to the assumption that $S, s_1 \models_T K\phi'$, $S, s_1' \models_T \phi'$. It follows from hypothetical assumption that $S, t_2 \models_T \phi'$. So we have that $S, s_2 \models_T K\phi'$.

Assume that $s_1 \leq s_2$ and $S, s_1 \models_F K\phi'$. It follows that there is a state t_1 such that $s_1K^-t_1$ and $S, t_1 \models_F \phi'$. So $(t_1, s_2) \in (K^- \circ \leq)$. Since S is an epistemic Belnap–Dunn structure, $(K^- \circ \leq) \subseteq (\leq \circ K^-)$ and hence $(t_1, s_2) \in (\leq \circ K^-)$. This implies that there is a state t'_1 such that $t_1 \leq t'_1$ and $t'_1 K^- s_2$. According to hypothetical induction, $S, t'_1 \models_F \phi'$. It follows from the fact that $t'_1 K^- s_2$ that $S, s_2 \models_F K\phi'$. □

Definition 3.21 *An epistemic Routley structure* is a tuple $S = \langle (S, \leq, g, v), K \rangle$, where

- $(S, \leq, g, v)$ is a Routley structure;
- K is an equivalence relation on S such that $(\leq \circ K) \subseteq (K \circ \leq)$.

For formulas of the form $K\phi$, the satisfaction relation is defined as follows.

$$S, s \models K\phi \quad \text{iff} \quad S, t \models \phi, \quad \text{for all } t \in S \text{ such that } sKt.$$

It is easy to check that the following lemma holds.

Lemma 3.22 *Let $\langle (S, \leq, g, v), K \rangle$ be an epistemic Routley structure. If $s_1 \leq s_2$, then, for any formula $\phi \in \Phi^K$,*

1. $S, s_1 \models \phi$ *implies* $S, s_2 \models \phi$;
2. $S, g(s_2) \models \phi$ *implies* $S, g(s_1) \models \phi$.

The above two epistemic structures are equivalent in the sense of the following proposition.

Proposition 3.23 *For any epistemic Routley structure S and world s in S, there is an epistemic Belnap–Dunn structure S' and world s' in S' such that for any formula $\phi \in \Phi^K$,*

$$S, s \models \phi \; \textit{iff} \; S', s' \models_T \phi \tag{11}$$

$$S, s \models {\sim}\phi \; \textit{iff} \; S', s' \models_F \phi \tag{12}$$

Conversely, for each epistemic Belnap–Dunn structure S' and world $s' \in S'$, there is an epistemic Routley structure S and world s such that (11) and (12) hold for each formula $\phi \in \Phi^K$.

Proof The interested reader may refer to the proof of a similar proposition Proposition 9.1. in Fagin et al. (1995). □

Now we consider probabilistic reasoning over an epistemic Routley structure $S = \langle (S, \leq, g, v), K \rangle$. Since K is an equivalence relation on S, it induces a partition Π on S and $\{ \Pi(s) \colon s \in S \}$ forms a basis for an algebra $\mathcal{A}$. It is easy to check that $\mathcal{A}$ is simply the set of subsets of the form $KE = \{ s \in S \colon \Pi(s) \subseteq E \}$. Those subsets are called *epistemic sets* of S. Probabilistic reasoning is performed on $\mathcal{A}$ according to a given probability function $Pr \colon \mathcal{A} \to [0, 1]$. A *Dempster–Shafer* structure (DS structure for short) M on the epistemic Routley structure is the tuple $M = \langle (S, \leq, g, v), K, Pr \rangle$. Let $F(S) = \{ I \colon I \text{ is a filter in } S \}$ denote the set of all filters in S. Note that $F(S)$ is a de Morgan lattice. Evidential reasoning is performed on the DS structure M through the following defined function: for any $E \in F(S)$,

$$Bel(E) := Pr(K(E)).$$

Note that such defined *Bel* is defined on $F(S)$ and is indeed a belief function on $F(S)$. *Bel* is called a *belief function on the epistemic Routely structure S*. If we want to make belief functions primary, we "hide" the epistemic part K and simply write the above DS structure M as $B = \langle S, \leq, g, v, Bel \rangle$.

4 Reasoning About Belief Functions for First Degree Entailments

In this section, we provide a sound and complete deductive system for reasoning about belief functions for first degree entailments and show that the satisfiability problem of a belief formula with respect to the corresponding class of Dempster–Shafer structures is *NP*-complete.

In this part, we adapt the deductive machinery from Fagin et al. (1990), Fagin and Halpern (1991) to provide a sound and complete axiomatization for reasoning about belief functions over Routley structures (Belnap–Dunn structures).

Definition 4.1 For the above given set Φ of formulas, a *term* is an expression of the form $a_1 bel(\phi_1) + a_2 bel(\phi_2) + \cdots + a_k bel(\phi_k)$, where $a_1, a_2, \ldots, a_k$ are integers, *bel* is the belief function symbol and $\phi_1, \phi_2, \ldots, \phi_k$ are formulas in Φ. *A basic belief formula* is one of the form $t \geq b$, where t is a term and b is an integer. A *belief formula* is a Boolean combination of basic belief formulas. We can always allow rational numbers in our formulas as abbreviations for the formula that would be obtained by clearing the dominator. Other derived relations such as $=$, $\leq$, $<$ and $>$ can be defined as usual.

Definition 4.2 Given a *DS*-structure $B = \langle S, \leq, g, v, Bel \rangle$ on a Routley structure $S := \langle S, \leq, g, v \rangle$ and a basic belief formula $f := a_1 bel(\phi_1) + a_2 bel(\phi_2) + \cdots + a_k bel(\phi_k) \geq b$, B *satisfies* f (denoted as $B \models f$) if

$$a_1 Bel([\![\phi_1]\!]_S) + a_2 Bel([\![\phi_2]\!]_S) + \cdots + a_k Bel([\![\phi_k]\!]_S) \geq b.$$

We then extend the above $\models$ in the obvious way to all belief formulas. Let $\mathcal{B}$ be a class of Dempster–Shafer structures. A belief formula f' is *satisfiable with respect to* $\mathcal{B}$ if it is satisfied in some $B \in \mathcal{B}$. It is *valid with respect to* $\mathcal{B}$ if $B \models f$, for all $B \in \mathcal{B}$.

The axiomatization $\mathcal{B}_{fde}$ of reasoning about belief functions for first degree entailments consists of three parts: the first degree entailments, reasoning about linear inequalities and reasoning about belief functions.

1. First degree entailments

- The complete system $\vdash_{\mathbf{R}_{fde}}$ of first degree entailment is provided in last section.

2. Reasoning about linear inequalities

(a) $a_1 bel(\phi_1) + a_2 bel(\phi_2) + \cdots + a_k bel(\phi_k) \geq b$ iff $a_1 bel(\phi_1) + a_2 bel(\phi_2) + \cdots + a_k bel(\phi_k) + a bel(\phi_{k+1}) \geq b$;
(b) $a_1 bel(\phi_1) + a_2 bel(\phi_2) + \cdots + a_k bel(\phi_k) \geq b$ iff $a_{j_1} bel(\phi_1) + a_{j_2} bel(\phi_2) + \cdots + a_{j_k} bel(\phi_k) \geq b$, where $j_1, j_2, \ldots, j_k$ is a permutation of $1, 2, \ldots, k$;
(c) $a_1 bel(\phi_1) + a_2 bel(\phi_2) + \cdots + a_k bel(\phi_k) \geq b$ iff $ca_1 bel(\phi_1) + ca_2 bel(\phi_2) + \cdots + ca_k bel(\phi_k) \geq cb$, where $c > 0$;
(d) $(a_1 + a_1')bel(\phi_1) + (a_2 + a_2')bel(\phi_2) + \cdots + (a_k + a_k')bel(\phi_k) \geq b + b'$ if $a_1 bel(\phi_1) + a_2 bel(\phi_2) + \cdots + a_k bel(\phi_k) \geq b$ and $a_1' bel(\phi_1) + a_2' bel(\phi_2) + \cdots + a_k' bel(\phi_k) \geq b'$;
(e) either $t \geq b$ or $t \leq b$, where t is a term;
(f) $t \geq b$ implies $t > b'$, where t is a term and $b' < b$.

Let AX_{Iq} denote this deductive reasoning system about linear inequalities, which is shown to be complete (Fagin et al. 1990).

3. Reasoning about belief functions

(a) $bel(\phi) \geq 0$, for all formulas $\phi \in \Phi$;
(b) $bel(\top) = 1$;
(c) $bel(\bot) = 0$;
(d) $bel(\phi_1 \vee \phi_2 \vee \cdots \vee \phi_n) \geq \sum_{I \subseteq \{1,2,\ldots,n\}} (-1)^{|I|+1} bel(\wedge_{i \in I} \phi_i)$;
(e) $bel(\phi) \leq bel(\psi)$ if $\phi \vdash_{\mathbf{R}_{fde}} \psi$.

Note that principle 3(e) is the connection of reasoning about belief functions to first degree entailments.

Theorem 4.3 *$\mathcal{B}_{fde}$ is a sound and complete axiomatization of belief formulas with respect to the class of DS-structures.*

Theorem 4.4 *The time complexity of deciding whether a belief formula is satisfiable with respect to the class of DS-structures is NP-complete.*

Proof One may refer to Zhou (2013) for the detailed proof. □

We may define logical implication for belief formulas as usual. The above theorem tells us that the complexity of the logical implication problem with respect to the class of *DS*-structures on Routley structures is the same as that with respect to the class of Routley structures (Theorem 3.18) and hence is not affected by the expansion of the propositional language with belief functions.

5 Related Work

This kind of integration of uncertainty measures such as belief functions and logics for knowledge representation is an important approach in reasoning under uncertain and imperfect knowledge in artificial intelligence (Parsons 1996). Logics including many non-classical logics play a central role in the task of knowledge representation in artificial intelligence (Nilsson 1991), and each of these logics was intended for some particular focus. On the other hand, uncertainty measures are usually employed to deal with uncertainty in information (Halpern 2005). However, non-classical logics are not expressive enough to capture uncertainty in a *gradual* way, and uncertainty measures such as belief functions are not enough to handling imperfect information. This is another motivation to combine uncertainty measures with non-classical logics in addition to that mentioned at the beginning of Sect. 3. Besnard and Lang applied possibility theory to non-classical logics especially paraconsistent logics and showed how to reason under uncertain and inconsistent information (Besnard and Lang 1994). Saffiotti proposed a formal framework to integrate logics for knowledge including first order logic and belief functions (Saffiotti 1990a, b, 1992).

However, none of these papers has touched any issue about the mathematical foundation and computational complexity behind the theory of belief functions, just as we have done in this paper. Here, we provided a sound and complete deductive system for reasoning about belief functions for a simple epistemic logic the first-degree-entailment fragment of relevance logic **R** through different duality theorems between algebraic semantics and logic. This axiomatization can be used to show how to deduce one belief of some events from beliefs of others. Moreover, we have given the complexity result of the satisfiability problem of belief formulas in this kind of non-classical settings. The deductive approach for belief functions for first degree entailments can be applied to other non-classical formalisms $\mathbb{L}$ that assume a setting of distributive lattices. The axiomatization $\mathcal{B}_{\mathbb{L}}$ for reasoning about belief functions on $\mathbb{L}$ is simply obtained from the axiomatization $\mathcal{B}_{fde}$ by replacing Part 1 of first degree entailments by $\mathbb{L}$ and the implication $\phi \vdash_{\mathbf{R}_{fde}} \psi$ in principle 3(e) in Part 3 by $\phi \vdash_{\mathbb{L}} \psi$. In particular, our deductive approach also covers the formalism developed by Besnard and Lang (1994).

Acknowledgments I would like to thank the two referees for their constructive comments to help improve this paper. Also I want to thank Katalin Bimbó for her patience with my asking for deadline extensions a couple of times.

Appendix A: Duality Theorem of de Morgan Lattices

In this part, we show that any finite de Morgan lattice can be represented as the concrete lattice of order ideals in some poset with an order-reversing involution and there is a one-to-one correspondence between de Morgan lattices and posets with order-reversing involutions. The following propositions are based on similar results in Białynicki-Birula and Rasiowa (1957), Dunn (1986), Urquhart (1979), Priestley (1970). Białynicki-Birula and Rasiowa (1957) and Dunn (1986) did not give a whole duality theory for de Morgan lattices; rather they proved representation theorems. Urquhart (1979) provided a duality theory for distributive lattices with a dual homomorphism operator instead for de Morgan lattices, which are distributive lattices with a dual homomorphism operator that is additionally an *involution*. Priestley (1970) presented a duality theory for distributive lattices by means of ordered Stone spaces, which is quite different from the form that we need to show the main theorems in this paper. So, according to our knowledge, our presentation of the duality theorem for finite de Morgan lattices here, which combines different techniques from the above mentioned papers, is *the first one* to represent de Morgan lattices in the same way as in Birkhoff (1967) and Stanley (1997) for distributive lattices. In this sense, this part of our paper is of independent interest.

Let $(P, \leq, g)$ be a poset with an order-reversing involution g, i.e., g is a function from P to P satisfying the following conditions: for any x and y in P,

1. $x \leq y$ implies $g(y) \leq g(x)$;
2. $g(g(x)) = x$.

It is easy to see that g is also one-to-one. $J(P)$ is defined to be the lattice of order ideals in P with the usual set operations $\cap$ and $\cup$. It is easy to check that $J(P)$ is a distributive lattice. According to g, we define $\sim$ as follows:

$$\sim I := P \setminus g(I) \text{ for any order ideal } I \in J(P).$$

It is easy to check that $\sim I$ is also an order ideal in $J(P)$. So $\sim$ is a unary operation on $J(P)$. We further show that $J(P)$ with this unary operation $\sim$ is a de Morgan lattice.

Theorem A.1 *The above defined $J(P)$ with the unary operation $\sim$ is a de Morgan lattice.*

Proof It suffices to show that g is an order-reversing involution on the set of order ideals of P.

1. First we show that it is order-reversing. Assume that $I_1 \subseteq I_2$ and $x \in \sim I_2$. It follows that $g(x) \notin I_2$ and hence $g(x) \notin I_1$. So we have that $I_1 \subseteq I_2$ implies $\sim I_2 \subseteq \sim I_1$.
2. Next we show that $\sim$ is an involution by the following chain of equivalences.

$$x \in \sim\sim I \Leftrightarrow x \in P \setminus g(P \setminus g(I)) \Leftrightarrow g(x) \notin P \setminus g(I) \Leftrightarrow g(x) \in g(I) \Leftrightarrow x \in I$$

So $I = \sim\sim I$ for any order ideal in $J(P)$. □

Next we show the converse to the above theorem: any de Morgan lattice can be represented as the lattice of order ideals in some poset with an order-reversing involution. Given a de Morgan lattice $(D, \wedge, \vee, \sim)$, P_D is defined as the sub-poset of join-irreducibles in D. In addition, we define, for any $a \in P_D$,

$$g(a) = \bigwedge \{x \in D : x \in D \setminus \sim[a)\}, \text{ where } \sim[a) = \{\sim x : x \in [a)\}$$

We won't distinguish the unary operation on D and the derived unary operation on P_D. The context will decide which we use. Similarly we have used the same notation $\sim$ for the unary operation on distributive lattices D and for the derived unary operation on $J(P)$.

Proposition A.2 *Let L be a finite distributive lattice. There is a one-to-one correspondence between join-irreducibles and prime filters in L in the following sense:*

1. *for any join-irreducible a in L,* $[a)$ *is a prime filter;*
2. *for any prime filter F,* $\bigwedge F$ *is a join-irreducible in L.*

Proof For the first part, assume that a is join-irreducible in L and $a \leq b \vee c$. It follows that $a \leq b$ or $a \leq c$. For the second part, assume that F is a prime filter and $a_F := \bigwedge F = b \vee c$. It follows that $b \leq a$ and $c \leq a$ and $b \vee c \in F$. Since F is a prime filter, $b \in F$ or $c \in F$, i.e., $a_F \leq b$ or $a_F \leq c$. So $a_F = b$ or $a_F = c$. That is to say, a_F is join-irreducible. □

Lemma A.3 *For any join-irreducible* $a \in P_D$, $g(a) \in P_D$, *i.e.,* $g(a)$ *is also join-irreducible.*

Proof Let a be join-irreducible in D. Assume that $g(a) = b \vee c$. We need to show that $g(a) = b$ or $g(a) = c$. Since a is join-irreducible in D. $[a)$ is a prime filter in D. We can further show that $D \setminus \sim[a)$ is also a prime filter. So $g(a) = \bigwedge\{x \in D : x \in D \setminus \sim[a)\}$ is a join-irreducible element in D. □

So the above defined g is a unary operation on P_D.

Theorem A.4 *g is an order-reversing involution on* P_D.

Proof First we show that g is order-reversing. Let a and b be two join-irreducibles in P_D such that $a \leq b$. The next series of implications holds.

$$a \leq b \Rightarrow [b) \subseteq [a) \Rightarrow \sim[b) \subseteq \sim[a) \Rightarrow D \setminus \sim[a) \subseteq D \setminus \sim[b) \Rightarrow g(b) \leq g(a)$$

Next we show that g is an involution. It suffices to show, by the following equivalences, that for any $a \in P_D$, $[a) = D \setminus \sim[g(a))$.

$$x \in [a) \Leftrightarrow \sim x \in \sim[a) \Leftrightarrow \sim x \notin D \setminus \sim[a) \Leftrightarrow g(a) \not\leq \sim x \Leftrightarrow$$
$$\sim x \notin [g(a)) \Leftrightarrow x \notin \sim[g(a)) \Leftrightarrow x \in D \setminus \sim[g(a)) \quad \square$$

Theorem A.5 *Let P be a poset with an order-reversing involution g. Then P is isomorphic to the sub-poset* $P_{J(P)}$ *of join-irreducibles in* $J(P)$ *which is the lattice of order ideals in P.*

Proof Let P be a poset with an order-reversing involution g. A function $h : P \to J(P)$ is defined as follows:

$$h(a) = (a] \text{ for any } a \in P.$$

First we show that h is actually a function from P to $P_{J(P)}$, i.e., $h(a)$ is join-irreducible in $J(P)$ for any $a \in P$. Assume that $a \in P$ and $(a] = I_1 \cup I_2$, where $I_1 \in J(P)$ and $I_2 \in J(P)$. It follows that $a \in I_1$ or $a \in I_2$. Either case implies that $I_1 = (a]$ or $I_2 = (a]$. So indeed $h(a)$ is join-irreducible in $J(P)$.

Next we show that h is one-to-one between P and $P_{J(P)}$. From the above, we only need to show that h is onto. Assume that $I \in P_{J(P)}$, i.e., I is join-irreducible in $J(P)$. Now we need to show that I is actually a principal order ideal in P. We prove this by contraposition. Suppose that I is not a principal order ideal in P. Let $I_{max} = \{x \in I : x$ is maximal in I in the sense that there are no other elements y in I such that $y \geq x\}$. It follows that $|I_{max}| \geq 2$. So $I_{max} = M_1 \cup M_2$ for some non-empty subsets M_1 and M_2. We define:

$$I_1 = \{x \in I : x \leq y \text{ for some } y \in M_1\}, \qquad I_2 = \{x \in I : x \leq y \text{ for some } y \in M_2\}.$$

It is easy to check that $I = I_1 \cup I_2$ but $I_1 \neq I$ and $I \neq I_2$. So I is not join-irreducible in $J(P)$.

It remains to show that h preserves the order and the operation g. It is easy to see that it does for the order. Now we show that it preserves g. For any $a \in P$,

$$\begin{aligned}
g(h(a)) &= x \in \bigcap\{I \in J(P) : I \in J(P) \setminus {\sim}[h(a))\} \\
&= \bigcap\{I \in J(P) : I \in J(P) \setminus {\sim}\{I \in J(P) : a \in I\}\} \\
&= \bigcap\{I \in J(P) : I \in J(P) \setminus \{P \setminus g(I) : I \in J(P), a \in I\}\} \\
&= \bigcap\{I \in J(P) : I \in J(P) \setminus \{J : a \in g(P \setminus J)\}\} \\
&= \bigcap\{I \in J(P) : a \notin g(P \setminus I)\} \\
&= \bigcap\{I \in J(P) : g(a) \notin P \setminus I\} \\
&= \bigcap\{I \in J(P) : g(a) \in I\} \\
&= (g(a)]
\end{aligned}$$

That is to say, $h(g(a)) = g(h(a))$. □

Theorem A.6 *Any finite de Morgan lattice D can be represented as the lattice $J(P_D)$ of order ideals in the sub-poset P_D of join-irreducibles with an order-reversing involution g.*

Proof Let D be a finite de Morgan lattice and P_D be its sub-poset of join-irreducibles with the order-reversing involution g. Now we need to show that D is isomorphic to the concrete de Morgan lattice $J(P_D)$. Define $h : D \to J(P_D)$ as $h(x) = \{y \in P_D : y \leq x\}$ for any $x \in D$. From the proof of Theorem 3.4.1 in Stanley (1997), we only need to show that h preserves negation. In order to prove this, it suffices to show that, for any $a \in D$, $h(\sim x) = P_D \setminus g(h(a))$. For any $x \in P_D$,

$$x \in h(\sim a) \Leftrightarrow x \leq \sim a \Leftrightarrow \sim a \in [x) \Leftrightarrow a \in \sim[x) \Leftrightarrow a \notin D \setminus \sim[x) \Leftrightarrow$$
$$g(x) \nleq a \Leftrightarrow g(x) \notin h(a) \Leftrightarrow x \notin g(h(a)) \Leftrightarrow x \in P \setminus g(h(a)) \Leftrightarrow x \in \sim h(a)$$

Note that the $\sim$ in the last line is the unary operation on $J(P_D)$. □

Corollary A.7 *There is a one-to-one correspondence between the class of de Morgan lattices and that of posets with order-reversing involutions.*

Proof This proposition follows from the above two theorems. This kind of correspondence is illustrated in the following diagram:

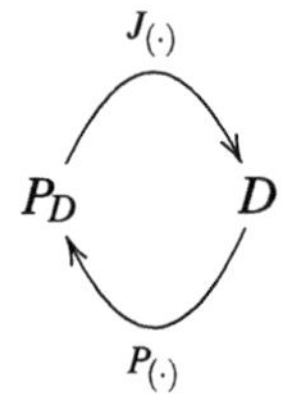

□

References

Anderson, A. R., & Belnap, N. D. (1975). *Entailment: The logic of relevance and necessity* (Vol. I). Princeton, NJ: Princeton University Press.

Arieli, O., & Avron, A. (1998). The value of the four values. *Artificial Intelligence*, *102*(1), 97–141.

Arieli, O., Avron, A., & Zamansky, A. (2011). What is an ideal logic for reasoning with inconsistency? *IJCAI* (pp. 706–711).

Barthélemy, J. P. (2000). Monotone functions on finite lattices: An ordinal approach to capacities, belief and necessaity functions. In J. Fodor, B. Baets & P. Perny (Eds.), *Preferences and decisions under incomplete knowledge* (pp. 195–208).

Belnap, N. D. (1977). A useful four-valued logic. In J. M. Dunn & G. Epstein (Eds.), *Modern uses of multiple-valued logic* (pp. 8–37). Dordrecht: Reidel.

Besnard, P., & Lang, J. (1994). Possibility and necessity functions over non-classical logics. In R. L. de Mántaras & D. Poole (Eds.), *UAI '94: Proceedings of the Tenth Annual Conference on Uncertainty in Artificial Intelligence* (pp. 69–76). Seattle, WA, USA: Morgan Kaufmann.

Białynicki-Birula, A., & Rasiowa, H. (1957). On the representation of quasi-Boolean algebras. *Bulletin de l'Académie Polonaise des Sciences*, *5*, 259–261.

Birkhoff, G. (1967). *Lattice theory, AMS colloquium publications* (3rd ed., Vol. 25). Providence, RI: American Mathematical Society.

Birkhoff, G., & von Neumann, J. (1936). The logic of quantum mechanics. *Annals of Mathematics*, *37*, 823–843.

Carnap, R. (1962). *Logical foundations of probability*. University of Chicago Press.

Dempster, A. P. (1967). Upper and lower probabilities induced by a multivalued mapping. *Annals of Mathematical Statistics*, *38*, 325–339.

Dubois, D., & Prade, H. (2008). An introduction to bipolar representations of information and preference. *International Journal of Intelligent Systems*, *23*(8), 866–877.

Dunn, J. M. (1966). The Algebra of Intensional Logics, PhD thesis, University of Pittsburgh, Ann Arbor (UMI).

Dunn, J. M. (1976a). Intuitive semantics for first-degree entailments and 'coupled trees'. *Philosophical Studies, 29*, 149–168.

Dunn, J. M. (1976b). A Kripke-style semantics for *R*-mingle using a binary accessibility relation. *Studia Logica, 35*, 163–172.

Dunn, J. M. (1986). Relevance logic and entailment. In D. Gabbay & F. Guenthner (Eds.), *Handbook of philosophical logic* (1st ed., Vol. 3, pp. 117–224). Dordrecht: D. Reidel.

Dunn, J. M. (1999). A comparative study of various model-theoretic treatments of negation: A history of formal negation. In D. M. Gabbay & H. Wansing (Eds.), *What is negation?* (pp. 23–51). Dordrecht: Kluwer.

Dunn, J. M. (2000). Partiality and its dual. *Studia Logica, 65*, 5–40.

Dunn, J. M. (2010). Contradictory information: Too much of a good thing. *Journal of Philosophical Logic, 39*, 425–452.

Dunn, J. M., & Zhou, C. (2005). Negation in the context of gaggle theory. *Studia Logica, 80*, 235–264.

Fagin, R., & Halpern, J. (1991). Uncertainty, belief, and probability. *Computational Intelligence, 7*, 160–173.

Fagin, R., Halpern, J., & Megiddo, N. (1990). A logic for reasoning about probabilities. *Information and Computation, 87*, 78–128.

Fagin, R., Halpern, J. Y., Moses, Y., & Vardi, M. (1995). *Reasoning about knowledge*. MIT Press.

Fitting, M. (1991). Bilattices and the semantics of logic programming. *Journal of Logic Programming, 11*(1–2), 91–116.

Fitting, M. (1994). Kleene's three valued logics and their children. *Fundamenta Informatica, 20*(1-2-3), 113–131.

Ginsberg, M. (1988). Multivalued logics: A uniform approach to reasoning in AI. *Computer Intelligence, 4*, 256–316.

Girard, J.-Y. (1987). Linear logic. *Theoretical Computer Science, 50*, 1–102.

Grabisch, M. (2009). Belief functions on lattices. *International Journal of Intelligent Systems, 24*(1), 76–95.

Halpern, J. (2005). *Reasoning about uncertainty*. MIT Press.

Jung, A., & Rivieccio, U. (2012). Priestley duality for bilattices. *Studia Logica, 100*(1–2), 223–252.

Jung, A., & Rivieccio, U. (2013). Kripke semantics for modal bilattice logic. In *28th Annual ACM/IEEE Symposium on Logic in Computer Science, LICS 2013* (pp. 438–447). New Orleans, LA, USA: IEEE Computer Society.

Kifer, M., & Lozinskii, E. L. (1992). A logic for reasoning with inconsistency. *Journal of Automated Reasoning, 9*(2), 179–215.

Konieczny, S., Marquis, P., & Besnard, P. (2008). Bipolarity in bilattice logics. *International Journal of Intelligent Systems, 23*(10), 1046–1061.

Kripke, S. A. (1965). Semantical analysis of intuitionistic logic I. In J. N. Crossley & M. A. E. Dummett (Eds.), *Formal Systems and Recursive Functions. Proceedings of the Eighth Logic Colloquium, Studies in Logic and the Foundations of Mathematics* (Vol. 40, pp. 92–130). North-Holland, Amsterdam.

Levesque, H. (1984). A logic of implicit and explicit belief, *AAAI* (pp. 198–202).

Lin, J. (1996). A semantics for reasoning consistently in the presence of inconsistency. *Artificial Intelligence, 86*(1), 75–95.

Mobasher, B., Pigozzi, D., Slutzki, G., & Voutsadakis, G. (2000). A duality theory for bilattices. *Algebra Universalis, 43*, 109–125.

Nilsson, N. (1991). Logic and artificial intelligence. *Artificial Intelligence, 47*(1–3), 31–56.

Parsons, S. (1996). Current approaches to handling imperfect information in data and knowledge bases. *IEEE Transactions on Knowledge and Data Engineering, 8*(3), 353–372.

Priest, G. (1979). The logic of paradox. *Journal of Philosophical Logic, 9*, 415–435.

Priestley, H. A. (1970). Representation of distributive lattices by means of ordered Stone spaces. *Bulletin of the London Mathematical Society, 2*, 186–190.

Reiter, R. (1978). On closed world database. In H. Gallaire & J. Minker (Eds.), *Logic and database* (pp. 55–76). New York, NY: Plenum Press.

Rota, G. C. (1964). On the foundations of combinatorial theory I: Theory of Möbius functions. *Zeitschrift für Wahrscheinlichkeitstheorie und Verwandte Gebiete*, *2*, 340–368.

Routley, R., & Routley, V. (1972). The semantics of first degree entailment. *Noûs*, *6*(4), 335–359.

Ruspini, E. H. (1987). Epistemic logics, probability, and the calculus of evidence. In J. P. McDermott (Ed.), *Proceedings of the 10th International Joint Conference on Artificial Intelligence* (pp. 924–931). Milan, Italy: Morgan Kaufmann.

Saffiotti, A. (1990a). A hybrid framework for representing uncertain knowledge. In H. E. Shrobe, T. G. Dietterich, & W. R. Swartout (Eds.), *Proceedings of the 8th National Conference on Artificial Intelligence* (pp. 653–658). Boston, MA, USA: AAAI Press and MIT Press.

Saffiotti, A. (1990b). Using Dempster-Shafer theory in knowledge representation, in P. P. Bonissone, M. Henrion, L. N. Kanal and J. F. Lemmer (Eds.), *UAI* (pp. 417–434). Elsevier.

Saffiotti, A. (1992). A belief-function logic. In W. R. Swartout (Ed.), *Proceedings of the 10th National Conference on Artificial Intelligence* (pp. 642–647). San Jose, CA, USA: AAAI Press and MIT Press.

Savage, L. (1972). *The foundations of statistics*. Dover Publications Inc.

Shafer, G. (1976). *A mathematical theory of evidence*. Princeton, NJ: Princeton University Press.

Smets, P. (1981). The degree of belief in a fuzzy event. *Information Science*, *25*(1), 1–19.

Smets, P. (1995). The canonical decomposition of a weighted belief. In *Proceedings of the Fourteenth International Joint Conference on Artificial Intelligence, IJCAI '95* (pp. 1896–1901). Montréal, QB, Canada: Morgan Kaufmann.

Smets, P., & Kennes, R. (1994). The transferable belief model. *Artificial Intelligence*, *66*(2), 191–234.

Stanley, R. (1997). *Enumerative combinatorics, Cambridge studies in advanced mathematics* (Vol. 1). Cambridge University Press.

Urquhart, A. (1979). Distributive lattices with a dual homomorphic operation, I. *Studia Logica*, *38*(2), 201–209.

Urquhart, A. (1990). The complexity of decision procedures in relevance logic. In J. M. Dunn & A. Gupta (Eds.), *Truth or consequences: Essays in honor of Nuel Belnap* (pp. 61–76). Amsterdam: Kluwer.

van Dalen, D. (2004). *Logic and structure*, University Texts in Mathematics (4th ed.). Springer.

Yen, J. (1990). Generalizing the Dempster-Shafer theory to fuzzy sets. *IEEE Transactions on Systems, Man and Cybernetics*, *20*(3), 559–570.

Ying, M. (2010). Quantum computation, quantum theory and AI. *Artificial Intelligence*, *174*(2), 162–176.

Zadeh, L. (1979). Fuzzy sets and information granularity, *Advances in Fuzzy Set Theory and Applications* (pp. 3–18).

Zadeh, L. (1988). Fuzzy logic. *IEEE Computer*, *21*(4), 83–93.

Zhou, C. (2012). Belief functions on distributive lattices. In J. Hoffmann & B. Selman (Eds.), *Proceedings of the Twenty-Sixth AAAI Conference on Artificial Intelligence*. Toronto, ON, Canada: AAAI Press.

Zhou, C. (2013). Belief functions on distributive lattices. *Artificial Intelligence*, *201*, 1–31. doi:10.1016/j.artint.2013.05.003.

Probabilistic Interpretations of Predicates

Janusz Czelakowski

Abstract In classical logic, any m-ary predicate is interpreted as an m-argument two-valued relation defined on a non-empty universe. In probability theory, m-ary predicates are interpreted as probability measures on the mth power of a probability space. m-ary probabilistic predicates are equivalently semantically characterized as m-dimensional cumulative distribution functions defined on $\mathbb{R}^m$. The paper is mainly concerned with probabilistic interpretations of unary predicates in the algebra of cumulative distribution functions defined on $\mathbb{R}$. This algebra, enriched with two constants, forms a bounded De Morgan algebra. Two logical systems based on the algebra of cumulative distributions are defined and their basic properties are isolated. Comparisons with the infinitely-valued Łukasiewicz logic and open problems are also discussed.

Keywords Consequence operation · Cumulative distribution function · De Morgan algebra · Predicate · Probability space · Random variable

1 Introduction

The reasons for writing this paper are multifold. Certainly, one of them is probability theory itself and its relationship with many-valued logics. But there is another reason, namely, the theory of distributoids and gaggles developed by J. Michael Dunn, originally in Dunn (1991, 1993). The seminal monograph (Dunn and Hardegree 2001) provides a uniform semantical approach to "a variety of non-classical logics." Professor Dunn's strategy is to adopt the framework of the Kripke-style semantics, using accessibility relations, to give truth-conditions for the connectives of many non-classical logics. This paper is concerned with the algebraic approach to probability based on the algebra of cumulative distribution functions, denoted as ***CDF***. This algebra, being the main semantic tool in our metalogical considerations, has a

J. Czelakowski (✉)
Institute of Mathematics and Informatics, Opole University, Opole, Poland
e-mail: jczel@math.uni.opole.pl
URL: http://www.math.uni.opole.pl/prac.php?id=97

K. Bimbó (ed.), *J. Michael Dunn on Information Based Logics*,
Outstanding Contributions to Logic 8, DOI 10.1007/978-3-319-29300-4_13

rather complicated intrinsic structure. However, it is not difficult to notice that ***CDF*** is a tonoid in the sense of Dunn (1993), Dunn and Hardegree (2001). Moreover, if one disregards the convolution operation, this algebra becomes a distributoid. Various subtle issues pertinent to the structure of ***CDF*** are not discussed at length here. This is a task for future work. But we remark that the methods worked out in Dunn and Hardegree (2001) enable one to develop Kripke-style semantics for various logical systems based on the algebra ***CDF***.

The author dedicates this paper to Professor Jon Michael Dunn with the hope that he finds it interesting.

From the perspective of classical logic, any m-ary predicate P is interpreted as an m-argument relation defined on a non-empty set A. The set A is called the *universe* of the pertinent model. Equivalently, an interpretation of P is an m-argument zero-one function F_P defined on A, that is, $F_P\colon A^m \to \{0, 1\}$. If $F_P(a_1, \ldots, a_m) = 1$, we say that F_P *holds* for the sequence $\langle a_1, \ldots, a_m\rangle$; otherwise, when $F_P(a_1, \ldots, a_m) = 0$, we say that F_P *does not hold* for $\langle a_1, \ldots, a_m\rangle$.

The *fuzzy interpretation* goes farther—truth-values range in degree between 0 and 1. Accordingly, each m-ary predicate P is evaluated as an m-argument function F_P from a universe A to the unit interval [0, 1]. A systematic elaboration of this idea leads to various many-valued interpretations of the predicate calculus.

At the foundations of frameworks classified as fuzzy and related logics (fuzzy logics, rough sets, etc.) lies the assumption that predicates (often taken from everyday language practice) display some blurry or fuzzy character; the dividing line between what irrefutably belongs to the range of a predicate's meaning and what is absent from it is unstable and vague.

In sentences like:

(1) *Mike is a better mathematician than Andrew.*
(2) *John is a good writer.*

there appear, respectively, the binary predicate *x is a better mathematician than y* and the unary predicate *x is a good writer*. Further examples of such vague formulas can be easily found.

We can (or even should) utter grammatically correct sentences of type (1) or (2), yet in fact, it is difficult to assign any logical value to such sentences (or utterances), that is, truth or falsity. We can positively or negatively justify (or refute) these sentences, but from justifying to assigning a logical value to them there is a long way. A pragmatic solution consists in adopting certain criteria of acceptance or rejection of sentences. These criteria may refer to (certain) populations of adult users of the English language and their opinions. In the case of type (2) sentences, the community of literary critics forms such a natural population and their opinions are treated as binding.

The mentioned pragmatic criteria may be distant from the classical definition of truth. It is easy to point in this context to various manipulative techniques (techniques of influence) shaping views of users of the language.

For general accounts of the theory of attributes see e.g., (Ganter et al. 2005) and (Hájek 1998).

The fuzzy logic approach "acts" with reference to certain fragments of the language used. It has turned out, however, to be convincing and has resulted in substantial applications of a purely practical nature.

According to the *probabilistic* interpretation, universes as well as their Cartesian powers A^2, A^3 etc., are treated as statistical populations, that is, collections of uniform objects which have some properties in common but these properties are not individuated. (Statistics distinguishes between *quantitative* and *qualitative* properties. The former are divided in turn into ratio and interval features, while the latter are divided into ordinal and nominal features.) In the simplest case, unary predicates are viewed as random variables, that is, certain real-valued functions defined for each element of the population. (The height of the inhabitants of New York City or their sex are examples of unary predicates.) But from the probabilistic perspective, each m-ary predicate over a statistical population is interpreted as an *m-dimensional cumulative distribution function* (CDF). Accordingly, if one is interested in the probabilistic description of the height of the inhabitants of New York City, one assigns to this predicate a suitably selected normal cumulative distribution function. The probabilistic interpretation does not treat predicates as 'place holders for individual variables' (because there are no individual variables which are quantified), but merely marks the arity of each predicate and assigns to each m-ary quantitative predicate an m-dimensional cumulative distribution function (CDF). Thus, from the probabilistic perspective, the m-dimensional cumulative distribution assigned to an m-ary predicate P fully encodes the quantitative characteristic of P in the population. In this way the *probabilistic* interpretation of predicates is established. In this paper, we are mainly concerned with unary predicates and, consequently, with unary CDFs.

From a more abstract perspective, we may disregard populations and isolate the set of unary predicates as an absolutely free algebra endowed with a finite set of operations (to be defined later on) and freely generated by a countably infinite set of unary predicate *variables*. The latter are viewed as sentential variables and the entire algebra of unary predicates is an example of a sentential language in the sense of formal logic. Unary probabilistic predicates are identified with sentential formulas belonging to this language. Interpretations of formulas are defined in terms of homomorphisms of the formula algebra in the algebra of unary cumulative distribution functions. This idea is elaborated in detail in Sect. 4.

More generally, for each positive integer m, one builds the absolutely free algebra of m-ary probabilistic predicates that is freely generated by the countably infinite set of m-ary probabilistic variables by way of emulating the above definition for unary predicates. One then arrives at the sentential language of m-ary predicate variables. By analogy, interpretations of this language are defined in terms of m-dimensional cumulative distribution functions.

The set *CDF* of unary cumulative distribution functions exhibits a definite algebraic structure, viz. the structure of a distributive lattice. However, this lattice is more complex. Extending the set *CDF* by means of two constants **0** and **1**, one arrives at the *De Morgan algebra* ***CDF*** of unary CDFs. (The constants **0** and **1** are not CDFs;

they are generalized functions.) The paper is mainly concerned with the structure of the algebra ***CDF*** and its interrelations with logic.

Let us first present the following simple case. Let P be a unary predicate and A a universe. A will be called a *temporary population*. The standard first-order logic interpretation assigns to P a unary zero-one function R defined on A. R partitions the universe A into two disjoint sets: A_1—consisting of the elements of A that possess the property R and A_0—comprising those which do not exhibit R.

On the other hand, the simplest probabilistic interpretation assigns to P a cumulative distribution function $\boldsymbol{F}_P$ that takes only two values 0 and 1. $\boldsymbol{F}_P$ determines a number p from the unit interval $[0, 1]$. p is the probability that P takes value 1 and $1 - p$ is the probability that P takes value 0.

From the point of view of a probabilistic interpretation, one does not define the satisfiability of m-argument relations as holding on m-tuples of elements of the population, because m-tuples are not individuated. While in fuzzy set theory it makes sense to assign a numerical value to each m-tuple $\langle a_1, \ldots, a_m \rangle$ of elements of A as a degree of a relation R 'holding' on $\langle a_1, \ldots, a_m \rangle$, the probabilistic interpretation does not do this. It merely provides a global probability distribution for the set A^m. In particular, in a probabilistic interpretation, one abandons the notation adopted for m-ary predicates, that is, $P(x_1, \ldots, x_m)$, where $x_1, \ldots, x_m$ are individual variables, because within the probabilistic framework one does not isolate the category of variables ranging over elements of populations. Nevertheless, the arity of predicates is preserved. As a result, there are no quantifiers binding individual variables. The equality predicate interpreted as the identity relation between pairs of elements of the population is no longer needed either.

The algebra that is defined below plays (to an extent) in the probabilistic approach the role analogous to the two-element Boolean algebra in classical predicate calculus or the infinitely valued Łukasiewicz algebra in fuzzy set theory (see Davey and Priestley 2002).[1] We are talking here about the classical probability theory. In quantum probability theory, one defines other models, originating from finite or infinite dimensional Hilbert spaces.

Returning to the above example, the predicate $\neg P$ is interpreted as the cumulative distribution function that is dual to $\boldsymbol{F}_P$ and denoted by $\boldsymbol{F}_P^d$. The distribution $\boldsymbol{F}_P^d$ reverses the probabilities defined by $\boldsymbol{F}_P$. Accordingly, $\neg P$ takes the value 1 with probability $1 - p$ and the value 0 with probability p.

Generally, we will be concerned with probability distributions on the set $\mathbb{R}$ of real numbers defining, for any numbers $a < b$, the probability

$$Pr_A(a < R \leqslant b)$$

of acquiring numerical values from the half-open interval $(a, b]$ by P.

---

[1] A full analogy is obtained through introduction of the probabilistic interpretation of predicates of arbitrary finite arity.

We will develop the above ideas in subsequent paragraphs. These ideas were outlined in a sketchy way in the article (Czelakowski 2012), published in Polish. The present paper is an extended and improved version of the above work.

2 Cumulative Distributions

$I := [0, 1] = \{x \in \mathbb{R} : 0 \leqslant x \leqslant 1\}$ is the unit interval.

Let m be a positive integer. A function $\boldsymbol{F}: \mathbb{R}^m \to I$ is an m-dimensional *cumulative distribution* if and only if it determines a probability measure $\mu_{\boldsymbol{F}}$ on the σ-field $B(\mathbb{R}^m)$ of Borel subsets of $\mathbb{R}^m$ such that

$$\mu_{\boldsymbol{F}}((a_1, b_1] \times \cdots \times (a_m, b_m]) = \boldsymbol{F}(b_1, \ldots, b_m) - \boldsymbol{F}(a_1, \ldots, a_m),$$

for any two m-tuples $a_1, \ldots, a_m$ and $b_1, \ldots, b_m$ of real numbers such that $a_i < b_i$ for $i = 1, \ldots, m$. (Here $(a, b]$ is a half-open interval with a and b being its endpoints and $(a_1, b_1] \times \cdots \times (a_m, b_m]$ is the Cartesian product of the indicated intervals.)

Intuitively, for any m-tuple $\underline{r} = \langle r_1, \ldots, r_m \rangle$ of real numbers,

(1) $\boldsymbol{F}(r_1, \ldots, r_m)$ is the probability that the numerical value of the m-ary predicate P belongs to the Cartesian product $(-\infty, r_1] \times \ldots \times (-\infty, r_m]$.

In order that a function $\boldsymbol{F}: \mathbb{R}^m \to I$ be an n-dimensional cumulative distribution, $\boldsymbol{F}$ must validate some conditions. They provide an intrinsic characterization of m-dimensional cumulative distributions. We shall not present them here in the general case. But it is relatively easy to formulate them in the one-dimensional case. The general case is much more intricate.

Let P be an m-ary predicate symbol. According to the probabilistic interpretation, one assigns to the predicate P an m-dimensional cumulative distribution $\boldsymbol{F}_P: \mathbb{R}^m \to I$. $\boldsymbol{F}_P$ is called an *interpretation of* P in the set of m-dimensional cumulative distribution functions.

In probability theory, one-dimensional cumulative distributions are *defined* as functions $\boldsymbol{F}: \mathbb{R} \to I$ that satisfy the following conditions:

(2a) $\boldsymbol{F}$ is non-decreasing;
(2b) $\boldsymbol{F}$ is right-continuous;
(2c) $\lim_{x \to -\infty} \boldsymbol{F}(x) = 0$ and $\lim_{x \to +\infty} \boldsymbol{F}(x) = 1$.

$\boldsymbol{F}$ is therefore a *càdlàg function* which means that for every real r, the left limit $\boldsymbol{F}(r^-)$ exists; further, the right limit $\boldsymbol{F}(r^+)$ exists and equals $\boldsymbol{F}(r)$.

One of the main theorems in probability theory states that every CDF $\boldsymbol{F}$, defined as above, determines a probability measure $\mu_{\boldsymbol{F}}$ on the σ-field $B(\mathbb{R})$ of Borel subsets of $\mathbb{R}$ such that $\mu_{\boldsymbol{F}}((a, b]) = \boldsymbol{F}(b) - \boldsymbol{F}(a)$ for any real numbers a, b, where $a < b$. (If $a = b$, then $\mu_{\boldsymbol{F}}(\{a\}) = \boldsymbol{F}(a) - \boldsymbol{F}(a^-)$, where $\boldsymbol{F}(a^-)$ is the left limit of $\boldsymbol{F}$ at a. If $\boldsymbol{F}$ is continuous at a, then $\mu_{\boldsymbol{F}}(\{a\}) = 0$.)

Moreover, every probability measure on $B(\mathbb{R})$ is determined by a unique CDF (see e.g., Billingsley 1995).

If $\boldsymbol{F}$ is a continuous CDF, the measure $\mu_{\boldsymbol{F}}$ takes value zero on one-element subsets of $\mathbb{R}$. Consequently, for any numbers a, b with $a < b$, it is the case that $\mu_{\boldsymbol{F}}((a, b)) = \mu_{\boldsymbol{F}}((a, b]) = \mu_{\boldsymbol{F}}([a, b)) = \mu_{\boldsymbol{F}}([a, b])$.

In what follows we shall mainly confine the discussion to unary predicates (i.e., attributes). Their probabilistic interpretations are formed by one-dimensional cumulative distribution functions.

3 The Algebra of Probabilistic Attributes

Let μ be a probability measure on the σ-field $B(\mathbb{R})$ of Borel subsets of $\mathbb{R}$. μ^d is the measure *dual* to μ. Thus

$$\mu^d(X) := \mu(-X), \tag{1}$$

for any set $X \in B(\mathbb{R})$, where $-X := \{-x \colon x \in X\}$. (If X is a Borel set, then so is $-X$.)

μ^d is a probability measure on $B(\mathbb{R})$. This directly follows from the equivalence that $A \cap B = \emptyset$ if and only if $-A \cap -B = \emptyset$, for any sets $A, B \subseteq \mathbb{R}$, and the fact that μ is a measure. μ^d agrees with μ on Borel sets A for which $A = -A$. Such sets A are called *symmetric*. It is also clear that $(\mu^d)^d = \mu$.

Let $\boldsymbol{F} \colon \mathbb{R} \to [0, 1]$ be a CDF. The cumulative distribution function *dual* to $\boldsymbol{F}$ is the function $\boldsymbol{F^d} \colon \mathbb{R} \to [0, 1]$ defined as follows. Let $\mu_{\boldsymbol{F}}$ be the probability measure on $B(\mathbb{R})$ corresponding to $\boldsymbol{F}$. $\boldsymbol{F^d}$ is, by definition, the cumulative distribution function that determines the dual measure μ^d. $\boldsymbol{F^d}$ is unambiguously defined. In fact,

$$\boldsymbol{F^d}(x) := \mu^d_{\boldsymbol{F}}((-\infty, x]), \tag{2}$$

for every $x \in \mathbb{R}$. Thus

$$\boldsymbol{F^d}(x) = \mu_{\boldsymbol{F}}([-x, +\infty)), \tag{3}$$

for every $x \in \mathbb{R}$. Since $\mu_{\boldsymbol{F}}([-x, +\infty)) = 1 - \mu_{\boldsymbol{F}}((-\infty, -x))$, we have that

$$\boldsymbol{F^d}(x) = 1 - \boldsymbol{F}((-x)^-), \tag{4}$$

where $\boldsymbol{F}((-x)^-)$ is the left limit of $\boldsymbol{F}$ at $-x$, for every $x \in \mathbb{R}$.

Lemma 3.1 $(\boldsymbol{F^d})^d = \boldsymbol{F}$.

Proof Let μ be the measure corresponding to $\boldsymbol{F}$ and let μ^d be the dual measure. μ^d is the measure corresponding to $\boldsymbol{F^d}$. Then $(\boldsymbol{F^d})^d(x) =$ (by (3)) $\mu^d([-x, +\infty)) = \mu(-[-x, +\infty)) = \mu((-\infty, x]) = \boldsymbol{F}(x)$, for all $x \in \mathbb{R}$. □

$\boldsymbol{F}$ is continuous at a point a if and only if the left limit of $\boldsymbol{F}$ at a is equal to $\boldsymbol{F}(a)$. It follows from (4) that if $\boldsymbol{F}$ is continuous at $-x$, then $\boldsymbol{F}^{\boldsymbol{d}}(x) = 1 - \boldsymbol{F}(-x)$. We thus obtain

Corollary 3.2 *If* $\boldsymbol{F}$ *is a continuous* CDF, *then* $\boldsymbol{F}^{\boldsymbol{d}}$ *is a continuous* CDF *as well. Moreover,*

$$\boldsymbol{F}^{\boldsymbol{d}}(x) = 1 - \boldsymbol{F}(-x),$$

for every $x \in \mathbb{R}$.

The corollary follows from the above remarks.

Suppose that $\boldsymbol{F}$ is a continuous CDF and it has density, i.e., there exists a measurable non-negative function $f\colon \mathbb{R} \to \mathbb{R}$ such that

$$\boldsymbol{F}(x) = \int\limits_{-\infty}^{x} f(t)\, dt, \tag{5}$$

for all $x \in \mathbb{R}$. The function $g\colon \mathbb{R} \to \mathbb{R}$ given by $g(x) := f(-x)$, $x \in \mathbb{R}$, is the density function of $\boldsymbol{F}^{\boldsymbol{d}}$. The graph of g is obtained by the reflection of the graph of f with respect to the y-axis.

CDF is the set of (unary) cumulative distribution functions. The order relation $\leqslant$ on *CDF* is defined pointwise as follows.

$$\boldsymbol{F} \leqslant \boldsymbol{G} \quad \Leftrightarrow_{df} \quad \boldsymbol{F}(x) \leqslant \boldsymbol{G}(x) \ \text{ for every real number } x. \tag{6}$$

Thus, in accordance to the meaning attached to cumulative distributions, $\boldsymbol{F} \leqslant \boldsymbol{G}$ states that for every real number x, the probability that the numerical value of the probabilistic attribute $\boldsymbol{F}$ belongs to the interval $(-\infty, x]$ is smaller or equal to the probability that a numerical value of $\boldsymbol{G}$ belongs to $(-\infty, x]$.

Other operations are also performable in the set *CDF*. Suppose $\boldsymbol{F}$ and $\boldsymbol{G}$ are cumulative distribution functions, not necessarily continuous. We define the functions $\boldsymbol{F} \wedge \boldsymbol{G}$ and $\boldsymbol{F} \vee \boldsymbol{G}$ as the minimum and the maximum of $\boldsymbol{F}$ and $\boldsymbol{G}$, that is,

$$(\boldsymbol{F} \wedge \boldsymbol{G})(x) := \min(\boldsymbol{F}(x), \boldsymbol{G}(x)),$$
$$(\boldsymbol{F} \vee \boldsymbol{G})(x) := \max(\boldsymbol{F}(x), \boldsymbol{G}(x)),$$

for all $x \in \mathbb{R}$. The operations $\wedge$ and $\vee$ are called the *conjunction* and the *disjunction*, respectively.

Lemma 3.3 *If* $\boldsymbol{F}$ *and* $\boldsymbol{G}$ *are cumulative distributions, then so are the functions* $\boldsymbol{F} \wedge \boldsymbol{G}$ *and* $\boldsymbol{F} \vee \boldsymbol{G}$. *The set CDF of cumulative distributions equipped with the operations* $\wedge$ *and* $\vee$ *forms a distributive lattice.*

Moreover, if $\boldsymbol{F}$ *and* $\boldsymbol{G}$ *are continuous, then so are* $\boldsymbol{F} \wedge \boldsymbol{G}$ *and* $\boldsymbol{F} \vee \boldsymbol{G}$.

The proof is easy and omitted.

The order relation $\leqslant$ defined in (6) is thus the order relation of the distributive lattice $\langle CDF, \wedge, \vee \rangle$.

A slightly less obvious fact is that the above distributive lattice together with the operation of dualization ^{d} satisfies De Morgan's laws.

Lemma 3.4 *For any cumulative distribution functions* $\boldsymbol{F}$ *and* $\boldsymbol{G}$,

$$(\boldsymbol{F} \wedge \boldsymbol{G})^d = \boldsymbol{F}^d \vee \boldsymbol{G}^d \quad \textit{and} \quad (\boldsymbol{F} \vee \boldsymbol{G})^d = \boldsymbol{F}^d \wedge \boldsymbol{G}^d.$$

Proof We shall show the first equality. The proof is restricted here to continuous CDFs. (The proof in the general case is a bit more involved.) Let x be a real number. In view of Corollary 3.2 we have:

$$(\boldsymbol{F} \wedge \boldsymbol{G})^d(x) = 1 - (\boldsymbol{F} \wedge \boldsymbol{G})(-x) = 1 - \min(\boldsymbol{F}(-x), \boldsymbol{G}(-x)) =$$
$$\max(1 - \boldsymbol{F}(-x), 1 - \boldsymbol{G}(-x)) = \max(\boldsymbol{F}^d(x), \boldsymbol{G}^d(x)) = (\boldsymbol{F}^d \vee \boldsymbol{G}^d)(x). \qquad \square$$

Thus ^{d} is an involution operation satisfying De Morgan's laws.

The algebra

$$\boldsymbol{CDF} = \langle CDF, \wedge, \vee, {}^d \rangle$$

satisfies the axioms of De Morgan algebras with one exception: it is not bounded as a distributive lattice; that is, it does not possess a bottom or a top element. We shall use, however, the suggestive term *the De Morgan algebra of cumulative distribution functions* as a proper name.

The distributive lattice $\langle CDF, \wedge, \vee \rangle$ is not complete because it lacks a top and a bottom element.

As we shall show later, one may extend in the standard way the universe CDF by augmenting it with two additional elements $\mathbf{0}$ and $\mathbf{1}$ so that one obtains a bounded distributive lattice (with zero $\mathbf{0}$ and unit $\mathbf{1}$) satisfying *all* conditions imposed on De Morgan algebras. The constants $\mathbf{0}$ and $\mathbf{1}$ are not functions defined on $\mathbb{R}$—they are distributions (or generalized functions) in the sense of Sobolev–Schwartz. Nevertheless, the lattice $\langle CDF, \wedge, \vee, \mathbf{0}, \mathbf{1} \rangle$ extended in such a way is not complete. To this end we define the following sequence $\boldsymbol{F}_n$, $n = 1, 2, \ldots$ of continuous cumulative distribution functions:

$$\boldsymbol{F}_n(x) := \begin{cases} 0 & \text{if } x \leqslant 0; \\ x^{1/n} & \text{if } 0 < x \leqslant 1; \\ 1 & \text{if } 1 < x. \end{cases}$$

The sequence $\{\boldsymbol{F}_n\}$ is pointwise convergent to the function F, where $F(x) = 0$ for $x \leqslant 0$, and $F(x) = 1$ for $x > 0$. F is not right-continuous at $x = 0$. Hence, F is not a CDF. On the other hand, we have that $\{\boldsymbol{F}_n\}$ is monotone, that is, $\boldsymbol{F}_1 \leqslant \boldsymbol{F}_2 \leqslant \cdots$ in the lattice $\langle CDF, \wedge, \vee \rangle$. It is then easy to see that $\sup\{\boldsymbol{F}_n \colon n \geqslant 1\}$ does not exist in $\langle CDF, \wedge, \vee \rangle$.

CCDF is the set of *continuous* cumulative distribution functions on $\mathbb{R}$. The system

$$\boldsymbol{CCDF} = \langle CCDF, \wedge, \vee, {}^d \rangle$$

is a subalgebra of $\boldsymbol{CDF}$. $\boldsymbol{CCDF}$ is called *the De Morgan algebra of continuous cumulative distributions*. The structure $\boldsymbol{CCDF}$ is also called *the algebra of continuous probabilistic attributes*.

Continuous cumulative distributions $\boldsymbol{F}$ such that $\boldsymbol{F} = \boldsymbol{F}^d$ determine symmetric probability measures with respect to the ordinate Oy. This means that $\mu_{\boldsymbol{F}}([-r, 0]) = \mu_{\boldsymbol{F}}([0, r])$ for all real numbers r. ($\mu_{\boldsymbol{F}}$ is the probability measure on the σ-field $B(\mathbb{R})$ corresponding to $\boldsymbol{F}$.) If $\boldsymbol{F}$ possesses a density function f, we see that f is an even function, that is $f(x) = f(-x)$ for all x, whenever $\boldsymbol{F} = \boldsymbol{F}^d$.

The algebraic structure of the set *CDF* of cumulative distribution functions is much richer. This set is endowed with the operation of *convolution*

$$(\boldsymbol{F} * \boldsymbol{G})(x) := \int_{-\infty}^{+\infty} \boldsymbol{F}(t)\boldsymbol{G}(x - t)\, dt, \tag{7}$$

for all $x \in \mathbb{R}$. The convolution of cumulative distributions is a CDF. The convolution operation preserves continuity of CDFs. As it is known, $*$ is associative and commutative. It is also distributive: $(\boldsymbol{F} + \boldsymbol{G}) * \boldsymbol{H} = (\boldsymbol{F} * \boldsymbol{H}) + (\boldsymbol{G} * \boldsymbol{H})$. (But the sum $\boldsymbol{F} + \boldsymbol{G}$ is not a CDF.) If $\boldsymbol{F}$ and $\boldsymbol{G}$ are cumulative distributions corresponding to independent random variables X and Y defined on a probabilistic space, then the convolution $\boldsymbol{F} * \boldsymbol{G}$ is the cumulative distribution of the sum $X + Y$ of these random variables.

The *bounded addition* of CDFs is the binary operation $\oplus$ defined as follows:

$$(\boldsymbol{F} \oplus \boldsymbol{G})(x) := \min(1, \boldsymbol{F}(x) + \boldsymbol{G}(x)),$$

for all $x \in \mathbb{R}$. The bounded sum $\boldsymbol{F} \oplus \boldsymbol{G}$ of CDFs $\boldsymbol{F}$ and $\boldsymbol{G}$ is a CDF. By an analogy to Łukasiewicz logics, the sum $\boldsymbol{F} \oplus \boldsymbol{G}$ is called the *weak disjunction* of cumulative distributions $\boldsymbol{F}$ and $\boldsymbol{G}$. Furthermore, if $\boldsymbol{F}$ and $\boldsymbol{G}$ are continuous CDFs, $\boldsymbol{F} \oplus \boldsymbol{G}$ is continuous as well.

By the *strong conjunction* (also in an analogy to Łukasiewicz logics) we shall understand the binary operation $\otimes$ defined as

$$(\boldsymbol{F} \otimes \boldsymbol{G})(x) := \max(0, \boldsymbol{F}(x) + \boldsymbol{G}(x) - 1),$$

for all $x \in \mathbb{R}$. The strong conjunction $\boldsymbol{F} \otimes \boldsymbol{G}$ of CDFs $\boldsymbol{F}$ and $\boldsymbol{G}$ is a cumulative distribution. If $\boldsymbol{F}$ and $\boldsymbol{G}$ are continuous CDFs, $\boldsymbol{F} \otimes \boldsymbol{G}$ is continuous too.

Lemma 3.5 *For any* $\boldsymbol{F}, \boldsymbol{G}, \boldsymbol{H} \in CDF$,

(1) $\boldsymbol{F} \oplus \boldsymbol{G} = \boldsymbol{G} \oplus \boldsymbol{F}$,
(2) $\boldsymbol{F} \oplus (\boldsymbol{G} \oplus \boldsymbol{H}) = (\boldsymbol{F} \oplus \boldsymbol{G}) \oplus \boldsymbol{H}$,

(3) $\boldsymbol{F} \otimes \boldsymbol{G} = (\boldsymbol{F}^d \oplus \boldsymbol{G}^d)^d$ *and* $\boldsymbol{F} \oplus \boldsymbol{G} = (\boldsymbol{F}^d \otimes \boldsymbol{G}^d)^d$,
(4) $\boldsymbol{F} \otimes \boldsymbol{G} \leqslant \boldsymbol{F} \wedge \boldsymbol{G} \leqslant \boldsymbol{F} \vee \boldsymbol{G} \leqslant \boldsymbol{F} \oplus \boldsymbol{G}$.

Proof The proofs of (1)–(3) are straightforward. We shall give a proof for (4).

Claim 1 $\max(0, \boldsymbol{F}(x) + \boldsymbol{G}(x) - 1) \leqslant \min(\boldsymbol{F}(x), \boldsymbol{G}(x))$, *for all* $x \in \mathbb{R}$.

The above inequality holds if $\max(0, \boldsymbol{F}(x) + \boldsymbol{G}(x) - 1) = 0$. We consider the case when $\max(0, \boldsymbol{F}(x) + \boldsymbol{G}(x) - 1) = \boldsymbol{F}(x) + \boldsymbol{G}(x) - 1$. But for any $x \in \mathbb{R}$ we have that $\boldsymbol{F}(x) + \boldsymbol{G}(x) - 1 \leqslant \boldsymbol{F}(x)$ and $\boldsymbol{F}(x) + \boldsymbol{G}(x) - 1 \leqslant \boldsymbol{G}(x)$, because cumulative distribution functions are bounded from above by 1. Hence $\max(0, \boldsymbol{F}(x) + \boldsymbol{G}(x) - 1) = \boldsymbol{F}(x) + \boldsymbol{G}(x) - 1 \leqslant \min(\boldsymbol{F}(x), \boldsymbol{G}(x))$. This proves the claim. □

Claim 2 $\max(\boldsymbol{F}(x), \boldsymbol{G}(x)) \leqslant \min(1, \boldsymbol{F}(x) + \boldsymbol{G}(x))$, *for all* $x \in \mathbb{R}$.

The above inequality holds if $\min(1, \boldsymbol{F}(x) + \boldsymbol{G}(x)) = 1$. In the other case when $\min(1, \boldsymbol{F}(x) + \boldsymbol{G}(x)) = \boldsymbol{F}(x) + \boldsymbol{G}(x)$, it suffices to notice that $\boldsymbol{F}(x) \leqslant \boldsymbol{F}(x) + \boldsymbol{G}(x)$ and $\boldsymbol{G}(x) \leqslant \boldsymbol{F}(x) + \boldsymbol{G}(x)$ for all x, because cumulative distributions are non-negative functions. Hence $\max(\boldsymbol{F}(x), \boldsymbol{G}(x)) \leqslant \boldsymbol{F}(x) + \boldsymbol{G}(x) = \min(1, \boldsymbol{F}(x) + \boldsymbol{G}(x))$. □

(4) follows from the above claims. □

The algebra $\langle CDF, \wedge, \vee, {}^d, *, \oplus, \otimes\rangle$ is called the *extended De Morgan algebra* of cumulative distributions. A similar name applies to the algebra $\langle CCDF, \wedge, \vee, {}^d, *, \oplus, \otimes\rangle$.

The list of operations which are performable on the set *CDF* is longer. We mention here convex combinations of finite sequences of cumulative distributions as well as translations along the x-axis. These operations have not been included in the list of primitive operations of $\langle CDF, \wedge, \vee, {}^d, *, \oplus, \otimes\rangle$.

Let $\boldsymbol{F}$ be a cumulative distribution. For each real number r, we define the function $\boldsymbol{F}_r$ by the following condition:

$$\boldsymbol{F}_r(x) := \boldsymbol{F}(x + r), \quad \text{for all } x \in \mathbb{R}.$$

$\boldsymbol{F}_r$ is a cumulative distribution. If $r > 0$, the graph of $\boldsymbol{F}_r$ is obtained from the graph of $\boldsymbol{F}$ by means of the translation r units to the left. Obviously, $\boldsymbol{F}_0 = \boldsymbol{F}$. We have:

Lemma 3.6 *For any* $a, b \in \mathbb{R}$, $a \leqslant b$ *if and only if* $\boldsymbol{F}_a \leqslant \boldsymbol{F}_b$.

Proof The implication "$\Rightarrow$" is obvious.

"$\Leftarrow$." We assume that $\boldsymbol{F}_a \leqslant \boldsymbol{F}_b$. So $\boldsymbol{F}(x + a) \leqslant \boldsymbol{F}(x + b)$ for all $x \in \mathbb{R}$. Suppose that $b < a$. Then $x + b \leqslant x + a$, and consequently $\boldsymbol{F}(x + b) \leqslant \boldsymbol{F}(x + a)$ for all $x \in \mathbb{R}$, by the fact that $\boldsymbol{F}$ is monotone. It follows that

$$\boldsymbol{F}(x + b) = \boldsymbol{F}(x + a), \quad \text{for all } x \in \mathbb{R}. \tag{a}$$

Let $\Delta := a - b$. So $a = b + \Delta$, Δ is a positive number and by (a),

$$\boldsymbol{F}(b + x) = \boldsymbol{F}(b + \Delta + x), \quad \text{for all } x \in \mathbb{R}. \tag{b}$$

Putting $x := 0$, we obtain that

$$\boldsymbol{F}(b) = \boldsymbol{F}(b + \Delta). \tag{c}$$

As $\boldsymbol{F}$ is monotone, it follows that $\boldsymbol{F}$ is a constant function throughout the interval $[b, b + \Delta]$ taking the value $\boldsymbol{F}(b)$.

Claim 1 *For any natural n,* $\boldsymbol{F}(b) = \boldsymbol{F}(b + n\Delta)$.

Proof (*of the claim*) Induction on n. The case $n = 0$ is trivial. The case $n = 1$ holds in virtue of (c). Assume $\boldsymbol{F}(b) = \boldsymbol{F}(b + n\Delta)$. Then substituting $x = n\Delta$ in (b), we have that $\boldsymbol{F}(b + n\Delta) = \boldsymbol{F}(b + (n + 1)\Delta)$. So $\boldsymbol{F}(b) = \boldsymbol{F}(b + (n + 1)\Delta)$. This proves the claim. □

Claim 2 *For any natural n,* $\boldsymbol{F}(b) = \boldsymbol{F}(b - n\Delta)$.

The proof is similar.

Claims 1–2 imply that $\boldsymbol{F}$, being a non-decreasing function, takes the constant value $\boldsymbol{F}(b)$ throughout the real line. This is excluded, because limits of $\boldsymbol{F}$ at $-\infty$ and $+\infty$ are 0 and 1. The obtained inconsistency proves the lemma. □

It follows from the lemma that the family of cumulative distributions $\{\boldsymbol{F}_r : r \in \mathbb{R}\}$ forms a chain in the poset $\langle CDF, \leqslant \rangle$ and the order type of this chain is equal to λ, the order type of the set $\mathbb{R}$. One may then say that the poset $\langle CDF, \leqslant \rangle$ has a rather complicated order structure, also due to the fact that $\mathbb{R}$ is unbounded. For example, the poset $\langle CDF, \leqslant \rangle$ contains neither maximal nor minimal elements.

4 The Logic of Unary Predicates

Let $\boldsymbol{L}$ be an absolutely free algebra built from a countably infinite set of sentential variables $Var = \{P_n : n = 1, 2, \ldots\}$ and endowed with binary connectives $\wedge, \vee, *, \oplus, \otimes$ and one unary connective ^{d}. Thus the universe L of $\boldsymbol{L}$ consists of all sentential formulas formed from Var and the above connectives in the well-known manner. $\boldsymbol{L}$ is called the *language of unary probabilistic predicates*.

The variables of Var range over one-dimensional cumulative distributions. We therefore define the notion of a valuation. A *valuation* of the language $\boldsymbol{L}$ in the extended algebra $\boldsymbol{CDF}$ is an arbitrary homomorphism $h : \boldsymbol{L} \to \boldsymbol{CDF}$. h is unambiguously determined by its values on the set Var. If $\alpha(P_1, \ldots, P_n)$ is a formula (and all its variables are displayed) and h is a valuation such that $\boldsymbol{F}_i = h(P_i)$ for $i = 1, \ldots, n$, then the cumulative distribution $h(\alpha)$ is denoted as $\alpha(P_1/\boldsymbol{F}_1, \ldots, P_n/\boldsymbol{F}_n)$, or shortly $\alpha(\boldsymbol{F}_1, \ldots, \boldsymbol{F}_n)$.

Both the values of the arguments of formulas, that is, the values of variables occurring in a formula as well as the values of formulas themselves are one-dimensional cumulative distributions. The language $\boldsymbol{L}$ thus represents the syntax of *unary* probabilistic predicates. The more general case of probabilistic predicates of *arbitrary* arity, although theoretically important, is only incidentally mentioned in the final paragraph.

A *model* for the language $\boldsymbol{L}$ of unary probabilistic predicates is a pair

$$\boldsymbol{M} = (Var, h),$$

where h is a mapping defined on the set of variables Var assigning to each predicate variable $P \in Var$ a unary cumulative distribution $h(P)$. When h is clear from context, the cumulative distribution $h(P)$ will be denoted by $\boldsymbol{F}_P$. h is then recursively extended to a homomorphism (valuation) from $\boldsymbol{L}$ to $\boldsymbol{CDF}$ in the well-known way.

We define the relation of *probabilistic entailment* $\models$ on $\boldsymbol{L}$. For any $n \geqslant 1$ and any formulas $\alpha_1, \ldots, \alpha_n, \beta$ of $\boldsymbol{L}$ we define:

$$\alpha_1, \ldots, \alpha_n \models \beta \quad \Leftrightarrow_{df} \quad (\forall h\colon \boldsymbol{L} \to \boldsymbol{CDF})\ h(\alpha_1) \wedge \ldots \wedge h(\alpha_n) \leqslant h(\beta).$$

If X is an infinite set of formulas, we assume that

$$X \models \beta \ \Leftrightarrow_{df}\ \alpha_1, \ldots, \alpha_n \models \beta \ \text{ for some } n \geqslant 1 \text{ and some formulas } \alpha_1, \ldots, \alpha_n \in X.$$

Moreover, it is assumed that $\emptyset \models \beta$ for no formula β. Thus the above probabilistic logic does not possess tautologies.

For each set of formulas X define:

$$C(X) := \{\beta \in L\colon X \models \beta\}.$$

C is an operation defined on the power set $\wp(L)$, assigning to each subset $X \subseteq L$ the set $C(X)$.

Theorem 4.1 *C is a finitary and structural consequence operation, that is, for all $X \subseteq L$:*

(C1) $X \subseteq C(X)$ (reflexivity)
(C2) $X \subseteq Y$ *implies* $C(X) \subseteq C(Y)$ (monotonicity)
(C3) $C(C(X)) \subseteq C(X)$ (idempotency)
(C4) $C(X) = \bigcup\{C(X_f)\colon X_f$ *is a finite subset of* $X\}$ (finitariness)
(C5) $eC(X) \subseteq C(eX)$ *for every endomorphism* $e\colon \boldsymbol{L} \to \boldsymbol{L}$. (structurality)

Moreover,

(C6) $C(\emptyset) = \emptyset$.

(The above conditions are not logically independent—(C2) follows from (C4).)

Proof A verification of (C1)–(C3) is straightforward. We shall check (C5). It suffices to prove (C5) for arbitrary finite sets $X = \{\alpha_1, \ldots, \alpha_n\}$.

We assume that $\alpha_1, \ldots, \alpha_n \models \beta$. Let $e\colon \boldsymbol{L} \to \boldsymbol{L}$ be an endomorphism. We claim that $e\alpha_1, \ldots, e\alpha_n \models e\beta$.

Let $h\colon \boldsymbol{L} \to \boldsymbol{CDF}$ be an arbitrary but fixed homomorphism. The composition $h \circ e$ is also a homomorphism from $\boldsymbol{L}$ to $\boldsymbol{CDF}$. As $\alpha_1, \ldots, \alpha_n \models \beta$, we therefore obtain that

$$(h \circ e)(\alpha_1) \wedge \cdots \wedge (h \circ e)(\alpha_n) \leqslant (h \circ e)(\beta) \tag{8}$$

in the algebra $\boldsymbol{CDF}$, by the definition of $\models$. But (8) means that

$$h(e\alpha_1) \wedge \cdots \wedge h(e\alpha_n) \leqslant h(e\beta). \tag{9}$$

Thus $e\alpha_1, \ldots, e\alpha_n \models e\beta$, because h is an arbitrary homomorphism. So (C5) holds. □

The monograph (Wójcicki 1988) contains a good exposition of the theory of consequence operations.

As C is a purely inferential consequence operation (there are no tautologies), it follows that C can be adequately characterized in terms of standard proper rules of inference. However, no inferential base for C is known thus far.

Problems. *Give an adequate inferential base for C. Give a description of C-filters on the algebra* $\boldsymbol{CDF}$.

We may also define (in a fully analogous way) the consequence operation C_{con} on the power set $\wp(L)$ in terms of valuations in the extended algebra $\boldsymbol{CCDF}$ of *continuous* cumulative distributions. As $\boldsymbol{CCDF}$ is a subalgebra of $\boldsymbol{CDF}$, it follows that C_{con} is stronger than C, that is, $C(X) \subseteq C_{con}(X)$ for any set $X \subseteq L$. A problem is whether these two consequences coincide on finite sets X.

5 Further Remarks

In a standard way, the lattice $\langle CDF, \wedge, \vee \rangle$ can be augmented with $\mathbf{0}$ (being the least element) and $\mathbf{1}$ (the greatest element). $\langle CDF, \wedge, \vee \rangle$ is thereby extended to a bounded distributive lattice. $\mathbf{0}$ and $\mathbf{1}$ are not cumulative distributions; they may be regarded as generalized functions in the sense of Sobolev–Schwartz. One may say that $\mathbf{1}$ is a "cumulative distribution" on the real line which is everywhere equal to 1, but $\lim_{x \to -\infty} \mathbf{1}(x) = 0$. Analogously, $\mathbf{0}$ is a "function" on $\mathbb{R}$ everywhere equal to 0, but $\lim_{x \to +\infty} \mathbf{0}(x) = 1$.

(Yet another option is to add two points at infinity, viz., $-\infty$ and $+\infty$, to $\mathbb{R}$ and define $\mathbb{R}^{\infty} := \mathbb{R} \cup \{-\infty, +\infty\}$. Each cumulative distribution $\boldsymbol{F}$ is extended onto $\mathbb{R}^{\infty}$ by adopting that $\boldsymbol{F}(-\infty) = 0$ and $\boldsymbol{F}(+\infty) = 1$. $\mathbf{0}$ and $\mathbf{1}$ are then treated as functions defined throughout $\mathbb{R}^{\infty}$ according to the formulas:

$$\mathbf{0}(x) = 0 \text{ for all } x \in \mathbb{R} \cup \{-\infty\}, \quad \text{and} \quad \mathbf{0}(+\infty) = 1,$$
$$\mathbf{1}(-\infty) = 0, \quad \text{and} \quad \mathbf{1}(x) = 1 \text{ for all } x \in \mathbb{R} \cup \{+\infty\}.$$

One may alternatively define ***CDF*** to be the algebra of such extended functions defined on $\mathbb{R}^{\infty}$.)

The operations ^{d}, $\oplus$, $\otimes$ on the set *CDF* are also extended on the universe $CDF_b := CDF \cup \{\mathbf{0}, \mathbf{1}\}$:

$$\mathbf{1}^d := \mathbf{0} \quad \text{and} \quad \mathbf{0}^d := \mathbf{1}.$$

Furthermore, we assume that

$$\mathbf{1} \oplus \boldsymbol{a} = \boldsymbol{a} \oplus \mathbf{1} = \mathbf{1} \quad \text{and} \quad \mathbf{0} \oplus \boldsymbol{a} = \boldsymbol{a} \oplus \mathbf{0} = \boldsymbol{a}$$

and

$$\mathbf{1} \otimes \boldsymbol{a} = \boldsymbol{a} \otimes \mathbf{1} = \boldsymbol{a} \quad \text{and} \quad \mathbf{0} \otimes \boldsymbol{a} = \boldsymbol{a} \otimes \mathbf{0} = \mathbf{0}$$

for all $\boldsymbol{a} \in CDF \cup \{\mathbf{0}, \mathbf{1}\}$.

It is unclear how to meaningfully combine the constants $\mathbf{0}$ and $\mathbf{1}$ with the convolution operation $*$. In the technical sense, one may assume that $\mathbf{1}$ is the two-sided unit (that is, the neutral element) for the convolution operation $*$ and $\mathbf{0}$ is the zero for $*$, i.e.,

$$\mathbf{1} * \boldsymbol{a} = \boldsymbol{a} * \mathbf{1} = \boldsymbol{a} \quad \text{and} \quad \mathbf{0} * \boldsymbol{a} = \boldsymbol{a} * \mathbf{0} = \mathbf{0}$$

hold for all $\boldsymbol{a} \in CCDF \cup \{\mathbf{0}, \mathbf{1}\}$. (But formula (7) does not apply to the above extension of $*$.)

As a result we obtain a (bounded) De Morgan algebra

$$\boldsymbol{CDF}_b := \langle CDF_b, \wedge, \vee, {}^d, *, \oplus, \otimes, \mathbf{0}, \mathbf{1}\rangle,$$

with the additional operations $*$, $\oplus$, $\otimes$ and the constants $\mathbf{0}$, $\mathbf{1}$.

$\boldsymbol{CDF}_b$ is called *the extended De Morgan algebra* of cumulative distributions.

In a fully analogous way, one defines the algebra

$$\boldsymbol{CCDF}_b := \langle CCDF_b, \wedge, \vee, {}^d, *, \oplus, \otimes, \mathbf{0}, \mathbf{1}\rangle,$$

the extended De Morgan algebra of *continuous* cumulative distribution functions.

6 Another System of Logic

The set of connectives of $\boldsymbol{L}$ is extended by adding two constants 0 and 1 (interpreted as $\mathbf{0}$ and $\mathbf{1}$ in $\boldsymbol{CDF}_b$, respectively). We denote the new language by $\boldsymbol{L}_b$.

The algebra $\boldsymbol{CDF}_b$ may be treated as a logical matrix for $\boldsymbol{L}_b$ with $\mathbf{1}$ being the designated element. One may then define yet another notion of *probabilistic entailment* $\models_b$ on $\boldsymbol{L}_b$. For a set of formulas X of $\boldsymbol{L}_b$ and a formula β of $\boldsymbol{L}_b$, we stipulate that

$$X \models_b \beta \;\Leftrightarrow\; (\forall h\colon \boldsymbol{L} \to \boldsymbol{CCDF}_b)\,(h(\beta) = \mathbf{1} \text{ whenever } h(\alpha) = \mathbf{1} \text{ for all } \alpha \in X).$$

For a set of formulas X of $\boldsymbol{L}_b$, we define:

$$C_b(X) := \{\, \beta \in L\colon X \models_b \beta \,\}.$$

C_b is a structural consequence operation, i.e., it satisfies the above conditions (C1)–(C3), (C5)–(C6) for all sets of formulas X, Y.

Formula 1 is a tautology in the sense of C_b. There are more tautologies of C_b but each tautology necessarily involves constants 0 or 1 as subformulas. Indeed, for any formula $\alpha(P_1, \ldots, P_n)$ of $\boldsymbol{L}$ (which does not contain 0 or 1), where $n \geqslant 1$, and any cumulative distribution functions $\boldsymbol{F}_1, \ldots, \boldsymbol{F}_n$, the element $\alpha(\boldsymbol{F}_1, \ldots, \boldsymbol{F}_n)$ is also a cumulative distribution; therefore it is not equal to the generalized function $\mathbf{1}$. Consequently, the formula $\alpha(P_1, \ldots, P_n)$ is not a tautology of the logic C_b.

It follows from the above remarks that C_b does not possess a (definable) implication connective which would determine the rule of detachment together with the law of identity, so that both rules should be valid in C_b. Thus C_b, together with the previously defined logics, is not protoalgebraic. It is also clear that the logic C_b is stronger than C on $\boldsymbol{L}$, that is for any set $X \subseteq L$ and any $\alpha \in L$, $\alpha \in C(X)$ implies that $\alpha \in C_b(X)$. (Note that every formula γ of L acquires neither the value $\mathbf{1}$ nor $\mathbf{0}$.)

No adequate set of (possibly infinitistic) rules of inference adequate for C_b is available thus far.

The algebra $\boldsymbol{CDF}$ is endowed with the operation $\to$, where

$$(\boldsymbol{F} \to \boldsymbol{G}) := \boldsymbol{F}^d \oplus \boldsymbol{G},$$

for all $\boldsymbol{F}, \boldsymbol{G} \in CDF$. It is clear that $\boldsymbol{F} \to \boldsymbol{G}$ is a cumulative distribution, that is, $\boldsymbol{F} \to \boldsymbol{G} \in CDF$ whenever $\boldsymbol{F}, \boldsymbol{G} \in CDF$. Moreover $\boldsymbol{F} \to \boldsymbol{G}$ is a continuous cumulative distribution whenever $\boldsymbol{F}, \boldsymbol{G} \in CCDF$.

We obviously have that

$$(\boldsymbol{F} \to \boldsymbol{G})(x) = \min(1, \boldsymbol{F}^d(x) + \boldsymbol{G}(x)) = \min(1, 1 - \boldsymbol{F}(-x) + \boldsymbol{G}(x)),$$

for any real number x. Moreover,

$$\boldsymbol{F} \to \boldsymbol{G} = \boldsymbol{G}^d \to \boldsymbol{F}^d, \quad \text{and} \tag{10}$$

$$\boldsymbol{G} \leqslant \boldsymbol{F} \to \boldsymbol{G} \tag{11}$$

for all $\boldsymbol{F}, \boldsymbol{G} \in CDF$.

Equation (10) directly follows from the definitions of the operations $\rightarrow$ and ^{d}. As to (11), suppose *a contrario* that $\boldsymbol{G}(x) > (F \rightarrow G)(x)$ for some $x \in \mathbb{R}$, that is,

$$\boldsymbol{G}(x) > \min(1, 1 - \boldsymbol{F}(-x) + \boldsymbol{G}(x)). \qquad (*)$$

We consider two cases.
Case 1. $1 = \min(1, 1 - \boldsymbol{F}(-x) + \boldsymbol{G}(x))$. Then, by $(*)$, $\boldsymbol{G}(x) > 1$, which is excluded.
Case 2. $1 - \boldsymbol{F}(-x) + \boldsymbol{G}(x)) = \min(1, 1 - \boldsymbol{F}(-x) + \boldsymbol{G}(x))$. Hence, by $(*)$, $\boldsymbol{G}(x) > 1 - \boldsymbol{F}(-x) + \boldsymbol{G}(x)$, which gives that $\boldsymbol{F}(-x) > 1$. This is also excluded.
So (11) follows.

Proposition 6.1 *Suppose $\boldsymbol{F}, \boldsymbol{G} \in CDF$ and $x \in \mathbb{R}$. Then the conditions $\boldsymbol{F}(-x) = 1$ and $(\boldsymbol{F} \rightarrow \boldsymbol{G})(x) = 1$ imply $\boldsymbol{G}(x) = 1$.*

Proof We assume that $\boldsymbol{F}(-x) = 1$ and $(\boldsymbol{F} \rightarrow \boldsymbol{G})(x) = 1$. We have $1 = (\boldsymbol{F} \rightarrow \boldsymbol{G})(x) = \min(1, 1 - \boldsymbol{F}(-x) + \boldsymbol{G}(x))$, which gives that $1 - \boldsymbol{F}(-x) + \boldsymbol{G}(x) \geqslant 1$. Hence $\boldsymbol{F}(-x) \leqslant \boldsymbol{G}(x)$. As $\boldsymbol{F}(-x) = 1$, we infer that $\boldsymbol{G}(x) = 1$. □

The property of the operation $\rightarrow$ expressed in Proposition 6.1 may be regarded as the validity of a version of the detachment rule.

Although the implication symbol is used to denote the above operation, it would be rather unnatural to attach the name 'implication' to the above function. The reason is in the fact that the operation $\rightarrow$ fails to satisfy the law of identity, relevant in metalogical consequences, that is, it is not the case that $(\boldsymbol{F} \rightarrow \boldsymbol{F})(x) = 1$ for all $x \in \mathbb{R}$. But we have:

$$(\boldsymbol{F} \rightarrow \boldsymbol{F})(x) = \begin{cases} 1 & \text{if } 0 \leqslant x, \\ 1 - \boldsymbol{F}(-x) + \boldsymbol{F}(x) & \text{otherwise,} \end{cases}$$

as one can easily check.

7 A Connection with Łukasiewicz Logics

We define:

$$\boldsymbol{\Delta} := \{\boldsymbol{F} \in CDF : \boldsymbol{F}(0) = 1\}.$$

Thus, if $\boldsymbol{F} \in \boldsymbol{\Delta}$, then $\boldsymbol{F}(x) = 1$, for all $x \geqslant 0$.

Theorem 7.1 *$\boldsymbol{\Delta}$ satisfies the following conditions, for all $\boldsymbol{F}, \boldsymbol{G} \in CDF$:*

(a) $\boldsymbol{F} \rightarrow \boldsymbol{F} \in \boldsymbol{\Delta}$;
(b) *if $\boldsymbol{F} \in \boldsymbol{\Delta}$ and $\boldsymbol{F} \leqslant \boldsymbol{G}$, then $\boldsymbol{G} \in \boldsymbol{\Delta}$;*
(c) *if $\boldsymbol{F} \in \boldsymbol{\Delta}$ and $\boldsymbol{G} \in \boldsymbol{\Delta}$, then $\boldsymbol{F} \otimes \boldsymbol{G} \in \boldsymbol{\Delta}$;*
(d) *if $\boldsymbol{F} \in \boldsymbol{\Delta}$ and $\boldsymbol{G} \in \boldsymbol{\Delta}$, then $\boldsymbol{F} \wedge \boldsymbol{G} \in \boldsymbol{\Delta}$;*
(e) *if $\boldsymbol{F} \in \boldsymbol{\Delta}$ and $\boldsymbol{F} \rightarrow \boldsymbol{G} \in \boldsymbol{\Delta}$, then $\boldsymbol{G} \in \boldsymbol{\Delta}$.*

Conditions (b) and (c) state that $\boldsymbol{\Delta}$ is a filter in the strong sense. According to (b) and (d), $\boldsymbol{\Delta}$ is also a "standard" lattice-theoretic filter. In turn, (a) and (e) state that $\boldsymbol{\Delta}$ validates the identity axiom and the detachment rule corresponding to $\boldsymbol{\rightarrow}$.

Proof Suppose $\boldsymbol{F}, \boldsymbol{G} \in CDF$.

(a) $(\boldsymbol{F} \boldsymbol{\rightarrow} \boldsymbol{F})(0) = \min(1, 1 - \boldsymbol{F}(-0) + \boldsymbol{F}(0)) = \min(1, 1 - \boldsymbol{F}(0) + \boldsymbol{F}(0)) = \min(1, 1) = 1$.

(b) is immediate.

(c) Assume $\boldsymbol{F}(0) = \boldsymbol{G}(0) = 1$. Then $\boldsymbol{F} \otimes \boldsymbol{G}(0) = \max(0, \boldsymbol{F}(0) + \boldsymbol{G}(0) - 1) = \max(0, 1 + 1 - 1) = \max(0, 1) = 1$.

(d) We have that $\boldsymbol{F} \otimes \boldsymbol{G} \leqslant \boldsymbol{F} \wedge \boldsymbol{G}$. Hence if $\boldsymbol{F}, \boldsymbol{G} \in \boldsymbol{\Delta}$, then $\boldsymbol{F} \wedge \boldsymbol{G} \in \boldsymbol{\Delta}$, by (b) and (c).

(e) Suppose $\boldsymbol{F} \in \boldsymbol{\Delta}$ and $\boldsymbol{F} \boldsymbol{\rightarrow} \boldsymbol{G} \in \boldsymbol{\Delta}$. Then $\boldsymbol{F}(0) = 1$ and $\min(1, 1 - \boldsymbol{F}(0) + \boldsymbol{G}(0)) = 1$. Hence $\min(1, 1 - 1 + \boldsymbol{G}(0)) = 1$, that is, $\min(1, \boldsymbol{G}(0)) = 1$. This gives that $\boldsymbol{G}(0) = 1$. □

On the unit interval $I = [0, 1]$, we define the operations $\rightarrow, \wedge, \vee, \otimes, \oplus, \neg$ by

$(\rightarrow)_Ł$ $a \rightarrow b := \min(1, 1 - a + b)$,
$(\wedge)_Ł$ $a \wedge b := \min(a, b)$,
$(\vee)_Ł$ $a \vee b := \max(a, b)$,
$(\otimes)_Ł$ $a \otimes b := \max(0, a + b - 1)$,
$(\oplus)_Ł$ $a \oplus b := \min(1, a + b)$,
$(\neg)_Ł$ $\neg a := 1 - a$.

They are called the *Łukasiewicz operations*. (To simplify the notation, the Łukasiewicz strong conjunction $\otimes$ and the weak disjunction $\oplus$ are denoted by the same symbols as in the algebra of cumulative distributions.)

$\boldsymbol{A}_c := \langle I, \rightarrow, \wedge, \vee, \oplus, \otimes, \neg \rangle$ is the infinite Łukasiewicz algebra. All the displayed operations are treated here as primitive operations of $\boldsymbol{A}_c$ but they are definable in terms of the operations $\rightarrow$ and $\neg$ in the well-known manner. (One may also take $\oplus$ and $\neg$ as primitive operations, because $a \rightarrow b = \neg a \oplus b$, $a \vee b = \neg(\neg a \oplus b) \oplus b$, $a \wedge b = \neg(a \vee \neg b)$, and $a \otimes b = \neg(a \oplus \neg b)$, for all $a, b \in I$.)

The pair $\langle \boldsymbol{A}_c, \{1\} \rangle$ is called the infinite *Łukasiewicz matrix*.

We now consider the extended algebra $\langle CDF, \boldsymbol{\rightarrow}, \boldsymbol{\wedge}, \boldsymbol{\vee}, \boldsymbol{\oplus}, \boldsymbol{\otimes}, {}^{d} \rangle$ of cumulative distribution functions augmented with the operation $\boldsymbol{\rightarrow}$ defined as in Sect. 5. (But the convolution operation is discarded.) To simplify the notation, we refer to this algebra by the same symbol $\boldsymbol{CDF}_0$. The algebra $\boldsymbol{CDF}_0 = \langle CDF, \boldsymbol{\rightarrow}, \boldsymbol{\wedge}, \boldsymbol{\vee}, \boldsymbol{\oplus}, \boldsymbol{\otimes}, {}^{d} \rangle$ is similar to $\boldsymbol{A}_c$.

The following theorem is an immediate consequence of the definitions of the above two algebras:

Theorem 7.2 *The mapping $h\colon CDF \rightarrow I$ given by $h(\boldsymbol{F}) := \boldsymbol{F}(0)$ is a homomorphism from the algebra $\boldsymbol{CDF}_0$ onto the Łukasiewicz algebra $\boldsymbol{A}_c$.*

Moreover, the filter $\boldsymbol{\Delta}$ is the pre-image of $\{1\}$ with respect to h, that is, $\boldsymbol{\Delta} = \{\boldsymbol{F} \in CDF : h(\boldsymbol{F}) = 1\}$.

Proof h is well defined. h is surjective, because for every number $r \in I$ there exists a cumulative distribution function $\boldsymbol{F}$ such that $\boldsymbol{F}(0) = r$.

Let $\boldsymbol{F}, \boldsymbol{G} \in CDF$. We set $a := \boldsymbol{F}(0)$ and $b := \boldsymbol{G}(0)$. Then we have $h(\boldsymbol{F} \to \boldsymbol{G}) = (\boldsymbol{F} \to \boldsymbol{G})(0) = \min(1, 1 - \boldsymbol{F}(-0) + \boldsymbol{G}(0)) = \min(1, 1 - a + b) = a \to b = \boldsymbol{F}(0) \to \boldsymbol{G}(0) = h(\boldsymbol{F}) \to h(\boldsymbol{G})$.

In a similar manner, one checks the remaining conditions imposed on h to be a homomorphism.

The second statement is immediate. □

Let $\boldsymbol{L}_0$ be the propositional language endowed with the binary connectives $\to, \wedge, \vee, \oplus, \otimes$ and the unary ^{d}. $\boldsymbol{L}_0$ differs from the language $\boldsymbol{L}$ we have defined by deleting the connective of convolution and, on the other hand, by adjoining the binary connective $\to$. $\boldsymbol{L}_0$ is thus the (full) language of Łukasiewicz logics.

We shall treat the filter $\boldsymbol{\Delta}$ as a designated subset of the algebra $\boldsymbol{CDF}_0$ and define the matrix $\boldsymbol{M} = \langle \boldsymbol{CDF}_0, \boldsymbol{\Delta} \rangle$. Let $C_{\boldsymbol{M}}$ be the consequence operation determined in the language $\boldsymbol{L}_0$ by $\boldsymbol{M}$ in the standard way.

Theorem 7.3 *The consequence operation $C_{\boldsymbol{M}}$ coincides with the consequence operation determined by the Łukasiewicz matrix $\langle \boldsymbol{A}_c, \{1\} \rangle$.*

Proof The consequence determined by a matrix and the consequence defined by any strict surjective homomorphic image of this matrix coincide—see e.g., (Wójcicki 1988, Lemma 3.1.8). (A strict homomorphism between matrices does not paste together designated elements with undesignated ones.) The above mapping h is a surjective, strict homomorphism from the matrix $\boldsymbol{M}$ onto $\langle \boldsymbol{A}_c, \{1\} \rangle$. □

The above theorem thus establishes the relationship between the infinitely-valued Łukasiewicz logic and the above consequence operations determined by the algebra of cumulative distribution functions.

In particular we obtain:

Corollary 7.4 *Let $\boldsymbol{F}, \boldsymbol{G}, \boldsymbol{H} \in CDF$.*

(f) $\boldsymbol{F} \to (\boldsymbol{G} \to \boldsymbol{F}) \in \boldsymbol{\Delta}$,

(g) $(\boldsymbol{F} \to \boldsymbol{G}) \to ((\boldsymbol{G} \to \boldsymbol{H}) \to (\boldsymbol{F} \to \boldsymbol{H})) \in \boldsymbol{\Delta}$.

(h) *More generally, if $\phi(P_1, \ldots, P_n)$ is a formula from the language $\boldsymbol{L}_0$ and $\phi(P_1, \ldots, P_n)$ is valid in the infinite Łukasiewicz matrix, then $\phi(\boldsymbol{F}_1, \ldots, \boldsymbol{F}_n) \in \boldsymbol{\Delta}$, for all $\boldsymbol{F}_1, \ldots, \boldsymbol{F}_n \in CDF$. Moreover, $\boldsymbol{\Delta}$ consists exactly of cumulative distribution functions of the form $\phi(\boldsymbol{F}_1, \ldots, \boldsymbol{F}_n)$, where $\phi(P_1, \ldots, P_n)$ is a tautology of the infinitely-valued Łukasiewicz logic.* □

8 Is There a Probability Logic?

The set CDF is endowed with a bunch of binary operations and the unary ^{d}. It is appropriate to look at the structure $\langle CDF, \to, \wedge, \vee, \oplus, \otimes, *, ^d \rangle$ (with the convolution operation $*$ included) from a more general universal algebraic perspective. We shall apply the terminology and notation adopted in Dunn and Hardegree (2001, p. 398).

A *tonoid* is a structure $\langle A, \leqslant, \{\circ_i\}_{i\in I}\rangle$ such that $\langle A, \leqslant\rangle$ is a poset, $\langle A, \{\circ_i\}_{i\in I}\rangle$ is an algebra with the property that each operation $f \in \{\circ_i\}_{i\in I}$ is either isotonic or antitonic, in each of its argument positions.

Theorem 8.1 *The structure* $\langle CDF, \leqslant, \rightarrow, \wedge, \vee, \oplus, \otimes, *, ^d\rangle$ *is a tonoid. More precisely, if* $\boldsymbol{F}, \boldsymbol{G}, \boldsymbol{H} \in CDF$, *then*

(i) *for any operation* $\bullet$ *from the list* $\oplus, \otimes, *$, *if* $\boldsymbol{F} \leqslant \boldsymbol{G}$, *then* $\boldsymbol{F} \bullet \boldsymbol{H} \leqslant \boldsymbol{G} \bullet \boldsymbol{H}$ *and* $\boldsymbol{H} \bullet \boldsymbol{F} \leqslant \boldsymbol{H} \bullet \boldsymbol{G}$;
(ii) *if* $\boldsymbol{F} \leqslant \boldsymbol{G}$, *then* $\boldsymbol{H} \rightarrow \boldsymbol{F} \leqslant \boldsymbol{H} \rightarrow \boldsymbol{G}$ *and* $\boldsymbol{G} \rightarrow \boldsymbol{H} \leqslant \boldsymbol{F} \rightarrow \boldsymbol{H}$,
(iii) *if* $\boldsymbol{F} \leqslant \boldsymbol{G}$, *then* $\boldsymbol{G}^d \leqslant \boldsymbol{F}^d$.

Proof (i) directly follows from the definitions and the commutativity of the operations $\oplus$, $\otimes$ and $*$. (iii) follows from Lemma 3.4. We shall check (ii).

Assume $\boldsymbol{F} \leqslant \boldsymbol{G}$ and let r be a real number. We compute

$$(\boldsymbol{H}\rightarrow\boldsymbol{F})(r) = \min(1, 1 - \boldsymbol{H}(-r) + \boldsymbol{F}(r)) \leqslant \min(1, 1 - \boldsymbol{H}(-r) + \boldsymbol{G}(r)) = (\boldsymbol{H}\rightarrow\boldsymbol{G})(r),$$

because $\boldsymbol{F}(r) \leqslant \boldsymbol{G}(r)$ and hence $1 - \boldsymbol{H}(-r) + \boldsymbol{F}(r) \leqslant 1 - \boldsymbol{H}(-r) + \boldsymbol{G}(r)$.

Similarly,

$$(\boldsymbol{G}\rightarrow\boldsymbol{H})(r) = \min(1, 1 - \boldsymbol{G}(-r) + \boldsymbol{H}(r)) \leqslant \min(1, 1 - \boldsymbol{F}(-r) + \boldsymbol{G}(r)) = (\boldsymbol{F}\rightarrow\boldsymbol{H})(r),$$

because $1 - \boldsymbol{G}(-r) \leqslant 1 - \boldsymbol{F}(-r)$. □

An abstract algebra $\langle A, \wedge, \vee, \{\circ_i\}_{i\in I}\rangle$ is called a *distributoid* (see Dunn and Hardegree 2001, p. 398) if $\langle A, \wedge, \vee\rangle$ is a distributive lattice, and each $f \in \{\circ_i\}_{i\in I}$ is a (finitary) operation on A that "distributes" in each of its argument places over $\wedge$ or $\vee$, leaving the lattice operation unchanged or switching it with its dual.

If we delete the convolution operation $*$ from the above list, we obtain the following stronger result.

Theorem 8.2 *The algebra* $\langle CDF, \wedge, \vee, \rightarrow, \oplus, \otimes, ^d\rangle$ *is a distributoid. More exactly, if* $\boldsymbol{F}, \boldsymbol{G}, \boldsymbol{H} \in CDF$, *then*

(i) $(\boldsymbol{F} \vee \boldsymbol{G}) \oplus \boldsymbol{H} = (\boldsymbol{F} \oplus \boldsymbol{H}) \vee (\boldsymbol{G} \oplus \boldsymbol{H})$,
(ii) $(\boldsymbol{F} \wedge \boldsymbol{G}) \otimes \boldsymbol{H} = (\boldsymbol{F} \otimes \boldsymbol{H}) \wedge (\boldsymbol{G} \otimes \boldsymbol{H})$,
(iii) $(\boldsymbol{F} \vee \boldsymbol{G}) \rightarrow \boldsymbol{H} = (\boldsymbol{F} \rightarrow \boldsymbol{H}) \wedge (\boldsymbol{G} \rightarrow \boldsymbol{H})$,
(iv) $\boldsymbol{H} \rightarrow (\boldsymbol{F} \wedge \boldsymbol{G}) = (\boldsymbol{H} \rightarrow \boldsymbol{F}) \vee (\boldsymbol{H} \rightarrow \boldsymbol{G})$,
(v) $(\boldsymbol{F} \vee \boldsymbol{G})^d = \boldsymbol{F}^d \wedge \boldsymbol{G}^d$ *and* $(\boldsymbol{F} \wedge \boldsymbol{G})^d = \boldsymbol{F}^d \vee \boldsymbol{G}^d$.

(We do not know if the convolution $*$ distributes in each argument.)

Proof We recall the following well-known facts.

Claim If a, b, c are real numbers, then

(a) $a + \max(b, c) = \max(a + b, a + c)$;
(b) $a + \min(b, c) = \min(a + b, a + c)$;
(c) $a - \max(b, c) = \min(a - b, a - c)$.

Fix a real number r. We then have:

$$
\begin{aligned}
\text{(i)}\quad ((\boldsymbol{F} \vee \boldsymbol{G}) \oplus \boldsymbol{H})(r) &= \min(1, (\boldsymbol{F} \vee \boldsymbol{G})(r) + \boldsymbol{H}(r)) \\
&= \min(1, \boldsymbol{H}(r) + \max(\boldsymbol{F}(r), \boldsymbol{G}(r))) \\
&= \min(1, \max(\boldsymbol{H}(r) + \boldsymbol{F}(r), \boldsymbol{H}(r) + \boldsymbol{G}(r))) \quad \text{(by Claim (a))} \\
&= \max(\min(1, \boldsymbol{H}(r) + \boldsymbol{F}(r)), \min(1, \boldsymbol{H}(r) + \boldsymbol{G}(r))) \\
&\qquad \text{(by distributivity of the chain } \langle \mathbb{R}, \leqslant \rangle) \\
&= ((\boldsymbol{F} \oplus \boldsymbol{H}) \vee (\boldsymbol{G} \oplus \boldsymbol{H}))(r)
\end{aligned}
$$

$$
\begin{aligned}
\text{(ii)}\quad ((\boldsymbol{F} \wedge \boldsymbol{G}) \otimes \boldsymbol{H})(r) &= \max(0, (\boldsymbol{F} \wedge \boldsymbol{G})(r) + \boldsymbol{H}(r) - 1) \\
&= \max(0, \min(\boldsymbol{F}(r), \boldsymbol{G}(r)) + \boldsymbol{H}(r) - 1) \quad \text{(by Claim (b))} \\
&= \max(0, \min(\boldsymbol{F}(r) + \boldsymbol{H}(r) - 1, \boldsymbol{G}(r) + \boldsymbol{H}(r) - 1)) \\
&= \min(\max(0, \boldsymbol{F}(r) + \boldsymbol{H}(r) - 1), \max(0, \boldsymbol{G}(r) + \boldsymbol{H}(r) - 1)) \\
&\qquad \text{(by distributivity of the chain } \langle \mathbb{R}, \leqslant \rangle) \\
&= \min((\boldsymbol{F} \otimes \boldsymbol{H})(r), (\boldsymbol{G} \otimes \boldsymbol{H})(r)) \\
&= ((\boldsymbol{F} \otimes \boldsymbol{H}) \wedge (\boldsymbol{G} \otimes \boldsymbol{H}))(r)
\end{aligned}
$$

$$
\begin{aligned}
\text{(iii)}\quad ((\boldsymbol{F} \vee \boldsymbol{G}) \to \boldsymbol{H})(r) &= \min(1, 1 - (\boldsymbol{F} \vee \boldsymbol{G})(-r) + \boldsymbol{H}(r)) \\
&= \min(1, 1 - \max(\boldsymbol{F}(-r), \boldsymbol{G}(-r)) + \boldsymbol{H}(r)) \\
&= \min(1, 1 + \boldsymbol{H}(r) - \max(\boldsymbol{F}(-r), \boldsymbol{G}(-r))) \\
&= \min(1, \min(1 + \boldsymbol{H}(r) - \boldsymbol{F}(-r), 1 + \boldsymbol{H}(r) - \boldsymbol{G}(-r))) \\
&\qquad \text{(by Claim (c))} \\
&= \min(1, \min(1 - \boldsymbol{F}(-r) + \boldsymbol{H}(r), 1 - \boldsymbol{G}(-r) + \boldsymbol{H}(r))) \\
&= \min(1, \min((\boldsymbol{F} \to \boldsymbol{H})(r), (\boldsymbol{G} \to \boldsymbol{H})(r))) \\
&= \min((\boldsymbol{F} \to \boldsymbol{H})(r), (\boldsymbol{G} \to \boldsymbol{H})(r)) \\
&= ((\boldsymbol{F} \to \boldsymbol{H}) \wedge (\boldsymbol{G} \to \boldsymbol{H}))(r)
\end{aligned}
$$

$$
\begin{aligned}
\text{(iv)}\quad (\boldsymbol{H} \to (\boldsymbol{F} \wedge \boldsymbol{G}))(r) &= \min(1, 1 - \boldsymbol{H}(-r) + \min(\boldsymbol{F}(r), \boldsymbol{G}(r))) \\
&= \min(1, \min(1 - \boldsymbol{H}(-r) + \boldsymbol{F}(r), 1 - \boldsymbol{H}(-r) + \boldsymbol{G}(r))) \\
&\qquad \text{(by Claim (b))} \\
&= \min(1, \min((\boldsymbol{H} \to \boldsymbol{F})(r), (\boldsymbol{H} \to \boldsymbol{G})(r))) \\
&= \min((\boldsymbol{H} \to \boldsymbol{F})(r), (\boldsymbol{H} \to \boldsymbol{G})(r)) \\
&= ((\boldsymbol{H} \to \boldsymbol{F}) \wedge (\boldsymbol{H} \to \boldsymbol{G}))(r)
\end{aligned}
$$

(v) This is established in Lemma 3.4. □

Note. The reduct $\langle \boldsymbol{CDF}, \wedge, \vee, \oplus, \otimes \rangle$ of the full algebra $\boldsymbol{CDF}$ is a subalgebra of the $\mathbb{R}$-power of the reduct $\langle I, \wedge, \vee, \oplus, \otimes \rangle$ of the infinite Łukasiewicz algebra $\boldsymbol{A}_c$. Therefore every identity valid in $\langle I, \wedge, \vee, \oplus, \otimes \rangle$ also holds in $\langle \boldsymbol{CDF}, \wedge, \vee, \oplus, \otimes \rangle$. Conditions (1)–(2) and (4) of Lemma 3.5 as well as conditions (i)–(ii) of Theorem 8.2 can be also deduced from this fact.

Not much more can be said about the algebraic structure of the set *CDF*. This is the topic which requires further scrutiny.

The sentential languages $\boldsymbol{L}$, $\boldsymbol{L}_b$ and $\boldsymbol{L}_0$ enable one to express in a uniform way the simplest logical interrelations holding between cumulative distribution functions. The system $\langle \boldsymbol{CDF}, \rightarrow, \wedge, \vee, \oplus, \otimes, *, {}^d \rangle$ and its fragments provide an algebraic semantics for these languages. This semantics determines the underlying consequence operations. The fact that the algebra $\langle \boldsymbol{CDF}, \rightarrow, \wedge, \vee, \oplus, \otimes, {}^d \rangle$ is a distributoid makes it possible to define truth-conditions for the pertinent connectives by means of applying the general approach to relational semantics elaborated by J.M. Dunn in Dunn (1991, 1993), Dunn and Hardegree (2001). Loosely speaking, a relational semantics for probability is available in a way which parallels the standard Kripke semantics for normal modal systems. This is due to the fact that every distributoid can be represented in terms of frames (see Definition 12.4.1 and Theorems 12.4.3 and 12.4.5 in Dunn and Hardegree 2001). It is however an open question how such truth-conditions are to be explicitly defined. The crucial problem consists then in providing the definition that a unary probabilistic variable (predicate) P is true at some world u.

The next question concerns the role of the mean value and the variance of a CDF $\boldsymbol{F}$ (provided that they exist). The *expected value* (or the mean value) of $\boldsymbol{F}$ is defined as the Riemann–Stjeltjes integral

$$E(\boldsymbol{F}) = \int_{-\infty}^{\infty} x \, d\boldsymbol{F}(x).$$

(The Cauchy distribution is an example of a CDF which has no mean and variance defined.) Can we use the mean values to attach meanings, and hence truth-values, to phrases such as e.g., "Men are taller than women" or "The Americans are richer than the Russians" in a consistent, uniform and precise way? How to combine the rules of assigning truth-values with inferential statistical analysis?

The general question that arises is: Can one make sense of *a probabilistic logic* whose natural algebraic semantics is constituted by the algebra $\boldsymbol{CDF}$ (or its expansion $\boldsymbol{CDF}_b$)? There are more questions here than answers in the affirmative. Below we shall discuss some other delicate issues that should be resolved beforehand.

The above definitions do not refer to the notions of a probabilistic space and of a random variable. The approach we presented is based on the notions of a probabilistic attribute and of a cumulative distribution function. Random variables are not even mentioned in the above definitions. However, such an approach simplifies matters, because it omits problems which are of vital importance to probability theory and statistics. It seems that the issues of stochastic independence and correlation are central in this context. More precisely, the well-known definition of stochastic independence is restricted to random variables and, as such, it requires prior introduction of probability spaces. (Random variables are defined on these spaces.) One may, of course, attempt to reformulate the classical notion of stochastic independence and redefine it uniformly in terms of cumulative distributions, thus disregarding random variables. The problem is that if one wants to base the notion of stochastic independence on cumulative distributions, the notion of probabilistic space *cannot be eliminated* from the discourse altogether, as we shall show. This and other issues are discussed in the final section, where we shall try to understand logical nuances involved in the problem of independence.

9 Binary Probabilistic Predicates

A mapping $\boldsymbol{F}\colon \mathbb{R}^2 \to I$ is a *two-dimensional* cumulative distribution function if it satisfies:

1. $\forall_{x\in\mathbb{R}} \lim_{y\to-\infty} \boldsymbol{F}(x,y) = 0, \qquad \forall_{y\in\mathbb{R}} \lim_{x\to-\infty} \boldsymbol{F}(x,y) = 0$
2. $\lim_{y\to\infty,\, x\to\infty} \boldsymbol{F}(x,y) = 1$
3. $\boldsymbol{F}$ is both non-decreasing and right-continuous with respect to both x and y.

Two-dimensional cumulative distribution functions are also called *binary* cumulative distributions.

Any two-dimensional cumulative distribution $\boldsymbol{F}$ determines a unique probability measure $\mu_{\boldsymbol{F}}$ on $B(\mathbb{R}^2)$. $\mu_{\boldsymbol{F}}$ is defined on the infinite sets $(-\infty, x] \times (-\infty, y]$ by means of the formula: $\mu_{\boldsymbol{F}}((-\infty, x] \times (-\infty, y]) := \boldsymbol{F}(x,y)$. It follows that for any half-closed rectangle $(x_1, x_2] \times (y_1, y_2]$, it is the case that $\mu_{\boldsymbol{F}}((x_1, x_2] \times (y_1, y_2]) = \boldsymbol{F}(x_2, y_2) - \boldsymbol{F}(x_2, y_1) - \boldsymbol{F}(x_1, y_2) + \boldsymbol{F}(x_1, y_1)$.

Conversely, any probability measure μ on $B(\mathbb{R}^2)$ is unambiguously defined by some two-dimensional cumulative distribution $\boldsymbol{F}$, namely, by the function given by $\boldsymbol{F}(x,y) := \mu((-\infty, x] \times (-\infty, y])$ for all $x, y \in \mathbb{R}$.

CDF_2 is the set of two-dimensional cumulative distribution functions. CDF_2 is partially ordered by $\leqslant$, where $\boldsymbol{F} \leqslant \boldsymbol{G}$ means that $\boldsymbol{F}(x,y) \leqslant \boldsymbol{G}(x,y)$ for all $x, y \in \mathbb{R}$. In fact, the poset $\langle CDF_2, \leqslant\rangle$ is a distributive lattice with $\max(\boldsymbol{F}, \boldsymbol{G})$ and $\min(\boldsymbol{F}, \boldsymbol{G})$ being the lattice operation of join and meet.

The problem arises: how to define the algebra $\boldsymbol{CDF}_2$ of binary cumulative distribution functions $\boldsymbol{F}\colon \mathbb{R}^2 \to I$ and, more generally, the algebra $\boldsymbol{CDF}_m$ of m-dimensional cumulative distribution functions, for any positive integer m? $\boldsymbol{CDF}_2$ is a distributive lattice. But there are more options concerning dualization operations. These options are determined by a group of symmetries attached to the system of Cartesian coordinates of the plane $\mathbb{R}^2$. Let $g_1, \ldots, g_5$ be the following transformations of the plane: g_1 is the reflection (of any subset of $\mathbb{R}^2$) against the y-axis, g_2 is the reflection against the x-axis, g_3 is the reflection in the origin of the coordinate system, g_4 is the reflection against the line $y = x$, and g_5 is the reflection against the line $y = -x$. (The identity transformation is omitted.) For example, g_1 determines the following operation on the power set of $\mathbb{R}^2$: given a set $A \subseteq \mathbb{R}^2$,

$$g_1 A := \{ (-a, b) \colon (a, b) \in A \}.$$

In an analogous way we define the remaining operations. Thus

$$\begin{aligned} g_2 A &:= \{ (a, -b) \colon (a, b) \in A \}, \\ g_3 A &:= \{ (-a, -b) \colon (a, b) \in A \}, \\ g_4 A &:= \{ (b, a) \colon (a, b) \in A \}, \\ g_5 A &:= \{ (-b, -a) \colon (a, b) \in A \}, \end{aligned}$$

for any set $A \subseteq \mathbb{R}^2$.

g_3 is the composition $g_2 \circ g_1$ of the group operations g_1 and g_2. (This composition commutes.) In turn, g_5 is equal to the composition $g_4 \circ (g_2 \circ g_1)$. The operations $g_1, \ldots, g_5$ together with the identity transformation form a 6-element Abelian group, in which all elements are idempotent.

Let μ be a probability measure on the σ-field $B(\mathbb{R}^2)$ of Borel subsets of the plane $\mathbb{R}^2$. For each g_i, $i = 1, \ldots, 5$, we define the probability measure $\mu \circ g_i$ on $B(\mathbb{R}^2)$:

$$(\mu \circ g_i)(A) := \mu(g_i A),$$

for any set $A \in B(\mathbb{R}^2)$. Let $\boldsymbol{F}$ be the two-dimensional cumulative distribution function that determines μ. The two-dimensional cumulative distribution function corresponding to the measure $\mu \circ g_i$ is marked as $\boldsymbol{F} \circ g_i$. Thus e.g., $(\boldsymbol{F} \circ g_1)((-\infty, x] \times (-\infty, y]) = \boldsymbol{F}([-x, +\infty) \times (-\infty, y])(= 1 - \boldsymbol{F}((-\infty, -x) \times (-\infty, y]))$, for all $x, y \in \mathbb{R}$.

It follows from the above remarks that the set CDF_2 is not only a distributive lattice, but it also becomes a (unbounded) De Morgan lattice with respect of each operation $\boldsymbol{g}_i \colon CDF_2 \to CDF_2$, where $\boldsymbol{g}_i(\boldsymbol{F}) := \boldsymbol{F} \circ g_i$, for all $\boldsymbol{F} \in CDF_2$ for $i = 1, \ldots, 5$. We thus arrive here at a new type of algebraic structures $\langle D, \wedge, \vee, d_1, \ldots, d_k \rangle$, viz., distributive lattices endowed with a finite number of unary operations $d_1, \ldots, d_k$ such that for every i, $\langle D, \wedge, \vee, d_i \rangle$ is a De Morgan algebra. Moreover, $d_1, \ldots, d_k$, together with the identity operation on D, form an Abelian idempotent group with respect to composition. The operations $d_1, \ldots, d_k$ are the actions of the above group on CDF_2.

The set CDF_2 is also equipped with the two-dimensional convolution operation:

$$(\boldsymbol{F} * \boldsymbol{G})(x, y) := \int_{-\infty}^{+\infty} \int_{-\infty}^{+\infty} \boldsymbol{F}(u, v) \boldsymbol{G}(x - u, y - v) \, du dv, \tag{12}$$

for all $x, y \in \mathbb{R}$. The convolution $\boldsymbol{F} * \boldsymbol{G}$ of any two-dimensional cumulative distributions $\boldsymbol{F}$ and $\boldsymbol{G}$ belongs to CDF_2.

As in the case of unary CDFs we define the *bounded addition* $\oplus$ of binary CDFs:

$$(\boldsymbol{F} \oplus \boldsymbol{G})(x, y) := \min(1, \boldsymbol{F}(x, y) + \boldsymbol{G}(x, y)),$$

for all $x, y \in \mathbb{R}$. The bounded sum $\boldsymbol{F} \oplus \boldsymbol{G}$ of binary CDFs $\boldsymbol{F}$ and $\boldsymbol{G}$ is a binary CDF. The sum $\boldsymbol{F} \oplus \boldsymbol{G}$ is also called the *weak disjunction* of cumulative distributions $\boldsymbol{F}$ and $\boldsymbol{G}$. Furthermore, if $\boldsymbol{F}$ and $\boldsymbol{G}$ are continuous binary CDFs, $\boldsymbol{F} \oplus \boldsymbol{G}$ is continuous as well.

By the *strong conjunction* (also in an analogy to Łukasiewicz logics) we shall understand the binary operation $\otimes$ defined as:

$$(\boldsymbol{F} \otimes \boldsymbol{G})(x, y) := \max(0, \boldsymbol{F}(x, y) + \boldsymbol{G}(x, y) - 1),$$

for all $x, y \in \mathbb{R}$. As in the one-dimensional case, the strong conjunction $\boldsymbol{F} \otimes \boldsymbol{G}$ of binary CDFs $\boldsymbol{F}$ and $\boldsymbol{G}$ is a binary cumulative distribution, too. If $\boldsymbol{F}$ and $\boldsymbol{G}$ are continuous CDFs, so is $\boldsymbol{F} \otimes \boldsymbol{G}$.

The algebra $\boldsymbol{CDF}_2 = \langle CDF_2, \wedge, \vee, \boldsymbol{g}_1, \ldots, \boldsymbol{g}_5, *, \oplus, \otimes \rangle$ is called the *extended De Morgan algebra* of binary cumulative distribution functions. (Thus $\boldsymbol{CDF}_2$ is a De Morgan algebra with respect to each unary operation $\boldsymbol{g}_i$, $i = 1, \ldots, 5$, separately.) A similar name applies to the algebra $\langle CCDF_2, \wedge, \vee, \boldsymbol{g}_1, \ldots, \boldsymbol{g}_5, *, \oplus, \otimes \rangle$ of continuous binary cumulative distributions.

The algebra $\boldsymbol{CDF}_2$ and the defined earlier algebra $\boldsymbol{CDF}$ of unary cumulative distribution functions are not altogether separated notions. They are linked by two operators assigning to each $\boldsymbol{F} \in \boldsymbol{CDF}_2$ the boundary cumulative distributions $\boldsymbol{F}_1 : \mathbb{R} \to I$ and $\boldsymbol{F}_2 : \mathbb{R} \to I$ corresponding to $\boldsymbol{F}$, respectively. More specifically, let $\boldsymbol{F} : \mathbb{R}^2 \to I$ be a two-dimensional cumulative distribution. One can assign to $\boldsymbol{F}$ two mappings $\boldsymbol{F}_1 : \mathbb{R} \to I$ and $\boldsymbol{F}_2 : \mathbb{R} \to I$ defined for any $x, y \in \mathbb{R}$ as follows:

$$\boldsymbol{F}_1(x) := \lim_{y \to +\infty} \boldsymbol{F}(x, y) \qquad \text{and} \qquad \boldsymbol{F}_2(y) := \lim_{x \to +\infty} \boldsymbol{F}(x, y).$$

$\boldsymbol{F}_1$ and $\boldsymbol{F}_2$ are one-dimensional cumulative distributions. They are called *boundary cumulative distributions corresponding to* $\boldsymbol{F}$. If $\boldsymbol{F}$ is continuous on $\mathbb{R}^2$, $\boldsymbol{F}_1$ and $\boldsymbol{F}_2$ are continuous on $\mathbb{R}$.

The following observation is immediate.

Lemma 9.1 *Suppose $\boldsymbol{F}, \boldsymbol{G} \in CDF_2$. If $\boldsymbol{F} \leqslant \boldsymbol{G}$, then $\boldsymbol{F}_1 \leqslant \boldsymbol{G}_1$ and $\boldsymbol{F}_2 \leqslant \boldsymbol{G}_2$, for the boundary distributions corresponding to $\boldsymbol{F}$ and $\boldsymbol{G}$.*

On the other hand, let $\boldsymbol{F}_1 : \mathbb{R} \to I$ and $\boldsymbol{F}_2 : \mathbb{R} \to I$ be arbitrary one-dimensional cumulative distributions (they may be identical). To the ordered pair $\langle \boldsymbol{F}_1, \boldsymbol{F}_2 \rangle$, one can assign the mapping

$$\boldsymbol{F}_1 \times \boldsymbol{F}_2 : \mathbb{R}^2 \to I$$

defined as follows:

$$(\boldsymbol{F}_1 \times \boldsymbol{F}_2)(x, y) := \boldsymbol{F}_1(x) \cdot \boldsymbol{F}_2(y),$$

for all pairs $(x, y) \in \mathbb{R}^2$. $\boldsymbol{F}_1 \times \boldsymbol{F}_2$ is a two-dimensional cumulative distribution called the *product* of $\boldsymbol{F}_1$ and $\boldsymbol{F}_2$. Thus $\times$ is a binary operation from CDF to CDF_2.

Let $\boldsymbol{L}_2$ be the absolutely free algebra built from a countably infinite set of predicate variables $Var_2 = \{P_n^2 : n = 1, 2, \dots\}$ and endowed with binary connectives $\wedge, \vee, *, \oplus, \otimes$ and five unary connectives $g_1, \dots, g_5$. The universe L_2 of $\boldsymbol{L}_2$ consists of all sentential formulas formed from Var_2 by means of the above connectives in the well-known manner. (The elements of Var_2 are therefore treated as sentential variables of $\boldsymbol{L}_2$.) $\boldsymbol{L}_2$ is called the *language of two-dimensional probabilistic predicates*.

The variables of Var_2 range over 2-dimensional cumulative distributions. A *valuation* of the language $\boldsymbol{L}_2$ in the algebra $\boldsymbol{CDF}_2$ is an arbitrary homomorphism $h: \boldsymbol{L}_2 \to \boldsymbol{CDF}_2$. h is unambiguously determined by its values on the set Var_2.

By an analogy to the one-dimensional case, we define a *model for* $\boldsymbol{L}_2$ to be the pair

$$\boldsymbol{M} = (Var_2, h),$$

where h is a mapping defined on the set Var_2 of variables of $\boldsymbol{L}_2$ and assigning to each binary predicate variable $P \in Var_2$ a 2-dimensional cumulative distribution $h(P)$ on $\mathbb{R}^2$. When h is clear from a context, the cumulative distribution $h(P)$ will be denoted by $\boldsymbol{F}_P$. h is then recursively extended to a homomorphism from $\boldsymbol{L}$ to $\boldsymbol{CDF}_2$ in the standard way.

The relation of *probabilistic entailment* $\models_2$ on $\boldsymbol{L}_2$ is defined in a similar way as in the one-dimensional case. For $n \geqslant 1$ and formulas $\alpha_1, \dots, \alpha_n, \beta$ of $\boldsymbol{L}_2$, we define

$$\alpha_1, \dots, \alpha_n \models_2 \beta \quad \Leftrightarrow_{df} \quad (\forall h: \boldsymbol{L}_2 \to \boldsymbol{CDF}_2)\ h(\alpha_1) \wedge \cdots \wedge h(\alpha_n) \leqslant h(\beta),$$

where $\leqslant$ is the above partial order on $\boldsymbol{CDF}_2$.

If X is an infinite set of formulas of $\boldsymbol{L}_2$, we assume that

$$X \models_2 \beta \ \Leftrightarrow_{df} \ \alpha_1, \dots, \alpha_n \models_2 \beta \ \text{ for some } n \geqslant 1 \text{ and some formulas } \alpha_1, \dots, \alpha_n \in X.$$

Moreover, it is assumed that $\emptyset \models_2 \beta$ for no formula β. Thus the above probabilistic logic does not possess tautologies.

For each set of formulas X of $\boldsymbol{L}_2$, we define:

$$C_2(X) := \{\beta \in L : X \models_2 \beta\}.$$

C_2 is an operation defined on the the power set $\wp(L_2)$, assigning the set $C_2(X)$ to each subset $X \subseteq L_2$.

In analogy with Theorem 4.1, it can be shown:

Theorem 9.2 *C_2 is a finitary and structural consequence operation.*

C_2 is called the *logic of binary probabilistic predicates*. C_2 is semantically defined in terms of models for $\boldsymbol{L}_2$.

10 Stochastic Independence and Related Issues

We recall that a probability space is a triple (Ω, F, μ) consisting of:

- the *sample space* Ω being an arbitrary non-empty set,
- the σ-field F of subsets of Ω; the elements of F are called *events*,
- the probability measure $\mu \colon F \to [0, 1]$; μ is a σ-additive measure such that $\mu(\Omega) = 1$.

Predicates are linguistic objects. Random variables may be regarded as interpreted predicates (and interpreted in a certain way). According to the classical theory of probability, each 1-dimensional random variable is a real valued function defined on a probability space (Ω, F, μ). (We shall however not change the terminology adopted here and we shall interchangeably speak of *populations* and *sample spaces*.) To each random variable X defined on Ω (and *a fortiori*—on the probability space (Ω, F, μ)) the cumulative distribution $\boldsymbol{F}_X \colon \mathbb{R} \to [0, 1]$ is assigned according to the formula: $\boldsymbol{F}_X(x) := \mu(\{a \in \Omega \colon X(a) \leqslant x\})$, for all $x \in \mathbb{R}$. (It is assumed that X is a measurable function in the sense of the σ-field F, that is, for every Borel subset $A \subseteq \mathbb{R}$, the pre-image $X^{-1}[A]$ belongs to F.)

In the approach presented here the situation is simpler. Probabilistic predicates are directly evaluated as cumulative distributions. In other words, to each unary predicate P a cumulative distribution $\boldsymbol{F}_P \colon \mathbb{R} \to [0, 1]$ is assigned. The number $\boldsymbol{F}_P(x)$ gives the probability that P takes values from the infinite interval $(-\infty, x]$.

The use of probability spaces as an intermediary notion is unnecessary in the definition of the logic C (or C_b). The burden of the approach presented so far rests on uniform interpretations h of predicate letters as cumulative distribution functions. What is lost here? We recall that a *model* for the language $\boldsymbol{L}$ of unary probabilistic predicates is the pair

$$\boldsymbol{M} = (Var, h),$$

where h is a mapping defined on the set of variables assigning to each predicate variable $P \in Var$ a unary cumulative distribution $h(P)$.

By a *standard model* for $\boldsymbol{L}$ we shall mean a model $\boldsymbol{M} = (Var, h)$ defined in terms of cumulative distributions functions of random variables. More specifically, given a probability space (Ω, F, μ), a sequence X_n, $n \in \omega$, of arbitrary random variables in the sense of (Ω, F, μ) is selected. Let $\boldsymbol{F}_n$ be the cumulative distribution of X_n, $n \in \omega$. Putting $h(P_n) := \boldsymbol{F}_n$, $n \in \omega$, we obtain a model (Var, h) for the language of probabilistic predicates. Thus any standard model is identified with an infinite sequence of the cumulative distributions corresponding to random variables (the latter being defined on a classical probability space).

Every standard model is obviously a model. But the converse also holds. The class of standard models is thus adequate for the systems C and C_b because

Lemma 10.1 *Every model is standard.*

In other words, we claim that for every infinite sequence $\{\boldsymbol{F}_n\}$ of unary cumulative distribution functions there exists a probability space (Ω, F, μ) and a sequence of F-measurable random variables X_n, $n \in \omega$, defined on Ω, such that $\boldsymbol{F}_n$ is the cumulative distribution of X_n, for all n.

Proof (*of the lemma*) We proceed as follows. Let μ_n be the probability measure on the σ-field $B(\mathbb{R})$ of Borel subsets of $\mathbb{R}$ determined by $\boldsymbol{F}_n$. (Thus $\mu_n((a, b]) = \boldsymbol{F}_n(b) - \boldsymbol{F}_n(a)$ for any interval $(a, b]$.) The triple $(\mathbb{R}, B(\mathbb{R}), \mu_n)$ is a probability space, for all $n \in \omega$. We then define (Ω, F, μ) to be the product of the countably infinite family $(\mathbb{R}, B(\mathbb{R}), \mu_n)$, $n \in \omega$. (The product of a family of probability spaces is a well-known construction applied in probability theory; see Billingsley (1995) for details.) Thus Ω is the Cartesian power $\Pi_{n\in\omega}\mathbb{R}$ of the real line, F is a σ-field of subsets of Ω, and μ is the probability measure on F. F is defined in a certain way and called a power of the σ-field $B(\mathbb{R})$. The measure μ is called the product of the family μ_n, $n \in \omega$. For each $n \in \omega$, the random variable $X_n \colon \Omega \to \mathbb{R}$ is defined as follows. For every sequence $\underline{x} = \langle x_k \colon k \in \omega\rangle \in \Omega$, we set $X_n(\underline{x}) := \boldsymbol{F}_n(x_n)$. X_n is a well-defined mapping. Moreover, X_n is F-measurable, because X_n is the composition of the projection of Ω onto the nth axis and of the distribution $\boldsymbol{F}_n$. X_n is therefore a random variable in the sense of (Ω, F). It is easy to see that the random variables X_n, $n \in \omega$, are stochastically independent in the sense of (Ω, F, μ).

Claim $\boldsymbol{F}_n$ *is the cumulative distribution function of* X_n, *for all* $n \in \omega$.

Proof (*of the claim*) Fix $r \in \mathbb{R}$ and n. Let π_n be the projection of Ω onto the nth axis. We have:

$$\begin{aligned}
\mu(\{\underline{x} \in \Omega \colon X_n(\underline{x}) \leqslant r\}) &= \mu(\{\underline{x} \in \Omega \colon \boldsymbol{F}_n(x_n) \leqslant r\}) \\
&= \mu(\{\underline{x} \in \Omega \colon \boldsymbol{F}_n(\pi_n(\underline{x})) \leqslant r\}) \\
&= \mu(\{\underline{x} \in \Omega \colon \pi_n(\underline{x}) \in \boldsymbol{F}_n^{-1}((-\infty, r])\}) \\
&= \mu_n(\{x \in \mathbb{R} \colon x \in \boldsymbol{F}_n^{-1}((-\infty, r])\}) \\
&\qquad\text{(by def. of the product of measures)} \\
&= \mu_n(\{x \in \mathbb{R} \colon \boldsymbol{F}_n(x) \leqslant r\}) \\
&= \boldsymbol{F}_n(r)
\end{aligned}$$

This proves the claim and the lemma. □

It therefore follows from the above considerations that the logic C is also determined by standard models.

In order to present the theory of stochastic independence of unary predicates in an undistorted way, it is necessary to take into account probabilistic predicates of higher arities as well as many-dimensional cumulative distributions. Here we shall restrict the discussion to the binary case.

One may attempt to reformulate the classical notion of stochastic independence and define it uniformly in terms of the corresponding cumulative distributions, thus disregarding random variables. The problem is that the notion of a probabilistic space *cannot be eliminated* from such a discourse altogether.

Let $(\Omega, \boldsymbol{F}, \mu)$ be a probability space and $X, Y\colon \Omega \to \mathbb{R}$ random variables (in the sense of this space). Let $\boldsymbol{F}$ and $\boldsymbol{G}$ be the (one-dimensional) cumulative distributions corresponding to X and Y. In the standard way one defines the product space

$$(\Omega \times \Omega, F \times F, \mu \times \mu).$$

We define the two-dimensional random variable $X \times Y\colon \Omega \times \Omega \to \mathbb{R} \times \mathbb{R}$ as follows: $(X \times Y)(a, b) := \langle X(a), Y(b)\rangle$, for every pair $(a, b) \in \Omega \times \Omega$. We then put:

$$\begin{aligned} \boldsymbol{H}(x, y) &:= (\mu \times \mu)(\{ (a, b) \in \Omega \times \Omega : X(a) \leqslant x \text{ and } Y(b) \leqslant y \}) \\ &= (\mu \times \mu)(\{ (a, b) \in \Omega \times \Omega : (X \times Y)(a, b) \in (-\infty, x] \times (-\infty, y] \}) \end{aligned}$$

$\boldsymbol{H}$ is a two-dimensional cumulative distribution. $\boldsymbol{H}$ is called the *product of the cumulative distributions* $\mathbf{F}$ *and* $\mathbf{G}$ *in the sense of the probability space* (Ω, F, μ) and denoted by $\boldsymbol{F} \times_\mu \boldsymbol{G}$.

Each probability space (Ω, F, μ) thus determines a binary, partial operation $\times_\mu$ on the set of cumulative distributions. The operation $\times_\mu$ is commutative and associative. (One may also define arbitrary finite as well as infinite μ-products of cumulative distributions; this aspect is omitted here.)

We recall that random variables X and Y are *stochastically independent* (in the space (Ω, F, μ)), if for all $x, y \in \mathbb{R}$,

$$\begin{aligned} &(\mu \times \mu)(\{ (a, b) \in \Omega \times \Omega : X(a) \leqslant x \text{ and } Y(b) \leqslant y \}) \\ &\qquad = \mu(\{ a \in \Omega : X(a) \leqslant x \}) \cdot \mu(\{ b \in \Omega : X(a) \leqslant y \}). \end{aligned}$$

We may reformulate the above definition in the terms of cumulative distributions and say that two cumulative distributions $\boldsymbol{F}$ and $\boldsymbol{G}$ are *independent in the sense of* $\times_\mu$ if

$$(\boldsymbol{F} \times_\mu \boldsymbol{G})(x, y) = \boldsymbol{F}(x) \cdot \boldsymbol{G}(y),$$

for all $x, y \in \mathbb{R}$. Note, however, that the definition of $\boldsymbol{F} \times_\mu \boldsymbol{G}$ makes sense only for cumulative distributions $\boldsymbol{F}$ and $\boldsymbol{G}$ of random variables over (Ω, F, μ), and not for *arbitrary* cumulative distributions. For instance, if Ω is countable, *no* continuous CDF $\boldsymbol{F}$ is determined by a random variable over (Ω, F). In this case, the product $\boldsymbol{F} \times_\mu \boldsymbol{G}$ is undefined whenever $\boldsymbol{F}$ and $\boldsymbol{G}$ are continuous CDFs.

The above definition of $\times_\mu$ thus explicitly refers to the space (Ω, F, μ). It is not clear how to formulate an "intrinsic" definition of $\times_\mu$, that is, how to characterize the partial operation $\times_\mu$ entirely in terms of properties of the set *CDF* of cumulative distributions without resorting to probability spaces.

We thus see that in order to adequately capture various probabilistic phenomena, as e.g., the stochastic independence, the algebra ***CDF*** must be supplemented with a bunch of binary partial operations of type $\times_\mu$ defined as above.

11 Convergence

Yet another aspect of probability theory that comes to light is that of convergence of cumulative distribution functions. The mapping assigning to each random variable X over (Ω, F, μ) its cumulative distribution $\boldsymbol{F}_X$ is not one-to-one. One identifies random variables X and Y relative μ according to the formula:

$$X =_\mu Y \quad \Leftrightarrow_{df} \quad \mu(\{ a \in \Omega : X(a) = Y(a) \}) = 1.$$

$=_\mu$ is an equivalence relation on the set of random variables over (Ω, F, μ) compatible with some arithmetic operations as, e.g., the sum and the product of random variables. One can prove that if $X =_\mu Y$, then the cumulative distributions $\boldsymbol{F}_X$ and $\boldsymbol{F}_Y$ coincide.

The part of probability theory which is preoccupied by pointwise convergence of (independent) random variables cannot be adequately rendered in terms of cumulative distribution functions—the presence of probability spaces is a necessary additional ingredient that enables us to express pertinent properties of cumulative distributions as, e.g., stochastic independence or convergence. For example, the central limit theorem (CLT) in its common form states that under some conditions, the mean of a sufficiently large number of iterates of independent and identically distributed (say, by $\boldsymbol{F}$) random variables with finite expected values and variance will be approximately normally distributed, regardless of the underlying cumulative distribution $\boldsymbol{F}$. More precisely, this theorem (in a bit restricted form) states that if X_n, $n \in \omega$, is a sequence of independent random variables over a probability space (Ω, F, μ) with the same cumulative distribution $\boldsymbol{F}$ with variance 1 and expected value 0, then the sequence $(X_0 + X_1 + \cdots + X_{n-1})/\sqrt{n}$ approaches a random variable with the standard normal distribution.

An infinite sequence $\boldsymbol{F}_n$, $n \in \omega$, of unary cumulative distribution functions is *convergent* to a distribution $\boldsymbol{F}$ if $\boldsymbol{F}(x) = \lim_{n\to\infty} \boldsymbol{F}_n(x)$, for all $x \in \mathbb{R}$. The phrase "to a distribution" is essential in this definition, because an infinite sequence of CDFs may turn out to be pointwise convergent to a function $\boldsymbol{F} : \mathbb{R} \to [0, 1]$ which is *not* a cumulative distribution (see the example following Lemma 3.4). It is known, however, that every sequence of cumulative distributions convergent to a *continuous* cumulative distribution $\boldsymbol{F}$ is uniformly convergent to $\boldsymbol{F}$. It follows that the pointwise limit of a uniformly convergent sequence of uniformly continuous cumulative distributions is a uniformly continuous CDF. This type of convergence is expressible in terms of the topological properties of the set *CCDF* without the need of introducing probability spaces. However, probability theory makes use of other forms of convergence of random variables. These forms of convergence of random variables are not captured in terms of convergence of cumulative distributions.

One defines yet another form of convergence of cumulative distributions, called weak convergence. An infinite sequence of cumulative distributions $\boldsymbol{F}_n$, $n \in \omega$, is *weakly convergent* to a CDF $\boldsymbol{F}$ if and only if $\boldsymbol{F}(x) = \lim_{n\to\infty} \boldsymbol{F}_n(x)$, for all points

$x \in \mathbb{R}$ at which $\boldsymbol{F}$ is continuous. The weak convergence plays a significant role in the formulation of the above central limit theorem.

Let X_n, $n \in \omega$, be a sequence of random variables defined on the same probability space (Ω, F, μ). Let $\boldsymbol{F}_n$ be the cumulative distribution of X_n, $n \in \omega$, and let X be a random variable on (Ω, F, μ), whose distribution is $\boldsymbol{F}$.

If $\boldsymbol{F}_n, n \in \omega$, is weakly convergent to $\boldsymbol{F}$, then according to the terminology adopted in probability theory, the variables X_n, $n \in \omega$, are also said to be *convergent in distribution* to the variable X and we write $X_n \Rightarrow X$. This type of convergence basically refers to properties of cumulative distributions and not to those of random variables.

However, there are other, stronger forms of convergence such as convergence in probability or almost sure convergence. These forms are defined in terms of sequences X_n, $n \in \omega$, of random variables over a space (Ω, F, μ).

It does not seem that these forms of convergence could be equivalently defined in terms of cumulative distributive functions without resorting to the space (Ω, F, μ). There is an even more basic point, though. Random variables X and Y can have the same distribution, yet $X \neq Y$ everywhere (or almost everywhere, if they are continuous). Clearly there is more information in X than in its CDF.

This supports the thesis that random variables cannot be eliminated from the probabilistic discourse and entirely replaced by cumulative distribution functions.

12 The General Case

Let m be a positive integer. The above definitions give clues to the characterization of m-dimensional cumulative distribution functions $\boldsymbol{F}\colon \mathbb{R}^m \to I$. In a fully analogous way to the two-dimensional case, we may define the algebra $\boldsymbol{CDF}_m$ of m-dimensional cumulative distribution functions. $\boldsymbol{CDF}_m$ is a distributive lattice furnished with some additional operations. The only difficulty is in isolating the group of symmetries of the n-dimensional Cartesian system of coordinates and the corresponding actions of the group on the set CDF_m. (We shall omit the details. We merely assume that $\boldsymbol{CDF}_m$ is a well-defined mathematical object.)

Let m be a positive integer. Let $\boldsymbol{L}_m$ be the absolutely free algebra built from a countably infinite set of predicate variables $Var_m = \{P^m_n \colon n = 1, 2, \dots\}$ and endowed with the set of connectives corresponding to the operations of the algebra $\boldsymbol{CDF}_m$. Thus the universe L_m of $\boldsymbol{L}_m$ consists of all sentential formulas formed from Var_m and the above connectives in the well-known manner. $\boldsymbol{L}_m$ is called the *language of m-ary probabilistic predicates*.

The variables of Var_m range over m-dimensional cumulative distributions. A *valuation* of the language $\boldsymbol{L}$ in the algebra $\boldsymbol{CDF}_m$ is an arbitrary homomorphism $h\colon \boldsymbol{L} \to \boldsymbol{CDF}_m$. h is unambiguously determined by its values on the set Var_m.

By an analogy to the one-dimensional case, a *model for* $\boldsymbol{L}_m$ is the pair

$$\boldsymbol{M} = (Var_m, h),$$

where h is a mapping defined on the set Var_m of variables of $\boldsymbol{L}_m$ and assigning to each m-ary predicate variable $P \in Var_m$ an m-dimensional cumulative distribution $h(P)$ on $\mathbb{R}^m$. When h is clear from context, the cumulative distribution $h(P)$ will be denoted by $\boldsymbol{F}_P$. h is then recursively extended to a homomorphism from $\boldsymbol{L}$ to $\boldsymbol{CDF}_m$ in the well-known way.

We return to the question of interpreting m-ary predicates as m-dimensional random variables.

Let (Ω, F, μ) be a probability space. By the mth *power* of (Ω, F, μ) we shall mean the probability space (Ω^m, F^m, μ^m), where $\Omega^m := \Omega \times \cdots \times \Omega$ (Ω occurs m-times), F^m is a σ-field of subsets of Ω defined in a certain way and called the mth power of the σ-field F (F^m is σ-generated by m-dimensional cuboids formed from the sets of F), and μ^m is the probability measure on F^m. The measure μ^m is called the mth power of μ and also denoted as $\mu \times \cdots \times \mu$. For a more technical aspects of the above definitions, the reader is advised to consult (Billingsley 1995).

Let μ be a probability measure on $B(\mathbb{R})$. The mth power of the probability space $(\mathbb{R}, B(\mathbb{R}), \mu)$ is usually identified with the space $(\mathbb{R}^m, B(\mathbb{R}^m), \mu^m)$.

Any mapping $X\colon \Omega^m \to \mathbb{R}^m$, which is F^m-measurable is called an *m-dimensional random variable over* the probability space (Ω, F, μ). The assumption that X is F^m-measurable means that for any Borel set $A \in B(\mathbb{R}^m)$, the pre-image $X^{-1}[A]$ belongs to F^m. But trivially, every m-dimensional random variable over (Ω, F, μ) is represented as an m-tuple $\langle X_1, \ldots, X_m \rangle$ composed of m unary random variables over (Ω, F, μ). Such an m-tuple $\langle X_1, \ldots, X_m \rangle$ is called a *random vector* consisting of m random variables over (Ω, F, μ). Following the common practice adopted in probability theory, one simply identifies m-dimensional random variables with random vectors of length m.

By a *standard m-dimensional model* (over a probability space (Ω, F, μ)) we shall mean the pair

$$\boldsymbol{M} = (Var_m, h),$$

where h is a mapping defined on the set Var_m and assigning to each m-ary predicate variable $P \in Var_m$ an m-dimensional cumulative distribution $h(P)$ on $\mathbb{R}^m$ corresponding to some random vector $\langle X_1, \ldots, X_m \rangle$ over (Ω, F, μ). In other words, the values of h are *not* arbitrary m-dimensional cumulative distributions but only those determined by random vectors of length m over (Ω, F, μ).

Suitably modifying the proof of Lemma 10.1, one can prove that

Every model $\boldsymbol{M} = (Var_m, h)$ *for* $\boldsymbol{L}_m$ *is a standard m-dimensional model (over some probability space* (Ω, F, μ)*).*

In other words, for every infinite sequence $\{\boldsymbol{F}_n\}$ of m-dimensional cumulative distribution functions there exists a probability space (Ω, F, μ) and a sequence $\underline{X}_n = \langle X_{1,n}, \ldots, X_{m,n} \rangle$ of random vectors over (Ω, F, μ) (each vector of length m) such that $\boldsymbol{F}_n$ is the m-dimensional cumulative distribution of the random vector $\underline{X}_n$, for $n = 1, 2, \ldots$.

The conclusion which can be drawn from the above remarks is that the standard interpretation of m-ary probabilistic predicates as vectors (of length m) of random variables m over probability spaces is fully legitimate and it gives all the models we have defined.

Acknowledgments The author warmly thanks the anonymous referees for their valuable suggestions that have shaped the final form of this paper.

References

Billingsley, P. (1995). *Probability and measure*. New York: Wiley.
Billingsley, P. (1999). *Convergence of probability measures*. New York: Wiley.
Czelakowski, J. (2012). O probabilistycznej interpretacji predykatów [Polish]. In A. Wójtowicz & J. Golińska-Pilarek (Eds.), *Identyczność znaku czy znak identyczności? [Identity of Sign or the Sign of Identity?]*. Warsaw: Warsaw University Press.
Davey, B. A., & Priestley, H. A. (2002). *Introduction to lattices and order* (2nd ed.). Cambridge: Cambridge University Press.
Dunn, J. M. (1991). Gaggle theory: An abstraction of Galois connections and residuation with applications to negation, implication, and various logical operators. In J. van Eijck (Ed.), *Logics in AI: European workshop JELIA '90, Lecture notes in computer science* (Vol. 478, pp. 31–51). Berlin: Springer.
Dunn, J. M. (1993). Partial gaggles applied to logics with restricted structural rules. In K. Došen & P. Schroeder-Heister (Eds.), *Substructural logics* (pp. 63–108). Oxford: Clarendon.
Dunn, J. M., & Hardegree, G. M. (2001). *Algebraic methods in philosophical logic, Oxford logic guides* (Vol. 41). Oxford: Oxford University Press.
Ganter, B., Stumme, G., & Wille, R. (Eds.). (2005). *Formal concept analysis: Foundations and applications, Lecture notes in artificial intelligence* (Vol. 3626). Berlin: Springer.
Hájek, P. (1998). *Metamathematics of fuzzy logics*. Dordrecht: Kluwer.
Wójcicki, R. (1988). *Theory of logical calculi. Basic theory of consequence operations*. Dordrecht: Kluwer.

Reasoning with Incomplete Information in Generalized Galois Logics Without Distribution: The Case of Negation and Modal Operators

Chrysafis Hartonas

Abstract We extend Dunn's treatment of various forms of negation developed in the context of his theory of generalized Galois logics (known as *gaggle theory*), by dropping the assumption of distribution. We also study modal operators of possibility and impossibility in a non-distributive context and in standard Kripke semantics, thus improving significantly over existing approaches developed in the last decade or so on the semantics of modalities when distribution of conjunction over disjunction and conversely is dropped. We prove representation and completeness theorems for the related logical calculi in appropriate Kripke frames. Without distribution, the points of the frame (we call them information sites) appear as possessing incomplete only information, supporting the truth of a disjunction $\varphi \vee \psi$ without necessarily supporting the truth of either φ or ψ. Our approach is based on and extends past results we have obtained on the (topological) representation (and Stone type duality) of non-distributive lattices with additional operators.

Keywords Impossibility · Modal lattice logic · Negation · Negation as impossibility · Non-distributive lattice logic · Possibility · Star and perp

1 Preliminaries

1.1 Generalized Galois Logics

Dunn's theory of Generalized Galois Logics (gaggles), motivated by the relational semantics for Relevance Logic of Routley and Meyer (1973) and the semantic analysis of orthologic of Goldblatt (1974), Goldblatt (1993) has been developed as an extension of the classical Jónsson and Tarski (1952) results on Boolean Algebras with Operators, where the latter is itself an extension of the topological representation of

C. Hartonas (✉)
Department of Computer Science & Engineering, University of Applied Sciences of Thessaly (TEI of Thessaly), Larissa, Greece
e-mail: hartonas@teilar.gr
URL: http://www.researchgate.net/profile/Chrysafis_Hartonas

K. Bimbó (ed.), *J. Michael Dunn on Information Based Logics*,
Outstanding Contributions to Logic 8, DOI 10.1007/978-3-319-29300-4_14

Boolean Algebras of Stone (1938). The theory is well-developed when distribution is assumed and, in fact, Dunn's original project presented in Dunn (1991) was restricted to the case of logical calculi whose Lindenbaum algebra is a distributive lattice, aiming at extending the Stone (1937–38), or Priestley (1970) topological representation of distributive lattices. A further generalization of gaggle theory introduced in Dunn (1991) is to extend the class of lattice operators from Jónsson and Tarski's additive operators (distributing over joins at each argument place) to the class of operators that either distribute or codistribute at each argument place over joins (turning them to meets, in the latter case), or over meets, and returning the same type of operator (a meet, or a join). For example, implication is not additive but it does fall within the class of operators treated by gaggle theory (it codistributes over joins in the first argument place, while it distributes over meets in the second argument place returning, in both cases, a meet). Each n-ary operator having a well-behaved distribution type, in the sense explained above, is then interpreted in a Kripke frame using an $(n + 1)$-ary relation that is canonically associated to it, generalizing on the Jónsson–Tarski idea of representing additive operators as the image operators of associated relations on their dual Stone space. The name of the theory derives from Dunn's observation that logical operators often come in adjoint pairs, forming either a Galois connection, or a residuated pair (the classical example being conjunction and implication), an observation that has been subsequently turned to a guiding principle, seeking to discover and explore the semantics of adjoint pairs of logical operators. Adjointness of operators is of course directly related to the (co)distribution properties of the operators, by well-known general facts in Category Theory. Dunn has developed the theory of generalized, distributive Galois logics in a series of papers, though Dunn (1991) remains the central reference.

Negation is but one example of a unary operator having a well-behaved distribution type (it codistributes over joins). Dunn (1999) runs through the history of formal negation and compares its various model-theoretic treatments, classifying the various types in what has been called Dunn's kite of negations, first introduced in Dunn (1996). The kite got subsequently slightly modified, in Dunn and Zhou (2005), where a thorough study of negation in the context of the theory of distributive generalized Galois logics is presented.

Distribution of meets over joins, and conversely, is natural in logic and a significant convenience, but not at all a necessity, explicitly rejected in the Full Lambek Calculus, Linear Logic, Orthologic and Quantum Logic. Dunn, having originally built distribution into the definition of a generalized Galois logic (a gaggle), has subsequently had a change of heart in the matter.

One way around non-distributivity is to ignore the lattice structure and focus only on the partial order generated by the consequence relation and this approach has been followed in Dunn (1993, 1996). But it is of course best to avoid this simplification and deal directly with non-distribution. The success of such an approach depends largely on expanding on an appropriate representation theorem for general lattices. By now, a number of related results have appeared in the literature: Urquhart (1978), Hartung (1992), Allwein and Hartonas (1993), Hartonas and Dunn (1997), Hartonas

(1997), Gerhke and Harding (2001), Gerhke and van Gool (2014), Moshier and Jipsen (2014a, b).

Our own approach in Hartonas and Dunn (1997) and Hartonas (1997) has been developed with the aim of making it possible to extend Dunn's gaggle theory to the case of non-distributive lattices. In particular, Hartonas (1997) explicitly extends the Jónsson–Tarski results on Boolean algebras with operators and Dunn's gaggle-theoretic results on Distributive Lattices with operators with well-behaved distribution types to the case of lattices with what we called in Hartonas (1997) *normal* operators (just another name for Dunn's family of operators with well-behaved distribution types).

The objective of this paper is to contribute to the study of non-distributive logical calculi, though restricting the scope and focusing only on non-distributive calculi with various notions of negation, as captured in Dunn's kite, and complementing in this respect the treatment of negation in distributive gaggle theory of Dunn and Zhou (2005). We also complement our study of negation as impossibility by introducing and studying a possibility operator, where we provide standard relational semantics, despite the lack of distribution, thus significantly improving over the results of Conradie and Palmigiano (2015), which follows the approach of Gehrke (2006) of so-called generalized Kripke frames, or of Suzuki (2010, 2012, 2014), as well as of Järvinen and Orlowska (2005). The systems we study are based on non-distributive lattice logic, variously extending it with negation operators (minimal negation, De Morgan or intuitionistic negation, negation as impossibility, etc.) and proving completeness theorems in related Kripke frames.

Dropping distribution may present a number of new technical difficulties in the semantic treatment of the related calculi, but it does also present itself as the natural approach to reasoning with incomplete information. Information sites (worlds, points of the underlying set of the Kripke frame) possess only partial knowledge of the world. It can happen that a site u supports the truth of a disjunction $\varphi \vee \psi$, without being in a position to distinguish which of the two is true. The various representation theorems for non-distributive lattices allow for two different understandings of the meaning of disjunction. Nearly all introduce a binary relation connecting information sites (Alasdair Urquhart's approach in Urquhart (1978) is slightly more complicated, introducing two relations), used to interpret disjunction, but they differ in whether this relation is to be understood as a relation of information extension, or of information incompatibility. The relation introduced generates a Galois connection on sets of information sites and, therefore, a closure operator (the composition of the Galois maps). Propositions in the frame, interpreting sentences, are then taken to be not just any sets of sites but only the *stable* ones (remaining unchanged under an application of the closure operator).

In this paper we shall follow the approach of Hartonas (1997), appropriately extending it for the needs of the task at hand. To interpret disjunction, we introduce *information neighborhood frames* (X, ν) where $\nu(x)$ is the information neighborhood of x, which can be thought of either negatively (the set of information sites with information incompatible with the site x), or positively (the set of informational expansions of x). A disjunction $\varphi \vee \psi$ is true at the information site x iff x is in the neighborhood of every site y whose neighborhood includes all sites z where at least one of φ, ψ is true.

According to Dunn (1999) the general gaggle-theoretic semantic treatment of negation is in terms of a binary (perp) relation, using perp-frames $(W, \perp)$, with the semantic clause

$$u \models \sim\varphi \quad \text{iff} \quad \forall v\ (v \models \varphi \ \text{implies}\ u \perp v) \tag{1}$$

first used by Goldblatt (1974) in his semantic analysis of Orthologic (the origin of the idea is perhaps to be found in Birkhoff and von Neumann 1936). The intuition here is that a world (information site) u satisfies $\sim\varphi$ iff it is "orthogonal" to every world that satisfies φ, where "orthogonality" is to be thought of as some sort of incompatibility of worlds. The meaning and properties of this notion of orthogonality are of course bound to be different, depending on the logical properties of the negation operator.

1.2 Topological Representation of Bounded Lattices

To get an insight on the semantics of disjunction for non-distributive lattice logic we briefly review in this section the results that have appeared in the literature on lattice representation and duality.

The first representation theorem for a class of non-distributive lattices is due to Goldblatt (1974), though the approach applies only to lattices equipped with an orthocomplementation operator $\neg$ in which case joins are definable as $a \vee b = \neg(\neg a \wedge \neg b)$ and the task essentially reduces to representing the meet-semilattice, but also and in addition, representing appropriately the orthocomplementation operator. The carrier set of the dual space of the ortholattice is the set X of its filters and an irreflexive, symmetric binary relation $\perp$ of orthogonality is defined on X by $x \perp y$ iff $\exists a \in x\ \ \neg a \in y$, inducing an antitone operation (in fact, a duality) on sets of filters $U^{\perp} = \{x : \forall u \in U\ u \perp x\}$. The set X is appropriately topologized and Goldblatt identifies the lattice of closed (in the topology) regular subsets of X, i.e., sets $A \subseteq X$ such that $A^{\perp\perp} = A$, as an isomorphic copy of the original lattice, with joins definable by

$$A \vee B = (A \cup B)^{\perp\perp} = \{x : \forall y\ (\forall z\ (z \in A \cup B \implies z \perp y) \implies x \perp y)\} \tag{2}$$

Urquhart (1978) was the first to produce a topological representation and an objects-only duality for bounded lattices that may drop the distribution law. The carrier set X of the dual space of a lattice is the set of maximal disjoint filter-ideal pairs, i.e., pairs $x = (u, v)$ where u is a filter and v is an ideal, $u \cap v = \emptyset$, u is maximal amongst the filters disjoint from v and similarly for v. Existence of maximal disjoint filter-ideal pairs is established by a use of Zorn's lemma, hence the proof is carried out in ZFC (Zermelo–Fraenkel set theory with the axiom of choice). An advantage of Urquhart's representation is that it specializes to the representation of Priestley (1970) when the lattice is distributive. The set of maximal disjoint filter-ideal pairs is doubly-ordered, $(X, \leq_1, \leq_2)$, with the coordinate-wise inclusion relations. The two relations induce a Galois connection from the family X^1 (actually, a lattice) of $\leq_1$-increasing subsets of X to the family X^2 (a lattice) of $\leq_2$-increasing subsets of X, defined on subsets $U \in X^1$, $V \in X^2$ by

$$\lambda U = \{x \in X : \forall y \in X\ (y \in U \Longrightarrow x \not\leq_1 y)\} \quad \text{and} \qquad (3)$$
$$\rho V = \{x \in X : \forall y \in X\ (y \in V \Longrightarrow x \not\leq_2 y)\}.$$

He then topologizes X and identifies an isomorphic copy of the original lattice consisting of the sets $A \subseteq X$ that are doubly closed (both A and ρA are closed sets in the topology) and stable ($\lambda\rho A = A$), with joins definable by

$$A \vee B = \lambda\rho(A \cup B) = \{x : \forall y\ (\forall z\ (z \in A \cup B \Longrightarrow y \not\leq_2 z) \Longrightarrow x \not\leq_1 y)\} \quad (4)$$

Allwein and Hartonas (1993) build on Urquhart's representation and, considering the set of all disjoint filter-ideal pairs, work in ZF and establish a full categorical duality. Joins of the dual lattice are again definable as in Eq. (4).

Hartung (1992) also builds on Urquhart's representation, considering as the dual space of a lattice L the triple $(F_0, I_0, \perp)$, where F_0 is the set of filters that are maximal disjoint with respect to some ideal y (not necessarily in I_0) and, similarly, I_0 is the set of ideals that are maximal disjoint with respect to some filter x, while $\perp \subseteq F_0 \times I_0$ is the relation $x \perp y$ iff $x \cap y \neq \emptyset$. Letting $\lambda \dashv \rho$ be the induced Galois connection and after imposing an appropriate topology on both sets F_0, I_0, Hartung identifies the collection of stable subsets of F_0 enjoying some additional topological properties as an isomorphic copy of the original lattice. Because of use of maximality, after Urquhart (1978), Hartung works in ZFC, while also modeling his construction as an instance of the notion of a concept lattice, building on ideas from Wille (1987). However, as far as our interest in this paper is concerned, we note that joins are defined in a way analogous to Eq. (2)

$$\begin{aligned} A \vee B &= \lambda\rho(A \cup B) \qquad (5)\\ &= \{x \in F_0 : \forall y \in I_0\ (\forall z \in F_0\ (z \in A \cup B \Longrightarrow y \perp z) \Longrightarrow x \perp y)\} \end{aligned}$$

Hartonas (1996) and Hartonas and Dunn (1997) work in ZF and build on the idea that even when a dualizing map, such as orthocomplementation, is not available on the lattice, the fact of order-duality of meets and joins should suffice to sustain a representation of the lattice. Indeed, they start with the observation that the identity map $\imath$ is a dual isomorphism of L with L^{op} (its opposite lattice): $L \rightleftarrows (L^{op})^{op}$. They view this as a diagram of meet-semilattices connected with a duality and represent both semilattices using their sets of filters (except that the filters of L^{op} are really the ideals of L). They also represent the duality the same way as orthonegation is represented by a perp relation. Thus the dual frames of lattices are triples $(X, Y, \perp)$ where X is the set of filters and Y is the set of ideals of L and $\perp$ is defined by $x \perp y$ iff $\exists a \in x\ \ \imath(a) \in y$ iff $x \cap y \neq \emptyset$. The Galois connection $\lambda \dashv \rho$ generated by the relation induces a closure operator and, after topologizing, a certain subclass of the family of stable sets of filters (to wit, the stable compact-opens) is recognized as an isomorphic copy of the original lattice. Joins are defined by a condition essentially identical to that in Eq. (5).

$$\begin{aligned} A \vee B &= \lambda\rho(A \cup B) \\ &= \{x \in X : \forall y \in Y\ (\forall z \in X\ (z \in A \cup B \Longrightarrow y \perp z) \Longrightarrow x \perp y)\} \end{aligned} \tag{6}$$

Hartonas (1997) also works in ZF, he regards the requirement that there should exist a duality induced isomorphism of the lattice L (with itself, or another (semi)lattice) as superfluous and uses order-duality to represent separately the meet $L^{\wedge}$ and join $L^{\vee}$ semilattices making up the lattice L. Subsequently, based on order-duality, Hartonas (1997) establishes a dual isomorphism of the image of $L^{\wedge}$ with the image of $L^{\vee}$ at the representation level. More specifically, the concrete semilattices are the families of sets of filters $X_a = \{x : x_a \leq x\}$ and $X^a = \{x : x \leq x_a\}$, where x is a filter, x_a is the principal filter generated by the element $a \in L$ and $\leq$ is inclusion of filters. The ordering relation $\leq$ on the set X of filters generates a Galois connection

$$\lambda U = \{x \in X : U \leq x\} \qquad\qquad \rho V = \{y \in X : y \leq V\}$$

where $U \leq x$ means that x is an upper bound of the elements of U and $y \leq V$ that y is a lower bound of the elements of V. After topologizing, a certain family of stable sets $A = \lambda\rho A$ (the stable compact-opens) is identified as an isomorphic copy of the original lattice and joins are defined by

$$A \vee B = \lambda\rho(A \cup B) = \{x : \forall y\ (\forall z\ (z \in A \cup B \Longrightarrow y \leq z) \Longrightarrow y \leq x)\} \tag{7}$$

Gerhke and Harding (2001), motivated by Hartung (1992), independently arrive at the same idea for a dual frame as in Hartonas and Dunn (1997), considering triples $(X, Y, \perp)$ where X, Y are the sets of filters and ideals, respectively, of X and the relation $x \perp y$ is again defined by the condition $x \cap y \neq \emptyset$. Their representation of joins then is precisely that given in Eq. (6). Other than investigating a number of interesting properties of their representation, Gerhke and Harding (2001) also address the issue of representing unary functions on the lattice L, a problem not raised in Hartonas and Dunn (1997).

Gerhke and van Gool (2014) revisit the lattice duality problem and they investigate a number of properties of interest. For our present concerns, however, we only need to note that the dual objects are of the form $(X, Y, \perp)$ where X, Y are the sets of filters and ideals of the lattice and $x \perp y$ is defined by $x \cap y \neq \emptyset$, hence no new insights on the definition of joins result.

Moshier and Jipsen (2014a, b) observe that all dualities for bounded lattices have proceeded by building, in one way or another, on the Priestley (1970) duality for distributive lattices, making use of spaces with an additional binary relation on them, in some cases explicitly an ordering relation, and they set out to prove a duality that is based instead on the original Stone (1937–38) representation of distributive lattices. They report a number of interesting facts relating to lattice duality and, as far as representing joins is concerned, they introduce directly a closure operator Γ (they call it $\texttt{fsat}$ and define $\texttt{fsat}(U) = \bigcap\{F \in OF(X)\colon U \subseteq F\}$, where $OF(X)$ is the set of subsets of X that are filters and open in the topology defined on X) and model joins accordingly, as closures of unions. They achieve their main objective on generalizing the Stone representation result for the case of bounded lattices, however, as far as our present task is concerned we note that it follows from general facts of category theory that a closure operator is induced by a Galois connection (though not a unique one). On the other hand, it is also known that any Galois connection on a powerset is induced by a binary relation on that set (by setting xRy iff $y \in \lambda(\{x\})$) and so it seems that a binary relation on the points of X is involved again in representing joins and, though we have not sorted out details, this relation is based on the ordering of X, hence their interpretation of joins falls within one of the above patterns.

Remark 1.1 (Notational Convention) In the sequel, we overload the use of $\leq$, leaving it to the context to disambiguate its use. More specifically,

1. we systematically use x, y, z, u, v, w for filters of a lattice and x_a etc. for a principal filter $x_a = \{b\colon a \leq b\}$;
2. for filters x, y of a lattice L, we write $x \leq y$ for their set-theoretic inclusion ($x \subseteq y$);
3. for a set U of filters and a filter x, we write $U \leq x$ as an abbreviation for $\forall u \in U\ u \leq x$ (x is an upper bound of the elements of U);
4. similarly, for a set of filters V and a filter x, we write $x \leq V$ as an abbreviation for $\forall v \in V\ x \leq v$ (x is a lower bound of the elements in V);
5. we also write $a \leq x$, for an element $a \in L$ of the lattice and a filter x as an abbreviation for $\forall e \in x\ \ a \leq e$ (a is a lower bound of the elements of x). Note that $a \leq x$ iff $x \leq x_a$ (the filter x is contained in the principal filter x_a).

2 Lattice Logic

2.1 Motivation, Syntactic and Proof-Theoretic Preliminaries

In this section we isolate the basic, minimal system of Lattice Logic, whose language includes no more than the connectives of conjunction and disjunction. The interest in this system lies with the semantics of disjunction in the absence of the distribution law (and of an orthonegation operator). In later sections we present extensions of the system, first with various notions of negation (Sect. 3), as presented in Dunn and Zhou (2005), and then we introduce Modal Lattice Logic (Sect. 4), with possibility and impossibility operators, with the study of the latter, in the context of distributive logic, originating in Došen (1986, 1999) and Vakarelov (1977, 1989) and, more recently, in Dunn and Zhou (2005). The presentation is based on this author's results on lattice representation and Stone type duality in Hartonas (1997).

The sentences of the language of lattice logic are generated from a set AtS of atomic sentences p, q, etc., using the connectives $\wedge$ and $\vee$.

$$\varphi := \; p \; (p \in AtS) \mid \top \mid \bot \mid \varphi \wedge \varphi \mid \varphi \vee \varphi$$

As in Dunn and Zhou (2005) we present its proof theory by means of a symmetric consequence relation (a single sentence to the left and right of the turnstile), with the following axioms and rules:

1. $\varphi \vdash \varphi \qquad\qquad \varphi \vdash \top \qquad\qquad \bot \vdash \varphi$

2. $\dfrac{\varphi \vdash \psi \quad \psi \vdash \vartheta}{\varphi \vdash \vartheta}$

3–4. $\varphi \wedge \psi \vdash \varphi$ and $\varphi \wedge \psi \vdash \psi \qquad\qquad \varphi \vdash \varphi \vee \psi$ and $\psi \vdash \varphi \vee \psi$

5–6. $\dfrac{\varphi \vdash \psi \quad \varphi \vdash \vartheta}{\varphi \vdash \psi \wedge \vartheta} \qquad\qquad \dfrac{\varphi \vdash \vartheta \quad \psi \vdash \vartheta}{\varphi \vee \psi \vdash \vartheta}$

Naturally, we omit the distribution axiom and we refer to this system as PLL, for Positive Lattice Logic. As usual we write $[\varphi]$ for the equivalence class of φ, under provability. We list the following immediate result, merely for reasons of completeness.

Lemma 2.1 *The Lindenbaum algebra of Lattice Logic is a bounded lattice.*

2.2 *The Semantics of Non-distributive Lattice Logic*

Dropping distribution may present a number of new technical difficulties in the semantic treatment of the related calculi [witness the difficulties in the semantics of necessity and possibility in any of Kamide (2002), Gehrke (2006), Conradie and Palmigiano (2015), Suzuki (2010, 2012, 2014), Järvinen and Orlowska (2005)], but it does also present itself as the natural approach to reasoning with incomplete information. Information sites (worlds, points of the underlying set of the Kripke frame) possess only partial knowledge of the world. It can happen that a site u supports the truth of a disjunction $\varphi \vee \psi$, without being in a position to distinguish which of the two is true. Thereby, the familiar semantic clause $w \models \varphi \vee \psi$ iff $w \models \varphi$, or $w \models \psi$ must be abandoned, as it directly relates to representing disjunction (lattice join) as a set-theoretic union. The various representation theorems for non-distributive lattices allow for two different accounts of the meaning of disjunction. Nearly all introduce a binary relation connecting information sites [the approach in Urquhart (1978) is slightly more complicated, introducing two relations], used to interpret disjunction, but they differ in whether this relation is to be understood as a relation of information extension, or of information incompatibility. In either case, the relation introduced generates a Galois connection on sets of information sites and, therefore, a closure operator (the composition of the Galois maps) is obtained. Propositions in the frame, interpreting sentences, are then taken to be not just any sets of sites but only the *stable* ones (remaining unchanged under an application of the closure operator). To provide an intuitive (though unfamiliar) semantics for disjunction in the non-distributive setting, we introduce information neighborhood frames.

Definition 2.2 (*Neighborhood frames*) *Information Neighborhood Frames* are structures (X, ν) where $\nu : X \longrightarrow \mathcal{P}(X)$ is the *neighborhood function* and X is a non-empty set of points, to be called *information sites.*

- If R is the relation generated by xRy iff $y \in \nu(x)$ and $\Gamma = \lambda\rho$ is the closure operator generated by the Galois connection $\lambda \dashv \rho$ on the powerset of X, where

$$\lambda U = \{x \colon \forall u \in U\ uRx\} \qquad\qquad \rho V = \{y \colon \forall v \in V\ yRv\},$$

  then the neighborhood function is recoverable by $\nu(x) = \Gamma(\{x\})$.
- The stable subsets A of X, $A = \Gamma A$, are closures of single points $A = \Gamma x^A$, where for simplicity of notation we write Γx for $\Gamma(\{x\})$ (hence the neighborhood function assigns a *stable* neighborhood to each point $x \in X$).
- There exists a subset $X_0 \subseteq X$ such that the stable sets generated by the points of X_0 form a bounded sublattice of the complete lattice of stable sets. We will refer to this sublattice as the *regular sublattice*, to its members as the *regular* subsets of X and to the points (information sites) of X_0 generating them as the *regular points* of X.
- The bounds $\mathbf{0}, \mathbf{1}$ of the lattice of regular subsets are the whole space $X = \mathbf{1}$, and a special singleton regular subset $\mathbf{0} = \{\omega\}$.
- The neighborhood function imposes a partial order on the carrier set of a frame by letting $x \leq y$ iff $\nu(y) \subseteq \nu(x)$, which we refer to as the *information ordering* of the frame.

We note that the information ordering on the points of the frame induced by the neighborhood function (equivalently, by the closure operator) has a largest element, *the inconsistent information site* ω, and a least informative site, *the trivial information site* to be denoted by 1.

Definition 2.3 A *neighborhood model* is a frame (X, ν) together with an interpretation function $\imath$ assigning to each atomic proposition p a regular subset $\imath(p)$ of the carrier set of the frame. The satisfaction relation is then defined on all sentences by the clauses

$x \models p$	iff $x \in \imath(p)$
$x \models \top$	iff $x \in X$
$x \models \bot$	iff $x = \omega$
$\omega \models \varphi$	(for any φ)
$x \models \varphi \wedge \psi$	iff $x \models \varphi$ and $x \models \psi$
$x \models \varphi \vee \psi$	iff $\forall y\ [(\forall z\ (z \models \varphi \text{ or } z \models \psi) \Longrightarrow z \in \nu(y)) \Longrightarrow x \in \nu(y)]$

In words:

$\varphi \vee \psi$ is true at the information site x iff x is in the neighborhood of every site whose neighborhood includes the sites where at least one of φ, ψ is true.

Lemma 2.4 *Letting $[\![\varphi]\!] = \{x : x \models \varphi\}$, the sets $[\![\varphi]\!]$ are regular subsets of X.*

Proof By induction on φ. The claim is true for the base cases (atomic sentence p and the special sentences $\top, \bot$) because of the requirements imposed on frames and interpretations. The rest follows from the fact that $[\![\varphi \wedge \psi]\!] = [\![\varphi]\!] \cap [\![\psi]\!]$ and $[\![\varphi \vee \psi]\!] = [\![\varphi]\!] \vee [\![\psi]\!]$, where the latter is the join in the regular sublattice, defined by $[\![\varphi]\!] \vee [\![\psi]\!] = \Gamma([\![\varphi]\!] \cup [\![\psi]\!]) = \lambda\rho([\![\varphi]\!] \cup [\![\psi]\!])$, given the definition of λ, ρ in Definition 2.2 and given the requirement that $\nu(x) = \Gamma x$ (which is an abbreviation for $\Gamma(\{x\})$) in neighborhood frames. □

For completeness, we first briefly present the representation theorem from Hartonas (1997).

Theorem 2.5 (Representation theorem) *For every lattice L, there is a neighborhood frame (X_L, ν_L) such that L is isomorphic to the regular sublattice of the frame.*

Proof For details we refer the reader to our proof in Hartonas (1997). Suffice it to indicate here that X_L is the space of filters of X, including the improper filter ω, ν_L delivers the upper closure under set-theoretic inclusion of a filter x, $\nu_L(x) = \{z : x \leq z\}$, the isomorphism is the representation map $H(a) = \{x : a \in x\} = \{x : x_a \leq x\}$ (where x_a is the principal filter generated by the lattice element a). Endowing X with the topology generated by the subbasis $\{H(a) : a \in L\} \cup \{-H(a) : a \in L\}$, the regular sublattice is identified in Hartonas (1997) as the lattice of stable compact-open subsets of X. Then the set $X_0 \subseteq X$ of regular points of X consists of the points generating the stable compact-opens, which are identified in Hartonas (1997) to be precisely the sets $H(a)$, with $a \in L$, and their generators are the principal filters x_a, $a \in L$. □

For the sake of thoroughness we list the following theorem.

Theorem 2.6 (Soundness and completeness) *Lattice logic is sound and complete in information neighborhood frames.*

Proof Soundness follows from Lemma 2.4. For completeness, represent the Lindenbaum algebra of the logic (a bounded lattice) in its filter space X following the approach of Hartonas (1997), as explained in Theorem 2.5, letting $H(a) = \{x : x \text{ is a filter and } a \in x\}$ and define the neighborhood of a filter x to be its upper closure $\nu(x) = \{u \in X : x \leq u\}$ (where $\leq$ is inclusion of filters). The binary relation associated to the neighborhood function is the relation $\leq$ of filter inclusion and it generates a Galois connection $\lambda \dashv \rho$ from which a closure operator $\Gamma = \lambda\rho$ is extracted. For simplicity of notation, we write Γx for an application of Γ to singletons, rather than $\Gamma(\{x\})$. Note that $\Gamma x = \{y : x \leq y\} = \nu(x)$. Joins are represented by Eq. (7), which we simply rewrite using the neighborhood function:

$$A \vee B = \lambda\rho(A \cup B) = \{x : \forall y \,(\text{if } \forall z\,(z \in A \cup B \text{ implies } z \in \nu(y)), \text{ then } x \in \nu(y))\}$$

By the results of Hartonas (1997), (X, ν) is an information neighborhood frame. The regular sublattice of the frame is the set of stable compact-opens of X, where the topology is generated by the subbasis $\{ H(a) \colon a \in L \} \cup \{ -H(a) \colon a \in L \}$. The satisfaction relation $\models$, defined by $x \models \varphi$ iff $x \in \{ x \colon [\varphi] \in x \}$ satisfies the required clauses for information neighborhood frames, as described above.

By the representation results of Hartonas (1997), the lattice is isomorphic to the lattice of stable compact-open sets of filters, i.e., the lattice of regular subsets of X. It then easily follows that $\varphi \vdash \psi$ iff $[\varphi] \leq [\psi]$ iff $\{ x \colon [\varphi] \in x \} \subseteq \{ x \colon [\psi] \in x \}$, where $[\varphi]$ is the equivalence class of the sentence φ under the provability relation. □

3 Lattice Logic with Negation Operators

Classical Propositional Logic interprets negation in a simple and straightforward way, $w \models \neg\varphi$ iff $w \not\models \varphi$, a semantic clause based on the fact that boolean negation is interpreted (represented) as set-theoretic complement. In Intuitionistic Logic, with sentences interpreted as open subsets of a topological space and negation defined in terms of implication, $\neg\varphi$ is interpreted as true at all points in the topological interior of the complement of the interpretation of φ. Staying within the bounds of logics adopting the distribution law, Relevance Logic interprets negation using either an operator $*$, with $x^* = \{ \varphi \colon \neg\varphi \notin x \}$ (and where x is a prime theory (filter)), or a relation $\perp$ on worlds, two approaches that have been shown to be equivalent in Dunn (1993). Weak notions of negation in the context of distributive logics have been also interpreted using $\perp$-semantics, witness Dunn and Zhou (2005).

With the axiom of distribution abandoned, perhaps the most significant case that has been studied in the literature is orthonegation Goldblatt (1974), using a notion of orthogonality, or perpendicularity, of worlds ($x \perp y$ iff $\exists a \in x \ \neg a \in y$), probably inspired by the semantic treatment of negation in Quantum Logic (see Birkhoff and von Neumann 1936; Dalla Chiara and Giuntini 2002, 2001; Dunn et al. 2013). Quantum Logic is originally interpreted in the lattice of closed linear subspaces of Hilbert spaces where points of the space are orthogonal when their inner product equals zero, $\mathbf{x} \perp \mathbf{y}$ iff $\mathbf{x} \cdot \mathbf{y} = 0$, a relation then inducing an orthogonality relation on closed linear subspaces in the obvious way.

This section is devoted to providing semantics and proving completeness theorems for a family of logical systems that drop the distribution law and come equipped with notions of negation of various strength, from pre-minimal negation to orthonegation, as classified by Dunn, thus doing for non-distributive logics what Dunn and Zhou (2005) do for the distributive case. In Fig. 1, we have annotated each system with an acronym, with LL standing for Lattice Logic and with the third letter of the acronym pointing to Dunn's minimal negation (system LLD), or Galois negation (system LLG) and similarly for each of LLJ (Lattice Logic with Johansson's minimal negation), LLM (Lattice Logic with a De Morgan negation), LLI (Lattice Logic with an intuitionistic negation) and LLO, standing for Lattice Logic with an Orthonegation, commonly referred to as Orthologic. We write LL? to refer to any of the above

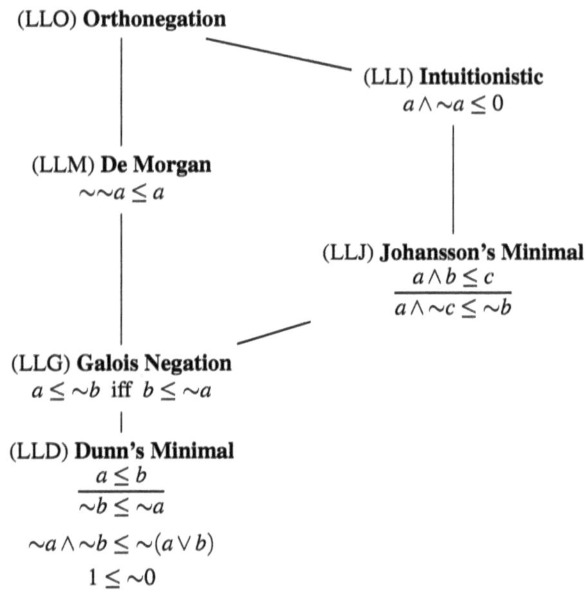

Fig. 1 Dunn's Kite of negations

systems. For the syntax and axiomatization of the systems we refer the reader to Dunn and Zhou (2005), though the axiomatization, besides the axioms for lattice logic (without distribution) that can be found in Sect. 2, can be easily also read off from Dunn's kite of negations in Fig. 1.

Remark 3.1 (Orthologic with Constructive Contraposition) Referring to Fig. 1, it should be noted that for the derivation of Johansson's constructive contraposition rule the assumption of an orthonegation is insufficient in itself. In fact, distribution is also required (see also Remark 4.1). Hence, in a non-distributive setting orthonegation as classically defined (satisfying antitonicity, double negation and the intuitionistic absurdity principle $a \wedge {\sim}a = 0$) is weaker than if we explicitly postulate constructive contraposition, as well. In other words, Orthologic in Goldblatt (1974) is weaker than Orthologic with Constructive Contraposition.

3.1 LL?-Algebras and Representation

Definition 3.1 An *LL?-algebra* $\langle L, \wedge, \vee, 0, 1, \sim \rangle$ is a non-distributive bounded lattice with an antitone unary operator $\sim : L \longrightarrow L^{op}$. We speak of LLD, LLG, LLJ, etc. algebras accordingly as the corresponding properties in Fig. 1 hold.

A *concrete LL?-algebra* is an LL?-algebra whose elements are sets, equipped with an antitone operator ${\sim}A$ defined on elements (sets) of the algebra and satisfying the corresponding properties in Fig. 1.

Note that the LLO-algebras are exactly the Ortholattices. For the sake of completeness we list the obvious result.

Lemma 3.2 *The Lindenbaum algebra of a logic system LL? is an LL?-algebra (where ? is, correspondingly, any of D, G, M, J, I or O).*

Theorem 3.3 (Representation theorem) *Every LL?-algebra is isomorphic to an LL?-algebra of sets.*

Proof The proof is an extension of our Lattice Representation Theorem 2.5, originally reported in Hartonas (1997). The negation operator, $\sim: L \longrightarrow L^{op}$, is a normal operator in the sense of Hartonas (1997) (it distributes over joins of L and delivers "joins" in L^{op}) and hence applying the technique we developed in Hartonas (1997) we define an operator $*$ on filters, then extend it to an operator ${-\!\!\wr}$ on stable sets, recalling that stable sets are the closures of single points:

$$x^* = \{b\colon \exists a\ (x \leq x_a \text{ and } {\sim}a \leq b)\} \qquad\qquad {-\!\!\wr}\Gamma x = \Gamma x^* \tag{8}$$

We leave it to the interested reader to verify that x^* is indeed a filter. Letting x_a be the principal filter generated by a and $H(a)$ be the representation map $H(a) = \{x\colon a \in x\} = \{x\colon x_a \leq x\} = \Gamma x_a$, we first list some basic facts.

Lemma 3.4 *The following hold.*

1. *The dual filter of x_a is $x_{\sim a}$: $(x_a)^* = x_{\sim a}$.*
2. *The operator $*$ is antitone on filters: $x \leq y$ implies $y^* \leq x^*$.*
3. *The operator ${-\!\!\wr}$ has the following properties:*
   (a) ${-\!\!\wr}H(a) = H({\sim}a)$;
   (b) $H(a) \leq H(b)$ *implies* ${-\!\!\wr}H(b) \leq {-\!\!\wr}H(a)$;
   (c) ${-\!\!\wr}(H(a) \vee H(b)) = {-\!\!\wr}H(a) \cap {-\!\!\wr}H(b)$.

Proof The first follows by a simple calculation.

$$(x_a)^* = \{b\colon \exists e \leq x_a\ {\sim}e \leq b\} = \{b\colon \exists e \leq a\ {\sim}e \leq b\} = \{b\colon {\sim}a \leq b\} = x_{\sim a}$$

For the second, assume $x \leq y$. Let $b \in y^*$. Then there exists some element e with $e \leq y$ such that ${\sim}e \leq b$. However, since $x \leq y$, a lower bound e of y is necessarily one of x, hence there exists $e \leq x$ such that ${\sim}e \leq b$ and this shows $b \in x^*$. This shows that if $x \leq y$, then $y^* \leq x^*$.

The proof of 3(a) is trivial, as it is a consequence of the first and of the definition of ${-\!\!\wr}$ in Eq. (8), given that $H(a) = \Gamma x_a$.

For 3(b), the assumption is equivalent to $\Gamma x_a \leq \Gamma x_b$, hence any filter z with $x_a \leq z$ will also satisfy $x_b \leq z$. In particular, $x_b \leq x_a$ follows. Then $b \in x_a$ and so $a \leq b$ and thereby ${\sim}b \leq {\sim}a$, from which we get $H({\sim}b) \leq H({\sim}a)$. This is equivalent to $\Gamma(x_b)^* \leq \Gamma(x_a)^*$, which is by definition equivalent to ${-\!\!\wr}\Gamma x_b \leq {-\!\!\wr}\Gamma x_a$, hence also ${-\!\!\wr}H(b) \leq {-\!\!\wr}H(a)$.

For 3(c), using the fact that $\neg H(a) = H(\sim a) = \Gamma x_{\sim a}$ and observing that $\Gamma x_{a \vee b} = \Gamma x_a \vee \Gamma x_b$ we obtain

$$
\begin{array}{rlrl}
\neg H(a) \cap \neg H(b) & = \Gamma x_{\sim a} \cap \Gamma x_{\sim b} & & = \{z : \sim a \in z \text{ and } \sim b \in z\} \\
& = \{z : \sim a \wedge \sim b \in z\} & & = \{z : \sim(a \vee b) \in z\} \\
& = \{z : x_{\sim(a \vee b)} \leq z\} & & = \Gamma x_{\sim(a \vee b)} \\
& = \neg \Gamma x_{a \vee b} & & = \neg(\Gamma x_a \vee \Gamma x_b) \\
& = \neg(H(a) \vee H(b)) & &
\end{array}
$$

This completes the argument for LLD-algebras. In the sequel we verify that the algebra of stable compact-opens of the filter space is an LLG, or an LLJ, etc. algebra, accordingly as the original Lindenbaum algebra $\langle L, \wedge, \vee, \sim \rangle$ is such an algebra, as well.

LLG In addition to the minimal negation properties, negation now forms a Galois connection with itself (in which case codistribution over joins is in fact derivable). By an elementary calculation, we show $H(a) \leq \neg H(b)$ iff $H(b) \leq \neg H(a)$, using the fact that $a \leq \sim b$ iff $b \leq \sim a$ holds in the Lindenbaum algebra.

$$
\begin{array}{lllll}
H(a) \leq \neg H(b) & \text{iff } \Gamma x_a \leq \Gamma x_{\sim b} & \text{iff } x_a \in \Gamma x_{\sim b} & & \\
& \text{iff } x_{\sim b} \leq x_a & \text{iff } a \leq \sim b & & \\
& & \text{iff } b \leq \sim a & \text{iff } x_{\sim a} \leq x_b & \\
& & \text{iff } x_b \in \Gamma x_{\sim a} & \text{iff } \Gamma x_b \leq \Gamma x_{\sim a} & \\
& & & \text{iff } H(b) \leq \neg H(a) &
\end{array}
$$

With the Galois property in place it easily follows that for any lattice elements a, b we have $x_b^* \leq x_a$ iff $x_a^* \leq x_b$.

LLM Given the Galois connection property, the identity $\sim\sim a = a$ holds in the system LLM. The proof that $\neg\neg H(a) = H(a)$ is elementary since the left side is identical to $\Gamma x_{\sim\sim a}$.

LLJ Having antitonicity and the Galois property, assume further that $H(a) \cap H(b) \leq H(c)$. Showing that $H(a) \cap \neg H(c) \leq \neg H(b)$ is trivial since the hypothesis is equivalent to $H(a \wedge b) \leq H(c)$, which is equivalent to $a \wedge b \leq c$ while the desired conclusion is equivalent to $a \wedge \sim c \leq \sim b$.

LLI, LLO Similar, by trivial calculations, based on the property of the representation that $a \leq b$ holds in the Lindenbaum algebra iff $H(a) \leq H(b)$ and the fact already proven that $\neg H(a) = H(\sim a)$. □

In the next two subsections, we explore two semantic approaches, namely, star semantics, familiar from the Routley and Routley (1972) semantic treatment of negation in Relevance Logic, and perp semantics, originating with Goldblatt's semantic treatment of Orthologic in Goldblatt (1974) and which is the standard gaggle-theoretic approach for distributive logics. The modal interpretation of negation (as impossibility), an approach initiated by Došen (1986, 1999) and Vakarelov (1977, 1989), and followed up in Dunn and Zhou (2005) will be explored in the next section for logics dropping the distribution law and equipped with both a possibility and an impossibility operator.

3.2 *Kripke Semantics for Negation Operators in* $*$*-Frames*

Star semantics for negation seems to have been first proposed in Białynicki-Birula and Rasiowa (1957), while Routley and Routley (1972) introduced star semantics for negation in Relevance Logic. The Routley approach amounts to defining an operator $*$ on prime filters x, delivering a prime filter $x^* = \{a : \neg a \notin x\}$ so that $\neg a \in x$ iff $a \notin x^*$, thereby interpreting relevant negation by the clause $x \models \neg\varphi$ iff $x^* \not\models \varphi$.

In the non-distributive case the semantics by means of a lattice representation result cannot use prime filters (else disjunction is interpreted as union), but uses all filters, in which case the Routley star cannot be shown to return a filter x^*, since the proof relies both on primality of the filter x and on the fact that relevant negation is a De Morgan negation. Consequently, a novel star operator needs to be invented, returning a filter, when applied to one. We will provide details in the course of the completeness proof, but first we introduce $*$-frames.

Definition 3.5 (*Star frames*) A *star frame* is a triple $\mathcal{F} = (X, \nu, *)$ where (X, ν) is a neighborhood frame in the sense of Definition 2.2 and $*: X \longrightarrow X^{op}$ is an antitone map on X (where the order on X is defined by $x \leq y$ iff $\nu(y) \subseteq \nu(x)$). For $x \in X$, we will refer to $x^* \in X$ as the *dual* of x. $\mathcal{F}$ is a *minimal*, or an *LLD-frame*, provided that

- the dual of a regular point is regular;
- for all regular points x, y, z, if $\nu(z)$ is the least upper bound (in the regular sub-lattice of the frame) of $\nu(x)$ and $\nu(y)$, then $\nu(z^*)$ is the greatest lower bound of $\nu(x^*), \nu(y^*)$;
- ω^* generates X (in the sense that $X = \nu(\omega^*) = \Gamma\omega^*$).

The following conditions classify LL?-frames:

LLG frames:	LLD + for all regular points x and y, $x^* \leq y$ iff $y^* \leq x$;
LLM frames:	LLG + $*$ is an involution on regular points: $x^{**} = x$;
LLJ frames:	LLG + for all regular points x, y, z if the neighborhood of z contains the intersection $\nu(x) \cap \nu(y)$, then the neighborhood of its dual contains the intersection $\nu(x) \cap \nu(y^*)$;
LLI frames:	LLJ + the neighborhoods of a regular point and its dual are disjoint: $\nu(x) \cap \nu(x^*) = \emptyset$;
LLO frames:	are the frames that are both LLM and LLI frames. □

Remark 3.2 (Representation and completion) A question of interest, given an LL?-algebra is whether it can be shown to be isomorphic to an LL?-subalgebra of a concrete and complete LL?-algebra. The question is both of an algebraic and of a semantic interest. As far as semantics is concerned, if the LL?-algebra cannot be embedded in a concrete and complete LL?-algebra, then the need for distinguishing in the frame a special set of stable subsets, or points that generate them, arises. Dealing with a variety of negation operators and lacking concrete completion results makes it necessary to resort to frames with a distinguished subfamily of stable sets, to serve as the propositions of the frame.

Lemma 3.6 *Let $(X, \nu, *)$ be a frame. Define an operator ${-\!\wr}$ on stable sets by setting ${-\!\wr}\Gamma x = \Gamma x^*$. Then ${-\!\wr}$ restricts to an operation on the regular subsets so that the algebra of regular subsets with the ${-\!\wr}$ operator is a concrete LL?-algebra, accordingly as the frame is an LL?-frame.*

Proof If x is a regular point, which is the same as saying that $\Gamma x = A$ is a regular subset of X, then since for LLD-frames it is already required that the dual of regular points are regular, it follows from the definition ${-\!\wr}\Gamma x = \Gamma x^*$ that ${-\!\wr}$ takes regular subsets to regular subsets of X. Furthermore, antitonicity of $*$ implies antitonicity of ${-\!\wr}$ on regular subsets.

Recall now that $\nu(x) = \Gamma x$, where the latter is an abbreviation of $\Gamma(\{x\})$. Given the second defining condition of LLD-frames and given the previous reminder and the definition of ${-\!\wr}$ we immediately obtain that for all regular points x, y, z we have ${-\!\wr}(\Gamma x \vee \Gamma y) = {-\!\wr}\Gamma z = {-\!\wr}\Gamma x \cap {-\!\wr}\Gamma y$, where Γz is assumed to be the least upper bound $\Gamma x \vee \Gamma y$, hence ${-\!\wr}$ codistributes over joins of regular subsets of X. The last requirement for an LLD-algebra, namely that $1 \leq {\sim}0$ is guaranteed by the fact that in LLD-frames we assume that $X = \Gamma\omega^*$.

Next assume that the frame is an LLG-frame, which is to say that for regular points x, y, $x^* \leq y$ iff $y^* \leq x$. But the assumption $x^* \leq y$, is equivalent to $\nu(y) = \Gamma y \leq \Gamma x^* = \nu(x^*) = {-\!\wr}\Gamma x$. Hence the frame condition translates to $\Gamma y \leq {-\!\wr}\Gamma x$ iff $\Gamma x \leq {-\!\wr}\Gamma y$, which is precisely the Galois condition for the operator ${-\!\wr}$.

If the frame is an LLM-frame then the involution property of $*$ on regular points and the definition of ${-\!\wr}$ by ${-\!\wr}\Gamma x = \Gamma x^*$ guarantees that ${-\!\wr}{-\!\wr}\Gamma x = \Gamma x$ and therefore the algebra of regular subsets is a concrete LLM-algebra.

For LLJ-frames and given the definition $\nu(x) = \Gamma x$ the condition for LLJ-algebras is that if $\Gamma x \cap \Gamma y \subseteq \Gamma z$, then $\Gamma x \cap {-\!\wr}\Gamma y \subseteq {-\!\wr}\Gamma z$, which is a direct translation of the frame condition.

Similarly for LLI-algebras, since the relevant condition $\Gamma x \cap {-\!\wr}\Gamma x$ is again a direct translation of the frame condition $\nu(x) \cap \nu(x^*) = \emptyset$.

For LLO-frames, there is nothing further to prove. □

The truth of the following claim has been demonstrated in the course of the proof of the previous lemma.

Corollary 3.7 *Let (X, ν) be a neighborhood frame. The following are equivalent:*

1. *There is a $*$ operator on the points of x such that the frame is an LL? frame (where ? is D, G, M, J, I or O).*
2. *There is an operator ${-\!\wr}$ on stable sets such that ${-\!\wr}$ preserves regular sets and the algebra of regular sets with the ${-\!\wr}$ operator is an LL? algebra (where ? is, correspondingly, D, G, M, J, I or O).* □

Definition 3.8 An LL?-model is an LL?-frame together with an interpretation function ι such that $\iota(p)$, for an atomic sentence p, is a regular subset of X. The satisfaction relation is required to satisfy the following clause for the negation operator

$$x \models {\sim}\varphi \quad \text{iff} \quad x \in {-\!\wr}\{y : y \models \varphi\} \tag{9}$$

in addition to the clauses for lattice logic. The model is an LLD, LLG, etc. model, accordingly as the frame is an LLD, LLG, etc. frame.

The following is an obvious consequence of the above.

Corollary 3.9 *If* $[\![\varphi]\!] = \{x : x \models \varphi\}$, *then* $[\![\varphi]\!]$ *is a regular subset of the carrier set* X *of the frame.*

We may now proceed to prove a soundness and completeness theorem.

Theorem 3.10 (Soundness and completeness in $*$-frames) *Each of the LL? logic systems is sound and complete for the respective LL?* $*$*-frames.*

Proof For soundness, every sentence is interpreted as a regular subset of the carrier set of the frame (Corollary 3.9) and the algebra of regular subsets is an LL? algebra accordingly as the frame is an LL? frame (Lemma 3.6). Hence every theorem of an LL? logical system is sound in the corresponding class of LL? frames.

Completeness is immediate from our representation theorem (Theorem 3.3). □

Remark 3.3 (A comparison with the Routley star) We have defined the satisfaction relation for negation by the clause

$$x \models \sim\varphi \quad \text{iff} \quad x \in {}^{-\wr}\{y : y \models \varphi\}.$$

Given the definition of the ${}^{-\wr}$ operator on stable sets, this can be easily seen to be equivalent to the following definition

$$x \models \sim\varphi \quad \text{iff} \quad \forall y\, (y \models \varphi \implies y^* \leq x). \tag{10}$$

In Dunn (1993), it is shown that the star and perp treatments of relevant negation are equivalent. In fact, they are related by the condition

$$\forall x, y\, (x \perp y \ \text{ iff } \ y \nleq x^*).$$

Using, as in Dunn (1993), the complement C of the perp relation the condition becomes

$$\forall x, y\, (xCy \ \text{ iff } \ y \leq x^*).$$

Reading xCy as "x is *compatible* with y," the site x^* is seen to be a maximal site compatible with x. Therefore, whereas the Routley star can be thought of as delivering a maximal information site compatible with x, so that $x \models \neg\varphi$ just in case its maximal compatible site x^* fails to satisfy φ, in the case of non-distributive logics an information site x satisfies $\sim\varphi$ just in case it contains the dual y^* of every site y where φ holds.

3.3 Kripke Semantics for Negation Operators In ⊥-Frames

The semantics of negation using an incompatibility relation follows the pattern

$$x \models \sim\varphi \quad \text{iff} \quad \forall y\ (y \models \varphi \text{ implies } y \perp x)$$

In words, $\sim\varphi$ holds at x just in case every information site satisfying φ is incompatible with x.

From Remark 3.3 comparing the Routley star with the star operator we have defined for the case of non-distributive logics and in particular from Eq. (10) it follows that the appropriate definition for non-distributive logics is $u \perp v$ iff $u^* \leq v$. In words, u is incompatible with v just in case v contains the dual of u. Hence u is always incompatible with its dual, $u \perp u^*$, and to any extension of it $v \geq u^*$. With this observation in hand, we can easily proceed to provide perp-semantics for Lattice Logic with various negation operators.

Definition 3.11 An *information neighborhood ⊥-frame* is a structure $(X, \nu, \perp)$ where (X, ν) is an information neighborhood frame in the sense of Definition 2.2 and $\perp$ is a binary relation on X. Moreover, writing $x \perp U$ for $x \perp u$, $\forall u \in U$, if U is regular, then so must be ${}^{-\wr}U = \{x : x \perp U\}$ and we may let u^* be the regular point that generates it, assuming u is the regular point that generates U. This gives rise to an associated $*$-frame and we may then classify ⊥-frames as LLD, LLG. etc. frames accordingly as their associated $*$-frames are LLD, LLG. etc. frames.

A model is a frame with an interpretation ι, assigning a regular subset of X to each atomic sentence p of the language. The satisfaction relation $\models$ from information sites to sentences is required to satisfy, in addition to the conditions for Lattice Logic, the following condition, familiar from the ⊥-semantics of negation in distributive gaggles.

$$x \models \sim\varphi \quad \text{iff} \quad \forall y\ (y \models \varphi \text{ implies } y \perp x)$$

It is an immediate consequence of the requirements in the above definition and by a straightforward inductive argument that $\{x : x \models \sim\varphi\}$ is regular, since $x \models \sim\varphi$ iff $x \perp \{z : z \models \varphi\}$ iff $x \in {}^{-\wr}\{y : y \models \varphi\}$.

Theorem 3.12 (Soundness and completeness in ⊥-frames) *Each of the LL? logic systems is sound and complete for the respective LL? ⊥-frames.*

Proof We may let ${}^{-\wr}A = \{x : x \perp A\}$, for regular sets A. If $A = \Gamma z$, for a regular point z, then using the associated $*$-frame we get ${}^{-\wr}\Gamma z = \Gamma z^*$. Then soundness follows by the same argument as for $*$-frames (Theorem 3.10).

For completeness, we turn the canonical $*$-frame of Theorem 3.10 (obtained from our representation result, Theorem 3.3) to a ⊥ frame by defining $x \perp z$ iff $x^* \leq z$, where x^* was defined by the equation $x^* = \{b : \exists a \leq x \ \ \sim a \leq b\}$. The following lemma is needed.

Lemma 3.13 *For any element a (of the Lindenbaum algebra of the logic) and filter* w *we have* $\sim a \in w$ *iff* $\forall x\ (a \in x \implies x \perp w)$.

Proof From left to right, assume $\sim\!a \in w$. It follows that $w \in \Gamma x_{\sim a} = -\!\!\wr\!\!-\Gamma x_a$, by Lemma 3.4. Assume now $a \in x$, for an arbitrary filter x. Then $x_a \leq x$, hence $\Gamma x \leq \Gamma x_a$, by antitonicity of Γ on singletons. Then $-\!\!\wr\!\!-\Gamma x_a \leq -\!\!\wr\!\!-\Gamma x$, by antitonicity of $-\!\!\wr\!\!-$ (Lemma 3.4). Hence $w \in -\!\!\wr\!\!-\Gamma x_a \leq -\!\!\wr\!\!-\Gamma x = \Gamma x^*$. This shows that $x^* \leq w$, hence by definition, $x \perp w$.

For the converse, assuming $\forall x\ (a \in x \Longrightarrow x \perp w)$ we have in particular $x_a \perp w$, i.e., $(x_a)^* \leq w$, which is equivalent to $x_{\sim a} \leq w$, i.e., $\sim\!a \in w$. □

Hence $-\!\!\wr\!\!-\Gamma x_a = H(\sim a) = \{w\colon \forall x\ (x \in H(a) \text{ implies } x \perp w)\}$ and the canonical frame is a $\perp$-frame.

Since the interpretation of sentences in the canonical $\perp$-frame coincides with that in the associated $*$-frame, the rest follows by the proof of Theorem 3.10. □

Remark 3.4 (A comparison with Goldblatt's orthogonality relation) If $\neg$ is an orthocomplementation operator, then the definition of the $*$ operator on filters becomes $x^* = \{a\colon \neg a \leq x\}$ (which is the same as $\{a\colon x \leq x_{\neg a}\}$). The star operator can be shown to be a Galois connection on X, $x \leq y^*$ iff $y \leq x^*$, and an involution on principal filters. Goldblatt (1974) defines a $\perp$ relation on filters by $x \perp y$ iff $\exists a \in x\ \neg a \in y$, an irreflexive and symmetric relation on proper filters. The orthogonality relation generates a Galois connection on subsets where, due to symmetry of $\perp$, the two Galois maps coincide and for a set of filters U we obtain $U^{\perp} = \{x\colon U \perp x\} = \{x\colon \forall u \in U\ u \perp x\}$. In particular, it can be verified that for a regular point x (a principal filter in the canonical frame) $\{x\}^{\perp} = \Gamma x^*$ where $x^* = \{a\colon \neg a \leq x\}$ is the star operator we introduced. Defining $x \perp_H y$ iff $y^* \leq x$ it can be verified that the two perp relations coincide on regular points (principal filters in the canonical frame). In this paper, we do not pursue this any further. We have done so, however, in a recent technical report Hartonas (2015), yet unpublished, where we demonstrate that, though in light of Goldblatt (1984) the subclass of Goldblatt's orthomodular orthoframes is not elementary, nevertheless, there does exist an elementary subclass of the orthoframes of Goldblatt (1974) for which a first-order condition can be specified to further characterize the subclass of orthomodular (quantum) frames.

4 Modal Lattice Logic: Possibility and Impossibility Operators

4.1 Axioms and Rules for Possibility and Impossibility

In Dunn and Zhou (2005) the authors address the issue of treating modal negation operators, previously introduced and studied by Došen (1986, 1999) and Vakarelov (1977, 1989). Negation, in this context, appears as an impossibility, or as an unnecessity operator. From the proof-theoretic point of view, the axiomatization of negation as impossibility can be given by any of the systems LL?, but the intended semantics is given in frames $(X, \nu, \leadsto_*)$.

$$x \models {\sim}\varphi \quad \text{iff} \quad \forall y\,(x \leadsto_* y \implies y \not\models \varphi) \tag{11}$$

where (X, ν) is an information neighborhood frame and $\leadsto_* \subseteq X \times X$ is an accessibility relation on the set of points of the frame.

In this section, we introduce Modal Lattice Logic, an extension of Lattice Logic with possibility and impossibility operators. Whereas Došen (1986, 1999), Vakarelov (1977, 1989), Dunn and Zhou (2005) focus on the study of a modal interpretation of negation as impossibility, we think that the interaction of possibility and impossibility operators deserves its own attention. The language of Modal Lattice Logic is generated by the following grammar, where AtS is a nonempty set of atomic sentences.

$$\varphi := p\ (p \in AtS) \mid \top \mid \bot \mid \varphi \wedge \varphi \mid \varphi \vee \varphi \mid {\sim}\varphi \mid \Diamond\varphi$$

We assume the axioms and rules for Lattice Logic, as presented in Sect. 2. The negation operator is interpreted by the semantic clause (11). The minimal system with the possibility operator will be denoted by MLL? and it includes the familiar axioms and rules for the possibility operator, shown in Eq. (12), and the respective axioms and rules for impossibility, accordingly as ? is D, G, M, J, I or O. The issue of interest now lies with proposing an axiomatization of the interaction between possibility and impossibility. For a minimal system we propose two axioms, a form of excluded middle and a contradiction principle, as shown in Eq. (13), amounting to the acknowledgement that nothing can be both possible and impossible and that it is always the case of anything that it is either possible, or impossible.

$$\frac{\varphi \vdash \psi}{\Diamond\varphi \vdash \Diamond\psi} \qquad \Diamond(\varphi \vee \psi) \vdash \Diamond\varphi \vee \Diamond\psi \qquad \Diamond\bot \vdash \bot \tag{12}$$

$$\top \vdash {\sim}\varphi \vee \Diamond\varphi \qquad \Diamond\varphi \wedge {\sim}\varphi \vdash \bot \tag{13}$$

Additional axioms can be imposed to strengthen the interaction and we first consider two weak forms of a T-like axiom, as follows:

($T^{\sim}$) If it is impossible that φ is possible, then φ is in fact impossible.
($T^{\Diamond}$) If it is impossible that φ is impossible, then φ is indeed possible.

Both are shown below, in (14). The reader can perhaps justify their naturalness by thinking of impossibility $\sim$ as necessary falsity $\Box\neg$.

$${\sim}\Diamond\varphi \vdash {\sim}\varphi \quad (T^{\sim}) \qquad {\sim}{\sim}\varphi \vdash \Diamond\varphi \quad (T^{\Diamond}) \tag{14}$$

We will also consider the converse of the principles in (14).

($S4^{\sim}$) If φ is impossible, then it is impossible that it is possible.
($S5^{\Diamond}$) If φ is possible, then it is impossible that it is impossible.

These are shown in (15) below.

$$\sim\varphi \vdash \sim\Diamond\varphi \quad (S4^{\sim}) \qquad\qquad \Diamond\varphi \vdash \sim\sim\varphi \quad (S5^{\Diamond}) \tag{15}$$

The first has the flavor of the S4 axiom and the second has the air of S5. Figure 2 presents the possible combinations, starting with the minimal system of Modal Lattice Logic (MLLD), extending LLD. In Fig. 2, we have not included combinations with either the T or the S4 axiom for possibility,

$$\varphi \vdash \Diamond\varphi \quad (T) \qquad\qquad \Diamond\Diamond\varphi \vdash \Diamond\varphi \quad (S4) \tag{16}$$

which we discuss in the context of Theorem 4.1. We have also left out the combinations that include both the $T^{\Diamond}$ and the $S5^{\Diamond}$ axioms, since in that case possibility becomes definable as double impossibility $\Diamond a = \sim\sim a$. This is indicated in Fig. 2 by enclosing this combination in a pair of large parentheses. Furthermore, the base system for negation is taken in Fig. 2 to be the weakest system LLD (Lattice Logic with Dunn's minimal negation). The case of stronger systems is discussed in Theorem 4.1. Note that, as it follows from Theorem 4.1, if the base system for negation is taken to be LLG (Lattice Logic where negation forms a Galois connection with itself), then

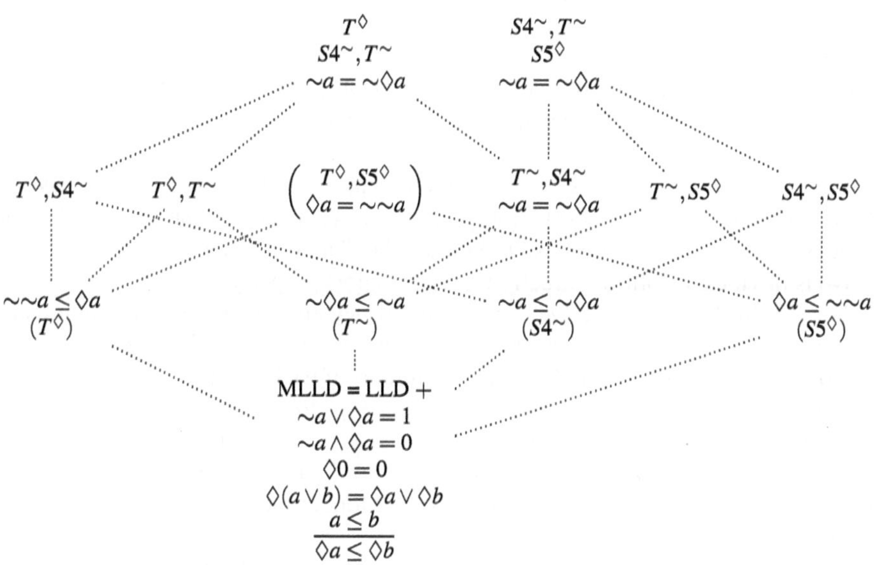

Fig. 2 LLD logics of possibility and impossibility

$S4^{\sim}$ and $S5^{\Diamond}$ are equivalent, hence some more of the combinations in Fig. 2 become uninteresting as negation becomes definable again in systems assuming both $T^{\Diamond}$ and $S4^{\sim}$ in that case. In the statement of the theorem and where A, B are logic systems, we write $B < A$ to indicate that A subsumes B.

Theorem 4.1 *The following hold:*

(1) *MLL? + $T^{\sim}$ < MLL? + T: If the T axiom $\varphi \vdash \Diamond\varphi$ is assumed for the possibility operator, then the axiom ($T^{\sim}$) in Eq. (14) becomes derivable.*
(2) *MLLG + $T^{\sim}$ < MLLG + T < MLLG + $T^{\Diamond}$: If negation is Galois and the $T^{\Diamond}$ axiom in Eq. (14) is assumed, then the T axiom for possibility becomes derivable.*
(3) *MLLM + $T^{\sim}$ = MLLM + T = MLLM + $T^{\Diamond}$: If negation is De Morgan (hence also if it is an orthonegation), then T, $T^{\sim}$ and $T^{\Diamond}$ are equivalent.*
(4) *MLLG + $S4^{\sim}$ = MLLG + $S5^{\Diamond}$: If negation is Galois, then $S4^{\sim}$ and $S5^{\Diamond}$ are equivalent.*
(5) *MLLM = MLLM + $S4^{\sim}$ + T = MLLM + $S5^{\Diamond}$ + T: If negation is De Morgan, then $S4^{\sim}$ and $S5^{\Diamond}$ are equivalent, by the previous case, and in addition if the T axiom is assumed, then the possibility operator is completely trivialized.*
(6) *In MLL? + $T^{\Diamond}$ + $S5^{\Diamond}$ possibility is identified with double negation (φ is possible iff it is impossible that it is impossible).*
(7) *In MLLM + $S4^{\sim}$ the intuitionistic principle $\varphi \wedge \sim\varphi \vdash \bot$ is derivable.*
(8) *In MLLM + $T^{\Diamond}$ (= MLLM + T = MLLM + $T^{\sim}$) the excluded middle principle $\top \vdash \sim\varphi \vee \varphi$ is derivable.*
(9) *MLLD + T + $S5^{\Diamond}$ = MLLG + T + $S5^{\Diamond}$: If both the T and the $S5^{\Diamond}$ are assumed, then negation is Galois, $T^{\sim}$ is derivable and $S4^{\sim}$ is equivalent to $S5^{\Diamond}$, by previous case of the Proposition.*

Proof
For (1), using the antitonicity rule $\dfrac{\varphi \vdash \Diamond\varphi}{\sim\Diamond\varphi \vdash \sim\varphi}$ of MLLD we obtain $T^{\sim}$ from T.

For (2), we get $\varphi \leq \sim\sim\varphi$ from the Galois property and then T is obtained using $T^{\Diamond}$ and Cut.

For (3), since $\varphi \equiv \sim\sim\varphi$ we obtain directly that $T^{\Diamond}$ and T are equivalent. By (1), $T^{\sim}$ is derivable from T. By the Galois property we get from $T^{\sim}$ that $\varphi \vdash \sim\sim\Diamond\varphi$ and since we assume that negation is De Morgan the T axiom is derived.

(4) is obvious, it follows directly from the Galois property.

For (5), by $S5^{\Diamond}$ we have $\Diamond\varphi \vdash \sim\sim\varphi$ and if negation is De Morgan, then using Cut we obtain $\Diamond\varphi \vdash \varphi$ and if the T axiom is also assumed then $\Diamond\varphi \equiv \varphi$.

(6) is obvious since $T^{\Diamond}$ and $S5^{\Diamond}$ are converses of each other.

For (7), using $S4^{\sim}$ we get $\sim\varphi \wedge \varphi \vdash \sim\Diamond\varphi \wedge \varphi$. When negation is De Morgan and using the excluded middle principle $\top \vdash \Diamond\varphi \vee \sim\varphi$ we further get $\sim\varphi \wedge \varphi \vdash \sim\sim\varphi \wedge \sim\Diamond\varphi \equiv \sim(\sim\varphi \vee \Diamond\varphi)$. But also $\sim(\sim\varphi \vee \Diamond\varphi) \vdash \sim\top$. Since $\top \equiv \sim\bot$, we get $\sim\varphi \wedge \varphi \vdash \sim\sim\bot$ and since negation is De Morgan the conclusion $\sim\varphi \wedge \varphi \vdash \bot$ follows.

For (8), use $\top \equiv \sim\bot$ and then from $\sim\varphi \wedge \Diamond\varphi \vdash \bot$ we obtain that $\top \equiv \sim(\sim\varphi \wedge \Diamond\varphi) \equiv \sim\sim\varphi \vee \sim\Diamond\varphi$. Eliminate the double negation and use $T^{\sim}$, i.e., $\sim\Diamond\varphi \vdash \sim\varphi$ to finally obtain $\top \vdash \varphi \vee \sim\varphi$.

For (9), T with $S5^{\Diamond}$ and Cut immediately give $\varphi \vdash {\sim}{\sim}\varphi$ from which the Galois property can be obtained. The rest has been proven in previous cases. □

The following is now an immediate consequence.

Corollary 4.2 *In either of MLLM* $+ S4^{\sim}$ *or MLLM* $+ T$ $(= MLLM + T^{\sim} = MLLM + T^{\Diamond})$ *negation satisfies the usual conditions by which orthonegation is defined, namely antitonicity* $\dfrac{a \leq b}{{\sim}b \leq {\sim}a}$*, the double negation principle* $a = {\sim}{\sim}a$ *and the intuitionistic principle* $a \wedge {\sim}a = 0$.

Proof For MLLM $+ S4^{\sim}$, the claim was proven in Theorem 4.1, case 7. For MLLM $+ T$ the claim follows directly from case 8 of Theorem 4.1, given that a De Morgan negation also satisfies codistribution over conjunction and disjunction. □

Remark 4.1 (Constructive orthonegation) In Remark 3.1, we pointed out that the derivation of Johansson's rule in a system with an orthonegation requires, in addition, an assumption of distribution. Indeed, assuming $a \wedge b \leq c$ we get by antitonicity and codistribution ${\sim}c \leq {\sim}a \vee {\sim}b$. Hence $a \wedge {\sim}c \leq a \wedge ({\sim}a \vee {\sim}b)$. Assuming distribution and using the intuitionistic principle $a \wedge {\sim}a = 0$, one gets from this that $a \wedge {\sim}c \leq (a \wedge {\sim}a) \vee (a \wedge {\sim}b) = a \wedge {\sim}b \leq {\sim}b$. This completes the derivation of the rule. However, without the assumption of distribution it is impossible to derive Johansson's rule from the hypotheses.

Orthologic (and therefore also Quantum Logic) as treated by Goldblatt (1974), assumes the usual definition of orthonegation and therefore fails to satisfy Johansson's constructive contraposition principle. There is then room in non-distributive logics for a stronger, constructive orthonegation notion, obtained by explicitly adding the rule $\dfrac{a \wedge b \leq c}{a \wedge {\sim}c \leq {\sim}b}$.

4.2 Modal Algebras and Frames

Definition 4.3 A structure $\langle L, \wedge, \vee, 0, 1, \sim, \Diamond\rangle$ is a Modal Lattice Algebra MLA? if $\langle L, \wedge, \vee, 0, 1, \sim\rangle$ is an LL?-algebra in the sense of Definition 3.1, with ? being one of D, G, etc., accordingly as the corresponding conditions for $\sim$ are assumed in the axiomatization. In addition, the $\Diamond$ operator is subject to the axioms and rules shown in Fig. 2 for the MLLD system. The algebra is a $T^{\sim}$-algebra if the $T^{\sim}$ axiom in Eq. (14) is assumed and similarly for $T^{\Diamond}$, T, $S4$, $S4^{\sim}$ or $S5^{\Diamond}$ algebras.

The proof of the following lemma is by a standard argument.

Lemma 4.4 *The Lindenbaum Algebra of Modal Lattice Logic is a Modal Lattice Algebra MLA? where ? is D, G, etc., accordingly as the logic is the system MLLD, MLLG, etc. Similarly when the logic assumes any combination of the axioms* $T^{\sim}$, $T^{\Diamond}$, T, $S4$, $S4^{\sim}$ *or* $S5^{\Diamond}$.

Turning now to Kripke semantics, we will always have $u \models {\sim}\varphi \vee \Diamond\varphi$, since $\top \vdash {\sim}\varphi \vee \Diamond\varphi$ is an axiom of the minimal system, as discussed above. In the minimal system we also expect it to be the case that $u \models {\sim}\varphi \Longrightarrow u \not\models \Diamond\varphi$ (unless $u = \omega$ is the inconsistent information site), which is equivalent to $u \models \Diamond\varphi \Longrightarrow u \not\models {\sim}\varphi$. What is dubious, at least in a non-distributive context, is that if $u \not\models \Diamond\varphi$, then it should be the case that $u \models {\sim}\varphi$. In a distributive setting and given that we always have $u \models {\sim}\varphi \vee \Diamond\varphi$ one is forced to accept that if $u \not\models \Diamond\varphi$, then it should be the case that $u \models {\sim}\varphi$ (because u is a prime filter in that case). However, lack of information at u that φ is possible does not automatically count as informational evidence at u that φ is, in fact, impossible. In a non-distributive setting, as we have discussed in Sect. 2 on Lattice Logic, it may well be that an information site u satisfies a disjunction, for example, ${\sim}\varphi \vee \Diamond\varphi$, without satisfying any of the two disjuncts. This is already true of course of ${\sim}\varphi \vee \varphi$ in the non-modal system and does not specifically relate to the possibility and impossibility operators. The issue is important when it comes to considering whether a single relation can and should be used to interpret both possibility and impossibility. If we assume that a single relation R is to be used to interpret the two operators, with the natural semantic clauses

$$
\begin{array}{lll}
u \models \Diamond\varphi & \text{iff} & \exists v\,(uRv \text{ and } v \models \varphi) \\
u \models {\sim}\varphi & \text{iff} & \forall v\,(uRv \text{ implies } v \not\models \varphi)
\end{array}
$$

then it follows that $u \models {\sim}\varphi$ iff $u \not\models \Diamond\varphi$. One half of this biconditional is unproblematic, as we discussed above, namely, we surely expect $u \models {\sim}\varphi \Longrightarrow u \not\models \Diamond\varphi$ (equivalently, $u \models \Diamond\varphi \Longrightarrow u \not\models {\sim}\varphi$), with the inconsistent site ω being the single exception to this rule. The other half, by our above discussion, is quite dubious in a non-distributive setting. In a sense, then, the decision to interpret both operators by the same relation seems to be based on a hidden distributivity assumption.

Consequently, we assume that information neighborhood frames (X, ν) come equipped with two relations $\rightsquigarrow_\Diamond$ and $\rightsquigarrow_*$ to interpret the possibility and the impossibility operators, respectively. Restrictions on the relation between $\rightsquigarrow_\Diamond$ and $\rightsquigarrow_*$ will need to be imposed, however, depending on the axiomatization of the considered system.

The reader may recall from Definition 2.2 that the neighborhood function ν returns a stable set $A = \Gamma A$ and that stable sets are the closures of single points $A = \Gamma(\{x^A\}) = \nu(x^A)$. She may also recall Definition 3.11 of $\perp$-frames. An accessibility frame for negation $(X, \nu, \rightsquigarrow_*)$ corresponds to a $\perp$-frame by interdefining the relations using the condition $x \rightsquigarrow_* y$ iff $y \not\perp x$. Thus, avoiding a straightforward but unnecessarily repetitive rephrasing we may define accessibility frames for negation by reducing the definition to that of $\perp$-frames, which in turn is reduced to that of $*$-frames for negation, as the reader may recall from Definition 3.11. The new feature for modal lattice frames $(X, \nu, \rightsquigarrow_*, \rightsquigarrow_\Diamond)$ is the relation $\rightsquigarrow_\Diamond \subseteq X \times X$, which induces an operation on all subsets of X definable by Eq. (17).

$$
V^\Diamond = \{u \colon \exists v\,(u \rightsquigarrow_\Diamond v \text{ and } v \in V)\} \tag{17}
$$

We further require that if V is stable then so must be $V^\Diamond$. Since stable sets in neighborhood frames are the closures of single points this further induces an operator $^\Diamond$ on information sites, where $x^\Diamond$ is the unique point generating the stable set $A^\Diamond$ defined by Eq. (17), assuming $A = \Gamma x$ is itself stable. Hence if $A = \Gamma x$, then $A^\Diamond = (\Gamma x)^\Diamond = \Gamma x^\Diamond$, where the latter abbreviates $\Gamma(\{x^\Diamond\})$.

Conversely, assuming an operator $^\Diamond$ on points, and recalling that the points of an information neighborhood frame are partially ordered by the induced ordering $x \leq y$ iff $\Gamma(\{y\}) = \nu(y) \leq \nu(x) = \Gamma(\{x\})$ a relation $\rightsquigarrow_\Diamond$ can be defined on X by setting $u \rightsquigarrow_\Diamond v$ iff $\nu(u) \leq \nu(v^\Diamond)$ iff $u \in \Gamma v^\Diamond$ iff $v^\Diamond \leq u$.

We then have the option of defining frames in alternate ways and, for consistency with our treatment of the accessibility relation $\rightsquigarrow_*$, we shall prefer to base the definition on frames with a point operator $\Diamond$, though both the relation and the point and stable sets $\Diamond$ operators will be used in the sequel, since in many occasions this simplifies the presentation of conditions on frames.

Definition 4.5 (*Pre-frames*) A structure $(X, \nu, \rightsquigarrow_*, \rightsquigarrow_\Diamond)$ is a *modal neighborhood pre-frame* (MLF?) iff (X, ν) is an information neighborhood frame in the sense of Definition 2.2, $\rightsquigarrow_*, \rightsquigarrow_\Diamond \subseteq X \times X$, and

1. defining $y \perp x$ iff $x \not\rightsquigarrow_* y$, the structure $(X, \nu, \perp)$ is a perp-frame in the sense of Definition 3.11;
2. there is a monotone operator $^\Diamond$ on points of the pre-frame such that $u \rightsquigarrow_\Diamond v$ iff $\nu(u) \leq \nu(v^\Diamond)$. The $^\Diamond$ operator induces an operation on stable sets where, if $A = \Gamma x$, then $A^\Diamond = \Gamma x^\Diamond$ $(= \Gamma(\{x^\Diamond\}))$. In addition,
    (a) the set of regular points of the pre-frame is closed under the $^\Diamond$ operation and $\omega^\Diamond = \omega$;
    (b) if x, y are regular points then Γz is the least upper bound of $\Gamma x, \Gamma y$, iff $\Gamma z^\Diamond$ is the least upper bound of $\Gamma x^\Diamond, \Gamma y^\Diamond$.
3. The MLF? pre-frame is an MLFD, MLFG, etc. pre-frame accordingly as its reduct $(X, \nu, \perp)$ is an LLD, LLG, etc. frame in the sense of Definitions 3.5 and 3.11.

A *modal neighborhood pre-model* is a pre-frame in the sense of this definition together with an interpretation function ι such that (X, ν, ι) is a neighborhood model in the sense of Definition 2.3 and the satisfaction relation is subject to the following requirements:

$$u \models \Diamond\varphi \quad \text{iff} \quad \exists v\ (u \rightsquigarrow_\Diamond v \text{ and } v \models \varphi)$$
$$u \models {\sim}\varphi \quad \text{iff} \quad \forall v\ (u \rightsquigarrow_* v \text{ implies } v \not\models \varphi)$$

in addition to those for neighborhood models.

The following result is now immediate.

Corollary 4.6 *The algebra of regular subsets of a pre-frame is an LL?-algebra in the sense of Definition 3.1 with a monotone operator $\Diamond$ which distributes over joins of regular sets and satisfies $\{\top\}^{\Diamond} = \{\top\}$, where $\top$ is the least informative site in the information ordering of the frame.* □

Lemma 4.7 *The following are equivalent for a pre-frame and its algebra of regular sets:*

1. *(frame condition) For all regular x, all y and any regular set A, if ${-\!\!\wr}A$, $A^{\Diamond}$ are both contained in the neighborhood $\nu(y)$ of y, then $x \in \nu(y)$.*
2. *(algebraic condition) In the algebra of regular subsets of the pre-frame the identity ${-\!\!\wr}A \vee A^{\Diamond} = X$ holds.*

If either of these equivalent conditions holds, then the frame validates the axiom $\top \vdash {\sim}\varphi \vee \Diamond\varphi$.

Proof Trivial, by definition of the semantics for disjunction in Lattice Logic. □

Lemma 4.8 *The following are equivalent for a pre-frame and its algebra of regular sets:*

1. *(frame condition) For all sites x and all regular sites y, if $x \rightsquigarrow_{\Diamond} y$ and $y \perp x$, then x is the inconsistent site ω.*
2. *(algebraic condition) In the algebra of regular subsets of the pre-frame the identity ${-\!\!\wr}A \cap A^{\Diamond} = \{\omega\}$ holds (where ω is the inconsistent site).*

If either of these equivalent conditions holds, then the axiom ${\sim}\varphi \wedge \Diamond\varphi \vdash \bot$ is valid.

Proof Assume the frame condition and let $x \in {-\!\!\wr}A \cap A^{\Diamond}$ where $A = \Gamma y$ is a regular set (so that, by definition, y is regular, too). Since $x \in A^{\Diamond} = (\Gamma y)^{\Diamond} = \Gamma y^{\Diamond}$ we obtain $y^{\Diamond} \leq x$, hence $x \rightsquigarrow_{\Diamond} y$ holds. But also $x \in {-\!\!\wr}A = {-\!\!\wr}\Gamma y = \Gamma y^*$ so that $y^* \leq x$, which is equivalent to $y \perp x$. By the frame condition $x = \omega$ is the inconsistent site, hence ${-\!\!\wr}A \cap A^{\Diamond} = \{\omega\}$. Conversely, assuming the algebraic condition, if $x \rightsquigarrow_{\Diamond} y$ and $y \perp x$, letting $A = \Gamma y$ we obtain $x \in {-\!\!\wr}A \cap A^{\Diamond}$, hence $x = \omega$, by the algebraic condition. Validation of the axiom ${\sim}\varphi \wedge \Diamond\varphi \vdash \bot$ is immediate. □

Definition 4.9 (*Neighborhood frames and models*) A pre-frame is a frame if either of the equivalent conditions in Lemmas 4.7 and 4.8 holds. A pre-model is a model if its pre-frame is a frame.

Lemma 4.10 *The following are equivalent for a frame and its algebra of regular sets:*

1. *(frame condition) For all sites u and all regular sites x, if $x^{\Diamond} \perp u$, then $x \perp u$.*
2. *(algebraic condition) In the algebra of regular subsets of the frame the inclusion ${-\!\!\wr}(A^{\Diamond}) \subseteq {-\!\!\wr}A$ holds.*

If either of the equivalent conditions holds, then the $T^{\sim}$ axiom ${\sim}\Diamond\varphi \vdash {\sim}\varphi$ is valid.

Proof Assume the frame condition and let $A = \Gamma x$ be regular. Then we have

$$-\!\!\wr(A^{\Diamond}) = \{u\colon A^{\Diamond} \perp u\} \subseteq \{u\colon A \perp u\} = -\!\!\wr A, \quad \text{or equivalently,}$$
$$\forall u\ (\Gamma x^{\Diamond} \perp u \implies \Gamma x \perp u).$$

Note that $\Gamma x \perp u$ iff $x \perp u$ (because $x \leq y$ implies $y^* \leq x^* \leq u$) and so the desired conclusion follows, given the frame identity. The converse is by essentially the same argument and validation of the $T^{\sim}$ axiom if any of the equivalent conditions holds is immediate. □

Lemma 4.11 *The following are equivalent for a frame and its algebra of regular sets:*

1. *(frame condition) For all sites u and all regular sites x, if $x \perp u$, then $x^{\Diamond} \perp u$.*
2. *(algebraic condition) In the algebra of regular subsets of the frame the inclusion $-\!\!\wr A \subseteq -\!\!\wr(A^{\Diamond})$ holds.*

If either of the equivalent conditions holds, then the $S4^{\sim}$ axiom $\sim\varphi \vdash \sim\Diamond\varphi$ is valid.

Proof Similar to the proof of Lemma 4.10. □

Lemma 4.12 *The following are equivalent for a frame and its algebra of regular sets:*

1. *(frame condition) For all sites u and all regular sites x, if for all y, $u \rightsquigarrow_* y \implies x \rightsquigarrow_* y$, then there exists an informational extension $v \geq x$ such that $u \rightsquigarrow_{\Diamond} v$.*
2. *(algebraic condition) In the algebra of regular subsets of the frame the inclusion $-\!\!\wr-\!\!\wr A \subseteq A^{\Diamond}$ holds.*

If either of the equivalent conditions holds, then the $T^{\Diamond}$ axiom $\sim\sim\varphi \vdash \Diamond\varphi$ is valid.

Proof Assume the frame condition and let $A = \Gamma x$. The required inclusion, using definitions, is

$$-\!\!\wr-\!\!\wr A = \{u\colon -\!\!\wr A \perp u\} = \{u\colon \Gamma x^* \perp u\} \subseteq \{u\colon \exists v\ (x \leq v \text{ and } u \rightsquigarrow_{\Diamond} v)\} = A^{\Diamond}$$

Since $\Gamma x^* \perp u$ is equivalent to $\forall y\ (x^* \leq y \implies y^* \leq u)$ or, contraposing and using the accessibility relation $\rightsquigarrow_*$, $\forall y\ (u \rightsquigarrow_* y \implies x \rightsquigarrow_* y)$ the frame condition allows us to conclude that there exists $v \geq x$, i.e., $v \in \Gamma x = A$ such that $u \rightsquigarrow_{\Diamond} v$ and therefore the inclusion holds. The converse, from the algebraic to the frame condition is along the same lines. It is also clear that if either of the equivalent conditions holds, then the frame validates the $T^{\Diamond}$ axiom $\sim\sim\varphi \vdash \Diamond\varphi$. □

Lemma 4.13 *The following are equivalent for a frame and its algebra of regular sets:*

1. *(frame condition) For all sites u and all regular sites x, if there exists an informational extension $v \geq x$ such that $u \rightsquigarrow_{\Diamond} v$, then for all y, $u \rightsquigarrow_* y \implies x \rightsquigarrow_* y$.*

2. *(algebraic condition) In the algebra of regular subsets of the frame the inclusion* $A^{\Diamond} \subseteq {-\!\wr\!-}{-\!\wr\!-}A$ *holds.*

If either of the equivalent conditions holds, then the S5$^{\Diamond}$ *axiom* $\Diamond\varphi \vdash {\sim\sim}\varphi$ *is valid.*

Proof Similar to the proof of Lemma 4.12. □

Lemma 4.14 *The following hold.*

1. *The* $\rightsquigarrow_{\Diamond}$ *relation is reflexive on regular points of the frame iff the inclusion* $A \subseteq A^{\Diamond}$ *holds in the algebra of regular sets of the frame. If either condition holds, then the T axiom* $\varphi \vdash \Diamond\varphi$ *is valid in the frame.*
2. *The condition* $\forall y\ (y \rightsquigarrow_{\Diamond} x \implies y^{\Diamond} \rightsquigarrow_{\Diamond} x)$ *holds in the frame for regular points* x *iff the inclusion* $A^{\Diamond\Diamond} \subseteq A^{\Diamond}$ *holds in the algebra of regular sets of the frame. If either condition holds, then the S4 axiom* $\Diamond\Diamond\varphi \vdash \Diamond\varphi$ *is valid in the frame.*

Proof The first is immediate by just recalling that regular sets A are closures of single points, $A = \Gamma x$, and given the definition of $A^{\Diamond}$ and that $x \rightsquigarrow_{\Diamond} x$ is equivalent to $x^{\Diamond} \leq x$.

For the second, the inclusion $A^{\Diamond\Diamond} \subseteq A^{\Diamond}$, where $A = \Gamma x$, is equivalent to $x^{\Diamond} \leq x^{\Diamond\Diamond}$, i.e., to $x^{\Diamond\Diamond} \rightsquigarrow_{\Diamond} x$. The latter can be easily seen by the reader to be equivalent to the frame condition $\forall y\ (y \rightsquigarrow_{\Diamond} x \implies y^{\Diamond} \rightsquigarrow_{\Diamond} x)$.

The validity of the axioms T and $S4$ in the respective cases above is then immediate. □

An immediate consequence of Corollary 4.6 and Lemmas 4.7–4.14 is a soundness result.

Theorem 4.15 (Soundness) *Every system MLL? of Modal Lattice Logic, perhaps with the addition of a combination of the axioms T, S4,* $T^{\sim}$, $T^{\Diamond}$, $S4^{\sim}$, $S5^{\Diamond}$, *is sound in the respective Modal Neighborhood Frames.*

4.3 Canonical Kripke Model and Completeness

Theorem 4.16 (Representation) *Every MLA? algebra is isomorphic to a concrete MLA? algebra.*

Proof The proof of a more general case has been given in Hartonas (1997). Here we only do the part that relates to modal lattice algebras. The construction extends the ones given for Lattice Logic and for the $*$ and $\perp$ interpretation of negation, as detailed in previous sections. The operators $*$ and $\Diamond$ are defined on filters as follows:

$$x^{\Diamond} = \{b \colon \forall a \leq x\ \ \Diamond a \leq b\} \qquad\qquad x^{*} = \{b \colon \exists a \leq x\ \ {\sim}a \leq b\}$$

In Theorem 3.3 we have verified that the canonical frame is a $\perp$-frame in the sense of Definition 3.11, hence the appropriate conditions for a frame with an accessibility

$\rightsquigarrow_*$ hold, given our definition of this relation in terms of $\perp$ (hence in terms of $*$). We now complete the argument for the representation of modal algebras.

To show that $H(a) \subseteq H(b)$ implies $H(\Diamond a) \subseteq H(\Diamond b)$, notice first that the hypothesis is equivalent to $a \leq b$, hence we obtain $\Diamond a \leq \Diamond b$. Hence, if $\Diamond a \in x$, for a filter x, we get $\Diamond b \in x$, i.e., $H(\Diamond a) \subseteq H(\Diamond b)$.

To show that $H(\Diamond(a \vee b)) = H(\Diamond a) \vee H(\Diamond b)$, recall that $H(\Diamond(a \vee b)) = \Gamma x_{\Diamond(a \vee b)} =$ (since $\Diamond(a \vee b) = \Diamond a \vee \Diamond b$ holds in the lattice) $= \Gamma x_{\Diamond a \vee \Diamond b} =$ [by the lattice representation theorem (Theorem 2.5) based on Hartonas (1997)] $= \Gamma x_{\Diamond a} \vee \Gamma x_{\Diamond b} = H(\Diamond a) \vee H(\Diamond b)$.

Observe also that $H(\Diamond 0) = \Gamma x_{\Diamond 0} =$ (since $\Diamond 0 = 0$ holds in the lattice) $= \Gamma x_0 = H(0)$. A similar proof applies to showing that $H(\sim a \vee \Diamond a) = H(1)$, since $1 = \sim a \vee \Diamond a$ holds in the MLA?-algebra.

Next, assume that $H(\sim a) \subseteq H(\Diamond b)$, which implies that $\sim a \leq \Diamond b$. We have that $H(\sim a \wedge \sim b) = \Gamma x_{\sim a \wedge \sim b}$. Since $\sim a \leq \Diamond b$ it follows by properties of the MLA?-algebra that $\sim a \wedge \sim b = 0$. Hence $H(\sim a \wedge \sim b) = \Gamma x_0 = H(0)$. Similarly, assuming that $H(\Diamond a) \subseteq H(\sim b)$, we obtain that $H(\Diamond a \wedge \Diamond b) = H(0)$.

By a similar argument, if the MLA?-algebra includes in its axiomatization any of the axioms in Eqs. (13)–(15), then the corresponding identity holds for the representation map. □

The following corollary is an immediate consequence of the representation Theorem 4.16 defining a diamond operator on stable sets as follows.

$$(\Gamma x)^{\Diamond} = \Gamma x^{\Diamond} \qquad (\text{where } x^{\Diamond} = \{b\colon \forall a \leq x \;\; \Diamond a \leq b\}) \tag{18}$$

Corollary 4.17 *The algebra of stable compact opens of the filter space with the operators* $\multimap\!\!\!\!\!\sim$ *and* $\Diamond$ *defined on stable compact opens by Eqs. (8), (18) is an MLA?-algebra in the sense of Definition 4.3.*

In the sequel we construct the canonical Kripke frame, based on the representation Theorem 4.16.

Define a structure $(X, \nu, \rightsquigarrow_{\Diamond}, \rightsquigarrow_*)$ letting X be the set of filters of L and $\lambda \dashv \rho$ the Galois connection detailed in the completeness Theorem 2.6 for Lattice Logic and setting $\Gamma = \lambda\rho$ and $\nu(x) = \Gamma(\{x\})$, as in Theorem 2.6. The relation $\rightsquigarrow_*$ is defined using $\perp$ and the $*$ operator, as in Sect. 3, and $\rightsquigarrow_{\Diamond}$ is defined by (20).

$$x \rightsquigarrow_* y \quad \text{iff} \quad y^* \not\leq x \quad (\text{where } y^* = \{a\colon \exists c \leq y \;\; \sim c \leq a\}) \text{ iff } y \not\perp x \tag{19}$$

$$x \rightsquigarrow_{\Diamond} y \quad \text{iff} \quad \text{for all } a, \text{ if } \forall e\; (y \leq x_e \text{ implies } a \in x_{\Diamond e}), \text{ then } a \in x \tag{20}$$

Recall that the operator $^{\Diamond}$ on filters is defined by setting $x^{\Diamond} = \{b\colon \forall a \leq x \;\; \Diamond a \leq b\}$ and the reader can easily verify that $x^{\Diamond}$ is indeed a filter. Recall that $y \perp x$ iff $y^* \leq x$ iff $x \not\rightsquigarrow_* y$ and note that $x \rightsquigarrow_{\Diamond} y$ iff $y^{\Diamond} \leq x$.

Lemma 4.18 *If $a \in y$, then $\Diamond a \in y^{\Diamond}$ and if $x \rightsquigarrow_{\Diamond} y$ and $a \in y$, then $\Diamond a \in x$.*

Proof Notice that $x \rightsquigarrow_{\Diamond} y$ iff $y^{\Diamond} \leq x$ and that $^{\Diamond}$ is a monotone operator on filters such that $x_a^{\Diamond} = x_{\Diamond a}$ (just like $x_a^* = x_{\sim a}$). Hence, if $a \in y$, i.e., $y_a \leq y$, then $y_a^{\Diamond} = y_{\Diamond a} \leq y^{\Diamond}$. Therefore, $a \in y \implies \Diamond a \in y^{\Diamond}$. □

Define an operation $^{\Diamond}$ on all stable sets by letting $^{\Diamond}\Gamma x = \Gamma x^{\Diamond}$ (which is an abbreviation for $\Gamma(\{x^{\Diamond}\})$). The representation map is $H(a) = \{x : a \in x\} = \Gamma x_a$. As we observed above, $x_a^{\Diamond} = x_{\Diamond a}$, hence $^{\Diamond}H(a) = {}^{\Diamond}\Gamma x_a = \Gamma x_a^{\Diamond} = \Gamma x_{\Diamond a} = H(\Diamond a)$.

Lemma 4.19 *The following hold:*

1. $\sim a \in x$ *iff* $\forall y\ (x \rightsquigarrow_{*} y$ *implies* $a \notin y)$;
2. $\Diamond a \in x$ *iff* $\exists y\ (x \rightsquigarrow_{\Diamond} y$ *and* $a \in y)$.

Proof The proof of the first follows from Lemma 3.13, given our definition of $\perp$ and $\rightsquigarrow_{*}$, using the $*$ operator. For the proof of the second claim of the lemma, if $\Diamond a \in x$, let $y = x_a$. Then the hypothesis is equivalent to $x_a^{\Diamond} = x_{\Diamond a} \leq x$, i.e., $x \rightsquigarrow_{\Diamond} x_a$ holds and of course $a \in x_a$. Conversely, If such a y with $a \in y$ and $y^{\Diamond} \leq x$ exists, we have $x_a \leq y$ and thereby $x_a^{\Diamond} = x_{\Diamond a} \leq y^{\Diamond} \leq x$ which implies $\Diamond a \in x$. □

It follows that

$$H(\Diamond a) = \{x : \exists y\ (x \rightsquigarrow_{\Diamond} y \text{ and } a \in y)\}$$
$$H(\sim a) = \{x : \forall y\ (x \rightsquigarrow_{*} y \text{ implies } a \notin y)\}$$

and therefore the satisfaction relation $x \models \varphi$ iff $x \in H([\varphi])$ satisfies the respective clauses for the connectives $\sim$ (as verified in Sect. 3) and $\Diamond$.

By the representation Theorem 4.16 the algebra of stable compact open sets of the space, which we define to be the algebra of regular subsets of the frame, is an MLA?-algebra such that if the logic satisfies any of the additional axioms T, $S4$, $T^{\sim}$, $T^{\Diamond}$, $S4^{\sim}$, $S5^{\Diamond}$, then so does the algebra of regular sets (stable compact opens). By Corollary 4.6 and Lemmas 4.7–4.14, this is equivalent to the corresponding frame condition and thereby the canonical frame satisfies the conditions corresponding to the additional axioms, if assumed. The reader can easily verify them directly, bearing in mind that the regular points of the canonical frame are precisely the principal filters.

We may then conclude by stating the completeness theorem, whose proof is contained in the above discussion and the results previously obtained that have been mentioned.

Theorem 4.20 (Completeness) *Every system MLL? of Modal Lattice Logic, perhaps including a combination of the additional axioms T, $S4$, $T^{\sim}$, $T^{\Diamond}$, $S4^{\sim}$, $S5^{\Diamond}$ is (sound, by Theorem 4.15, and) complete in modal neighborhood frames.*

Remark 4.2 (Generalized Kripke frames and bi-approximation semantics) The approaches to the semantics of modal extensions of non-distributive logics developed over the last decade or so Kamide (2002), Gehrke (2006), Suzuki (2010, 2012,

2014), Conradie and Palmigiano (2015), Düntsch et al. (2004), invariably depart from the standard Kripke semantics for the modal operators in important ways. First and because of lack of distribution distinct accessibility relations for each of the necessity and possibility operators seem to be forced. More importantly, however, even for a single modal operator (box, or diamond), the familiar semantic clauses for boxes and diamonds are abandoned in favor of some notion or other of generalized semantics. This typically involves a pair of accessibility relations for the same modal operator and therefore a non-standard semantic clause for the interpretation of modal sentences, violating well established intuitions and resulting in an awkward and rather obscure interpretation.

For example, proposed frames in Conradie and Palmigiano (2015) are structures $(X, Y, R, R_\Box, R_\Diamond)$, where (X, Y, R) is the polarity arising from the underlying lattice representation approach and the accessibility relations connect (in converse direction) worlds (in X) and co-worlds (in Y): $R_\Box \subseteq X \times Y$, $R_\Diamond \subseteq Y \times X$. Furthermore, the semantic clauses for the modal operators mark a clear departure from the standard ones, which are shown in (21) and (22) below.

$$x \models \Box\varphi \quad \text{iff} \quad \forall x' \; (R^{\Box}xx' \implies x' \models \varphi) \tag{21}$$

$$x \models \Diamond\varphi \quad \text{iff} \quad \exists x' \; (R^{\Diamond}xx' \text{ and } x' \models \varphi) \tag{22}$$

while the intuitions behind the new clauses remain rather obscure, to this author at least. The underlying technical necessity, of course, is that modal sentences, too, are to be modeled by *stable* sets (remaining unchanged under an application of the closure operator involved in the underlying lattice representation). For example, in Conradie and Palmigiano (2015), the following is the proposed satisfaction clause for the possibility operator

$$X \ni x \Vdash \Diamond\varphi \quad \text{iff} \quad \forall y \in Y \; \big(\forall z \in X \; (z \Vdash \varphi \implies yR_\Diamond z) \implies x \leq y\big)$$

where $x \leq y$ is defined after Gehrke's (2006) using the polarity relation $R \subseteq X \times Y$. Aside from the satisfaction relation $\Vdash$, a relation of co-satisfaction $\succ$, or refutation, is defined

$$Y \ni y \succ \Diamond\varphi \quad \text{iff} \quad \forall z \in X \; (z \Vdash \varphi \implies yR_\Diamond z)$$

The idea of the two relations $\Vdash$ and $\succ$ originates in Gehrke's contribution on Generalized Kripke Frames Gehrke (2006).

Despite differences, the main ideas in Suzuki (2010, 2012, 2014) for the semantics of possibility and necessity are quite similar to those in Conradie and Palmigiano (2015) and Suzuki's bi-approximation semantics and Gehrke's generalized Kripke frames seem to be variants of each other. Suzuki, overloading the use of $\Vdash$, uses it for both relations of satisfaction of sentences at worlds and refutation of sentences at co-worlds. The semantics for possibility proposed is the following:

$$X \ni x \Vdash \Diamond\varphi \;\text{ iff }\; \forall y \in Y\,(y \Vdash \Diamond\varphi \implies x \leq y)$$
$$Y \ni y \Vdash \Diamond\varphi \;\text{ iff }\; \forall x \in X\,(x \Vdash \varphi \implies xRy)$$

Regarding the obvious divergence from the standard semantics, Suzuki (2010) explicitly acknowledges that "this is because it is essential to set up our interpretation to return Galois stable sets."

In Düntsch et al. (2004) the authors study monomodal systems, hence separately each of $\Box$ and $\Diamond$. They base their approach on Urquhart's representation theorem for bounded lattices Urquhart (1978). Therefore, the points on the frames are maximal disjoint filter-ideal pairs (x, y), with a double-order $\leq_1, \leq_2$ on the points of the frame and this forces again two accessibility relations $R_\Box$ and $S_\Box$ (and similarly $R_\Diamond, S_\Diamond$ for the diamond operator) on the frame and thereby the familiar semantics for modal operators is lost.

By contrast to the approaches briefly discussed above we have presented in this paper a solution that is in keeping with the familiar semantic clause for possibility. In recent research we have extended our approach to systems with both a necessity and a possibility operator, maintaining the familiar semantic clauses in Eqs. (21) and (22), while dropping the assumption of distribution. Moreover, it is evident that the approaches briefly discussed above arrive at a complete impasse when considering temporal logics, where evidently a single time-flow relation must interpret all modal (temporal) operators. In our opinion, the difficulties encountered by these authors have their source in their choice of representation theorem for lattices that underlies their semantic approach. By contrast, in our recent research, extending our representation results for lattices with operators in Hartonas (1997) we have extended our work to Tense Logic on a non-distributive propositional basis, without being forced to resort to some awkward interpretation of time-flows.

5 Conclusions

Non-distributive logics are better suited for reasoning with incomplete information. This is particularly clear when the interpretation of disjunction is considered, but it affects the semantics of other logical operators, as well. This paper is an extension to the non-distributive setting of results produced in Dunn and Zhou (2005), while also introducing a standard semantic treatment of modal operators in logics over a non-distributive propositional basis. In particular, we first studied Lattice Logic with various notions of negation of increasing strength, from Dunn's minimal negation to orthonegation. We have pointed out that in the non-distributive setting orthonegation, as usually defined, does not satisfy Johansson's rule of constructive contraposition and there are therefore two variants of it in the non-distributive setting. Furthermore, extending to the non-distributive setting the study of negation as impossibility of Došen (1986), Vakarelov (1977, 1989) and Dunn and Zhou (2005) we initiated the study of Modal Lattice Logic, with possibility and impossibility operators, exploring natural choices for their axiomatization. Interpreting the two operators by the same relation has a hidden semantic assumption of distributivity, hence the natural venue in the non-distributive setting is to use distinct relations and impose conditions on their connection, depending on the axiomatization of the logic that we adopt.

We expect that our results will prove useful in studying variants and extensions of non-distributive logics, such as Quantum Logic, and they can be further extended, enriching the study of non-distributive logical calculi.

References

Allwein, G., & Hartonas, C. (1993). Duality for bounded lattices, Preprint series, IULG-93-25, Indiana University Logic Group.

Białynicki-Birula, A., & Rasiowa, H. (1957). On the representation of quasi-Boolean algebras. *Bulletin de l'Académie Polonaise des Sciences, 5*, 259–261.

Birkhoff, G., & von Neumann, J. (1936). The logic of quantum mechanics. *Annals of Mathematics, 37*, 823–843.

Conradie, W., & Palmigiano, A. (2015). Algorithmic correspondence and canonicity for non-distributive logics. *Journal of Logic and Computation*. Forthcoming.

Dalla Chiara, M. L., & Giuntini, R. (2001). Qantum logics. arXiv:quant-ph/0101028.

Dalla Chiara, M. L., & Giuntini, R. (2002). Quantum logics. In D. M. Gabbay & F. Guenthner (Eds.), *Handbook of philosophical logic* (pp. 129–228). Dordrecht: Kluwer Academic Publishers.

Došen, K. (1986). Negation as a modal operator. *Reports on Mathematical Logic, 20*, 15–27.

Došen, K. (1999). Negation in the light of modal logic. In D. Gabbay & H. Wansing (Eds.), *What is negation?* (pp. 77–86). Dordrecht: Kluwer.

Dunn, J. M. (1991). Gaggle theory: An abstraction of Galois connections and residuation with applications to negation, implication, and various logical operators. In J. van Eijck (Ed.), *Logics in AI: European workshop JELIA '90, Lecture notes in computer science* (Vol. 478, pp. 31–51). Berlin: Springer.

Dunn, J. M. (1993). Star and perp: Two treatments of negation, *Philosophical Perspectives 7*, 331–357. (Language and Logic, J. E. Tomberlin (Ed.)).

Dunn, J. M. (1996). Generalised ortho negation. In H. Wansing (Ed.), *Negation: A notion in focus* (pp. 3–26). New York, NY: Walter de Gruyter.

Dunn, J. M. (1999). A comparative study of various model-theoretic treatments of negation: A history of formal negation. In D. M. Gabbay & H. Wansing (Eds.), *What is negation?* (pp. 23–51). Dordrecht: Kluwer.

Dunn, J. M., Moss, L. S., & Wang, Z. (2013). The third life of quantum logic: Quantum logic inspired by quantum computing. *Journal of Philosophical Logic, 42*, 443–459.

Dunn, J. M., & Zhou, C. (2005). Negation in the context of gaggle theory. *Studia Logica, 80*, 235–264.

Düntsch, I., Orlowska, E., Radzikowska, A. M., & Vakarelov, D. (2004). Relational representation theorems for some lattice-based structures, Preprint CS-04-07.

Gehrke, M. (2006). Generalized Kripke frames. *Studia Logica, 84*(2), 241–275.

Gerhke, M., & Harding, J. (2001). Bounded lattice expansions. *Journal of Algebra, 238*, 345–371.

Gerhke, M., & van Gool, S. J. (2014). Distributive envelopes and topological duality for lattices via canonical extensions. *Order, 31*(3), 435–461.

Goldblatt, R. (1974). Semantic analysis of orthologic. *Journal of Philosophical Logic, 3*, 19–35.

Goldblatt, R. (1984). Orthomodularity is not elementary. *Journal of Symbolic Logic, 49*, 401–404.

Goldblatt, R. (1993). *Mathematics of modality, CSLI lecture notes* (Vol. 43). Stanford, CA: CSLI Publications.

Hartonas, C. (1996). Order-duality, negation and lattice representation. In H. Wansing (Ed.), *Negation: A notion in focus* (pp. 27–36). W. de Gruyter.
Hartonas, C. (1997). Duality for lattice-ordered algebras and for normal algebraizable logics. *Studia Logica*, *58*, 403–450.
Hartonas, C. (2015). Elementary (first-order) frames for orthomodularity and the semantics of orthomodular quantum logic, Preprint CS-2015-3, University of Applied Sciences of Thessaly (TEI of Thessaly).
Hartonas, C., & Dunn, J. M. (1997). Stone duality for lattices. *Algebra Universalis*, *37*, 391–401.
Hartung, G. (1992). A topological representation for lattices. *Algebra Universalis*, *29*, 273–299.
Järvinen, J., & Orlowska, E. (2005). Relational correspondences for lattices with operators. In I. D. Wendy MacCaull & M. Winter (Eds.), *Relational Methods in Computer Science*. Amsterdam.
Jónsson, B., & Tarski, A. (1952). Boolean algebras with operators, II. *American Journal of Mathematics*, *74*(1), 127–162.
Kamide, N. (2002). Kripke semantics for modal substructural logics. *Journal of Logic, Language and Information*, *11*, 453–470.
Moshier, M. A., & Jipsen, P. (2014a). Topological duality and lattice expansions, I: A topological construction of canonical extensions. *Algebra Universalis*, *71*, 109–126.
Moshier, M. A., & Jipsen, P. (2014b). Topological duality and lattice expansions, II: Lattice expansions with quasi-operators. *Algebra Universalis*, *71*, 221–234.
Priestley, H. A. (1970). Representation of distributive lattices by means of ordered Stone spaces. *Bulletin of the London Mathematical Society*, *2*, 186–190.
Routley, R., & Meyer, R. K. (1973). The semantics of entailment. In H. Leblanc (Ed.), *Truth, syntax and modality. Proceedings of the Temple University Conference on Alternative Semantics* (pp. 199–243). North-Holland, Amsterdam.
Routley, R., & Routley, V. (1972). The semantics of first degree entailment. *Noûs*, *6*(4), 335–359.
Stone, M. H. (1937–38). Topological representations of distributive lattices and Brouwerian logics. *Časopis pro pěstování matematiky a fysiky, Čast matematická*, *67*, 1–25.
Stone, M. H. (1938). The representation of Boolean algebras. *Bulletin of the American Mathematical Society*, *44*, 807–816.
Suzuki, T. (2010). Bi-approximation semantics for substructural logic at work. *Advances in Modal Logic*, *8*, 411–433.
Suzuki, T. (2012). Morphisms on bi-approximation semantics. *Advances in Modal Logic*, *9*, 494–515.
Suzuki, T. (2014). On polarity frames: Applications to substructural and lattice-based logics. *Advances in Modal Logic*, *10*, 533–552.
Urquhart, A. (1978). A topological representation theorem for lattices. *Algebra Universalis*, *8*, 45–58.
Vakarelov, D. (1977). Theory of Negation in Certain Logical Systems: Algebraic and Semantic Approach, PhD thesis, University of Warsaw.
Vakarelov, D. (1989). Consistency, completeness and negations. In G. Priest, R. Routley, & J. Norman (Eds.), *Paraconsistent logic: Essays on the inconsistent* (pp. 328–368). Munich: Philosophia Verlag.
Wille, R. (1987). Bedeutungen von Begriffsverbänden. In R. W. B. Ganter & K. E. Wolff (Eds.), *Beiträge zur Begriffsanalyse* (pp. 161–211). Mannheim: B. L.-Wissenschafts Verlag.

Mereocompactness and Duality for Mereotopological Spaces

Robert Goldblatt and Matt Grice

Abstract Mereotopology studies relations between regions of space, including the contact relation. It leads to an abstract notion of Boolean contact algebra which has been shown to be representable as an algebra of regular closed subsets of a compact topological space. Here we define mereotopological spaces and their mereomorphisms, and construct a dual equivalence between the category of Boolean contact algebras and a category of mereotopological spaces that have a property we call mereocompactness, strictly stronger than ordinary compactness. This is a further illustration of the kind of duality that has been widely used in the semantic analysis of propositional logics, and which has been a significant theme in the research of J. Michael Dunn.

Keywords Boolean algebra · Clan · Compact · Contact relation · Duality · Mereocompact · Mereotopology · Regular closed · Ultrafilter

1 Introduction

Duality in the semantic analysis of propositional logics has been a significant theme in the research of J. Michael Dunn. It is involved in his gaggle theory, whose development motivated the construction of a new topological duality for general lattices (Hartonas and Dunn 1997). It underlies the framework of a number of topics he has worked on, including: the representation of quasi-Boolean algebras (Dunn 1982) and positive modal algebras (Dunn 1995); the modelling of negation using information states (Dunn 1993); the representation of relation algebras over Routley–Meyer structural models for relevant logics (Dunn 2001); the relational semantics for linear

R. Goldblatt (✉) · M. Grice
Victoria University of Wellington, Wellington, New Zealand
e-mail: rob.goldblatt@msor.vuw.ac.nz
URL: http://www.sms.vuw.ac.nz/~rob

M. Grice
e-mail: gricematt@myvuw.ac.nz

K. Bimbó (ed.), *J. Michael Dunn on Information Based Logics*,
Outstanding Contributions to Logic 8, DOI 10.1007/978-3-319-29300-4_15

logic and other substructural logics (Allwein and Dunn 1993; Dunn et al. 2005). There are chapters on duality in his two most recent books (Dunn and Hardegree 2001; Bimbó and Dunn 2008).

This notion of duality is a certain relationship between two kinds of model: *algebraic* and *structural*. In an algebraic model, propositional formulas denote elements of an algebra whose fundamental operations interpret logical connectives. In a structural model, formulas denote subsets of some background set that carries relational and/or topological structure. The elements of the set are viewed as possible worlds/situations/information states/temporal instants etc., and certain of its subsets are taken to be propositions. The structure gives rise to connective-interpreting operations on propositions, so a structural model S has an associated algebra S^+ of propositions. S^+ is the *dual* of S.

In the opposite direction, representation theorems are applied to show that an algebra A has a *dual structure* A_+ such that A is isomorphic to the algebra $(A_+)^+$ of propositions of A_+. Topological properties may be used to characterise the propositions of A_+ and to identify which structures are (isomorphic to) the duals of algebras, or equivalently which structures S are isomorphic to their double dual structure $(S^+)_+$. For example, the duals of Boolean algebras are the Stone spaces, and the propositions of such spaces are the clopen (closed-and-open) subsets. The duals of distributive lattices can be described as the spectral spaces, with their propositions being the compact open subsets; or as the Priestley spaces, with clopen down-sets as propositions.

Ultimately, duality is a category-theoretic notion, taking the form of a pair of contravariant functors that constitute a *dual equivalence* between a category of algebras and a category of structures.[1]

The purpose of the present paper is to add another brick to the pyramid of ideas on duality, in the context of *mereotopology*. This is an approach to the abstract geometry of space, based on *regions* rather than points, in which there is a primitive relation of *contact* between regions. It originates in philosophical work in the early 20th century by de Laguna (1922) on postulates for a "can connect" relation between "solids," and by Whitehead (1929) on an "extensive connection" relation between regions. Its name derives from the word "mereology," devised by Leśniewski in the 1920s to refer to his theory of the part-whole relation. In more recent times the study of such theories has received impetus from theoretical computer science, since they provide a framework for qualitative spatial reasoning, as embodied in the Region Connection Calculus (Randell and Cohn 1989; Randell et al. 1992) that was built on an axiomatisation of Whitehead's theory by Clarke (1981).[2]

---

[1]For categories **A** of algebras and **S** of structures, if the assignments $A \mapsto A_+$ and $S \mapsto S^+$ extend to contravariant functors from **A** to **S** and vice versa, then this constitutes a dual equivalence when the composition of these functors in either order gives functors that are naturally isomorphic to the identity functors on **A** and **S**, respectively. This perspective on models of propositional logic was introduced (for modal algebras and Kripke frames) in the first author's thesis (Goldblatt 1974). The concept of natural isomorphism is explained in the present article at the end of Sect. 4.

[2]Further work on the Region Connection Calculus is surveyed in (Cohn et al. 1997).

In a topological interpretation, regions can be taken to be *regular closed* subsets of a topological space, with such sets being in contact if they have a non-empty intersection. Alternatively, regions may be *regular open* sets, in contact when their closures intersect. These two kinds of regular sets form Boolean algebras that have distinct operations but are isomorphic.

Düntsch and Winter (2005) studied an axiomatic notion of Boolean contact algebra and showed that such an algebra could be represented by embedding it into the regular-closed-subset algebra of a T_1 topological space. Dimov and Vakarelov (2006a) extended this by dropping an extensionality requirement on the contact relation and representing the resulting algebras in spaces that are T_0, compact and semi-regular (the latter meaning that the regular closed sets form a closed basis for the topology).[3] They gave an example in (Dimov and Vakarelov 2006b) to show that those three topological conditions do not suffice to characterise the dual spaces of contact algebras.

Our aim here is to lift these results to a full categorical duality, making Boolean contact algebras into a category and identifying a suitable dual category of "mereotopological" spaces with "mereomorphisms" between them. To characterise the duals of contact algebras we define a new notion of *mereocompactness* (see Sect. 5) which is strictly stronger than ordinary topological compactness.

We will work through the details of this programme according to the following steps, which serve as a summary of the paper.

- Define the category **BCA** of Boolean contact algebras whose morphisms are the Boolean homomorphisms that *reflect contact* (or equivalently, preserve non-tangential inclusion).
- Define a mereotopological space S as an ordinary topological space with a distinguished Boolean sub-algebra of its regular-closed-set algebra that is a closed basis for the topology, and which forms the dual contact algebra S^+ of S.
- Define the category **MS** of mereotopological spaces, whose *mereomorphisms* $\theta\colon S_1 \to S_2$ are functions whose pullback action provides a Boolean algebra homomorphism $S_2^+ \to S_1^+$.
- Construct a contravariant functor $\Phi\colon \mathbf{MS} \to \mathbf{BCA}$ having $\Phi(S) = S^+$.
- Adapt the representation theory of (Düntsch and Winter 2005; Dimov and Vakarelov 2006a) to associate with each contact algebra A a T_0 mereotopological space A_+ such that A is isomorphic in **BCA** to $(A_+)^+$.
- Define *mereocompactness* for a mereotopological space. Show that the dual space A_+ of any contact algebra is mereocompact, and that an arbitrary space S is isomorphic in **MS** to its double dual $(S^+)_+$ iff it is mereocompact and T_0.
- Construct a contravariant functor $\Theta\colon \mathbf{BCA} \to \mathbf{MS^*}$ having $\Theta(A) = A_+$, where $\mathbf{MS^*}$ is the category of mereocompact T_0 spaces.

[3] See the Introductions of papers (Düntsch and Winter 2005; Dimov and Vakarelov 2006a) for an overview of the background literature on region-based theories of space.

- Show that the categories **BCA** and **MS*** are dually equivalent, by showing that $\Phi^* \circ \Theta$ is naturally isomorphic to the identity functor on **BCA**, while $\Theta \circ \Phi^*$ is naturally isomorphic to the identity functor on **MS***, where $\Phi^*\colon \mathbf{MS^*} \to \mathbf{BCA}$ is the restriction of Φ to **MS***.

In the final section we explore alternative versions and consequences of the notion of mereocompactness.

2 Contact Algebras

We use the notation $(B, +, \cdot, -, 0, 1)$ for an abstract Boolean algebra on a set B, with operations $+$ of join, $\cdot$ of meet and $-$ of complement; and least element 0 and greatest element 1 under the partial ordering $\leq$ that has $x \leq y$ iff $x + y = y$ iff $x \cdot y = x$. We may denote this algebra by its underlying set B.

A *contact relation* on a Boolean algebra is a binary relation C on B satisfying the following axioms.

C1. xCy implies $x, y \neq 0$.
C2. xCy implies yCx.
C3. $xC(y+z)$ iff xCy or xCz.
C4. $x \neq 0$ implies xCx.

Such a C is always *monotonic* in each variable: if xCy, $x \leq x'$ and $y \leq y'$, then $x'Cy'$. Each Boolean algebra has a smallest contact relation $\{\,(a, b)\colon a \cdot b \neq 0\,\}$ and a largest one $\{\,(a, b)\colon a \neq 0 \neq b\,\}$.

A *Boolean contact algebra*, or BCA, is a pair $A = (B_A, C_A)$ with C_A a contact relation on Boolean algebra B_A. We may also denote such an algebra in the form $A_i = (B_i, C_i)$ where i is some suitable label, or as $A' = (B', C')$ etc.

We define a *homomorphism* $f\colon A \to A'$ of contact algebras, or *BCA-morphism*, to be a homomorphism $f\colon B \to B'$ of Boolean algebras such that, for all $x, y \in B$,

$$f(x)C'f(y) \text{ implies } xCy.$$

Thus a BCA-morphism *reflects contact*. Equivalently it preserves separation in the sense that if elements are not in contact in A, then their f-images are not in contact in A'.

A relation $\ll$ of *non-tangential inclusion* is defined on any contact algebra by putting $x \ll y$ iff not $xC(-y)$. A BCA-morphism can then be characterised as a Boolean algebra homomorphism that preserves non-tangential inclusion in the sense that

$$x \ll y \text{ implies } f(x) \ll' f(y).$$

The axioms (C1)–(C4) for a contact relation can be equivalently formulated entirely as properties of the relation $\ll$ (Dimov and Vakarelov 2006a, p. 214).

It is readily seen that the functional composition of two BCA-morphisms is a BCA-morphism, and that the identity function on a contact algebra is a BCA-morphism. Thus the Boolean contact algebras and their morphisms are the objects and arrows of a concrete category, which we denote **BCA**.

Category theory gives us a definition of an *isomorphism* of contact algebras: a BCA-morphism $f\colon A \to A'$ is an isomorphism when there exists a BCA-morphism $g\colon A' \to A$, the *inverse* of f, such that each of the compositions $g \circ f$ and $f \circ g$ is the identity morphism on its domain.

Theorem 2.1 *A BCA-morphism $f\colon A \to A'$ is an isomorphism if, and only if, it is bijective and preserves contact in the sense that xCy implies $f(x)C'f(y)$.*

Proof Let $f\colon A \to A'$ be an isomorphism, with inverse BCA-morphism $g\colon A' \to A$ as above. Since $g \circ f$ and $f \circ g$ are identity functions it follows that f is bijective. If xCy, then since $x = g(f(x))$ and $y = g(f(y))$, it follows that $f(x)C'f(y)$ as g reflects contact.

Conversely, suppose $f\colon A \to A'$ is a bijective BCA-morphism preserving contact. As a bijection, f has an inverse $g\colon B_{A'} \to B_A$. It is a fact of universal algebra that *the inverse of a bijective homomorphism of algebras is itself a homomorphism.*[4] So in this case g is a Boolean algebra homomorphism. If $g(x)C_A g(y)$, then $f(g(x))C_{A'} f(g(y))$ as f preserves contact, hence $xC_{A'}y$. This shows that g reflects contact and so is a BCA-morphism $A' \to A$, providing the inverse in **BCA** that ensures f is an isomorphism. □

3 Mereotopological Spaces

Let (X, τ) be a topological space, comprising a topology τ on set X. We typically denote the space just as X. Let cl_X and int_X be the closure and interior operators induced on subsets of X by its topology. A subset a of X is *regular closed* if it is equal to the closure of its interior: $a = \mathrm{cl}_X(\mathrm{int}_X(a))$. The set $\mathrm{RC}(X)$ of all regular closed subsets of X forms a Boolean algebra in which $a + b = a \cup b$, $a \cdot b = \mathrm{cl}_X(\mathrm{int}_X(a \cap b))$, $-a = \mathrm{cl}_X(X\backslash a)$, $0 = \emptyset$ and $1 = X$.

There is a contact relation C_X on $\mathrm{RC}(X)$ defined by putting aC_Xb iff $a \cap b \neq \emptyset$.[5] Thus $(\mathrm{RC}(X), C_X)$ is a Boolean contact algebra in which 'in contact' means to have a non-empty intersection. The non-tangential inclusion relation on this algebra has $a \ll b$ iff $a \subseteq \mathrm{int}(b)$.

By a *mereotopological space* we mean a pair $S = (X_S, M_S)$ where X_S is a topological space and M_S is a subalgebra of the Boolean algebra $\mathrm{RC}(X_S)$ of regular closed subsets of X_S, such that M_S is a closed basis for X_S. This last condition on M_S means that every closed subset of X_S is an intersection of a collection of members of M_S.

[4] Henkin et al. 1971, 0.2.9.

[5] Note that $a \cdot b \subseteq a \cap b$ for regular closed a and b, so $a \cap b \neq \emptyset$ is a weaker assertion here than $a \cdot b \neq 0$.

That M_S is a subalgebra of $\mathrm{RC}(X_S)$ means that it is closed under the Boolean algebra operations of $\mathrm{RC}(X_S)$, and hence is itself a Boolean algebra under these operations. Kontchakov et al. (2008) call such a pair S a *closed mereotopology*.

Remark 3.1 Mereotopology can equivalently be approached from the point of view of sets a that are *regular open* in the sense that $a = \mathrm{int}(\mathrm{cl}(a))$. A set is regular open iff its complement $X \backslash a$ is regular closed. The set $\mathrm{RO}(X)$ of regular open subsets of X is a Boolean algebra in which $a + b = \mathrm{int}(\mathrm{cl}(a \cup b))$, $a \cdot b = a \cap b$ and $-a = \mathrm{int}(X \backslash a)$. Its natural contact relation D_X has $a D_X b$ iff $\mathrm{cl}(a) \cap \mathrm{cl}(b) \neq \emptyset$ iff there is a point that is *close to* both a and b. The map $a \mapsto \mathrm{cl}(a)$ is a **BCA**-isomorphism from $(\mathrm{RO}(X), D_X)$ onto $(\mathrm{RC}(X), C_X)$ (Dimov and Vakarelov 2006a, Example 2.1). Pratt-Hartmann (2007, Definition 2.5) defines a *mereotopology* over a topological space X as a Boolean sub-algebra of $\mathrm{RO}(X)$ that is a basis for the topology.

A *semiregular* topological space is one that has a basis of regular open sets, or equivalently has a closed basis of regular closed sets. For instance, the real line $\mathbb{R}$ with its standard topology is semi-regular since its intervals (x, y) are regular open and form a basis. If a space X is semiregular, then $(X, \mathrm{RC}(X))$ is a mereotopological space as defined here.

Contact algebras, especially those of the form $(\mathrm{RC}(X), C_X)$, can be used to model logics based on propositional languages with a binary connective C and possibly other connectives, including modalities. Work in this direction can be found in (Lutz and Wolter 2006; Kontchakov et al. 2008; Nenov and Vakarelov 2008; Vakarelov 2007) as well as in some chapters of the *Handbook of Spatial Logic* (Aiello et al. 2007).

Now we define mereomorphisms. If $S_1 = (X_1, M_1)$ and $S_2 = (X_2, M_2)$ are mereo-topological spaces, a *mereomorphism* $\theta : S_1 \to S_2$ is a function $\theta : X_1 \to X_2$ whose pullback action on members of M_2 is a Boolean algebra homomorphism from M_2 to M_1. This means that for each subset $a \subseteq X_2$ with $a \in M_2$, the pre-image $\theta^{-1}(a) = \{x \in X_1 : \theta(x) \in a\}$ belongs to M_1, and the map $M_2 \to M_1$ acting by $a \mapsto \theta^{-1}(a)$ is a Boolean algebra homomorphism.

Lemma 3.1 *Every mereomorphism is continuous.*

Proof Let θ be a mereomorphism as above, and b a closed subset of X_2. Then $b = \bigcap_{i \in I} a_i$ for some $a_i \in M_2$, since M_2 is a closed basis for the space X_2. So $\theta^{-1} b = \bigcap_{i \in I} \theta^{-1} a_i$, with each $\theta^{-1} a_i$ belonging to M_1 and hence being (regular) closed in X_1. Therefore $\theta^{-1} b$ is closed.

This shows that under $\theta : X_1 \to X_2$, pre-images of closed sets are closed, implying that θ is continuous. □

Remark 3.2 The map $a \mapsto \theta^{-1}(a)$ always preserves Boolean joins (=unions), so for it to be a Boolean homomorphism it is sufficient that it preserve Boolean complements: $\theta^{-1}(-a) = -\theta^{-1}(a)$. But for $\theta : X_1 \to X_2$ to be a mereomorphism, it is not sufficient in general that it be continuous and pull back members of M_2 to members of M_1, as we show next.

Example 3.3 Let S be the mereotopological space $(\mathbb{R}, \mathrm{RC}(\mathbb{R}))$, where $\mathbb{R}$ is the real line with its standard topology. Let θ be the constant function having $\theta(x) = 0$ for all $x \in \mathbb{R}$. Then $\theta\colon \mathbb{R} \to \mathbb{R}$ is continuous, and if $a \in \mathrm{RC}(\mathbb{R})$ then $\theta^{-1}(a)$ is either $\mathbb{R}$ or $\emptyset$ accordingly as $0 \in a$ or not, so $\theta^{-1}(a) \in \mathrm{RC}(\mathbb{R})$.

However θ is not a mereomorphism, because the map $a \mapsto \theta^{-1}(a)$ does not preserve the Boolean complement operation on $\mathrm{RC}(\mathbb{R})$. For example, let a be the regular closed interval $[0, \infty)$. Then $-a = (-\infty, 0]$, so $\theta^{-1}(-a) = \mathbb{R}$. But also $\theta^{-1}(a) = \mathbb{R}$, so $-\theta^{-1}(a) = \emptyset \neq \theta^{-1}(-a)$.

Observe also that the map $a \mapsto \theta^{-1}(a)$ does not preserve Boolean meets. With $a = [0, \infty)$ as above we have $\theta^{-1}(a \cdot -a) = \theta^{-1}(\emptyset) = \emptyset$, whereas $\theta^{-1}(a) \cdot \theta^{-1}(-a) = \mathbb{R} \cdot \mathbb{R} = \mathbb{R}$.

The identity function on a mereotopological space is a mereomorphism, and the functional composition of two mereomorphisms is a mereomorphism. Thus we have a category **MS** of mereotopological spaces and mereomorphisms.

In the next result we use the notation $\theta[a]$ for the direct image $\{\theta(x) : x \in a\}$ of a subset a of the domain of θ. If θ is a bijection with inverse σ, then $\theta[a] = \sigma^{-1}(a)$.

Theorem 3.2 *A mereomorphism $\theta\colon S_1 \to S_2$ is an isomorphism in the category* **MS** *if, and only if, it is a bijection that has $\theta[a] \in M_2$ for all $a \in M_1$.*

Proof Let θ be an isomorphism. This means that there is an inverse mereomorphism $\sigma\colon S_2 \to S_1$ such that each of the compositions $\sigma \circ \theta$ and $\theta \circ \sigma$ is the identity morphism on its domain. The existence of σ ensures that θ is a bijection. For each $a \in M_1$ we have $\theta[a] = \sigma^{-1}(a) \in M_2$ as σ is a mereomorphism.

Conversely, assume θ is a bijective mereomorphism having $\theta[a] \in M_2$ for all $a \in M_1$. As a bijection, θ has an inverse function $\sigma\colon X_2 \to X_1$. For each $a \in M_1$ we have $\sigma^{-1}(a) = \theta[a] \in M_2$, showing that the map $a \mapsto \sigma^{-1}(a)$ pulls members of M_1 back to members of M_2. Since θ is a mereomorphism, the map $b \mapsto \theta^{-1}(b)$ is a Boolean algebra homomorphism from M_2 to M_1. But this map is bijective, with inverse $a \mapsto \sigma^{-1}(a)$, since $\sigma^{-1}(\theta^{-1}(b)) = b$ and $\theta^{-1}(\sigma^{-1}(a)) = a$. As the inverse of a bijective homomorphism of algebras is itself a homomorphism, it follows that $a \mapsto \sigma^{-1}(a)$ is a Boolean algebra homomorphism.

This shows that σ is a mereomorphism $S_2 \to S_1$ and provides the inverse in **MS** that ensures θ is an isomorphism. □

An isomorphism in **MS** might well be called a *mereo-isomorphism*. By Lemma 3.1 such a map is a *homeomorphism*, i.e., is a continuous bijection with a continuous inverse. However, a mereomorphism that is a homeomorphism need not be a mereo-isomorphism:

Example 3.4 Let $\mathbb{Q}$ be the set of rational numbers, and for each $p, q \in \mathbb{Q}$ with $p < q$, let a_{pq} be the regular closed subset $(-\infty, p] \cup [q, \infty)$ of $\mathbb{R}$. Put $M_0 = \{a_{pq} : p, q \in \mathbb{Q}\}$, and let M be the Boolean subalgebra of $\mathrm{RC}(\mathbb{R})$ generated by M_0. Since M_0 is countable, so too is M, and therefore M is a proper subset of $\mathrm{RC}(\mathbb{R})$.

Now every open subset of $\mathbb{R}$ is a union of open intervals (p, q) with rational endpoints, so every closed subset is an intersection of members of M_0. Thus M is a closed basis for the standard topology on $\mathbb{R}$. Hence $(\mathbb{R}, \mathrm{RC}(\mathbb{R}))$ and $(\mathbb{R}, M)$ are distinct mereotopological spaces based on the same topological space $\mathbb{R}$.

Let θ be the identity function on $\mathbb{R}$. Then θ is a homeomorphism, and is a mereomorphism because the map $a \mapsto \theta^{-1}(a)$ is the inclusion homomorphism of M into $\mathrm{RC}(\mathbb{R})$. However θ is not a mereo-isomorphism, by Theorem 3.2, as there are (uncountably many) elements $a \in \mathrm{RC}(\mathbb{R}))$ such that $\theta[a] = a \notin M$.

We now define the dual of a mereotopological space S by putting $S^+ = (M_S, C_S)$ where C_S is the intersect relation on the Boolean algebra M_S, i.e., aC_Sb iff $a \cap b \neq \emptyset$, for all $a, b \in M_S$. Then S^+ is a Boolean contact algebra.

For each mereomorphism $\theta\colon S_1 \to S_2$, define $\theta^+\colon M_2 \to M_1$ by putting $\theta^+(a) = \theta^{-1}(a)$. The definition of mereomorphism ensures that $\theta^+(a) \in M_1$ for all $a \in M_2$, and that θ^+ is a Boolean algebra homomorphism. Moreover, if $\theta^+(a)C_{S_1}\theta^+(b)$, then $\theta^{-1}(a) \cap \theta^{-1}(b) \neq \emptyset$, hence $a \cap b \neq \emptyset$ and so $aC_{S_2}b$. Thus θ^+ reflects contact as well, making it a BCA-morphism $S_2^+ \to S_1^+$.

Now given a pair of composable mereomorphisms

$$S_1 \xrightarrow{\theta_1} S_2 \xrightarrow{\theta_2} S_3,$$

we obtain the composable BCA-morphisms

$$S_1^+ \xleftarrow{\theta_1^+} S_2^+ \xleftarrow{\theta_2^+} S_3^+,$$

for which it can be shown that $\theta_1^+ \circ \theta_2^+ = (\theta_2 \circ \theta_1)^+$ (because $\theta_1^{-1} \circ \theta_2^{-1} = (\theta_2 \circ \theta_1)^{-1}$). Also, if θ is the identity mereomorphism on S, i.e., the identity function on X_S, then θ^+ is the identity function on M_S, hence is the identity BCA-morphism on S^+.

Thus the assignments $\Phi(S) = S^+$ and $\Phi(\theta) = \theta^+$ form a *contravariant functor* $\Phi\colon$ **MS** $\to$ **BCA** from the category of mereotopological spaces to the category of Boolean contact algebras.

Our next task is to construct a functor in the opposite direction.

4 Representation by Clans

If $A = (B_A, C_A)$ is a Boolean contact algebra, then a *clan* of A is a non-empty subset Γ of B_A such that:

K1. $0 \notin \Gamma$.
K2. $x \in \Gamma$ and $x \leq y$ implies $y \in \Gamma$.
K3. $x + y \in \Gamma$ implies $x \in \Gamma$ or $y \in \Gamma$.
K4. $x, y \in \Gamma$ implies xCy.

A non-empty Γ satisfying K1–K3 is called a *grill*.[6] So a clan is a grill for which any two members are in contact. It is readily seen that any ultrafilter of B_A is a clan (Dimov and Vakarelov 2006a). For, it is standard that an ultrafilter satisfies K1–K3, so is a grill. For K4, if x and y belong to an ultrafilter, then $x \cdot y \neq 0$, so $x \cdot y \, C_A \, x \cdot y$ by C4. Since $x \cdot y \leq x, y$ and C_A is monotonic in each variable, it follows that $x C_A y$.

Let X_A be the set of all clans of A. For each $x \in B_A$ let $f_A(x) = \{ \Gamma \in X_A : x \in \Gamma \}$. The function f_A is injective, for if $x \neq y$, then say $x \nleq y$, so there is an ultrafilter U of B_A that contains x but not y. Then U is a clan belonging to $f_A(x)$ but not $f_A(y)$, showing that $f_A(x) \neq f_A(y)$.

Let $M_A = \{ f_A(x) : x \in B_A \}$. Now as Lemma 5.1(i) of (Dimov and Vakarelov 2006a) states, f_A has the properties $f_A(0) = \emptyset$, $f_A(1) = X_A$, $f_A(x + y) = f_A(x) + f_A(y)$. So M_A contains $\emptyset$ and X_A and is closed under finite unions. This is enough to ensure that M_A is a closed basis for a topology on X_A whose closed subsets of X_A are the intersections of collections of members of M_A. We now view X_A as a space under this topology.

It is proved in (Dimov and Vakarelov 2006a) that X_A is compact and T_0. We will show in Sect. 5 that compactness follows from our stronger mereocompactness property (see Theorem 5.2). The T_0 separation property is that for any pair of distinct points there is an open neighbourhood of one that excludes the other. To show this for X_A, let Γ and Δ be distinct clans of A. Then there is an element of one that does not belong to the other, say $x \in \Gamma$ and $x \notin \Delta$. Then $X_A \setminus f_A(x)$ is an open neighbourhood of Δ that excludes Γ.

By Lemma 5.3(ii) of (Dimov and Vakarelov 2006a), each set $f_A(x)$ is regular closed in the space X_A, so f_A maps B_A into the Boolean algebra $\mathrm{RC}(X_A)$. Moreover, Lemma 5.3(i) of (Dimov and Vakarelov 2006a) shows that $f_A(-x) = \mathrm{cl}_{X_A}(X_A \setminus f_A(x)) = -f_A(x)$ in $\mathrm{RC}(X_A)$. So together with its above listed properties, we see that f_A is a Boolean algebra homomorphism into $\mathrm{RC}(X_A)$, making its image M_A a subalgebra of $\mathrm{RC}(X_A)$.

Thus the structure $A_+ = (X_A, M_A)$ is a mereotopological space. This is the dual space of the algebra A.

Theorem 4.1 *A is isomorphic to the contact algebra $(A_+)^+$ in the category* ***BCA***.

Proof By definition, $(A_+)^+ = (M_A, C_{A_+})$, where C_{A_+} is the intersect relation on the Boolean set algebra M_A. We have already observed that f_A is an injective Boolean algebra homomorphism, and it maps B_A onto M_A. By (Dimov and Vakarelov 2006a, Proposition 3.3(i)) we have that $x C_A y$ iff there is a clan Γ of A with $x, y \in \Gamma$, which is equivalent to $f_A(x) \cap f_A(y) \neq \emptyset$, i.e., to $f_A(x) \, C_{A_+} f_A(y)$. Hence f_A preserves and reflects contact.

Altogether this shows that $f_A \colon B_A \to M_A$ is a bijective BCA-morphism preserving contact, so is an isomorphism from A to $(A_+)^+$ by Theorem 2.1. □

Now for any BCA-morphism $f \colon A \to A'$, define a function f_+ on $X_{A'}$ by putting, for each clan Γ of A', $f_+(\Gamma) = f^{-1}(\Gamma) = \{ x \in B_A : f(x) \in \Gamma \}$.

[6]Grills originate with Choquet (1947) and clans with Thron (1973).

Theorem 4.2 *f_+ is a mereomorphism from A'_+ to A_+.*

Proof First we need that f_+ is a function from $X_{A'}$ to X_A, i.e., that $f^{-1}(\Gamma)$ is a clan of A when Γ is a clan of A'. First, $f^{-1}\Gamma$ is non-empty because $f(1_A) = 1_{A'} \in \Gamma$ and so $1_A \in f^{-1}\Gamma$. Next, the grill properties K1–K3 lift from Γ to $f^{-1}\Gamma$ because f preserves least elements, the partial orders $\leq$, and joins. For K4, if $x, y \in f^{-1}\Gamma$, then $f(x)C'f(y)$ as Γ has K4, hence xCy as f reflects contact. Thus $f^{-1}(\Gamma)$ is indeed a clan.

Also we require that pulling back along f_+ gives a Boolean homomorphism from M_A to $M_{A'}$. For an arbitrary element $f_A(x)$ of M_A, we have, for any $\Gamma \in X_{A'}$, that $\Gamma \in (f_+)^{-1}(f_A(x))$ iff $x \in f^{-1}(\Gamma)$ iff $\Gamma \in f_{A'}(f(x))$. This shows that for any $x \in B_A$,

$$(f_+)^{-1}(f_A(x)) = f_{A'}(f(x)), \tag{1}$$

confirming that $(f_+)^{-1}$ maps M_A into $M_{A'}$. Then Eq. (1) and the fact that f, f_A and $f_{A'}$ are all Boolean homomorphisms allow us to infer that $(f_+)^{-1}$ preserves Boolean complements, because

$$(f_+)^{-1}(-f_A(x)) = (f_+)^{-1}(f_A(-x)) = f_{A'}(f(-x)) = -f_{A'}(f(x)) = -(f_+)^{-1}(f_A(x)).$$

As already noted in Remark 3.2, that suffices to ensure that $(f_+)^{-1}$ is a Boolean homomorphism. □

Now given a pair of composable BCA-morphisms

$$A_1 \xrightarrow{f_1} A_2 \xrightarrow{f_2} A_3,$$

we obtain the composable mereomorphisms

$$A_{1+} \xleftarrow{f_{1+}} A_{2+} \xleftarrow{f_{2+}} A_{3+},$$

for which it can be shown that $f_{1+} \circ f_{2+} = (f_2 \circ f_1)_+$. Also, if f is the identity BCA-morphism on A, then f_+ is the identity mereomorphism on A_+.

Thus the assignments $\Theta(A) = A_+$ and $\Theta(f) = f_+$ form a contravariant functor Θ : **BCA** → **MS**. For any BCA-morphism $f : A \to A'$, Eq. (1) implies that the following diagram commutes.

$$\begin{array}{ccc} A & \xrightarrow{f_A} & (A_+)^+ = \Phi(\Theta(A)) \\ {\scriptstyle f}\big\downarrow & & \big\downarrow{\scriptstyle (f_+)^+ = \Phi(\Theta(f))} \\ A' & \xrightarrow{f_{A'}} & (A'_+)^+ = \Phi(\Theta(A')) \end{array}$$

This means, by definition, that the assignment $A \mapsto f_A$ for all BCA's A constitutes a *natural transformation* from the identity functor on **BCA** to the functor

$\Phi \circ \Theta$: **BCA** $\to$ **BCA** that assigns to each contact algebra A its "double dual" $(A_+)^+$. The morphisms f_A are the *components* of this natural transformation. In general, a natural transformation is called a *natural isomorphism* if its components are isomorphisms (i.e., invertible morphisms) in their ambient category. Thus in our present situation, as the components f_A are all mereo-isomorphisms (Theorem 4.1), it follows that $\Phi \circ \Theta$ is *naturally isomorphic* to the identity functor.[7]

5 Mereocompactness

The functor $\Theta \circ \Phi$: **MS** $\to$ **MS** is not naturally isomorphic to the identity functor on **MS**, because a mereotopological space S need not be isomorphic to its double dual $(S^+)_+$. For instance, $(S^+)_+$ can be of higher cardinality that S. As an example, let $S = (X, \mathrm{RC}(X))$ where X is a discrete space of any infinite cardinality κ. Then $\mathrm{RC}(X)$ is the powerset algebra of X, of cardinality 2^κ, having 2^{2^κ} ultrafilters. Since ultrafilters are clans, it follows that $(S^+)_+$ is of cardinality 2^{2^κ}.

Note that the question of whether S is isomorphic to its double dual is equivalent to the question of whether it is isomorphic to the dual of something. For if $S \cong A_+$, then $S^+ \cong (A_+)^+ \cong A$ (Theorem 4.1), and so $(S^+)_+ \cong A_+ \cong S$.

We now explore conditions under which a space $S = (X_S, M_S)$ *is* isomorphic to $(S^+)_+$. For each $x \in X_S$, define $\rho_S(x) = \{a \in M_S : x \in a\}$.

Theorem 5.1 (1) *$\rho_S(x)$ is a clan of the algebra $S^+ = (M_S, C_S)$, hence a point of the space X_{S^+}.*
(2) *$\rho_S : X_S \to X_{S^+}$ is a mereomorphism $S \to (S^+)_+$.*
(3) *ρ_S is injective if, and only if, X_S is T_0.*
(4) *For any mereomorphism $\theta : S \to S'$, the following diagram commutes:*

$$\begin{array}{ccc} S & \xrightarrow{\rho_S} & (S^+)_+ \\ {\scriptstyle\theta}\downarrow & & \downarrow{\scriptstyle(\theta^+)_+} \\ S' & \xrightarrow{\rho_{S'}} & (S'^+)_+ \end{array} \tag{2}$$

Proof (1) That $\rho_S(x)$ satisfies K1–K3 is routine. For K4 recall that C_S is the intersect relation, and note that if $a_1, a_2 \in \rho_S(x)$, then $x \in a_1 \cap a_2$, so $a_1 C_S a_2$.
(2) First we need to have ρ_S^{-1} pulling back members of $M_{S^+} = \{ f_{S^+}(a) : a \in M_S \}$ to members of M_S. But for any $a \in M_S$ we have

$$\rho_S^{-1}(f_{S^+}(a)) = a, \tag{3}$$

hence $\rho_S^{-1}(f_{S^+}(a)) \in M_S$ as required. Equation (3) holds since $x \in \rho_S^{-1}(f_{S^+}(a))$ iff $\rho_S(x) \in f_{S^+}(a)$ iff $a \in \rho_S(x)$ iff $x \in a$.

[7] See (Mac Lane 1998, I.4) for the theory of natural transformations and isomorphisms.

With the help of (3) for $-a$ and for a we reason that

$$\rho_S^{-1}(-f_{S^+}(a)) = \rho_S^{-1}(f_{S^+}(-a)) = -a = -\rho_S^{-1}(f_{S^+}(a))$$

so ρ_S^{-1} preserves Boolean complements. That is enough to make it a Boolean homomorphism (Remark 3.2), completing the proof that ρ_S is a mereomorphism.

(3) Suppose X_S is T_0. Then if $x, y \in X_S$ with $x \neq y$, there is an open set containing one but not the other, hence its complement is a closed set containing one but not the other. Since M_S is a closed basis for X_S there must then be a member a of M_S containing one but not the other, so a belongs either to $\rho_S(x)\backslash\rho_S(y)$ or to $\rho_S(y)\backslash\rho_S(x)$. In either case $\rho_S(x) \neq \rho_S(y)$, showing ρ_S is injective.
For the converse, let ρ_S be injective. If $x \neq y$, there is some $a \in M_S$ such that a belongs either to $\rho_S(x)\backslash\rho_S(y)$ or to $\rho_S(y)\backslash\rho_S(x)$. Then the complement of a is an open set containing one of x and y but not the other. This shows that distinct points of X_S do not have the same open neighbourhoods, which is the T_0 property.

(4) For each $x \in X_S$ we have $\rho_{S'}(\theta(x)) = \{a \in M_{S'} : \theta(x) \in a\}$, while

$$(\theta^+)_+(\rho_S(x)) = (\theta^+)^{-1}(\rho_S(x)) = \{a \in M_{S'} : \theta^+(a) \in \rho_S(x)\}.$$

But $\theta^+(a) = \theta^{-1}(a) \in \rho_S(x)$ iff $x \in \theta^{-1}(a)$ iff $\theta(x) \in a$. So $\rho_{S'}(\theta(x)) = (\theta^+)_+ (\rho_S(x))$ as required for the diagram to commute. □

Now define a *mereocompact space* to be a mereotopological space S satisfying the following property:

For every $\Gamma, \Delta \subseteq M_S$ with Γ a *clan* of S^+, if $\bigcap \Gamma \subseteq \bigcup \Delta$ then $\Gamma \cap \Delta \neq \emptyset$.

Theorem 5.2 (1) *S is mereocompact iff $\rho_S \colon X_S \to X_{S^+}$ is surjective.*
(2) *Every mereocompact space is compact.*
(3) *If A is any Boolean contact algebra, then A_+ is mereocompact.*

Proof (1) Let S be mereocompact. Take any $\Gamma \in X_{S^+}$, i.e., Γ is a clan of the algebra S^+. Put $\Delta = M_S\backslash\Gamma$. Then $\Gamma \cap \Delta = \emptyset$ so by mereocompactness $\bigcap \Gamma \nsubseteq \bigcup \Delta$. Hence there is some $x \in \bigcap \Gamma\backslash \bigcup \Delta$. Then $\rho_S(x) = \Gamma$. This shows ρ_S is surjective.
Conversely, suppose ρ_S is surjective. Let $\Gamma, \Delta \subseteq M_S$ with Γ a clan, and $\bigcap \Gamma \subseteq \bigcup \Delta$. Then $\Gamma = \rho_S(x)$ for some $x \in S$, hence $x \in \bigcap \Gamma$. Thus there is some $\delta \in \Delta$ with $x \in \delta$. Hence $\delta \in \rho_S(x)$, so $\delta \in \Gamma \cap \Delta \neq \emptyset$. This shows S is mereocompact.

(2) Let S be mereocompact. For compactness of X_S it suffices to show that any collection of closed sets with the *finite intersection property* has non-empty intersection. (Recall that collection M has the finite intersection property if each finite subcollection of M has non-empty intersection.) But since M_S is a closed basis for X_S, it is enough to prove this for subcollections of M_S. So take any $M \subseteq M_S$ with the finite intersection property. Then M extends to an ultrafilter

U of the powerset algebra $\mathcal{P}(X_S)$ of all subsets of X_S. Let $\Gamma = U \cap M_S$. Then Γ is a clan of S^+: the fact that U is a grill of $\mathcal{P}(X_S)$ ensures that Γ is a grill of M_S, and if $a, b \in \Gamma$, then $a, b \in U$ and so $a \cap b \neq \emptyset$, i.e., $aC_S b$.

Now put $\Delta = \emptyset$ in the definition of mereocompactness of S. Since Γ is a clan and $\Gamma \cap \emptyset = \emptyset$, it follows that $\bigcap \Gamma \neq \bigcup \emptyset = \emptyset$. Since $M \subseteq \Gamma$, this implies $\bigcap M \neq \emptyset$ as required.

(3) Recall that $A_+ = (X_A, M_A)$ with X_A the set of clans of A and $M_A = \{ f_A(x) : x \in B_A \}$. Take subsets Γ, Δ of M_A with Γ a clan of $(A_+)^+ = (M_A, C_{A_+})$. Put $J = f_A^{-1}\Gamma = \{ x \in B_A : f_A(x) \in \Gamma \}$. Then J is a clan of A by the proof of Theorem 4.2, so $J \in X_A$.

Now for any $f_A(x) \in \Gamma$ we have $x \in J$ and so $J \in f_A(x)$. Thus $J \in \bigcap \Gamma$. So if $\bigcap \Gamma \subseteq \bigcup \Delta$ then there is some $f_A(y) \in \Delta$ such that $J \in f_A(y)$. But then $y \in J$, implying $f_A(y) \in \Gamma$. Hence $f_A(y) \in \Gamma \cap \Delta$, showing that $\Gamma \cap \Delta$ is non-empty. This proves mereocompactness of A_+. □

Mereocompactness is a strictly stronger property than compactness:

Example 5.1 A topological space is said to be *strongly compact* (Rasiowa and Sikorski 1963, p. 101) if it is not covered by open proper subsets, i.e., if every open cover of the space must include the space itself as a member. Equivalently, this means that the intersection of any set of non-empty closed subsets is non-empty, which is a much stronger condition than compactness. Any topological space X has a *one-point strong compactification* (Rasiowa and Sikorski 1963, p. 102) obtained by adding a new point π to X and declaring that the open subsets of $X \cup \{ \pi \}$ are $X \cup \{ \pi \}$ itself and all the open subsets of X. Then the closed subsets of $X \cup \{ \pi \}$ are $\emptyset$ and all sets of the form $b \cup \{ \pi \}$ with b a closed subset of X. Thus π belongs to every non-empty closed set in $X \cup \{ \pi \}$, ensuring strong compactness. The regular closed subsets of $X \cup \{ \pi \}$ are $\emptyset$ and all sets $b \cup \{ \pi \}$ with b a *non-empty* regular closed subset of X.

Now let X be the three-element set $3 = \{ 0, 1, 2 \}$ with the discrete topology, and put $S = (X_S, M_S)$ with $X_S = X \cup \{ \pi \}$ and

$$M_S = \mathrm{RC}(X \cup \{ \pi \}) = \{ \emptyset \} \cup \{ b \cup \{ \pi \} : \emptyset \neq b \subseteq 3 \}.$$

$S^+ = (M_S, C_S)$ is an eight-element Boolean contact algebra (see Fig. 1) in which any two non-empty members are C_S-related, i.e., intersect, since they contain π. For each $x \in 3$, the *point-clan* $\rho_S(x) = \{ a \in M_S : x \in a \}$ is the principal ultrafilter of S^+ generated by the atom $\{ x, \pi \}$, and contains the four elements $\{ x, \pi \}$, $\{ x, y, \pi \}$, $\{ x, z, \pi \}$, $\{ x, y, z, \pi \}$, where y, z are the two elements of 3 other than x. $\rho_S(\pi)$ is the seven-element set $\{ b \cup \{ \pi \} : \emptyset \neq b \subseteq 3 \}$.

Let $\Gamma = \rho_S(0) \cup \rho_S(1)$. Then Γ is a clan of S^+ and is in fact the six-element set $M_S \setminus \{ \{ 2, \pi \}, \emptyset \}$. Thus $\Gamma \neq \rho_S(w)$ for all $w \in 3 \cup \{ \pi \}$, so the function $\rho_S : X_S \to X_{S^+}$ is not surjective. Hence while X_S is strongly compact, S is not mereocompact by Theorem 5.2(1).

The idea of this counter-example was prompted by Example 4.2 of (Dimov and Vakarelov 2006b), which exhibited a compact semiregular T_0 space X whose RC-algebra has a clan that is not equal to any point-clan by taking $X = \mathbb{R} \cup \{ \pi \}$ (with

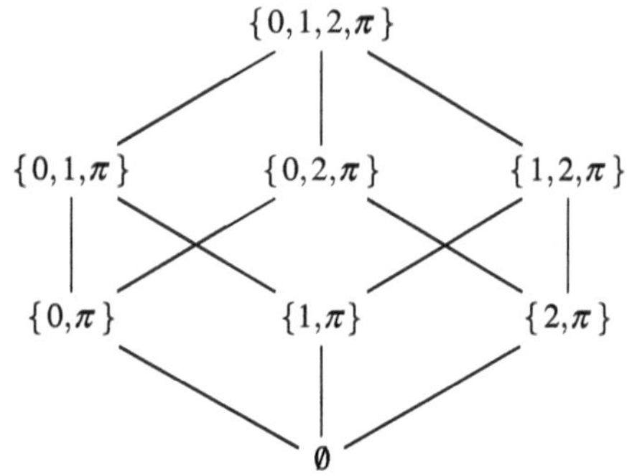

Fig. 1 The algebra $\mathrm{RC}(X \cup \{\pi\})$ with $X = \{0, 1, 2\}$

a different description of its topology and without the discussion of strong compactness). We make use of the finiteness of our example in the final section below.

Although mereocompactness implies compactness of the underlying topology, it is independent of *strong* compactness. To see this, let S_1 be any mereotopological space containing non-empty regular closed sets a, b that are disjoint. E.g., $S_1 = (\mathbb{R}, \mathrm{RC}(\mathbb{R}))$ with a, b any two disjoint closed intervals. Let $A = S_1^+$. Then a and b are not in contact in A, so A_+ is a mereocompact space in which $f_A(a)$ and $f_A(b)$ are non-empty closed subsets of X_A that are disjoint. Thus A_+ is not strongly compact.

We now establish the conditions under which a space is isomorphic to its double dual:

Theorem 5.3 *The mereomorphism $\rho_S\colon S \to (S^+)_+$ is a mereo-isomorphism if, and only if, S is mereocompact and T_0.*

Proof By Theorems 5.1 (3) and 5.2(1), ρ_S is a bijection from X_S onto X_{S^+}, the set of all clans of S^+, iff S is mereocompact and T_0.

Now let S be mereocompact and T_0. To prove that the bijection ρ_S is a mereo-isomorphism, it suffices by Theorem 3.2 to prove that for each $a \in M_S$, the direct image $\rho_S[a]$ belongs to M_{S^+}.

But by Theorem 4.1 with $A = S^+$, the BCA-isomorphism f_{S^+} between S^+ and its double dual maps M_S onto M_{S^+}, with $f_{S^+}(a)$ being the set of all clans of S^+ that contain a. Since ρ_S is surjective, any clan of S^+ is equal to $\rho_S(x)$ for some $x \in X_S$. Thus

$$f_{S^+}(a) = \{\, \rho_S(x)\colon x \in X_S \text{ and } a \in \rho_S(x) \,\} = \{\, \rho_S(x)\colon x \in a \,\} = \rho_S[a].$$

So $\rho_S[a] = f_{S^+}(a) \in M_{S^+}$ as required. □

Now let **MS*** be the full subcategory of **MS** whose objects are the mereocompact T_0 spaces. For each Boolean contact algebra A, the dual space $\Theta(A) = A_+$ is mereocompact and T_0, so we can view Θ as a functor from **BCA** into **MS***. In the opposite direction, let $\Phi^*\colon$ **MS*** $\to$ **BCA** be the restriction of functor Φ to **MS***.

From our earlier analysis, $\Phi^* \circ \Theta$: **BCA** $\to$ **BCA** is naturally isomorphic to the identity functor on **BCA**.

The commuting diagram (2) in Theorem 5.1 shows that the mereo-isomorphisms ρ_S for all **MS***-objects S form the components of a natural isomorphism between the identity functor on **MS*** and the functor $\Theta \circ \Phi^*$: **MS*** $\to$ **MS*** that assigns to each mereocompact T_0 space S its double dual $(S^+)_+$. These properties of Θ and Φ^* establish that the category of Boolean contact algebras is *dually equivalent* to the category of mereocompact T_0 spaces. That is the principal result of this paper.

There are further results in the topological representation of BCA's, for instance concerning the notion of an *extensional contact algebra* (ECA) as a BCA satisfying

$$\forall z\,(xCz \text{ iff } yCz) \text{ implies } x = y.$$

By restricting the points of the representing space to be *maximal* clans, it was shown in (Düntsch and Winter 2005; Dimov and Vakarelov 2006a) that any ECA is embeddable into the RC-algebra of a space that is compact, T_1 and weakly regular, the later meaning that the space is semi-regular and any non-empty open set a has a non-empty open subset b with $\mathrm{cl}(b) \subseteq a$. It is left to the interested reader to extend this result to a full duality for ECA's, and to do likewise for other classes of BCA's discussed in the literature.

6 Variations on Mereocompactness

We conclude by giving an alternative formulation of mereocompactness, and exploring some consequences that have been used in other duality theories to characterise dual spaces of algebras.

Let the notation $\Gamma \subseteq_f \Gamma'$ mean that Γ is a *finite* subset of Γ'. Consider the property

(μ_0) For every $\Gamma, \Delta \subseteq M_S$ with Γ a *clan* of S^+, if $\bigcap \Gamma \subseteq \bigcup \Delta$ then there exists a $\gamma \in \Gamma$ and a $\Delta_0 \subseteq_f \Delta$ such that $\gamma \subseteq \bigcup \Delta_0$.

(μ_0) is equivalent to mereocompactness. For if $\gamma \in \Gamma$ and $\Delta_0 \subseteq_f \Delta$ with $\gamma \subseteq \bigcup \Delta_0$, then $\bigcup \Delta_0 \in \Gamma$ by K2 for Γ, and so by K3 there is some $\delta \in \Delta_0$ with $\delta \in \Gamma$, hence $\delta \in \Gamma \cap \Delta \neq \emptyset$. Conversely, if there is a $\gamma \in \Gamma \cap \Delta$, then $\gamma \subseteq \bigcup\{\gamma\}$ and $\{\gamma\} \subseteq_f \Delta$.

Next consider

(μ_1) For every $\Gamma, \Delta \subseteq M_S$ with $\bigcap \Gamma \subseteq \bigcup \Delta$, there exist sets $\Gamma_0 \subseteq_f \Gamma$ and $\Delta_0 \subseteq_f \Delta$ such that $\bigcap \Gamma_0 \subseteq \bigcup \Delta_0$.

This property holds when M_S is the dual algebra of clopen subsets of the Stone space of a Boolean algebra, and is a consequence of, indeed equivalent to, the compactness of that Stone space. μ_1 also holds when M_S is the algebra of compact open subsets of the dual space of a distributive lattice, and has been used as one of the characterising properties of such spaces (Balbes and Dwinger 1974, Chap. IV).

In our mereotopological setting, μ_1 follows from mereocompactness of S. To see why, suppose that $\bigcap \Gamma_0 \nsubseteq \bigcup \Delta_0$ for all sets $\Gamma_0 \subseteq_{\mathrm{f}} \Gamma$ and $\Delta_0 \subseteq_{\mathrm{f}} \Delta$. Then $\Gamma \cup \{ X_S \backslash \delta : \delta \in \Delta \}$ has the finite intersection property and so extends to an ultrafilter U of the powerset algebra $\mathcal{P}(X_S)$ that includes Γ and is disjoint from Δ. Then $\Gamma' = U \cap M_S$ is a clan of S^+ that includes Γ. But if $\bigcap \Gamma \subseteq \bigcup \Delta$, then $\bigcap \Gamma' \subseteq \bigcap \Gamma \subseteq \bigcup \Delta$, so from mereocompactness of S we infer $\Gamma' \cap \Delta \neq \emptyset$, contradicting $U \cap \Delta = \emptyset$. Hence $\bigcap \Gamma \nsubseteq \bigcup \Delta$, confirming that μ_1 holds for S.

μ_1 is in fact weaker than mereocompactness. It holds trivially whenever M_S is finite, so it holds in the finite space S of Example 5.1, which is not mereocompact.

Now we modify μ_1 to the statement

(μ_2) For every $\Gamma, \Delta \subseteq M_S$ with $\bigcap \Gamma \subseteq \bigcup \Delta$, there exist sets $\Gamma_0 \subseteq_{\mathrm{f}} \Gamma$ and $\Delta_0 \subseteq_{\mathrm{f}} \Delta$ such that $\bigwedge \Gamma_0 \subseteq \bigcup \Delta_0$,

where $\bigwedge \Gamma_0 = \mathrm{cl}(\mathrm{int}(\bigcap \Gamma_0))$ is the Boolean meet of Γ_0 in S^+. Since $\bigwedge \Gamma_0 \subseteq \bigcap \Gamma_0$, it is evident that ($\mu_2$) follows from ($\mu_1$).

Property μ_2 is itself equivalent to requiring that

(μ_3) for all ultrafilters Γ of S^+, for all $\Delta \subseteq M_S$, $\bigcap \Gamma \subseteq \bigcup \Delta$ implies $\Gamma \cap \Delta \neq \emptyset$.

Proof Assume μ_2 and take any $\Gamma, \Delta \subseteq M_S$ such that Γ is an ultrafilter and $\bigcap \Gamma \subseteq \bigcup \Delta$. Then there exist sets $\Gamma_0 \subseteq_{\mathrm{f}} \Gamma$ and $\Delta_0 \subseteq_{\mathrm{f}} \Delta$ such that $\bigwedge \Gamma_0 \subseteq \bigcup \Delta_0$. As a filter, Γ is closed under finite meets and closed upwards under $\subseteq$, so then $\bigcup \Delta_0 \in \Gamma$. Since Γ satisfies K3 it follows that $\delta \in \Gamma$ for some $\delta \in \Delta_0$. Hence $\delta \in \Gamma \cap \Delta \neq \emptyset$, proving μ_3.

Conversely assume μ_3, take any $\Gamma, \Delta \subseteq M_S$ and suppose that $\bigwedge \Gamma_0 \nsubseteq \bigcup \Delta_0$ for all $\Gamma_0 \subseteq_{\mathrm{f}} \Gamma$ and $\Delta_0 \subseteq_{\mathrm{f}} \Delta$. Then $\Gamma \cup \{ -\delta : \delta \in \Delta \}$ has the *finite meet property*, i.e., each of its finite subsets has non-zero meet in S^+—which means non-empty meet. Hence $\Gamma \cup \{ -\delta : \delta \in \Delta \}$ extends to an ultrafilter Γ' of S^+ that is disjoint from Δ. By μ_3, since $\Gamma' \cap \Delta = \emptyset$ we immediately get $\bigcap \Gamma' \nsubseteq \bigcup \Delta$. But $\bigcap \Gamma' \subseteq \bigcap \Gamma$, so then $\bigcap \Gamma \nsubseteq \bigcup \Delta$. This proves μ_2. □

μ_3 is in turn equivalent to the condition

(μ_4) Each ultrafilter of S^+ is equal to $\rho_S(x)$ for some $x \in X$.

The equivalence of μ_3 and μ_4 follows by the same reasoning that shows that mereocompactness is equivalent to the surjectivity of ρ_S (Theorem 5.2).

μ_4 has the immediate consequence that every ultrafilter of S^+ has non-empty intersection. This consequence is weaker than μ_4, because it is also implied by topological compactness whereas μ_4 is not. In fact μ_4 is not even implied by strong compactness.

Example 6.1 We show that μ_4 fails in $S = (X, \mathrm{RC}(X))$ where X is the strong compactification $\mathbb{R} \cup \{ \pi \}$ of $\mathbb{R}$ (see Example 5.1). Hence $\mu_0 - \mu_3$ also fail in this space.

For each real number r, let $a_r = [r, \infty) \cup \{\pi\} \in \mathrm{RC}(X)$. Put $\Gamma = \{a_r : r \in \mathbb{R}\}$. Then $a_r \subseteq a_s$ iff $s \leq r$, so if $\emptyset \neq \Gamma_0 \subseteq_{\mathrm{f}} \Gamma$, then $\bigwedge \Gamma_0 = a_r \neq \emptyset$ where a_r is the $\subseteq$-least member of Γ_0. Hence Γ has the finite meet property and so extends to an ultrafilter U of S^+.

Now $\bigcap \Gamma = \{\pi\}$, since $s \notin a_r$ for any real $s < r$. Hence $\bigcap U = \{\pi\}$. If we had $U = \rho_S(x)$ for some x, then $x \in \bigcap U$ and so $x = \pi$ and $U = \rho_S(\pi)$. But this is impossible as $\rho_S(\pi)$ is $\mathrm{RC}(X)\backslash\{\emptyset\}$ and is not an ultrafilter. Indeed it is not even a filter since it contains both a_r and its complement $-a_r = (-\infty, r] \cup \{\pi\}$ but does not contain their meet $a_r \wedge -a_r = \emptyset$. So U violates μ_4.

We can also see directly that U violates μ_3: since π belongs to every non-empty member of $\mathrm{RC}(X)$ we have $\bigcap U \subseteq \bigcup(\mathrm{RC}(X)\backslash U)$ while $U \cap (\mathrm{RC}y(X)\backslash U) = \emptyset$.

It seems possible that μ_2 could be weaker than μ_1. To show this it would suffice to exhibit a mereotopological space S that satisfies any of μ_2–μ_4 and has a subset of M_S with the finite intersection property but empty intersection.

References

Aiello, M., Pratt-Hartmann, I., & van Benthem, J. (2007). *Handbook of spatial logics*. Springer.

Allwein, G., & Dunn, J. M. (1993). Kripke semantics for linear logic. *The Journal of Symbolic Logic*, *58*(2), 514–545.

Balbes, R. & Dwinger, P. (1974). *Distributive lattices*. University of Missouri Press.

Bimbó, K., & Dunn, J. M. (2008). *Generalized Galois logics: Relational semantics of nonclassical logical calculi, vol. 188 of CSLI lecture notes*. CA: CSLI Publications.

Choquet, G. (1947). Sur les notions de filtre et de grille. *Comptes Rendus de l'Académie des Sciences Paris*, *224*, 171–173.

Clarke, B. L. (1981). A calculus of individuals based on 'connection'. *Notre Dame Journal of Formal Logic*, *22*(3), 204–218.

Cohn, A. G., Bennett, B., Gooday, J., & Gotts, N. M. (1997). Qualitative spatial representation and reasoning with the region connection calculus. *GeoInformatica*, *1*(3), 275–316.

Dimov, G., & Vakarelov, D. (2006a). Contact algebras and region-based theory of space: A proximity approach - I. *Fundamenta Informaticae*, *74*, 209–249.

Dimov, G., & Vakarelov, D. (2006b). Contact algebras and region-based theory of space: A proximity approach - II. *Fundamenta Informaticae*, *74*, 251–282.

Dunn, J. M. (1982). A relational representation of quasi-boolean algebras. *Notre Dame Journal of Formal Logic*, *23*, 353–357.

Dunn, J. M. (1993). Star and perp: Two treatments of negation. *Philosophical perspectives* (Vol. 7, pp. 331–357). (Language and Logic, J. E. Tomberlin (Ed.)).

Dunn, J. M. (1995). Positive modal logic. *Studia Logica*, *55*, 301–317.

Dunn, J. M. (2001). A representation of relation algebras using Routley-Meyer frames. In C. A. Anderson & M. Zelëny (Eds.) *Logic, Meaning and Computation. Essays in Memory of Alonzo Church* (pp. 77–108). Kluwer Academic Publishers.

Dunn, J. M. & Hardegree, G. M. (2001). *Algebraic methods in philosophical logic, Vol. 41 of Oxford logic guides*. Oxford: Oxford University Press.

Dunn, J. M., Gehrke, M., & Palmigiano, A. (2005). Canonical extensions and relational completeness of some substructural logics. *The Journal of Symbolic Logic*, *70*(3), 713–740.

Düntsch, I., & Winter, M. (2005). A representation theorem for Boolean contact algebras. *Theoretical Computer Science*, *347*, 498–512.

Goldblatt, R. (1974). Metamathematics of Modal Logic, PhD thesis, Victoria University, Wellington. Included in (Goldblatt 1993).
Goldblatt, R. (1993). *Mathematics of modality, vol. 43 of CSLI lecture notes*. CA: CSLI Publications.
Hartonas, C., & Dunn, J. M. (1997). Stone duality for lattices. *Algebra Universalis*, *37*, 391–401.
Henkin, L., Monk, J. D. & Tarski, A. (1971). *Cylindric algebras I*. North-Holland.
Kontchakov, R., Pratt-Hartmann, I., Wolter, F. & Zakharyaschev, M. (2008). Topology, connectedness, and modal logic, In Goldblatt, R., & Areces, C. (Eds.) *Advances in Modal Logic* (Vol. 7, pp. 151–176). College Publications. www.aiml.net/volumes/volume7/.
de Laguna, T. (1922). Point, line, and surface, as sets of solids. *Journal of Philosophy*, *19*(17), 449–461.
Lutz, C., & Wolter, F. (2006). From varieties of algebras to covarieties of coalgebras. *Logical Methods in Computer Science*, *2*(2).
Mac Lane, S. (1998). *Categories for the working mathematician* (2nd ed.). Springer.
Nenov, Y. & Vakarelov, D. (2008). Modal logics for mereotopological relations. In R. Goldblatt & C. Areces (Eds.) *Advances in Modal Logic* (Vol. 7, pp. 249–272). College Publications. www.aiml.net/volumes/volume7/.
Pratt-Hartmann, I. (2007). First-order mereotopology. In M. Aiello, I. Pratt-Hartmann & J. van Benthem (Eds.) *Handbook of Spatial Logics* (pp.13–97). Springer.
Randell, D. A., & Cohn, A. G. (1989). Modelling topological and metrical properties of physical processes. In R. Brachman, H. Levesque & R. Reiter (Eds.) *Proceedings 1st International Conference on the Principles of Knowledge Representation and Reasoning* (pp. 55–66). Morgan Kaufman.
Randell, D. A., Cui, Z., & Cohn, A. G. (1992). A spatial logic based on regions and connection. In B. Nebel, C. Rich & W. Swartout (Eds.) *Proceedings 3rd International Conference on the Principles of Knowledge Representation and Reasoning* (pp. 165–176). Morgan Kaufman.
Rasiowa, H., & Sikorski, R. (1963). *The Mathematics of Metamathematics*. Warsaw: PWN-Polish Scientific Publishers.
Thron, W. J. (1973). Proximity structures and grills. *Mathematische Annalen*, *206*, 35–62.
Vakarelov, D. (2007). Region-based theory of space: Algebras of regions, representation theory, and logics. In D. M. Gabbay, S. S. Goncharov & M. Zakharyaschev (Eds.) *Mathematical Problems from Applied Logic II* (pp.267–348). Springer.
Whitehead, A. N. (1929). *Process and Reality*. Cambridge University Press.

Distributed Modal Logic

Gerard Allwein and William L. Harrison

Abstract Modal logics typically have only one domain of discourse—i.e., the collection of worlds or states. For distributed computing systems, however, it makes sense to have several collections of worlds and to relate one domain's local worlds to another's using either relations or special maps. To this end, we introduce distributed modal logics. Distributed modal logics lift the distribution structure of a distributed system directly into the logic, thereby parameterizing the logic by the distribution structure itself. Each domain supports a "local logic" (which can itself be a modal logic). The connections between local logics are realized as "distributed modal connectives" where these connectives take propositions in one logic to propositions in another. Weak distributed logic systems require neighborhood semantics and, hence, the connection between domains becomes a neighborhood map linking each world in one domain to a collection of neighborhoods in another domain. In sufficiently strong distributed logic systems, the maps may be Kripke relations linking worlds from two different domains. We briefly illustrate distributed modal logics with the outline of a security verification for a hardware distributed system (i.e., a system-on-a-chip) with components that must be woven into proofs of security statements. Distributed modal logics also support probabilistic systems using stochastic relations.

Keywords Hilbert systems · Kripke · Modal algebras · Modal logic · Simulations

G. Allwein (✉)
Naval Research Laboratory, Code 5543, Washington, USA
e-mail: gerard.allwein@nrl.navy.mil
URL: http://www.nrl.navy.mil/itd/chacs/5543

W.L. Harrison
Department of Computer Science, University of Missouri, Columbia, USA
e-mail: harrisonwl@missouri.edu
URL: http://engineering.missouri.edu/person/harrisonw/

K. Bimbó (ed.), *J. Michael Dunn on Information Based Logics*,
Outstanding Contributions to Logic 8, DOI 10.1007/978-3-319-29300-4_16

1 Introduction

Logic in 20th century had many parents. It settled upon a very linguistic base and many logical investigations concern exploring logical notions as represented via this linguistic base. The reasoning that we can perform using logic in this manner is filtered through this linguistic base. This has the tendency to force some notions to be expressed (if even possible) in higher abstract formal machinery and more complicated semantics than is desired if we intend for logics to be used by humans for reasoning, as opposed to machines. The problems encountered are not to be considered as artificially imposed via the linguistic base, but rather there is much more that could be represented directly of the world about which we use logic to reason. It is in this sense that we present distributed logics, i.e., as an attempt to capture more of the work of reasoning that we need logic to support. The received linguistic syntactic structure should not be seen as paradigmatic for logic but rather a first attempt at coming to terms with logical reasoning. Distributed systems are commonplace in computing and engineering, yet they have been rather less so in the philosophical world. Distributed logic extends the notion of what is to be considered as logical, and yet we still rely heavily on the hard work of our predecessors in logic.

Much of the background in distributed logic owes a debt to J. Michael Dunn for his work in Gaggle Theory (Dunn 1991; Dunn and Hardegree 2001)—Gaggle Theory's notion of residuation is essentially a notion of distribution, for example. Gaggle theory can be used to relate two different algebraic systems and it is but a short step to view logics through algebraic eyes as do most algebraic logicians (of which the first author considers himself). Another precursor to distributed logics is Barwise and Seligman (1997). The colloquial term used is *channel theory* and channel theory is billed as the logic of distributed systems. We have done work in channel theory (Allwein 2005) and, indeed, spent quite a bit of time learning how its notion of distribution is used in a logical setting. The notion of a *local logic* stems from channel theory. Channel theory itself relies heavily on classical logic.

The direct precursor to distributed logic is partially ordered modalities (Allwein et al. 2010). The partial order among modalities is generally sparse in any application and is modeled via a partial order on relations. Of course, there is, at least, a complete lattice of relations on a set. However, there are few relations in most applications and, consequently, the entire complete lattice is mostly noise, i.e., most of the relations have no realistic counterpart in an application. It was our desire to generalize partially ordered modalities that led directly to distributed logics. We concentrated on modal logics because we were attempting to generalize a modal base. In subsequent work, we will modify the modal distributed logics to intensional distributed logics in an analogous sense to how relevance logic modifies modal logic.

A traditional approach to distributed systems is Markov transition systems. Here, the notion of measurement is prominent. There has been some recent work (see Doberkat 2010 and his references) showing how stochastic relations (in place of Kripke relations) can be used in measuring Kripke systems expressed as coalgebras. We only present the notion here to show how distributed modal logics are appropriate logic systems for Markov transition systems. Modal systems as coalgebras require a single local logic, and, hence, do not really provide an adequate logical framework for Markov transition systems and stochastic relations. Our work in distributed logic did not arise from stochastic relations. However, in retrospect, the match is very tight and we can view the work on stochastic relations as giving us a continuous mathematics interpretation for distributed modal logic.

A *distributed modal logic* is a collection of *local modal logics* linked together by *distributed modal connectives*, each of which takes formulas in one logic and returns formulas in a different logic. Semantically, each local logic is interpreted over a collection of worlds. Let this collection be called the *local collection* for this local logic. A *local neighborhood (nbd) map* takes each world to a set of neighborhoods taken from the local collection and is used to interpret the modal connectives of the local logic. The distributed modal connectives are also interpreted using nbd maps; here, the nbd maps take worlds from a local collection of worlds to nbds of worlds from a different local collection.

Extra properties, via logical axioms and rules, can be imposed on the interpreting nbd maps. This is precisely analogous to imposing conditions on Kripke relations or nbd maps in traditional modal logic. Many of the usual conditions (e.g., normality or functionality) can be generalized from their traditional counterparts. The selection of axioms reflects the model theory one needs for an application. If one adds enough axioms to force the distributed modal connectives to be normal modal connectives (even though they map from one logic to another), the interpreting nbd maps can be defined to be Kripke relations that, here, span between local collections.

There are other approaches to locality in logic: we have already mentioned channel theory (Barwise and Seligman 1997; Allwein 2005). Institutions (Goguen and Burstall 1985) and Chu spaces (van Benthem 2000) are others. There are also multi-agent logic systems (Fagin et al. 1995). What distinguishes distributed logics from these are that the morphisms—i.e., the nbd maps—have been lifted into the logic and hence are given properties via logical axioms and rules.

The methods of fusion, fibring and algebraic fibring (Thomason 1984; Gabbay 1999; Sernadas et al. 1999) are mainly concerned with gluing logics together while preserving the logics or constructing a minimal combination of the logics. These methods sometimes introduce new logical properties, but these are a by-product of ironing out technical details to make the methods work. They are not primarily concerned with making modal logic more expressive and certainly are very far removed from engineering applications. The motivation for distributed modal logic was to lift model theoretic notions into modal connectives and to provide a more expressive way of reasoning about distributed engineered systems. Combining logics by means of multi-graphs (Rasga et al. 2010) comes somewhat closer in that new connectives are introduced to manage importing of one logic into another. However, the goal there is to map one logic into another while the goal here is not that one modal logic is expressed within another but rather the two can be connected in such a way as to produce a new logic without necessarily any sort of embedding. One could conceive of the distributed connectives as providing an embedding but only if the underlying semantics has enough properties to force the distributed connectives to have such

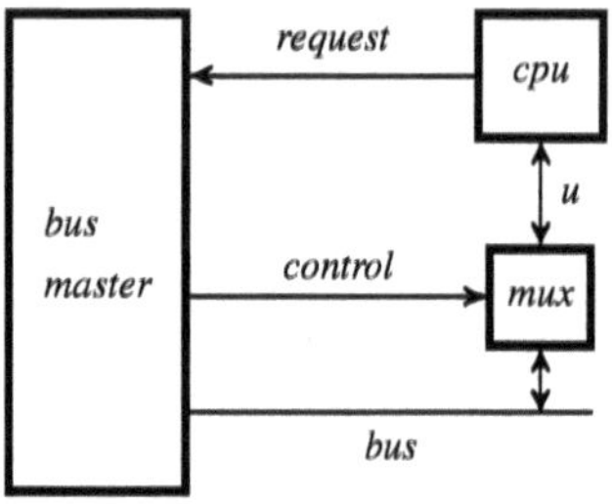

Fig. 1 The bus master example

power. Also, that embedding in distributed modal logic would be controlled or better, specified, by logic axioms.

A planned companion paper to the current paper will use Grothendieck fibrations (see Jacobs 1999 and his sources). The notion of distribution matches well with the notion of distribution in these fibrations, namely through the eyes of a total category over a base category. The stalk over an object of the base category represents a local logic. The pullback and other associated functors become distributed modal operators. Some of the conditions such as Beck-Chevalley and Frobenius are expressed as modal axioms.

An Application to Computer Security. The obvious practical question is "What are distributed logics good for?" Consider Fig. 1; this is a simplified view of an actual system.

The *cpu* issues a request to the *bus master* to read from the bus. The *mux* either connects line u to the bus or leaves it undefined as a "tri-state value," $\perp$, which will be used as a predicate in the security specification below. The control line tells the *mux* when to make the connection. The formulas are distributed logic statements that hold of the *bus master*:

$$(control = 0) \supset [c](\perp(u)), \quad (control = 1) \supset [c](bus = u)$$

Note that without the modality $[c]$, the statement $bus = u$ would mean the mux was not mediating the connection between u and bus.

This simplified view of a hardware bus system illustrates how reasoning in distributed logic supports formal verification of distributed computing systems. The *bus master* does not have access to the line u and, hence, u cannot be part of the *bus master*'s state. The two statements hold of any state in the *bus master* since the *control* line is either 0 or 1. Every state in the *bus master* is related to at least one state of the *cpu-mux* via the *control* line; this co-occurrence relation, which will be called C, is used in interpreting the (necessity) distributed modal connective $[c]$.

Let σ be a state in the *bus master*'s worlds where *control* = 0. The evaluation of the first statement is then

$$\begin{aligned} &\sigma \models_{bus\ master} (control = 0) \supset [c](\bot(u)) \\ &\therefore\ \sigma \models_{bus\ master} [c](\bot(u)) \\ &\therefore\ \text{for all } \tau \in cpu\text{-}mux\ (\mathcal{C}\sigma\tau \ \text{ implies } \tau \models_{cpu\text{-}mux} \bot(u)) \end{aligned}$$

Note how the appellation of the semantic turnstile changes from *bus master* to *cpu-mux* as the formula is evaluated.

More abstractly, some security properties of distributed systems can be expressed using these forms of logic statements. Distribution prevents the necessity of taking large cross products of states which tend to degrade the performance of model checking algorithms beyond reasonable levels. Intuitively, although space prevents us from explicating it here, distributed logic statements can be paired with a process algebra where the terms yield something like a tensor product of states of the components.

There is another use for distributed logics in testing systems. The situation frequently arises where one is tasked with producing a distributed system for a system-on-a-chip where what is known as "foreign IP (intellectual property)" must be used. While in one state of a known component, tests are made to a foreign IP component. The tests generate neighborhoods which are not neighborhoods of the state in the known component but rather neighborhoods of the foreign IP component. In sufficiently weak (undistributed) modal systems, neighborhoods need not contain the point of which they are neighborhoods. A distributed logic is merely an extreme example where the neighborhoods are not even in the same state space as the state of the known component. The situation is similar to the non-normal diagram in the next section. The worlds are the states and the $\mathcal{R}$ neighborhood map indicates tests for each state (world).

2 The Logic

A distributed logic starts with a directed graph where every node constitutes a *local logic*. Each node is a (possibly null) extension of a classical propositional logic with a set of modal connectives, and any axioms and rules to govern behavior. The graph makes apparent the structure of the collection of the local logics. Using an arc for every modal connective can get a bit "noisy" due to classical negation and defining possibility from necessity or vice versa. Instead, arcs specify semantic maps that must exist in any interpretation. Each arc is then a bit of abstract syntax which, in an interpretation, will be turned in for a nbd map.

The collection of distributed modal connectives is specified in the axioms. These axioms can be mixed and matched depending upon the properties desired for the domain of discourse being modeled. (Recall that we use the expression "domain of discourse" in a general sense of the term, not in its technical sense as it is used in first-order logic.) One should look at one's axiom set as a control panel of switches and knobs which select the properties of the underlying nbd maps or Kripke relations. The distributed structure is typically lifted from the universe of discourse and is generally small. It is certainly possible to define meta-linguistically a very large graph of local logics and distributed modal connectives. We do not do so in this paper to keep the level of abstraction to a minimum.

2.1 Conventions

The intuitive picture for models of two local logics h and k semantically connected by either a nbd map $\mathcal{R}$ or a relation $\mathcal{R}$ is in Fig. 2.

As depicted in the diagrams, the arrows labeled $\mathcal{R}$ are morphisms in a category, not functions. The $\langle r\rangle$ and $[r]$ are *forward looking* modal connectives in that their interpretation by the neighborhood map $\mathcal{R}$ looks forward along $\mathcal{R}$ from head to tail. The $\langle\cdot r\cdot\rangle$ and $[\cdot r\cdot]$ are backwards looking modal connectives. Let x be a world for h and y be a world for k, then in the first diagram, $\mathcal{H}x$, $\mathcal{R}x$, and $\mathcal{K}y$ are each a collection of neighborhoods.

One can add axioms for the distributed modal connectives to force the nbd maps to be simulation relations in the normal case and to respect a simulation condition for neighborhoods in the non-normal case. Other axioms can require that the relations be functions. Using both simulation and function axioms requires that the relations be p-morphisms, and the resulting logic is simulation logic (Allwein et al. 2014). We simplify a bit and allow the indices h and k to refer to a local logic as well as indexing the local logic's modal connectives, and we also assume there are only the modal connectives $[k]$, $\langle k\rangle$ in the logic for k and similarly for h. There are no problems adding more modal connectives and axioms and rules to govern their behavior. In particular, one can add conditions expressing the interaction between local modal connectives and distributed modal connectives. We use the simulation axiom

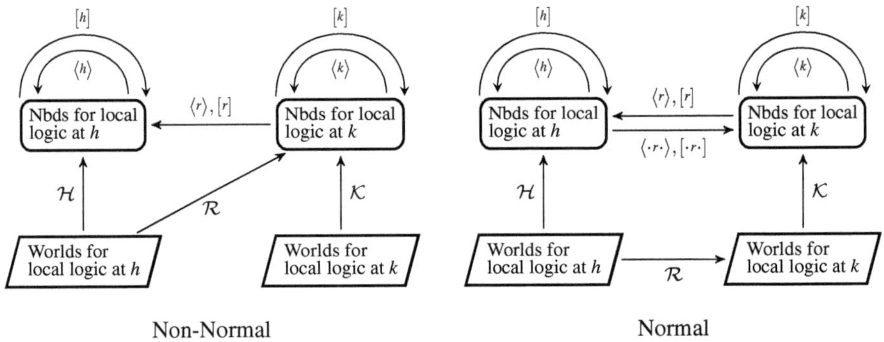

Fig. 2 Intuitive picture of distributed modal logics

(see axiom F1 below) to illustrate this. There are a wealth of choices that are driven by the particular distributed system about which a distributed logic is desired.

As mentioned previously, in sufficiently weak modal systems, it is not necessary that a point be a member of its neighborhoods. Here, it is almost a requirement or the notion of distribution is not present. Model theoretically, $\mathcal{R}$ relates two different neighborhood systems. These neighborhood maps, as morphisms, compose and there is an identity for each domain of worlds. In the normal case, the morphisms can be represented as relations with suitable modifications of the definitions.

The notation dom(r) refers to the domain or source of the arc r in a graph and cod(r) refers to the codomain or target of the arc, $r\colon \mathrm{dom}(r) \curvearrowright \mathrm{cod}(r)$. We use the locution $\langle h \rangle \in \mathrm{dom}(r)$ to refer to a modal connective in the logic associated with the node which is the source for the arc $r\colon h \curvearrowright k$. The symbol $\equiv$ is used for *bi-implication*, i.e., $P \equiv Q$ stands for $(P \supset Q) \wedge (Q \supset P)$. We use the following letter conventions:

Entity	Description
h, k, l	Nodes and endo-arcs in a graph $\mathfrak{G}$
$\langle h \rangle, [h], \langle k \rangle, [k]$	Local modal connectives at nodes h and k
r, s	Arcs in a graph $\mathfrak{G}$
$\langle r \rangle, [r], \langle s \rangle, [s]$	Forward-looking modal connectives for arcs r and s
$\langle \cdot r \cdot \rangle, [\cdot r \cdot]$	Backward-looking modal connectives for arc r
H, K, L	Sets of worlds in interpretations for logics at h, k
$\mathbb{H}, \mathbb{K}, \mathbb{L}$	Sets of sets of worlds for interpretations at h, k, and l
$\mathcal{H}, \mathcal{K}, \mathcal{L}$	Interpret modal connectives for endo-arcs at h, k
$(H, \mathcal{H}, \mathbb{H}), (K, \mathcal{K}, \mathbb{K}), (L, \mathcal{L}, \mathbb{L})$	Neighborhood frames for the logics at h, k, and l
$\mathcal{R}, \mathcal{S}$	Relations to interpret modal connectives for arcs r and s

We will assume, without loss of generality, that the modal connectives of local logic can be interpreted with a single neighborhood map. Hence, the node and its endo-arc can share the same label with use disambiguating meaning. This allows us to equate a node usually labeled h or k with the modal logic at that node.

2.2 Axioms and Rules

A local logic is "local" in that it is associated with one node in the graph. In this paper, the accompanying notion of a global logic does not entail formulas "spanning" two local logics in the sense of P in one logic implying Q in another where implying is reified as an implication connective (and similarly with other two place connectives). Each formula is entirely within a single local logic although it may contain subformulas from others.

The distributed logic graphs we use have *endo-diagrams*, each of which is a labeled node and a single endo-arc (self-arc). Each endo-arc will be translated into an endo-morphism. Each node is required to have at least one endo-diagram whose arc will be translated in an interpretation into an identity morphism. This is necessary since the models for the logic will be a category. The graph specifies which local logics there are to be, which morphisms are to appear in any model, and force identity morphisms to exist. Each local logic may have its own propositional atoms and local modal connectives. The **S** specification and **A** and **B** axioms are not optional.

Graph Specification S:

S1. A graph $\mathfrak{G}$ of nodes and arcs
A set $\mathfrak{D}$ of endo-diagrams

S2. An endo-diagram with an arc i for each node in $\mathfrak{G}$

Axiom Schemes A: For each node in $\mathfrak{G}$,

A1. all (two-valued) truth functional theorems of propositional logic

A2. Modal axioms for a logic at this node

Each node h must contain an endo-diagram for each class of modal connectives in its local logic. A class is a subset of the collection $\{\, [h], \langle h \rangle \,\}$ if the local logic is non-normal and a subset of $\{\, [h], \langle h \rangle, [\cdot h \cdot], \langle \cdot h \cdot \rangle \,\}$ if the local logic is normal.

Axiom Schemes B: These axioms force arcs to be interpreted as morphisms in a category. For arcs $r : h \curvearrowright k$ and $s : k \curvearrowright l$,

B1. $P \equiv [i]\, P$

B2. $[r]\,[s]\, P \equiv [s \circ r]\, P$

Axiom Schemes C: Taken all together these axioms would force the distributed modal connectives to be normal. Each may be optionally added.

C1. $[r]\, P \wedge [r]\, Q \supset [r](P \wedge Q)$

C2. $[r](P \wedge Q) \supset [r]\, P \wedge [r]\, Q$

C3. $\top \supset [r]\top$

The Axiom Schemes **C** should be present to specify simulation logic (Allwein et al. 2014); they also allow the specification of backward looking connectives residuated with their forward looking counterparts (see Dunn and Hardegree 2001). Simulation logic could also be built on a non-normal basis using the same main simulation axiom. However, the semantic conditions then involve neighborhoods, not relations.

Definition of Possibility: $\langle m \rangle\, P \stackrel{def}{=} \neg\, [m] \neg P, \quad m \in \{\, k, r \,\}$

Rules A: For each local logic k,

$$\frac{\vdash_k P \qquad \vdash_k P \supset Q}{\vdash_k Q} \qquad\qquad \frac{\vdash_k (P_1 \wedge \cdots \wedge P_n) \equiv P}{\vdash_k ([k]\, P_1 \wedge \cdots \wedge [k]\, P_n) \equiv [k]\, P}$$

where the appellation of $\vdash$ indicates the local logic to which the proof sign attaches.

Rule B: For each $r : h \curvearrowright k$ arc in $\mathfrak{G}$,

$$\frac{\vdash_k (P_1 \wedge \cdots \wedge P_n) \equiv P}{\vdash_h ([r]\, P_1 \wedge \cdots \wedge [r]\, P_n) \equiv [r]\, P}$$

Note in Rule **B**, the appellation of the $\vdash$ changes from premise to conclusion against the direction of arc $r : h \curvearrowright k$.

We will only be concerned with the forward versions of necessity and possibility connectives since the backwards versions are so similar and can easily be added in when necessary for a particular application. The backward versions are only present for normal systems.

2.3 Options

Axiom Schemes D: The **D** axioms are examples of extra properties to be enforced on the interpreting morphisms. Other axioms can be added as well. A further example is in Sect. 7. We use Axioms **D** as paradigm examples:

D1. $[{}^{r}]\, P \supset \langle{}^{r}\rangle\, P$ D2. $\langle{}^{r}\rangle\, P \supset [{}^{r}]\, P$

In non-normal systems, the axiom D1 specifies consistency and the axiom D2 specifies maximality, both with respect to the collection of neighborhoods about any world when the world is in the source of the nbd map used in interpreting $[{}^{r}]$ and $\langle{}^{r}\rangle$. In normal systems, the first specifies that the interpreting relation be total on its domain and the second that it act functionally.

Axiom Schemes E: The axiom E1 is only necessary if you wish the classical propositional logic at $\mathrm{dom}(r)$ to be included in the logic at $\mathrm{cod}(r)$. This condition is part of the definition of simulation in Blackburn et al. (2000) although it is not strictly necessary in that it can be removed without damaging the logic.

For all propositional letters p,

E1. $p \supset [{}^{r}]\, p$

From now on, a distributed logic contains at least the specification **S** and axiom schemes **A** and **B**, and the Definition of Possibility, and the rules **A** and **B**. Normal distributed logics include the non-normal axioms and rules as well as the axiom schemes **C**. The latter can also be added individually rather than en masse if only a subset of the properties of normality are desired. The axiom schemes **D** are of interest and we have modeling conditions for them. The axiom scheme **E** must be handled quite separately in the semantics. Other axioms can be added, we stop with the list chosen for the purposes of this presentation.

Axiom Scheme F: Simulation logic in Allwein et al. (2014) requires for an arc $r : h \curvearrowright k$ in $\mathfrak{G}$, and modal connectives $[{}^{h}] \in \mathrm{dom}(r)$, $[{}^{k}] \in \mathrm{cod}(r)$,

F1. $\langle{}^{r}\rangle\, [{}^{k}]\, P \supset [{}^{h}]\, \langle{}^{r}\rangle\, P$

In normal distributed logics, the axiom F1 forces the arcs in the graph to be interpreted as simulation relations and B2 forces composition of relations to hold, where a simulation relation is one "half" of a bisimulation in Sangiorgi (2012). One common use of the simulation relation is when the interpretation of $\langle r \rangle$ via a relation $\mathcal{R}$ is a p-morphism. To force this, add the Axiom Schemes **C** and **D** to the simulation axiom.

3 Frames and Algebras

In keeping with our simplifications, assume there is only one local modality per frame, including both a $\Box$ and $\Diamond$ since they are inter-definable. More modal connectives can be added if needed by the particular distributed system under consideration.

3.1 Frames

Definition 3.1 A *neighborhood frame* is a structure $\mathcal{H} = (H, \mathcal{H}, \mathbb{H})$ such that H is a collection of worlds. $\mathcal{H}\colon H \to \mathcal{P}\mathbb{H}$ is a nbd map taking every world of H into a collection of neighborhoods. We use the same symbol for the frame and its nbd map, and let use disambiguate what is meant. $\mathbb{H}$ is a collection of *neighborhoods* which are subsets of H and the entire collection is closed under the Boolean operations and under the operations $[h]$, $\langle h \rangle : \mathbb{H} \to \mathbb{H}$ given by:

$$[h]\, C \stackrel{def}{=} \{x \in H : C \in \mathcal{H}x\}, \qquad \langle h \rangle\, C \stackrel{def}{=} \{x \in H : \; -C \notin \mathcal{H}x\},$$

where $-C$ is the set complement of C in H.

Each node in a distributed logic's graph has a local logic associated with it. That local logic, in turn, must have a neighborhood frame associated with it.

Definition 3.2 Let $\mathcal{H}$ and $\mathcal{K}$ be neighborhood frames. A *nbd map* $\mathcal{R}\colon \mathcal{H} \to \mathcal{K}$ is a map (also using the symbol $\mathcal{R}$) $\mathcal{R}\colon H \to \mathcal{P}\mathbb{K}$ such that for any $C \in \mathbb{K}$,

$$[r]\, C \stackrel{def}{=} \{x \in H : C \in \mathcal{R}x\} \in \mathbb{H}, \qquad \langle r \rangle\, C \stackrel{def}{=} \{x \in H : \; -C \notin \mathcal{R}x\} \in \mathbb{H}.$$

Let $\mathcal{R}\colon \mathcal{H} \to \mathcal{K}$ and $\mathcal{S}\colon \mathcal{K} \to \mathcal{L}$ be morphisms. The identity morphism $I\colon \mathcal{H} \to \mathcal{H}$ and the composition $\mathcal{S} \circ \mathcal{R}\colon \mathcal{H} \to \mathcal{L}$ are defined with $(x \in H)$

$$Ix \stackrel{def}{=} \{C \in \mathbb{H} \colon x \in C\}, \qquad (\mathcal{S} \circ \mathcal{R})_x \stackrel{def}{=} \{C \in \mathbb{L} \colon \{y \colon C \in \mathcal{S}y\} \in \mathcal{R}_x\}.$$

Each arc $r\colon h \curvearrowright k$ of the graph must be associated with a *semantic morphism* in the interpretation. The semantic morphisms are *neighborhood maps* $\mathcal{R}\colon H \to \mathcal{P}\mathbb{K}$ where $\mathbb{K}$ is the collection of neighborhoods, i.e., the $\mathbb{K}$ in $(K, \mathcal{K}, \mathbb{K})$. In the normal case, the neighborhood maps can be replaced with relations. These relations are derivable in the usual way (Chellas 1980), i.e., $\mathcal{R}xy$ iff $y \in \bigcap \mathcal{R}x$; that is, take intersection of all the neighborhoods at x under $\mathcal{R}$.

Note that the definition for composition can be rewritten as

$$(\mathcal{S} \circ \mathcal{R})_x \stackrel{def}{=} \{C \in \mathbb{L}\colon [s]C \in \mathcal{R}_x\}$$

using the Definition 3.2 for $[s]C$. The definition is found in Manes (1976) for the Kleisli category of the double power set monad. Our models are always in the category of neighborhood frames.

Each node representing a distinct local logic must be mapped to a distinct frame object in any interpretation. This informal way of restricting interpretations is the result of treating the graph, $\mathfrak{G}$ as not defining everything in a distributed logic, but the alternative would make the logic impenetrable.

The corresponding Kripke frame conditions for the logical axioms are

Frame Conditions S:

FS1. A category, with underlying graph $\mathfrak{G}$, of *local neighborhood frames* and neighborhood maps

FS2. An identity morphism for each i arc in $\mathfrak{G}$

Frame Conditions A: For each node in $\mathfrak{G}$,

FA1. A set of classical worlds

FA2. Frame conditions for a local logic at this node

Frame Conditions B: For $I\colon H \to \mathbb{H}$, $\mathcal{R}\colon H \to \mathbb{K}$ and $\mathcal{S}\colon K \to \mathbb{L}$

FB1. $I_x = \{C \in \mathbb{H}\colon x \in C\}$

FB2. $(\mathcal{S} \circ \mathcal{R})_x = \{C \in \mathbb{L}\colon [s]C \in \mathcal{R}_x\}$

Frame Conditions C:

FC1. $B, C \in \mathcal{R}_x$ implies $B \cap C \in \mathcal{R}_x$

FC2. $B \in \mathcal{R}_x$ and $B \subseteq C$ implies $C \in \mathcal{R}_x$

FC3. $\top \in \mathcal{R}_x$

Frame Conditions D:

FD1. $C \in \mathcal{R}_x$ implies $-C \notin \mathcal{R}_x$

FD2. $C \notin \mathcal{R}_x$ implies $-C \in \mathcal{R}_x$

Frame Condition F:

FF1. $-\{y\colon C \in \mathcal{K}y\} \notin \mathcal{R}x$ implies $\{z\colon -C \notin \mathcal{R}z\} \in \mathcal{H}x$

with the convention that the nbd maps that use upper case script relation letters will interpret modal connectives that use the corresponding lower case Roman letters. Each distributed frame category interpreting a distributed logic will have the conditions matching the axioms. The frame conditions **S**, **A**, and **B** are always assumed, the others are required if the corresponding axioms are present in the modeled local logic.

Slightly different frames are used for the axiom E1; the local frames will contain functions to interpret constants, one for every atomic proposition of the local logic for which the local frame provides a model.

The following proposition allows for the use of one neighborhood frame per local logic.

Proposition 3.3 *There are no provable instances of formulas of the form $P \bullet Q$ for $\bullet \in \{\supset, \wedge, \vee\}$ with P in one local logic and Q in different local logic.*

The proof is an easy induction on the axiom schemes and rules. The consequence is that no formula in the logic has a binary connective between formulas in two different local logics.

Note that we stated the above proposition in terms of formula "instances" rather than formulas because it is possible to attach a local logic to more than one node in the graph. In effect, this would give more than one instance of the logic in the entire distributed logic.

Using the semantic conditions, it is easy to show that

$$x \models_{\mathcal{H}} \neg [r] \neg P \quad \text{iff} \quad x \models_{\mathcal{H}} \langle r \rangle P,$$

hence the definition of $\langle r \rangle$ in terms of $[r]$ makes sense. A distributed category model has neighborhood frames for every node with a valuation for each node. The morphisms are neighborhood maps.

Definition 3.4 A *distributed category model* is a neighborhood frame category with a valuation and a local frame for each local logic. The local frame and its valuation are called a *local model*. A valuation specifies a collection of points in the local frame where the atomic propositions are true.

3.2 Algebras

We rely on heterogeneous (multisorted) algebras from Birkhoff and Lipson (1968) for the free algebra construction. The categorical version is most easily accessible in Adámek and Rosický (1994) who attribute the multisorted (non-categorical) case to Birkhoff and Lipson (1968).

Definition 3.5 (Birkhoff and Lipson 1968) A heterogeneous algebra is a system $A = [\mathcal{L}, F]$ in which

1. $\mathcal{L} = \{ S_i \}$ is a family of non-void sets S_i of different types of elements, each called a *phylum* of the algebra A. The phyla S_i are indexed by some set I; i.e., $S_i \in \mathcal{L}$ for $i \in I$ (or are called by appropriate names).
2. $F = \{ f_\alpha \}$ is a set of finitary operations, where each f_α is a mapping

$$f_\alpha : S_{i(1,\alpha)} \times S_{i(2,\alpha)} \times \cdots \times S_{i(n(\alpha),\alpha)} \to S_{p(\alpha)}$$

for some non-negative integer $n(\alpha)$, function $i_\alpha\colon j \to i(j,\alpha)$ from $n(\alpha) = \{1, 2, \ldots, n(\alpha)\}$ to I, and $p(\alpha) \in I$. The operations f_α are indexed by some set Ω; i.e., $f_\alpha \in F$ for $\alpha \in \Omega$ (or are called by appropriate names).

Definition 3.6 A *distributed algebra* appropriate for a distributed logic is a heterogeneous algebra with a modal algebra, called a *local modal algebra*, for each node of a graph, identity modal operators for each node, and *distributed operators* $\langle r \rangle$ and $[r]$ for every arc r of the graph. For $r : h \curvearrowright k$ in the graph,

- $[r]\,[s]\,a = [s \circ r]\,a$;
- $[i]\,a = a$, for the i arc in an endo-diagram;
- if the Axiom Schemes **C** are used
  - $[r]\,a \wedge [r]\,b \leq [r](a \wedge b)$;
  - $[r](a \wedge b) \leq [r]\,a \wedge [r]\,b$;
  - $\top_{\mathbb{H}} = [r]\,\top_{\mathbb{K}}$, for $\top$ the top of a Boolean lattice;
- if Axiom Schemes **D** are used
  - $[r]\,a \leq \langle r \rangle\,a$;
  - $\langle r \rangle\,a \leq [r]\,a$;
- $\langle r \rangle\,[k]\,a \leq [h]\,\langle r \rangle\,a$, if Axiom Scheme **F** is used.

The axiom E1 will be handled in the next subsection where we must add constant operations and functions to help interpret the propositional atoms.

Appropriate distributed algebras give a "localization" view of heterogeneous algebras which is isomorphic to the definition given above. Each phylum S_i with operators defined only upon S_i is a local modal algebra. The operations associated with $r : h \curvearrowright k$ of the graph map from a local modal algebra to a local modal algebra. This stratifies the heterogeneous distributed algebra and treats every local modal algebra as an object in the surrounding distributed algebra.

Algebraic versions of soundness and completeness depend on the Lindenbaum–Tarski (LT) algebra. We must first show that the operators all respect the congruence of bi-implication induced on the local word algebras by the local logics. The only operators not already covered in previous modal algebraic work are the distributed operators.

Lemma 3.7 *The distributed operators respect bi-equivalence.*

The connective $[r]$ respects bi-equivalence because of the Rule **B**. Using Boolean negation, it is easy to show that $\langle r \rangle$ does as well.

Next, we must show that the LT algebra is actually a distributed algebra. The only operators that are at issue are the distributed operators.

Lemma 3.8 *The LT distributed operators satisfy the required properties for a distributed algebra.*

The equivalence classes for the LT algebras are defined (as usual) with $[\![P]\!] = \{Q\colon \vdash_{\mathcal{H}} P \equiv Q\}$. The operators are defined inductively, e.g., $[\![P]\!] \wedge [\![Q]\!] = [\![P \wedge Q]\!]$, $[r][\![P]\!] = [\![[r]\,P]\!]$, etc.

Corollary 3.9 *The LT heterogeneous algebra is a distributed algebra.*

Proof (Proof Outline) The free heterogeneous algebra is the usual algebra of equivalence classes of terms in the variables as generators. One runs the induction procedure to get the word algebras over all the local logics simultaneously (Birkhoff and Lipson 1968), then divide out by the equalities in each algebra. Proposition 3.3 shows that no additional sorts over and above the local modal algebra carrier sets are necessary. Lemma 3.7 shows that the replacement property for the bi-implication congruence holds for each operator. Finally, Lemma 3.8 shows each of LT operators satisfy the distributed algebra axioms. □

Theorem 3.10 *Distributed Logic is sound with respect to the algebraic and distributed frame category models.*

Proof (Proof Outline) Soundness over the algebraic models is an induction starting with a valuation into a distributed algebra and then using the fact that the LT algebra is a free algebra for the heterogeneous class of distributed algebras. From this, it is easy to see that $\supset$ interprets to $\leq$ in the algebra. The axioms of the LT algebra clearly interpret to the axioms of the logic, and the rules of the logic preserve truth in the algebra. The free heterogeneous algebras are then used to generate the universal morphism for any interpretation into a heterogeneous modal algebra thus validating the axioms and rules.

The Frame Conditions FS1, FS2, FB1 and FB2, given the work in Manes (1976) on the double power set monad restricted to neighborhoods, show that the neighborhood maps are the Kleisli morphisms and hence form a category, so the identity and associative laws of categories are met. In the presence of the normal axioms, the previous prescription for manufacturing relations from neighborhood maps shows these frame conditions ensure the maps act like Kleisli morphisms for the power set monad restricted to neighborhoods.

The rest of the axioms and rules are easily checked. □

The *canonical frame* is generated by the LT algebra; the frame's neighborhoods are the output of a representation function for the LT algebra. The representation function β is defined by

$$\beta a = \{x\colon a \in x \text{ and } x \text{ is a maximal filter}\}.$$

Let $\mathrm{MA}(h)$, $\mathrm{MA}(k)$ stand for the local modal algebras and $\mathrm{CF}(h)$, $\mathrm{CF}(k)$ stand for the canonical frames at h and k, respectively. To get a frame category from the LT modal algebra requires that one take the (dual) Stone space containing all the maximal filters of each local algebra and define the local neighborhood maps with:

$$\beta a \in \mathcal{H}x \quad \text{iff} \quad [h]\,a \in x.$$

Since $[h]$ and $\langle h \rangle$ are De Morgan duals of each other and β is a homomorphism,

$$-\beta a \notin \mathcal{H}x \text{ iff } \beta\neg a \notin \mathcal{H}x \text{ iff } [h]\neg a \notin x \text{ iff } \neg[h]\neg a \in x \text{ iff } \langle h \rangle a \in x.$$

These same definitions work for the canonical relation $\mathcal{R}$ for $r: h \frown k$ where now $a \in \mathrm{MA}(k)$, $[r]a$, $\langle r \rangle a \in \mathrm{MA}(h)$, $x \in \mathrm{CF}(h)$, and $\mathcal{R}x \subseteq \mathbb{K}$ for $\mathbb{K}$ the neighborhoods of $\mathrm{CF}(k)$.

It is not hard to show that $\beta[h]a = [h]\beta a$ and $\beta\langle h \rangle a = \langle h \rangle \beta a$. Set union, intersection, and set complement interpret the classical logic connectives $\vee$, $\wedge$, and $\neg$. The only question is the status of $\langle r \rangle$, $[r]$ for $r: h \frown k$.

Lemma 3.11 *For* $a \in \mathrm{MA}(k)$ *and* $\langle r \rangle a \in \mathrm{MA}(h)$,

$$\beta[r]a = [r]\beta a \quad \textit{and} \quad \beta\langle r \rangle a = \langle r \rangle \beta a.$$

Proof $x \in \beta[r]a$ iff $[r]a \in x$ iff $\beta a \in \mathcal{R}x$ iff $x \in [r]\beta a$. The proof for $\langle r \rangle$ is similar. □

The modal completeness argument is the usual algebraic argument (Dunn and Hardegree 2001) using contraposition and the frame argument uses the canonical frame derived from a representation theorem (Allwein and Dunn 1993; Dunn and Hardegree 2001). The modal representation theorem represents a modal algebra as an algebra of sets using the canonical frame (Stone space) of the algebra. One defines the 1–1 homomorphism β on the distributed algebra for each carrier set and the operations using the above prescriptions.

Theorem 3.12 *Distributed Logic is complete with respect to the distributed algebras and the distributed category models.*

Proof From Proposition 3.3, we need only concern ourselves with formula (instances) which sit entirely within a single local logic. So one presents the formula instance at issue and then picks the local logic for which it must be determined whether it is a theorem. The argument is a contraposition argument using the LT heterogeneous algebra and its canonical frame category.

Note that any theorem without an implication as the main connective can be outfitted with one because $\vdash P$ iff $\vdash T \supset P$ where T is the truth constant in a local logic. Hence we need only to check implications. Suppose $\nvdash P \supset Q$, then $[\![P]\!] \nleq [\![Q]\!]$ in the LT algebra where $[\![P]\!]$, $[\![Q]\!]$ are the bi-implicational equivalence classes. This along with Corollary 3.9 is enough for algebraic completeness.

For frame completeness, there is a maximal separating filter x such that $[\![P]\!] \in x$ and $[\![Q]\!] \notin x$, i.e., $x \in \beta[\![P]\!]$ and $x \notin \beta[\![Q]\!]$, so $x \models P$ and $x \not\models Q$. Therefore there is a local model falsifying the non-theorem, and hence a distributed category model falsifying the non-theorem.

Taking the contrapositive in the algebraic and frame cases yields the required result. □

3.3 The Axiom Schemes E

The Axiom E1 requires some special treatment. The algebra will now have a collection of constant operators, one for each propositional atom in the language. The axiom does not merely necessitate a finite collection of operators but rather requires one for each propositional atom. This is in contradistinction to the usual prescription that a Hilbert-style axiom imposes on an algebraic interpretation. The usual prescription imposes a finite collection of operators and a finite number of properties.

Definition 3.13 An *E local modal algebra* is a local modal algebra with a collection of (local) constant operations. In symbols, if p is a propositional atom, then its constant, nullary operation, σ_p, is such that $\sigma_p = p$ in the word algebra of the logic and $\sigma_p = [\![p]\!]$ in the LT algebra. In addition to any axioms necessary for the local modal logic, we add the axiom

$$\sigma_p \leq [r]\,\sigma_p$$

for an arc r in the diagram to another node. We also require the logic at cod(r) to contain at least the same propositional atoms as those at dom(r).

Note that two constant operations, being functions, can point to the same element of the local modal algebra. The Lindenbaum–Tarski E local modal algebra has each constant operation pointing out the equivalence class of the propositional atom to which it is attached.

Definition 3.14 An *E neighborhood frame* is a neighborhood frame with a collection of constant functions, f_p, one for each propositional atom. A constant function selects an element of the set algebra, i.e., a local neighborhood.

Fix a distributed algebra with any desired E local modal algebras. Modal valuations vary over what gets assigned to the propositional atoms. Here, the valuations must be consistent with the nullary operations associated with each atom. We get the variation necessary for valuations by choosing different algebras which agree on everything except the nullary operations. So the variation gets satisfied at a slightly higher level. A similar statement holds for E neighborhood frames. The inductive definition generating interpretations from valuations remains the same and hence the restriction on valuations gets transferred to interpretations.

Definition 3.15 An *E local algebra valuation*, $[\![-]\!]$, must take every propositional atom to an element of the carrier set pointed to by the nullary operation for that atom, i.e., if $\sigma_p = a$, then $[\![p]\!] = a$. Similarly, for an E local neighborhood frame and valuations $[\![-]\!]$, we demand $[\![p]\!] = C$ if $f_p = C$. Also, we demand that for $r : h \curvearrowright k$, the r interpreting relation $\mathcal{R}$ must respect the constant functions in the sense that $x \in f_p$ at the neighborhood frame for h and $f_p \in \mathcal{R}x$ at the neighborhood frame for k.

The axiom $\sigma_p \leq [r]\,\sigma_p$ effectively forces $[\![p]\!] \leq [r][\![p]\!]$ for any interpretation $[\![-]\!]$. For the LT algebra, $\sigma_p = p$ in the word algebra forces $\sigma_p = [\![p]\!]$ in the LT algebra. The result is that we get the same LT algebra as we would have without the nullary constants. The universal property of the free algebra with respect to unique maps to the other E local modal algebras are unaffected since the restriction on interpretations will force the unique maps to choose the same elements of the algebras to which the nullary operations point for the respective propositional atoms. In the freeness diagram below, p indicates some propositional atom in the language, FA_h is the carrier set of the local modal logic for h inside of the free algebra $\mathcal{A}$. The algebra $\mathcal{B}$ is some other appropriate distributed algebra, and γ is an induced interpretation from the freeness property of $\mathcal{A}$, $U\mathcal{A}$ is the forgetful functor U (from algebras to sets) applied to the algebra $\mathcal{A}$ and returns the carrier sets (or sorts) of $\mathcal{A}$, and similarly for $U\mathcal{B}$. Ug is the underlying set function of the homomorphism g. SL is the set of atoms of the source language, and η maps (injects) them into the proper elements of the object $U\mathcal{A}$.

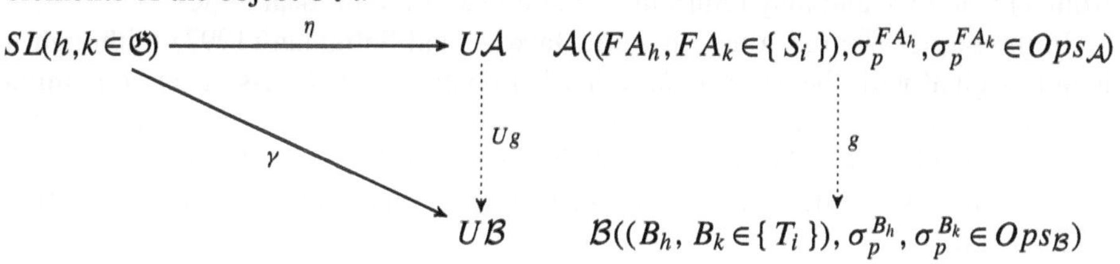

The algebra $\mathcal{B}$ has no notion of propositional atoms. The σ_p, being operations, are preserved by g. Hence, $\eta(p) = \sigma_p^{FA_h}$ and $g(\eta(p)) = g(\sigma_p^{FA_h}) = \sigma_p^{FB_h}$. Since the diagram commutes, $\gamma(p) = \sigma_p^{FB_h}$.

The extension to distributed algebras and distributed category models are called **E** distributed algebras and **E** distributed category models.

Theorem 3.16 *Distributed logics with the* **E** *axioms are sound and complete with respect to* **E** *distributed algebras and* **E** *distributed category models.*

4 Cheap Entailment Arrows

We use the word "cheap" because we embed the entailment arrow into a distributed modal logic. This is in contradistinction to treating the entailment arrows as more like relevance logic's entailment. In later work, we will show how to distribute relevance logic's entailment arrow; for now we stick to modal logic due to its simplicity and that its models are relatively free of auxiliary partial orders such as is necessary for relevance logic.

Gödel's embedding of intuitionistic entailment into S4 leaves open the possibility that an intuitionistic entailment might have a distributed counterpart. Barwise and Seligman (1993) used a similar encoding but into S5 for an entailment. In their setup, worlds are essentially three-valued; they use a partial function from a world and a proposition into $\{+, -\}$. S4 and S5 suffer from the problem that the Kripke relations used in an interpretation are required to be reflexive and transitive which would cause too much of the distributed structure of the interpreting category to collapse. However, one can still use the general idea of modal embedding.

We take the property of residuation to be the primary feature of an entailment arrow. The residuation partner of an entailment we will term (following Dunn) *fusion*, which is an intensional conjunction. The care and feeding of fusion dictates the distributed nature of the entailments.

In the sequel, we sometimes use superscripts over connectives to indicate to which logic they belong, i.e., $\overset{h}{\supset}$ is classical implication for the local logic at node h in the graph of a distributed logic. The general rule of thumb is that classical logical connectives take their arguments and return their result entirely within a single logic whereas distributed intensional and modal connectives each may take an argument from a local logic and may return their result in a different local logic.

In the sequel, we use a notion from Barwise and Seligman (1997) (although it is not original with them) of a *classification* containing two sets, a set containing the logic *over* a set containing the models or worlds. They are connected with a satisfaction relation $\models$. We use the terminology *classification of h* to refer to the classification at a node h in a distributed logic's graph and $\models_h$ is the satisfaction relation at h.

4.1 Gödel Entailment and Fusion

Gödel entailment and fusion connectives arise from insisting that the interpretation of the intuitionistic entailment retain the form of its first order logic evaluation, although dropping the restrictions of S4 on the interpreting relation. Referring to Fig. 3, we use $\Rightarrow$ and $\Leftarrow$ for the entailment connectives. Later it will turn out these are identical, but for now we will leave them as separate connectives. The $\circledcirc$ and $\circledcirc'$ connectives are

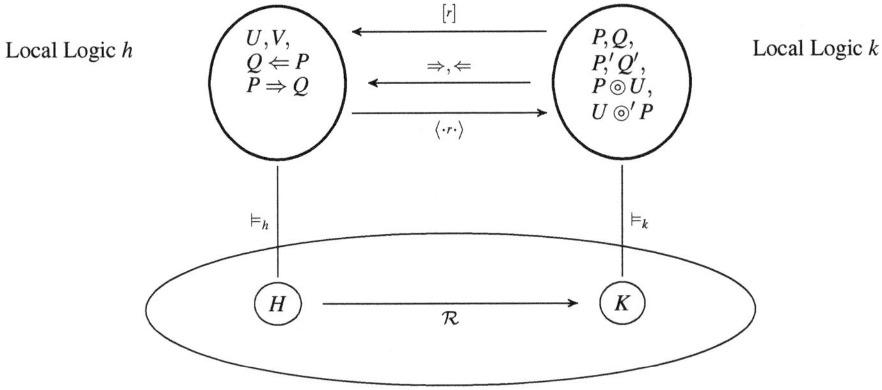

Fig. 3 Gödel entailment and fusion

distributed intensional conjunctions. The following diagram shows the logic (top) layer for two logics h and k and the model (bottom) layer of their respective sets of worlds H and K. The arrows between the logics are logical connectives. As before, $[r]$ is the forward looking necessity interpreted with the relation $\mathcal{R}$ and $\langle \cdot r \cdot \rangle$ is the backwards looking possibility connective interpreted with $\mathcal{R}$.

The evaluation condition for $\circledcirc$ is

$$y \models_k P \circledcirc U \quad \text{iff} \quad \exists x\, (\mathcal{R}xy \text{ and } x \models_h U \text{ and } y \models_k P).$$

The evaluation condition for $\Rightarrow$ is

$$x \models_h P \Rightarrow Q \quad \text{iff} \quad \forall y\, (\mathcal{R}xy \text{ and } y \models_k P \text{ implies } y \models_k Q).$$

Note that these modeling conditions are precisely the modeling conditions for the coding

$$P \Rightarrow Q \stackrel{def}{=} [r](P \supset Q).$$

and the conditions for $\circledcirc$ generate the encoding

$$P \circledcirc U \stackrel{def}{=} P \wedge \langle \cdot r \cdot \rangle\, U.$$

Theorem 4.1 *The two residuation axioms*

$$U \stackrel{h}{\supset} \big(P \Rightarrow (P \circledcirc U)\big), \qquad \big(P \circledcirc (P \Rightarrow Q)\big) \stackrel{k}{\supset} Q,$$

which dictate into which logics the formulas fall, are valid.

One adds the following monotonicity rules:

$$\frac{U \vdash_h V \qquad P \vdash_k Q}{P \circledcirc U \vdash_k Q \circledcirc V}\ \circledcirc\textit{-monotonicty} \qquad\qquad \frac{P \vdash_k Q \qquad P' \vdash_k Q'}{Q \Rightarrow P' \vdash_h P \Rightarrow Q'}\ \Rightarrow\textit{-monotonicty}$$

From the axioms and rules, the usual (bidirectional) form of residuation is derivable:

$$\frac{U \stackrel{h}{\supset} (P \Rightarrow Q)}{(P \circledcirc U) \stackrel{k}{\supset} Q}\ \textit{residuation}$$

There is a second fusion connective with the evaluation condition

$$y \models_k U \circledcirc' P \quad \text{iff} \quad \exists x\, (\mathcal{R}xy \text{ and } x \models_h U \text{ and } y \models_k P),$$

One might conjecture there is a second entailment $\Leftarrow$ connective with the evaluation condition

$$x \models_h Q \Leftarrow P \quad \text{iff} \quad \forall y\,(\mathcal{R}xy \text{ and } y \models_k P \text{ implies } y \models_k Q).$$

However, $\Leftarrow$ is the same as $\Rightarrow$ simply because there is no additional freedom to alter it given the Gödel evaluation form we are following and where the formulas must fall in the distribution. This is not true of the $\circledcirc'$ connective because one of the subformulas is evaluated using the same position in the relation as the result of the connective.

The $\Leftarrow$ and $\circledcirc'$ have the following two residuation properties

$$U \overset{h}{\supset} ((U \circledcirc' P) \Leftarrow P), \qquad ((Q \Leftarrow P) \circledcirc' P) \overset{k}{\supset} Q,$$

which again dictate into which logics the formulas fall. Residuation in the following form holds:

$$\frac{U \overset{h}{\supset} (Q \Leftarrow P)}{(U \circledcirc' P) \overset{k}{\supset} Q}\ \textit{residuation}$$

It is then easy to show that $P \circledcirc U \equiv U \circledcirc' P$.

If one were to treat the two entailment and two fusion connectives as native rather than defined, then one needs to add the axioms

$$(P \circledcirc U) \overset{k}{\supset} (U \circledcirc' P) \qquad (U \circledcirc' P) \overset{k}{\supset} (P \circledcirc U)$$

and

$$(P \Rightarrow Q) \overset{h}{\supset} (Q \Leftarrow P) \qquad (Q \Leftarrow P) \overset{h}{\supset} (P \Rightarrow Q).$$

Syntactically, $\Rightarrow$ and $\Leftarrow$ are the same connective because the arguments come from the same local logic. $\circledcirc$ and $\circledcirc'$ cannot be the same because the arguments come from different local logics.

It is necessary to have an intensional conjunction, $\circledcirc$, rather than an extensional conjunction, $\wedge$, because in the statement of residuation, (rewritten) the lower premise $P \wedge U \overset{k}{\supset} Q$ would require that P and U be in the same classification for $\wedge$ to make sense and also that Q be in that same classification for $\overset{k}{\supset}$ to make sense.

4.2 Simple Entailment and Fusion

A simple version of relevant-like entailment is definable. Let there be the set up in Fig. 4.

It turns out that there is no distinction between $\rightarrow$ and $\leftarrow$ mainly because $\circ$ must be commutative (Fig. 4). This latter is so because the evaluation condition is

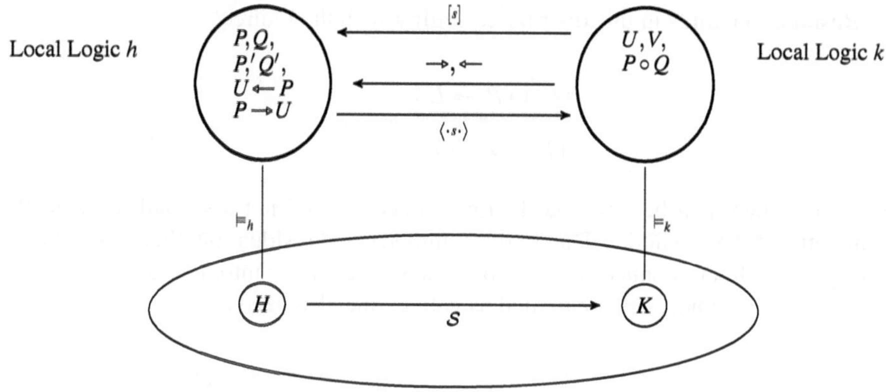

Fig. 4 Simple entailment and fusion

$$y \models_k P \circ Q \quad \text{iff} \quad \exists x\, (\mathcal{S}xy \text{ and } x \models_h P \text{ and } x \models_h Q).$$

The $\twoheadrightarrow$ connective has the following evaluation condition:

$$x \models_h P \twoheadrightarrow U \quad \text{iff} \quad \forall y\, (\mathcal{S}xy \text{ and } x \models_h P \text{ implies } y \models_k U).$$

Since x is a free variable and $x \models_h P$ does not rely on the $\forall y$ quantifier, we can rewrite this as

$$x \models_h P \twoheadrightarrow U \quad \text{iff} \quad x \models_h P \text{ implies } \forall y\, (\mathcal{S}xy \text{ implies } y \models_k U).$$

The two evaluation conditions show the following definitions can be made

$$P \twoheadrightarrow U \stackrel{def}{=} P \stackrel{h}{\supset} [s]\, U, \qquad P \circ Q \stackrel{def}{=} \langle \cdot s \cdot \rangle (P \wedge Q).$$

The following monotonicity rules must be added:

$$\frac{P \vdash_h P' \qquad Q \vdash_h Q'}{P \circ P' \vdash_k Q \circ Q'}\ \circ\text{-monotonicty} \qquad\qquad \frac{P \vdash_h Q \qquad U \vdash_k V}{Q \twoheadrightarrow U \vdash_h P \twoheadrightarrow V}\ \twoheadrightarrow\text{-monotonicty}$$

Theorem 4.2 *The two residuation properties*

$$Q \stackrel{h}{\supset} (P \twoheadrightarrow (P \circ Q)), \qquad (P \circ (P \twoheadrightarrow U)) \stackrel{k}{\supset} U,$$

which dictate into which logics the formulas fall, are valid.

Residuation holds in this distributed setting with the bidirectional rule:

$$\frac{Q \overset{h}{\supset} (P \twoheadrightarrow U)}{(P \circ Q) \overset{k}{\supset} U}\ \textit{residuation}$$

It is clear there can be only one fusion connective and not a second $\circ'$ given the symmetry of the semantics. Due to the symmetry and residuation, there can be only a single entailment connective and so $\twoheadrightarrow$ and $\twoheadleftarrow$ collapse into a single entailment. Also, if these connectives are treated as native, one also needs

$$(P \circ Q) \overset{k}{\supset} (Q \circ' P) \qquad (Q \circ' P) \overset{k}{\supset} (P \circ Q)$$

and

$$(P \twoheadrightarrow U) \overset{h}{\supset} (U \twoheadleftarrow P) \qquad (U \twoheadleftarrow P) \overset{h}{\supset} (P \twoheadrightarrow U)$$

since the semantics will identify the respective connectives.

4.3 Preservation Conditions

We work first with Gödel Entailment. Assume the confluent diagram of Fig. 5 appropriating the categorical notation of Freyd and Scedrov (1990) (although the diagram is not category theoretic) where one reads from left to right the Simulation Property,

for all x, y, u such that $\mathcal{R}xu$ and $\mathcal{H}xy$, there exists a z such that $\mathcal{K}uz$ and $\mathcal{R}yz$.

The relation $\mathcal{R}$ is the Simulation Relation. The schema G in Chellas (1980) for $i = j = m = n = 1$ is also known as the Geach axiom (left):

$$\Diamond^i \Box^j P \supset \Box^m \Diamond^n P, \qquad \langle r \rangle\, [k]\, P \supset [h]\, \langle r \rangle\, P.$$

where the i, j, m, n refer to repetitions of their respective connectives, i.e., $\Diamond^i$ stand for i instances of $\Diamond$ concatenated together, and so on for the rest. From Simulation Logic in Allwein et al. (2014), the condition is the first-order Simulation Property for modal logics. The axiom on the right is the Simulation Axiom. The Simulation Property validates the axiom. The modal connectives $\langle r \rangle$, $[k]$, and $[h]$ are interpreted by the relations $\mathcal{R}$, $\mathcal{K}$, and $\mathcal{H}$ respectively in the Simulation Property. Again, we allow the indices h and k to refer to a local logic as well as indexing the local logic's modal connectives, and we also assume there are only the modal connectives $[k]$, $\langle k \rangle$ in the logic at k and similarly at h. The intuitive picture of two local logics h and k semantically connected by a simulation $\mathcal{R}$ is Fig. 6.

The two entailments, $\overset{h}{\Longrightarrow}$ and $\overset{k}{\Longrightarrow}$ are similar to the Gödel entailments of the previous section but result in formulas in the same local logics as their arguments. The labels are meant as a reminder of this fact and that they are not to be confused with the previous distributed Gödel entailments.

The confluent condition underwrites the Gödel entailment preservation in the form of the following axiom replacing the Simulation Logic axiom:

$$\langle r \rangle (P \overset{k}{\Longrightarrow} Q) \overset{h}{\supset} \big([r]\, P \overset{h}{\Longrightarrow} \langle r \rangle\, Q \big)$$

The following is a derived rule by virtue of residuation and the fact that the $\langle r \rangle$ connective is monotone:

$$\frac{\langle r \rangle (P \overset{k}{\Longrightarrow} Q) \overset{h}{\supset} \big([r]\, P \overset{h}{\Longrightarrow} \langle r \rangle\, Q \big)}{\big([r]\, U \overset{h}{\circledcirc} \langle r \rangle\, V \big) \overset{h}{\supset} \langle r \rangle (U \overset{k}{\circledcirc} V)}$$

where $\overset{h}{\circledcirc}$ and $\overset{k}{\circledcirc}$ are similar to the previous Gödel fusion connective $\circledcirc$ except restricted to a single local logic.

One can also use residuation to take the conclusion of this derived rule as an axiom and derive the premise.

Now for the simple entailment preservation of $\overset{h}{\multimap}$. In the Freyd–Scedrov diagram of Fig. 7, $\mathcal{H}^{\smile}$ and $\mathcal{K}^{\smile}$ refer to the converse of $\mathcal{H}$ and $\mathcal{K}$, and are used because $\circ$ is a backwards looking connective. This condition validates

$$\langle \cdot h \cdot \rangle\, [r]\, P \leq [r]\, \langle \cdot k \cdot \rangle\, P.$$

The confluent condition also underwrites the simple fusion preservation in the form of the following axiom:

$$\big([r]\, P \overset{h}{\circ} [r]\, Q \big) \overset{h}{\supset} [r](P \overset{k}{\circ} Q)$$

The following is a derived rule by virtue of residuation and the fact that the $[r]$ connective is monotone:

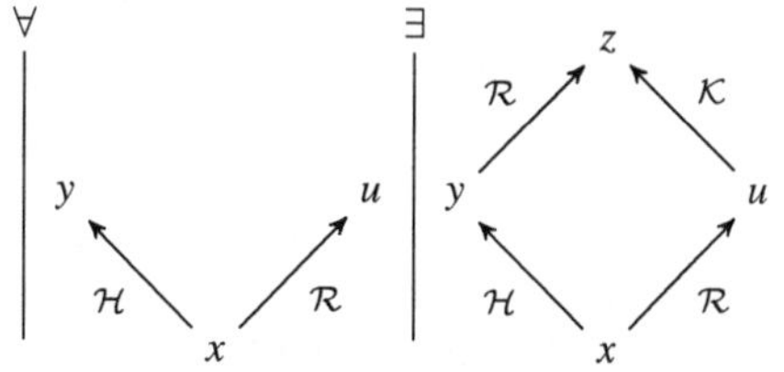

Fig. 5 Simulation property for Gödel entailment

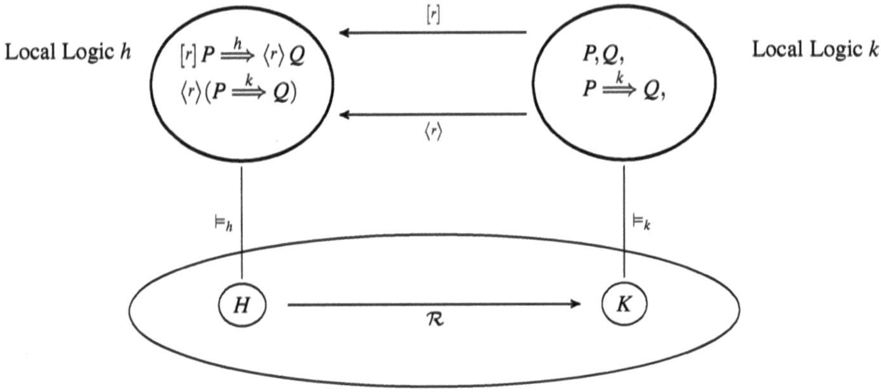

Fig. 6 Preserving Gödel entailment

Fig. 7 Simulation property for simple entailment

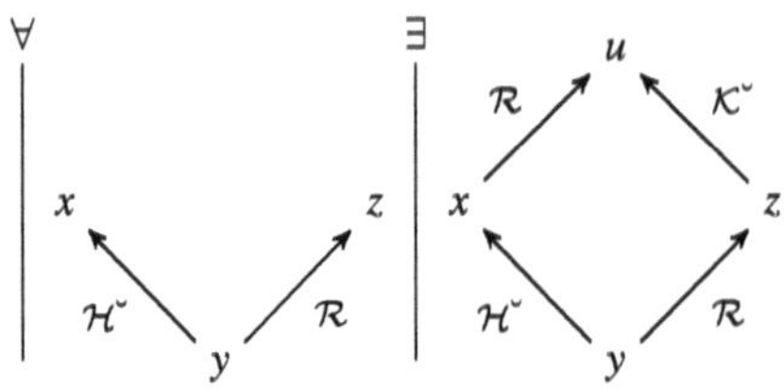

$$\frac{([r]\,P \overset{h}{\circ} [r]\,Q) \overset{h}{\supset} [r](P \overset{k}{\circ} Q)}{[r](P \overset{k}{\longrightarrow} Q) \overset{h}{\supset} ([r]\,P \overset{h}{\longrightarrow} [r]\,Q)}$$

As before, one can also use residuation to take the conclusion of this derived rule as an axiom and derive the premise.

5 Noninterference as a Simulation

High level security properties are generally expressed informally using distributed notions. However, when coerced into formal models, they frequently lose their distribution and the notion of a cross-product of system states replaces the notion of distribution. This has the effect of making the analysis complicated because then the distribution structure has to be disentangled from the combined system. Distributed logic cuts through the encode-the-distribution and subsequent decode-the-distribution steps. Noninterference (Goguen and Meseguer 1982) is a security property that is frequently imposed on state-based systems that process a combination of high security and low security data. Note that we refer to data here and not information. Information requires a more sophisticated typing scheme beyond mere values. To go one step further, knowledge requires a certain relationship between an agent and information. So we are concerned here with the most basic form of processing. In this section, we show how to view noninterference in a distributed setting.

A *covert channel* is a stream of data that rides atop an actual stream of data. An example is a message where the number of "*a*"s relative to the number of "*e*"s in a sentence could indicate a 0 or 1. Taken over the entire message, the covert channel is a sequence of 0s and 1s while the actual channel is the contents of the message. One cannot simply use the overt flow of data from high to low using Shannon communication theory. All the covert channels must be taken into account. It is hard to account for all of them because there is no way to predict their existence. Also, it is possible through the measurement process to pass as much data as you like through a channel of capacity zero (Moskowitz et al. 2002). This occurs because capacity is defined via a mathematical limiting process.

A high security process (one that processes high security data) influences or interferes with a low security process just when the behavior of the low security process changes in the presence of the high security process. The high and low processes are run together and compared against running the low process with high's output deleted. The reason for running the high and low processes together for the comparison is to account for any covert channels from high to low that are not accounted for merely by looking at high's output. The output will contain the overt channel's data as well as any covert channels residing therein. We need to look at low's output while running with high and then again while running low by itself. Noninterference is then expressed using a particular simulation relation which formulates what low can see in terms of the deletion of high's outputs. So in effect, we are choosing a canonical simulation relation for the combined high-low system; this is a simulation relation intimately tied to noninterference.

Let there be a High and Low system k which is the combination $h + l$ of the high and low system running together as a single system and a Low system l. If h noninterferes with l then there is a simulation relation so that every move that k can make is simulated by l. However, this is not enough. It could be that l is simply reading h's output and that would be interfering. So it is not enough to claim that noninterference means the existence of a simulation relation. The simulation relation must have some other properties.

Goguen and Meseguer (1982) state that noninterference exists when a particular equation holds. We let v be a user, and b be an executable instruction. The pair is denoted by (b, v), and is called a command. A finite sequence w of commands may be issued by both High and Low users. Assuming a start state, $[[w]]$ represents executing the commands in w. This results in a state that contains both High and Low data. Let A be a collection of commands. A function $S_{G,A}$ strips out or purges commands from A in w which contain a user in G. Stripping is denoted by $S_{G,A}(w)$. We will let G be the collection of High users. Let *out* be an output function, then

$$out\Big([[w]], v, b\Big)$$

is read "the output of the command b executed by user v after commands in w are run." Similarly,

$$out\Big([[S_{G,A}(w)]], v, b\Big)$$

is read "the output of the command b executed by user v after the commands in the stripped w are run."

Let v be a Low user, b a command that can be issued by v, and A a collection of commands which we deem to be security critical. High *noninterferes* with Low just when the following equation holds for all w, v, and b:

$$out\Big([[w]], v, b\Big) = out\Big([[S_{G,A}(w)]], v, b\Big).$$

5.1 Derivation of the Simulation Relation

Let $\mathcal{K}$ be the next state relation on the combination High and Low system k and $\mathcal{L}$ be the next state relation on the low system l. The next state relation is a collection of state pairs derived from all the commands a particular system can produce. Since we are abstracting over all commands, it is sufficient for the effect of a set of commands to be represented by a relation, i.e., we do not care about any one particular command.

The simulation axiom is

$$\langle r\rangle\,[l]\,Q \supset [k]\,\langle r\rangle\,Q$$

for all properties Q of l. (Note the $[k]$ is now in a different position than the previous use of this axiom due to nomenclature issues.)

Satisfaction in k for the left hand side is

$$x \models_k \langle r\rangle\,[l]\,Q \quad \text{iff} \quad \exists y \in l\,(\mathcal{R}xy \text{ and } \forall z \in l\,(\mathcal{L}yz \text{ implies } z \models_l Q))$$

and for the right hand side is

$$x \models_k [k]\,\langle r\rangle\,Q \quad \text{iff} \quad \forall u \in k\,(\mathcal{K}xu \text{ implies } \exists z \in l\,(\mathcal{R}uz \text{ and } z \models_l Q)).$$

We will use $\mathcal{R}$ to stand for stripping out High's state data since it relates k states with l states. If x is a state in k, then it must contain both High and Low data. y is a state in l and hence can contain only Low's data.

Now we attempt to rewrite Goguen and Meseguer's definition of noninterference. Recall

$$out\Big([[w]], v, b\Big) = out\Big([[S_{G,A}(w)]], v, b\Big),$$

Here, v is more or less a useless parameter, we assume the user is some v. So we can drop v. Every command must have start and stop states. Note that $[[w]]$ relates start and stop states. Hence we can rewrite with $[[w]]$ standing for a relation in infix notation:

$$x\ [[w]]\ u \text{ and } out(b(u)) \quad \text{iff} \quad y\ [[S_{G,A}(w)]]\ z \text{ and } out(b(z)).$$

where b reads the state and *out* produces some output value. The left side involves the states x and u of k and the right side involves the states y and z of l. So we will need to restructure the statement.

To restructure the statement, note that we are only interested in properties of l. The expression $out(b(u))$ can be rewritten in terms of a property of l if we relate u to some state z' in the l system. A stripper function will do precisely what we want so we use the relation $\mathcal{R}$ to represent stripping a k state of its high component and returning its low component. Hence, we rewrite the statement to be

$$x\ [[w]]\ u \text{ and } \mathcal{R}uz' \text{ and } out(b(z')) \quad \text{iff} \quad y\ [[S_{G,A}(w)]]\ z \text{ and } out(b(z)).$$

The $out(b(z'))$ expression is awkward and represents a property evaluated at the state z'. The fact that the read instruction b is used is merely parametric at this point. We abstract the output value into some predicate, Q, and assume the predicate is being held true or false of an l state. Hence $out(b(z'))$ will be represented as $Q(z')$. Rewriting, we get

$$x\ [[w]]\ u \text{ and } \mathcal{R}uz' \text{ and } Q(z') \quad \text{iff} \quad y\ [[S_{G,A}(w)]]\ z \text{ and } Q(z).$$

Both computations should start out in the same state, and they would were they both in the same state space. However, they are not, and we use $\mathcal{R}$ again to strip out h's data from x:

$$x\ [[w]]\ u \text{ and } \mathcal{R}uz' \text{ and } Q(z') \quad \text{iff} \quad \mathcal{R}xy \text{ and } y\ [[S_{G,A}(w)]]\ z \text{ and } Q(z).$$

All the commands in k together can be abstracted into a next state relation $\mathcal{K}$ and similarly for l using $\mathcal{L}$ (note, $S_{G,A}$ need not be a surjective function from k command sequences to l command sequences):

$$\mathcal{K}xu \text{ and } \mathcal{R}uz' \text{ and } Q(z') \quad \text{iff} \quad \mathcal{R}xy \text{ and } \mathcal{L}yz \text{ and } Q(z).$$

If $z \neq z'$ then l would be interfered with, so we make them equal:

$$\mathcal{K}xu \text{ and } \mathcal{R}uz \text{ and } Q(z) \quad \text{iff} \quad \mathcal{R}xy \text{ and } \mathcal{L}yz \text{ and } Q(z).$$

Since $Q(z)$ is parametric and appears on both sides of the iff, we can remove it:

$$\mathcal{K}xu \text{ and } \mathcal{R}uz \quad \text{iff} \quad \mathcal{R}xy \text{ and } \mathcal{L}yz.$$

This statement expresses the confluent diagram of Fig. 8

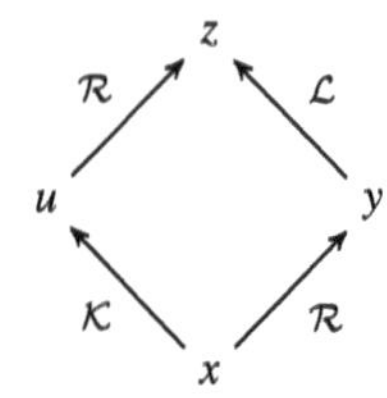

Fig. 8 Noninterference confluent diagram

We have ignored quantification over states. Certainly the confluent diagram does not hold for any u, x, y, and z, only the states in the proper relations. u is restricted by $\mathcal{K}$ and y by $\mathcal{R}$. z is determined by $\mathcal{L}$ and $\mathcal{R}$. The first is of the form "for any x and u such that $\mathcal{K}xu$," and the latter by "for any x and y such that $\mathcal{R}xy$," even if $\mathcal{R}$ appears as a stripper function. z is determined by an existential which is hidden by using a function, namely being a state determined by a stripper function but satisfying $\mathcal{L}yz$.

Now we must motivate the reformulation in terms of a simulation. We need only be concerned with whether k can make a move that l possibly cannot mimic. There are two parts, k must make a move recorded by $\mathcal{K}xu$ and the beginning state for the move, x, must be strippable into a state y of l. If k can make no such move or the beginning state is not strippable into a state for l, then we have "do not care" situation. The equality is really masking the fact that the statement should read

> Were k to execute the sequence of commands w, then l could not detect the difference from executing the sequence of w with k commands stripped out.

This situation is neatly handled by using a conditional

$$\mathcal{K}xu \text{ and } \mathcal{R}xy \text{ implies } \exists z\, (\mathcal{L}yz \text{ and } \mathcal{R}uz).$$

which is the Simulation Property. Now we can bring back the parametric $Q(z)$. Since z is a world, and the worlds are being abstracted away in modal logic, what is left over is just the proposition Q. Here there is a choice. In order to be a logic, Q should be a metalinguistic variable and range over all propositions. However, there is no need to go this far, Q can be restricted to range only over propositions that are critical to a particular implementation. Either way, the axiom

$$\langle r\rangle\, [l]\, Q \supset [k]\, \langle r\rangle\, Q$$

is certainly apropos in the analysis since the Simulation Property is necessary to validate this axiom. The situation can be diagrammed as illustrated in Fig. 9.

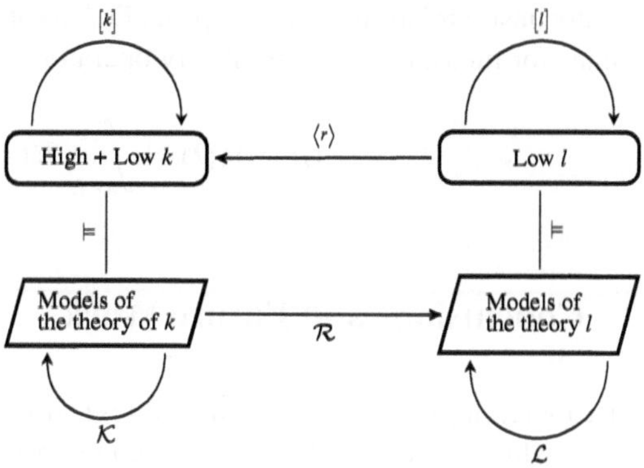

Fig. 9 Noninterference simulation

6 Stochastic Relations

Distributed logics lend themselves to being measured. This is important because many security properties will never hold either entirely or not at all of any particular system. By measuring the system over which noninterference is desired, the goal is to measure the security via noninterference.

The relation $\mathcal{R}$ as a set-valued map returns a set of all the points y to which an x is related. If instead there is some ambiguity about whether y is in $\mathcal{R}x$, then the notion of map must be relaxed much in the same way as that presented with neighborhood maps. The prescription is then

$$\mathcal{R}(x)(Q) \in [0, 1]$$

where Q is a neighborhood at k and $[0, 1]$ is the continuum from 0 to 1.

From Doberkat (2010), the following, with nomenclature changes to match this paper, defines stochastic relations. The models are now promoted to measurable spaces. In particular, the clopen sets of the Stone topology are now measurable and that topology is promoted to a σ-algebra. $\mathcal{H} = (H, \mathbb{H})$ and $\mathcal{K} = (K, \mathbb{K})$ are now measurable spaces and $\mathbb{H}$ and $\mathbb{K}$ are promoted to measurable relations as below:

Definition 6.1 A *stochastic relation* $\mathcal{R}: \mathcal{H} \to \mathcal{K}$ is a measurable map $\mathcal{H} \to \mathcal{G}(\mathcal{K})$ where $\mathcal{G}(\mathcal{K})$ is the collection of subprobability measures and the initial σ-algebra on the subprobability measures.

The following proposition from Doberkat (2010) characterizes stochastic relations:

Proposition 6.2 *Given measurable spaces $\mathcal{H}$ and $\mathcal{K}$, the following are equivalent:*

(i) *$\mathcal{R}: \mathcal{H} \to \mathcal{K}$ is a stochastic relation.*
(ii) *$\mathcal{R}(x)$ is a subprobability measure on $\mathcal{K}$ for each $x \in H$ such that the map $x \mapsto \mathcal{R}(x)(Q)$ is $\mathcal{H}$ measurable for each measurable set $Q \in \mathbb{K}$.*

Stochastic relations also compose, Doberkat (2010). Stochastic relations give the means for measuring the possibility operator $\langle r \rangle$ as

$$(\mu \langle r \rangle)(Q) = \int_{x \in H} \mathcal{R}(x)(Q) d\mu(x).$$

7 Conclusions and Future Work

Distributed logic is best viewed as a logical toolbox for integrating many different logics which are themselves configured by axioms. One specifies the "connectivity" of local logics as a graph structure and then configures these "connections" with axioms and rules based upon a particular application. Many of the common modal axioms can be altered to fit distributed modal connectives. The simulation axiom shows this. As a further example, consider the Euclidean axiom (in a normal modal logic) $\langle h \rangle P \supset [h] \langle h \rangle P$ and its validating condition $\mathcal{H}xy$ and $\mathcal{H}xz$ implies $\mathcal{H}yz$. In distributed form for $r : h \curvearrowright k$ in Fig. 10, this becomes $\langle r \rangle P \supset [r] \langle k \rangle P$ and the condition becomes ($\mathcal{R}xy$ and $\mathcal{R}xz$) implies $\mathcal{K}yz$.

Figure 10 represents a common situation: the relation $\mathcal{R}$ between domain h and k is an artifact of the model and as such, deserves to be represented in a logic over the model. This is the sense in which distributed logic could be considered a model theoretic logic (Barwise and Feferman 1985). One must make choices up front before parts of the toolbox come together for a logic; the choices are made because models of a particular kind are needed for an application.

More philosophically speaking, modal logics come with a model theory which includes morphisms between models. The logic is abstracted over the model theory giving valid axioms and rules for reasoning about the models. Since morphisms are used in the model theory to describe critical aspects of the model, the obvious question is why these aspects are not formalized in the logics? The work in this paper (and its predecessor Allwein et al. 2014) represents the first steps in this direction.

Part of the challenge of including morphisms in a logic is deciding which morphisms to include and how the included morphisms should be structured. Category theory presents us with the theory of morphisms, and considering modal logic, one

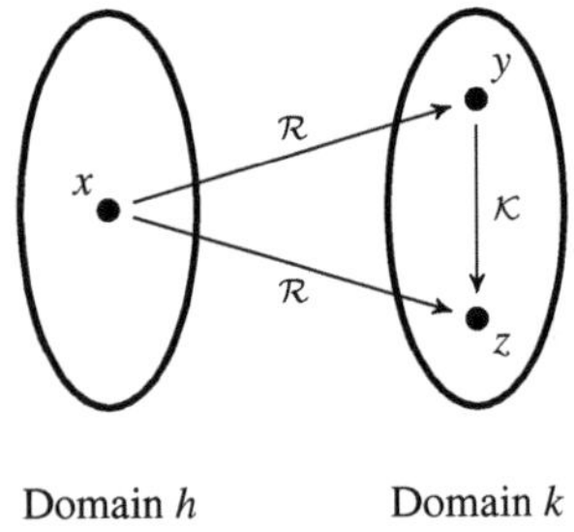

Fig. 10 Distributed Euclidean axiom

could have started with p-morphisms. The approach we have taken is to generalize the notion of what should be considered a model theoretic morphism and then use logical axioms to give the morphisms the properties desired. In effect, we are choosing logical morphisms that preserve only some desired structure (but not all structure). The axiom system is then used as an array of control switches to configure distributed logics. In addition, the morphisms can be fine-tuned between some local logics but not imposed between all local logics within a distributed logic. This accords well with the notion that distributed logics should be useful for representing reasoning about distributed computing systems where there is much variation and nuance that must be represented formally.

Space prevents us from also covering two-place intensional connectives such as entailment in relevance logic. That, too, has a pleasant reconstruction in distributed logic, although the three place relations require an extended notion of categorical morphism. Distributed logic was originally formulated with relations and consideration of testing for externally defined components in system-on-a-chip designs required the use of neighborhood systems. The ease of modification of distributed logic forced by two place intensional connectives and weak modal connectives requiring a neighborhood semantics is part of a larger theme for distributed logic: many model theoretic notions are "orthogonal" to distribution in that they do not seem to cause any significant hurdles to their re-expression in a distributed logic. Some model theoretic notions, such as morphism, are inherently distributed. Some, such as, Kripke relations, can be re-expressed as distributed notions. The bounds of what is possible seems to be related to the question of what is modality.

A good source of applications which require distributed reasoning are the security guarantees necessary for system-on-a-chip (SoC) architectures. In on-going and future work, we are expanding the use of distributed logics to provide a programming logic for a hardware specification language called ReWire in Procter et al. (2015). Formal logic for SoCs almost demands a distributed logic. The sub-components are scattered across the chip and each is a small universe of internal states or worlds. One sub-component's connections with other sub-components can be either tight or very loose. Distributed Kripke relations provide the right kind of flexibility in this environment for interpreting logical properties of the SoC.

Most systems in the engineering world have some kind of distribution, whether it is concretely in space or abstractly as a mathematical distribution of modeling conditions. As is typical of the real world, many properties are not binary (i.e., that either the system or component has the property or it does not) but, rather, only admit to a probability of holding. Much of the goal of engineering is to have systems perform to a certain tolerance. Chip companies recognize this and have produced designs with error correction circuitry although, even then, they realize that they cannot achieve perfection. The advantage distributed logic holds, given the structural similarity with stochastic relations and Markov modeling, is that these mathematical modeling techniques can now be seen as a direct weakening of logical properties.

References

Adámek, J., & Rosický, J. (1994). *Locally presentable and accessible categories*, London Mathematical Society. Lecture Note Series 189.

Allwein, G. (2005). A qualitative framework for Shannon information theories. In *Proceedings of the New Security Paradigms Workshop, 2004* (pp. 23–31). ACM Press.

Allwein, G., & Dunn, J. M. (1993). Kripke semantics for linear logic. *Journal of Symbolic Logic, 58*(2), 514–545.

Allwein, G., & Harrison, W. (2010). Partially-ordered Modalities. Proceedings of the Advances in Modal Logic Conference, *8*, 1–21.

Allwein, G., Harrison, W., & Andrews, D. (2014). Simulation logic. *Logic and Logical Philosophy, 23*(3), 277–299.

Barwise, J., & Feferman, S. (Eds.). (1985). *Model-theoretic logics*. Springer.

Barwise, J., & Seligman, J. (1993). Imperfect information flow, *Logic in Computer Science, 1993. LICS '93* (pp. 252–260). IEEE.

Barwise, J., & Seligman, J. (1997). *Information flow: The logic of distributed systems*. CUP. Cambridge Tracts in Theoretical Computer Science 44.

Birkhoff, G., & Lipson, J. D. (1968). Heterogeneous algebras. *Journal of Computational Theory, 8*, 115–133.

Blackburn, P., de Rijke, M., & Venema, Y. (2000). *Modal logic*. Cambridge, UK: Cambridge University Press.

Chellas, B. F. (1980). *Modal logic: An introduction*. Cambridge University Press.

Doberkat, E. -E. (2010). *Stochastic coalgebraic logic*. Springer.

Dunn, J. M. (1991). Gaggle theory: An abstraction of Galois connections and residuation with applications to negation, implication, and various logical operators. In J. van Eijck (Ed.), *Logics in AI: European workshop JELIA '90, Lecture notes in computer science* (Vol. 478, pp. 31–51). Berlin: Springer.

Dunn, J. M., & Hardegree, G. M. (2001). *Algebraic methods in philosophical logic, Oxford logic guides* (Vol. 41). Oxford, UK: Oxford University Press.

Fagin, R., Halpern, J. Y., Moses, Y., & Vardi, M. (1995). *Reasoning about knowledge*. MIT Press.

Freyd, P. J., & Scedrov, A. (1990). *Categories, allegories*. North-Holland, Amsterdam.

Gabbay, D. (1999). *Fibring logics*. Oxford University Press.

Goguen, J. A., & Burstall, R. M. (1985). Institutions: Abstract model theory for specification and programming. *CSLI Research Reports, 85–30*, 1–73.

Goguen, J. A., & Meseguer, J. (1982). Security policies and security models. In *Proceedings of the 1982 IEEE Symposium on Security and Privacy* (pp. 11–20). IEEE Press.

Jacobs, B. (1999). *Categorical logic and type theory*. Springer.

Manes, E. G. (1976). *Algebraic theories*. Springer.

Moskowitz, I. S., Chang, L., & Newman, R. E. (2002). Capacity is the wrong paradigm. In *Proceedings of New Security Paradigms Workshop*, Sept. 23–26 (pp. 114–126). ACM Press.

Procter, A., Harrison, W. L., Graves, I., Becchi, M., & Allwein, G. (2015). Semantics driven hardware design, implementation, and verification with ReWire. In *ACM SIGPLAN/SIGBED Conference on Languages, Compilers, Tools and Theory for Embedded Systems (LCTES)* (pp. 1–10).

Rasga, J., Sernadas, A., & Sernadas, C. (2010). Importing logics. *Studia Logica, 100*, 545–581.

Sangiorgi, D. (2012). *Introduction to bisimulation and coinduction*. Cambridge University Press.

Sernadas, A., Sernadas, C., & Caleiro, C. (1999). Fibring of logics as a categorical construction. *Journal of Logic and Computation, 9*, 149–179.

Thomason, R. (1984). Combinations of tense and modality. *Handbook of philosophical logic* (pp. 135–165). Kluwer Academic Publishers.

van Benthem, J. (2000). Information transfer across Chu spaces. *Logic Journal of the IGPL, 8*(6), 719–731.

Tracking Information

Johan van Benthem

Abstract Depending on a relevant task at hand, information can be represented at different levels, less or more detailed, each supporting its own appropriate logical languages. We discuss a few of these levels and their connections, and investigate when and how information growth at one level can be tracked at another. The resulting view has two intertwined forms of logical dynamics for informational agents: one of update and one of representation. Mike Dunn has been a lifelong pioneer in the study of logic and information, with seminal contributions to relevant and resource logics, including their semantic, algebraic and proof-theoretic dimensions. I offer the thoughts to follow as an academic fellow-traveler.

Keywords Information · Logic · Level · Update · Tracking

1 Introduction: Information and Logic

Connections between logic and information have been a lifelong interest of Michael Dunn, witness his seminal contributions that will be highlighted in this volume. I have long been intrigued by this interface and its many dimensions, but my offering in this volume in Mike's honor concerns just one special topic: the dynamics of information-driven agency over time.[1]

A first major issue in this setting is one that every logician studying the area encounters sooner or later: information is not one single notion, but a content that can be represented at *different levels* of detail, rougher or finer (van Benthem and Martinez 2008). Each level supports natural "attitudes" that agents can have, not in any concrete psychological sense, but in the sense of different attunements to information. In this paper, I will start from perhaps the roughest level, that of semantic information as a

[1]More specific points of contact between our interests will be found at the end of this paper.

J. van Benthem (✉)
Institute for Logic, Language and Computation, University of Amsterdam,
Stanford University, and Tsinghua University, Amsterdam, The Netherlands
e-mail: J.vanBenthem@uva.nl
URL: http://staff.fnwi.uva.nl/j.vanbenthem

K. Bimbó (ed.), *J. Michael Dunn on Information Based Logics*,
Outstanding Contributions to Logic 8, DOI 10.1007/978-3-319-29300-4_17

range of possibilities, and then discuss richer views that are prominent in the recent literature: plausibility order and belief, evidence, and eventually also some levels higher up, such as prioritized evidence or probability.

A second major issue is the pervasive phenomenon of *information dynamics* (van Benthem 2011). Agents do not have fixed information once and for all: it changes all the time when receiving new informational signals, hard or soft. Moreover, their attitudes tag along in a systematic manner, leading to dynamic knowledge update, belief revision, or whatever term fits the current representation level. This second issue is not unrelated to the first, since in any serious study of information, the choice of a representation level cannot be made in isolation from the dynamic actions that one wants to understand (Adriaans and van Benthem 2008).

However, my main aim in this paper is not to propose new representation levels or new dynamic update actions. Rather, I want to investigate how the two themes interact, resulting in the problem of "*tracking*" information dynamics at different levels. I will state a few new results, and on that basis, raise some general problems for the logical analysis of information in its proper generality. I will not develop a fully general theory, but Sects. 10 and 11 will contain some thoughts on that further step, including links with Mike Dunn's work on abstract information structure.[2]

2 Semantic Information

2.1 Basic Epistemic Logic

Perhaps the roughest form of representing information is the common-sense picture of a set of still live possible candidates for the actual world, an "epistemic range" that shrinks when new information comes in. These sets are models for the language of epistemic logic whose key operator $K_i\varphi$ says that φ holds in all accessible alternative worlds for agent i, or in other terminology: the agent has the "semantic information" that φ is true.[3]

When such models are used to describe some informational scenario taking place in reality, one assumes there is a unique true state of affairs, that can be marked as the "actual world" in the model—even though no agent needs to know which possibility is in fact the actual one.

The valid principles of reasoning with semantic information are those of the well-known modal logic $S5$ for each separate agent i. Note that this setting has no non-trivial valid laws that relate the information of different agents: any significant

[2]In this paper, reviews of standard material will be brief, making reference to the literature. New notions and new observations are marked as definitions, facts, or theorems. Also, I will be thinking mostly of *finite models* in what follows, not as a point of principle, but to avoid standard complexities in lifting simple intuitions to more complex infinite settings.

[3]Each agent i gets an equivalence relation $\sim_i$ whose clusters encode its information ranges.

dependencies must come from information channels or other forms of alignment that can hold between agents.

It is easy to criticize this setting for its extreme simplicity, but a bare range of options represents a well-chosen mathematical abstraction that occurs across the sciences and even philosophy.

We make even further simplifications. In this paper, we will consider just one agent, and hence indices i will be dropped—although many paradigmatic informational scenarios essentially involve many agents.[4] Another issue that we will leave aside in this paper is the connection between semantic information in the sense of epistemic logic and the usual philosophical notions of knowledge: see (Holliday 2012) for a sophisticated modern treatment.

2.2 Dynamics of Hard Information

The simplest informational events in this setting are "hard announcements" or observations $!\varphi$ of the fact that the proposition φ is true in the actual world. What this new information does is it retains those worlds in the current epistemic model that satisfy φ, while it eliminates those worlds that do not currently satisfy φ. This is arguably the simplest common-sense picture of obtaining new information, and it can be modeled technically as transforming the current model $\mathbf{M}, s$ into the new restricted model $\mathbf{M} \mid \varphi, s$.

Updates of this sort can be tricky if the new information φ is not just factual (being just a Boolean combination of atomic facts), but contains epistemic modalities.[5] However, there is a complete logic for this dynamics of semantic information, which can be brought out in a suitable two-tiered syntax. We introduce *announcement actions* $!\varphi$ for each formula: one can also think of these as public observations, or other ways of receiving hard information. In constructing formulas, we then allow Boolean operations, the K-modality, but now also a new dynamic modality $[!\varphi]\psi$ saying that after a truthful update with φ, ψ is the case in the updated model.

Theorem 2.1 *The dynamic epistemic logic of update with hard information is completely axiomatizable.*

Proof This result is easy to prove. The heart of the axiomatization is a "recursion law" describing what new knowledge obtains after new hard information has been received:

$$[!\varphi]K\psi \leftrightarrow (\varphi \rightarrow K[!\varphi]\psi)$$

[4]In this simple case, we essentially identify the K-operator with the "universal modality" that ranges over all worlds in the model.

[5]A well-known example are so-called "Moore sentences." If an agent lacks information if p, but p is the case (that is, $\neg Kp \wedge p$ is true), then announcing this fact will make $\neg Kp \wedge p$ false, as Kp has become true in the updated model given its restriction to p-worlds.

Such recursion laws, here and in other systems to follow, are the basic principles describing the stepwise dynamics of information change. □

This may suffice as a first level of representing information. Knowledge grows as agents receive successive new inputs and make matching updates. In some scenarios, this may zoom in to just the actual world, representing a state of common knowledge for the agents about what the world is like. However, one can also study infinite processes of endless learning. For further information on this dynamic-epistemic methodology, including delicate private informational actions with more complex updates, one can consult (van Ditmarsch et al. 2007) or (van Benthem 2011).

3 Plausibility and Belief

3.1 Plausibility Ordering

Mere semantic ranges ignore the fact that often, not all candidates for the actual world are on a par. Say, in planning my upcoming trip, I may consider some possibilities more plausible than others, and may well confine attention to the most plausible cases. This can be modeled in an enrichment of the earlier models with additional structure. We now expand the earlier epistemic models $(W, \sim, V)$ to *epistemic plausibility models* $(W, \sim, \leq, V)$, where $\leq$ is an ordering of "relative plausibility." For the purposes of this paper, we will assume that $\leq$ is a reflexive transitive order (i.e., a "pre-order"), not necessarily connected, leaving room for genuinely incomparable worlds. Moreover, to avoid technicalities that are not germane to our main theme, we assume that the plausibility order is the same at each world. I.e., in the sense of semantic information, the agent knows her own plausibility ordering, and the beliefs that she has, based on it.

This fine-structure of information at once suggests a richer repertoire of attitudes for agents. In particular, it makes sense to look only at the most plausible worlds in the current range, and define *belief* $B\varphi$ as truth of φ in all of these. This is a sort of less-demanding semantic information "to the best of one's knowledge." As for a logic of belief construed in this way, it is better to also introduce a further notion of *conditional belief* $B^{\psi}\varphi$, that is, a belief in φ conditional on already being in the set of ψ-worlds. Conditional belief satisfies exactly the principles of conditional logic over Lewis-style models, with the proviso that the order need not be connected.

Plausibility order may look like a technical device, but there are interesting issues in interpreting it. At the start of a process of inquiry, a plausibility order may be viewed as a "prior," a set of expectations, including conditional beliefs that state what we would believe were certain new information to arrive.[6] But over time, the plausibility order can be modified by informational events (see below), and hence

[6]In other settings, this prior ordering amounts to a "learning method" telling the agent how to respond to new information (Baltag et al. 2011).

we can also think of the current order as a sort of rough record of past experience. Finally, it is this order we tend to use for deciding on new actions, so plausibility models are at the same time a record of the past and a guide to the future.

3.2 Plausibility Dynamics

What sort of informational events can affect epistemic plausibility models? Public announcements still make sense, and validate a complete logic.

Theorem 3.1 *The dynamic logic of belief change under hard information is completely axiomatizable.*

Proof In particular, the key recursion laws for announcement are as follows:

$$[!\varphi]B\psi \leftrightarrow (\varphi \rightarrow B^{\varphi}[!\varphi]\psi)$$
$$[!\varphi]B^{\alpha}\psi \leftrightarrow (\varphi \rightarrow B^{\varphi\wedge[!\varphi]\alpha}[!\varphi]\psi)$$

Note how conditional beliefs after announcement modalities serve here to describe beliefs formed after receiving new information. □

Events of hard information can interact with beliefs in surprising ways.

Example Misleading with a truth.

Consider a model with an actual world 1 plus two more possible worlds 2, 3, ordered as follows qua plausibility: $1 \leq 2 \leq 3$. Let the atomic proposition p be true in 1 and 3, but not in 2, and let q be true at 3 only. In this model, the agent believes that p, because p is true in the most plausible world 3. Now announce the true fact that $\neg q$. This will eliminate world 3, leaving an ordering $1 \leq 2$, where the agent now falsely believes that not p. Despite this potentially surprising feature of update, a logic with the above laws will keep reasoning about such scenarios straight.

However, when we have more structure, there is often a richer repertoire of relevant actions. In particular, in addition to the "hard information" in events $!\varphi$, there is "soft information" that does not rule out any possibility, but changes the plausibility order for the existing possibilities.

A well-known example is *radical upgrade* $\Uparrow\varphi$, with the following effect. We put all φ-worlds above all $\neg\varphi$-worlds, on top in the ordering—while inside these two zones, the old plausibility order is retained. Reasoning gets more complex than with public announcement, but yields to techniques like before.

Theorem 3.2 *The dynamic logic of belief change under radical upgrade is completely axiomatizable.*

Proof This time, the key recursion laws are more involved. Perhaps the most complex principle is the valid equivalence for the new conditional beliefs formed under radical upgrade. It reads as follows:

$$[\Uparrow\varphi]B^{\alpha}\psi \leftrightarrow (\Diamond(\varphi \wedge [\Uparrow\varphi]\alpha) \wedge B^{\varphi\wedge[\Uparrow\varphi]\alpha}[\Uparrow\varphi]\psi)$$
$$\vee (\neg\Diamond(\varphi \wedge [\Uparrow\varphi]\alpha) \wedge B^{[\Uparrow\varphi]\alpha}[\Uparrow\varphi]\psi))$$

with $\Diamond$ an existential modality "somewhere in the current epistemic range." □

But softness comes in different kinds. A less radical way of taking new information is the *suggestion* $\#\varphi$, whose effect is merely to remove any links that run from a φ-world to some more plausible $\neg\varphi$-world. This forces us to take best worlds inside the φ-zone seriously in our beliefs, though formerly best $\neg\varphi$-worlds are still in play, too. Again, the dynamic logic of belief change is completely axiomatizable, but we will not state its key recursion law here.

There is a wide spectrum of order-changing operations behind these three specific examples of update actions, that can be found in many places: not just as plausibility change, but also as operations that change preferences (Liu 2011) or relevance (van Benthem 2015).[7] Indeed, there exist general methods for defining updates and deriving recursion laws in all these cases, for which we refer the reader to Baltag and Smets (2006), van Benthem and Liu (2007), Girard et al. (2012), and van Benthem and Smets (2015).

Remark Factual or epistemic-doxastic propositions.

One reason why the above recursion laws get complex syntactically is having to deal with knowledge and belief operators inside incoming new information, which can lead to surprising truth value changes under update. While this "higher information" is realistic in actual communication, it is also something of a technical side issue in this paper. Therefore, we make a sweeping simplification, namely, *we will restrict attention to factual propositions only*. Or stated differently, we decontextualize propositions denoted by our epistemic-doxastic syntax, by thinking of them merely as absolute subsets of the model. This huge simplification will play everywhere in what follows, and the reader should remain aware of it. Of course, a final version of our account should, and can, deal with the more sophisticated full version.

3.3 Interplay of Statics and Dynamics

It may look as if dynamic events of information flow and their matching update acts are just additions to a given static base logic of attitudes, that is usually on the shelf

[7]Even hard information can be taken as an order change operation, witness so-called "link cutting" versions of public announcement where the φ- and $\neg\varphi$-zones are made disjoint.

already in philosophical logic. But the dynamic component can also affect the design of the static base. Here are two examples.

Our earlier examples of misleading with the truth suggests a notion of robust or *safe belief* $SB\varphi$ that remains stable under whatever true new information comes in. It is easy to see that this amounts to the following notion, at least in models with *connected* plausibility orders[8]:

$$\mathbf{M}, s \models SB\varphi \text{ iff } \textit{for all } t \textit{ with } s \leq t,\ \mathbf{M}, t \models \varphi$$

On connected plausibility orders, safe belief is a new attitude for agents, in between ordinary belief and knowledge qua informational strength.

In addition to its intuitive attraction, safe belief also has the technical advantage of allowing us to define absolute and conditional belief. This is shown in the following two valid equivalences, that also work on arbitrary pre-orders, where K is the earlier knowledge modality over the whole epistemic range, and $\langle SB\rangle$ is the existential dual modality of SB:

Fact 3.1 *The following laws hold for knowledge, belief, and safe belief:*

$$B\varphi \leftrightarrow K\langle SB\rangle SB\varphi$$
$$B^{\psi}\varphi \leftrightarrow K(\psi \rightarrow \langle SB\rangle(\psi \wedge SB(\psi \rightarrow \varphi)))$$

Given these definitions, it is often easier to state recursion laws for informational actions and safe belief, since others will be derivable. As an illustration, here are laws for the three operations that we discussed in the above:

Fact 3.2 *The following recursion laws hold for factual propositions:*

$$[!\varphi]SB\psi \leftrightarrow (\varphi \rightarrow SB(\varphi \rightarrow \psi))$$
$$[\Uparrow\varphi]SB\psi \leftrightarrow (\neg\Diamond\varphi \wedge SB\psi) \vee (\varphi \wedge SB(\varphi \rightarrow \psi)) \vee (\neg\varphi \wedge \Diamond\varphi \wedge SB(\neg\varphi \rightarrow \psi) \wedge K(\varphi \rightarrow \psi))$$
$$[\#\varphi]SB\psi \leftrightarrow (\varphi \wedge SB(\varphi \rightarrow \psi)) \vee (\neg\varphi \wedge SB\psi)$$

Another example of how the dynamic component can influence the static base language of attitudes is the notion of a *conditional*, so central in much of Mike Dunn's work on logic and information (Dunn 1971, 1976).

When we say that "if φ, then ψ," there is an issue about the force of the "if." How are we to imagine the hypothetical situation that is introduced? The standard view of making a hypothesis fits a public announcement $!\varphi$: we restrict attention to the φ-worlds and work inside that restricted space. But sensitized to the above dynamic distinctions, we can also make the hypothesizer "if" weaker in force, letting it just promote the φ-worlds to top positions (as in radical upgrade $\Uparrow\varphi$), or even making all of them relevant cases for inspection, as in the above operation of suggestion $\#\varphi$.

---

[8] This is just the standard universal modality over a binary order. However, on pre-orders that allow incomparable worlds, safe belief in our dynamic sense refers to all worlds that do not strictly precede the current world in the ordering (van Benthem and Pacuit 2011). We will ignore this technicality in this paper, as it does not affect our main concerns.

This gives us three conditionals that we will read as saying that after the relevant way of assuming φ, the agent believes that the conclusion ψ is true:

$$\varphi \to^! \psi \qquad \varphi \to^{\Uparrow} \psi \qquad \varphi \to^{\#} \psi$$

These three notions give us more refined ways of thinking about conditional reasoning, and more generally, of generalized notions of consequence. It is not our aim to develop this line in depth here, but a few observations may help illustrate the attraction of this richer perspective.[9]

Fact 3.3 *The three conditionals validate different laws.*

Proof We have $\varphi \to^! K\varphi$, but not for the other two conditionals. We have $\varphi \to^{\Uparrow} B\varphi$, but the latter does not hold for the suggestion conditional. □

Next, consider structural rules as in standard proof theory (Bimbó 2015).

Fact 3.4 *The given three conditionals validate the same structural rules.*

Proof It is easy to see that the first two conditionals refer to the same worlds, since the maximal worlds within the φ-zone are the same as the maximal worlds overall when the φ-zone lies at the top in the model. Thus, their structural rules refer to the same zones in models.

To see that the third conditional complies with the same structural rules, note that $\varphi \to^{\#} \psi$ refers to maximal worlds that can be of two kinds: maximal within the φ-zone, or maximal worlds overall that are $\neg\varphi$. It says that all of these satisfy ψ. But this can be stated equivalently as follows:

$$(\varphi \to^! \psi) \wedge (T \to^! \psi)$$

Now it is easy to check that the notion defined by this conjunction satisfies the structural laws of the basic conditional logic. □

4 Connecting the Two Levels

Epistemic models as mere semantic ranges and plausibility models are two different ways of representing information, one richer than the other. Although this extension is intuitively clear, we give a brief discussion of some general issues that will return later on in this paper.

First, there are systematic connections between the two levels. Moving from poorer to richer, we can *embed* epistemic models into the realm of plausibility models as special cases where all worlds are equiplausible, or equivalently for our present purpose: incomparable qua plausibility. Let us call this the functor *equi*(**M**) that takes

[9]In what follows, again, we only consider factual propositions to avoid some complexities.

epistemic models to plausibility models.[10] Going in the opposite direction, there is also a natural notion of *projection*, by a functor *forg*(**M**) that forgets the plausibility ordering and just returns the bare epistemic domain.

The embedding and projection functors have this obvious connection:

$$\mathit{forg}(\mathit{equi}(\mathbf{M})) = \mathbf{M}$$

What does not hold is, for plausibility models **N** is the equality stating that $\mathit{equi}(\mathit{forg}(\mathbf{N})) = \mathbf{N}$. Structure that has been lost cannot be retrieved faithfully by some uniform stipulation.

Corresponding to these structural maps, there are also *translations* between the languages of the two levels for representing information. These satisfy the typical "adjointness" scheme for translations in logic:

$$\mathbf{M} \models \mathit{trans}(\varphi) \ \text{ iff } \ F(\mathbf{M}) \models \varphi, \text{ for all models } \mathbf{M} \text{ and formulas } \varphi$$

Here the map F is a model transformation, and *trans* is a matching translation between languages. In our specific case, the language translation corresponding to the functor *equi* is as follows:

> we leave atomic propositions, Boolean operations, and the epistemic modality K the same, and replace the doxastic modality B by K.

It is easy to show that this yields the above equivalence, since all worlds in plausibility models of the form *equi*(**M**) are maximal. But what this really says of course is that belief does not mean anything new in such special models. The same trivialization extends to safe belief: it, too, collapses into knowledge when the ordering is uniform equiplausibility.

The extra richness of the plausibility level shows, as we have seen earlier, in the dynamics. It is possible to also translate informational actions from the epistemic to the plausibility level, since we can use public announcements on plausibility models just as on epistemic models, as we have seen. Thus, the translation also extends to modalities $[!\varphi]$.

Things are more complex on the other side: not every natural transformation on plausibility models has an exact counterpart on bare epistemic models. Operations like $\Uparrow\varphi$, $\#\varphi$ will normally turn models of the form *equi*(**M**) into models that are not of this form. Stated in other terms, we cannot faithfully translate formulas $[\Uparrow\varphi]\psi$ into purely epistemic formulas.

Next consider projection from the richer to the poorer level. What translation takes the epistemic language into the doxastic one according to the above scheme? As fits a case of straightforward extension of models, this translation is just the *identity of formulas* in the dynamic-epistemic language of public announcement. However, this cannot be extended further.

---

[10]Here we use category-theoretic terminology such as "functor" in a loose sense, though a precise formal development in category-theoretic terms is outside the scope of this paper.

Fact 4.1 *There is no translation for belief via the projection functor forg.*

Proof Consider the same world in two plausibility models that have the same worlds and atomic valuation but differ in their plausibility structure, making Bp true in one model and false in the other. The functor *forg* will take these models to the same epistemic model, and so, the translations of Bp and $\neg Bp$ cannot differ in truth value, whereas they should. □

Next consider dynamics at the level of plausibility models, say, a soft informational change such as a radical upgrade $\Uparrow\varphi$. It is easy to see that, this time, there is a matching transformation at the epistemic level, namely, just the identity map on models. But this trivial harmony shows precisely what is going on: the epistemic level cannot detect mere plausibility changes. Thus, the latter reordering acts are genuine "internal" operations in the doxastic realm, leaving no significant "external" traces that can be tracked epistemically.

None of the preceding observations are deep, but they set the scene for the more interesting comparisons across informational levels to be made later.

5 Evidence

Plausibility order between worlds does not record which reasons determined these relative differences among the available candidates. A more fine-structured approach is that of van Benthem and Pacuit (2011), where evidence is modeled as a family of subsets of the domain, viewed as information obtained from various sources that may be consistent, but could also contradict each other. Such an array of evidence allows agents to form beliefs, but it also gives them more structure to work with when they have to supply reasons, or give up beliefs. In this section, we survey evidence modeling as our third richer level for representing information.

5.1 Evidence Models

An *evidence model* is a set of possible worlds, viewed as an epistemic range for a K-modality as before, but now with an added family $\mathcal{E}$ of non-empty subsets.[11] We assume for simplicity that evidence sets are uniformly available in the model, not depending on particular worlds. This may be considered a very special case of "neighborhood models" for modal logic. The most straightforward modality in this setting is then interpreted as follows:

$$\mathbf{M}, s \models \Box\varphi \quad \text{iff} \quad \text{there exists a set } E \in \mathcal{E} \text{ with } \mathbf{M}, t \models \varphi \text{ for all } t \in E.$$

[11] It would also make sense to model sources of evidence, but we ignore this aspect here.

The modality $\Box\varphi$ expresses the existence of evidence available to the agent that supports the proposition φ. As we make no special assumptions on the available evidence, it is easy to see that this modality is only upward monotonic, while it distributes neither over conjunctions nor over disjunctions. For instance, having $(\Box\varphi \wedge \Box\psi) \rightarrow \Box(\varphi \wedge \psi)$ valid would assume that all available evidence has already been "processed" to include combinations.

One notion that assumes such processing is belief, viewed as what we can safely conclude by combining our evidence to the utmost. Let a *maximal body of evidence* $\mathcal{X}$ be a family of sets in $\mathcal{E}$ for which all finite intersections are non-empty, and that cannot be extended with further evidence to retain this property. In finite models, we then look at the intersection of the whole family $\mathcal{X}$ as what follows on the basis of this body of evidence. Let us define

$\mathbf{M}, s \models B\varphi$ iff $\mathbf{M}, t \models \varphi$ for all worlds t in all intersections of maximal bodies of evidence.[12]

This can be generalized to conditional belief $B^{\psi}\varphi$ in a fitting manner using maximally consistent sets with respect to including the set $[\![\psi]\!]$. Further attitudes make sense as well, such as "*entertaining* φ" in the sense of φ being true throughout the intersection of at least one maximal body of evidence.[13] Yet other evidence-based attitudes will be mentioned below.

Belief as defined here satisfies the axioms of a normal modality, and in particular, we do have the validity of $(B\varphi \wedge B\psi) \rightarrow B(\varphi \wedge \psi)$. It is not trivial to axiomatize the resulting logic that combines both normal and non-normal monotonic attitudes, witness the completeness proof in van Benthem et al. (2012). Such technical details will not concern us here.

5.2 *Evidence Dynamics*

As with earlier levels for representing information structure, evidence supports a natural dynamics of change. To explore this, we start with public announcements $!\varphi$ restricting a current model to the subset of all φ-worlds.

Theorem 5.1 *The dynamic logic of public announcement over evidence models is completely axiomatizable.*

Proof The key recursion law for the belief modality is straightforward, though it requires also dealing with conditional belief, as we did on plausibility models. Moreover, the recursion law for $[!\varphi]\Box\psi$ needs a new notion of "conditional evidence" $\Box^{\varphi}\psi$ that we do not spell out here. □

---

[12] In infinite models with infinite sets E, this stipulation must and can be modified.

[13] This is not the existential dual $\neg B \neg\varphi$ of belief as defined above. Note also that, in models with conflicting evidence, we can "entertain" contradictory propositions at the same time.

Of greater interest are new operations that are typical for an evidence setting. One striking pilot example is *evidence addition* $+\varphi$. What this model transformation does is it adds the set $[\![\varphi]\!] = \{ t \in W : \mathbf{M}, t \models \varphi \}$ as a new set to the current evidence family $\mathcal{E}$ to obtain a new model $\mathbf{M}+\varphi$.

Theorem 5.2 *The dynamic logic of evidence and conditional belief under evidence addition is completely axiomatizable.*

Proof This time we give a bit more detail, since more is involved here than in the dynamics of simpler structures such as plausibility models. We start with the evidence modality:

$$[+\varphi]\Box\psi \leftrightarrow (\Box\psi \vee K(\varphi \rightarrow \psi))$$

The rationale for this will be clear, though this particularly simple form only holds for factual propositions. For the belief modality, things get more complex. After we have added $[\![\varphi]\!]$ as a new piece of evidence, there are two sorts of maximal bodies of evidence. The first consists of a family of evidence sets in the old model that is maximally consistent with respect to adding $[\![\varphi]\!]$, which is exactly the basis for conditional belief. The second sort is maximal bodies of evidence in the old model that also satisfy the condition that their intersection is disjoint from $[\![\varphi]\!]$. Now we get the following recursion law, again modulo our restriction to factual propositions:

$$[+\varphi]B\psi \leftrightarrow (B^{\varphi}\psi \wedge B(\neg\varphi \rightarrow \psi))$$

It is easy to see that this reduces, much as in our discussion of static suggestion-conditionals $\varphi \rightarrow^{\#} \psi$, to the following equivalence:

$$[+\varphi]B\psi \leftrightarrow (B^{\varphi}\psi \wedge B\psi)^{14}$$

The analogy of this formula with the behavior of suggestion update on plausibility models is no coincidence, and it will return below. □

But there are many further natural and appealing operations that affect our current evidence. Indeed, in the present setting, what used to be the basic notion of public announcement $!\varphi$ can be deconstructed into two intuitively independent operations: (a) adding evidence that φ is the case, and (b) removing all old evidence that supported $\neg\varphi$. The latter notion can be defined by itself as the following natural operation on evidence.

Deleting evidence $-\varphi$ transforms the current model as follows: all old evidence sets $E \in \mathcal{E}$ that are included in the set $[\![\neg\varphi]\!]$ will be removed. Of course, there can be many reasons for such a removal: one can think of "retraction" as in belief revision

[14]The recursion laws stated in van Benthem and Pacuit (2011) for these and the following operations in this section are more complex syntactically because they are meant to hold also for non-factual propositions. They have to deal with maximal sets of old evidence that are consistent with some given propositions while excluding others. In a full treatment of our themes, we would work in this more complex framework.

theory (van Benthem and Smets 2015), or of "forgetting" in some other sense, perhaps like cognitive decay.

For a final example of a natural operation on evidence, consider the K-axiom for conjunction that failed for the evidence modality. One might say that, given consistent evidence for φ and evidence for ψ, we have implicit evidence for $\varphi \wedge \psi$, that we can make explicit just by combining the two pieces of evidence. Now there are a few technical difficulties in making this work, but clearly, there is a natural operation $\cap$ of *intersecting* all mutually consistent evidence sets to form new evidence sets.

Of interest to our later discussion of tracking is that intersection is a sort of *internal* processing or re-arrangement of evidence, unlike addition or deletion. We will give a more precise sense to this notion later on. Intuitively, internal operations are one more illustration of what can be done at a richer level that need not show up at coarser levels of information structure.

5.3 Dynamic-Static Interactions Once More

As with plausibility models, the dynamics of evidence suggests new notions at the level of static attitudes as well. Van Benthem and Pacuit (2011) have several new notions of evidence-based conditional belief, for which axiomatizing the resulting richer doxastic base language is still an open problem.[15]

6 Representation and Translation

Again there are connections between our levels for representing information, one running from finer to coarser, and one from coarser to finer.

6.1 Projection

Evidence structure induces plausibility structure via a natural stipulation that can be found in many areas, from topology to Chu spaces (van Benthem 2000). We simply require that more plausible worlds satisfy all available evidence that is satisfied by less plausible ones:

$s \leq t$ iff for all $E \in \mathcal{E}$, if $s \in E$, then $t \in E$.[16]

[15]New modal logics for such generalized conditional evidence have recently been proposed and developed in van Benthem et al. (2015).

[16]It is instructive to see this work in concrete cases. The reader, when following the proofs to come, might draw some set diagrams and their induced plausibility orders.

We will call this the map *ord*(**M**) taking evidence models to plausibility models, where we do not change domains or propositional valuations.

There is a natural matching translation here for logical languages. As before, it runs in the opposite direction, from the language of plausibility models to that of evidence models. Actually, the main clauses are simple:

Knowledge K goes to knowledge K,
Belief B goes to belief B, and the same is true for conditional belief.

One may find the latter clause unsurprising, as our notion of evidence-based belief mirrored standard maximality notions in relational models. Indeed, any projection map allows us to "*retract*" notions from the plausibility level to the evidence level in a systematic manner.

For another case of such retraction or borrowing of notions, what about the earlier notion of safe belief? This now gets related to a natural notion of what may be called "reliable evidence" in the following sense. For any world s, let E_s be the family of sets $\{ E \in \mathcal{E} : s \in E \}$. This is a consistent family with a non-empty intersection. We now define *evidence-based safe beliefs* φ at s as the propositions φ true at each world in the intersection of $\{ E \in \mathcal{E} : s \in E \}$. Of course, agents need not know the world they are in, so what is reliable in this objective local sense may be unknown to them.

Now we can look at notions at the evidence level and see if they have plausibility counterparts. For the most obvious example, this fails.

Fact 6.1 *The evidence modality has no plausibility match via the map ord.*

Proof Consider a universe $W = \{ 1, 2, 3 \}$ with two evidence sets $\{ 1, 2 \}$, $\{ 2, 3 \}$, where only the world 2 satisfies the proposition p. In this model, the formula $\Box p$ does not hold. Now consider the same model with one evidence set added, viz. $\{ 2 \}$. This additional intersection of earlier evidence does not change the induced plausibility order. However, in the new model, $\Box p$ is true, and so cannot have been definable in the plausibility language. □

6.2 Embedding

There is also a natural map *evi*(**M**) running in the opposite direction, sending plausibility models to evidence models. A moment's reflection will show that it should work as follows (one might compare this stipulation with the structure of propositions in the semantics of intuitionistic logic):

the evidence sets are all upward closed in the plausibility ordering $\leq$.

This embeds plausibility models as special evidence models. In particular, the intersection of any two upward closed sets is upward closed, so the evidence sets are automatically closed under intersections.[17] This also shows in what happens when we repeat the level maps:

$$ord(evi(\mathbf{M})) = \mathbf{M}, \qquad \text{but not } evi(ord(\mathbf{M})) = \mathbf{M}.$$

Still, $evi(ord(\mathbf{M}))$ is just the closure of **M** under non-empty intersections.

Again there is a syntactic translation matching the embedding *evi*. It is correct for the identity translation of K and B from the language of evidence models to that of plausibility models. Given these translations, there is a natural issue whether they extend to dynamic modalities for changing the plausibility order or the evidence structure. This kind of correlation will be the topic of our next section, albeit with a slightly different emphasis.

7 Tracking Evidence via Plausibility

One particular interest we stated at the beginning was how the dynamics of information flow at one level might be tracked at another level. We will look a bit more closely into this now, finding both positive and negative results in our richer realm, and we will discuss what these findings mean.

7.1 Tracking Diagrams

Let us start with a case of harmony between operations.[18] Here, $ord(\mathbf{M}) \mid \varphi$ refers to the model transformation associated with public announcement.

Fact 7.1 *For all evidence models* $\mathbf{M}$ *and factual propositions* φ*, we have that* $ord(\mathbf{M} \mid \varphi) = ord(\mathbf{M}) \mid \varphi$.

Proof This follows immediately from the natural definition of public announcement on evidence models: we restrict the domain to the worlds that satisfy φ, and we restrict all evidence sets to this subdomain (though dropping the empty set if it arises in this manner). □

We can picture this situation in the following "tracking diagram."[19]

$$\begin{array}{ccc} ord(\mathbf{M}) & !\varphi & ord(\mathbf{N}) \\ \uparrow & & \uparrow \\ \mathbf{M} & !\varphi & \mathbf{N} \end{array}$$

---

[17] In this section, we sidestep a few technicalities with the empty set.

[18] The first two harmony results to follow were stated for preference order in Liu (2011).

[19] Again we assume for simplicity that propositions are subsets of models here, disregarding technical issues of translation between complex formulas. In a more detailed treatment, we would compare correlated updates of the form $!\varphi$ and $!translation(\varphi)$.

In a diagram like this, we say that the update map at the poorer upper level *tracks* the lower one at the richer level.[20]

One might think that this harmony holds here only because public announcement works in much the same way at both levels. But here is a less straightforward example. Evidence addition is tracked by our earlier operation of "suggestion" on plausibility models. Again, in what follows, operator notations refer to the model transformations introduced earlier for plausibility orders and evidence models.

Fact 7.2 *For all evidence models* $\boldsymbol{M}$ *and factual propositions* φ, *we have that* $ord(\boldsymbol{M} + \varphi) = ord(\boldsymbol{M})\#\varphi$.

Proof The essential observation is that, given our definition of the induced order, adding the denotation of φ as an evidence set, the order changes precisely as described in the operation $\#\varphi$. Points where φ holds now satisfy evidence that is not shared by any $\neg\varphi$-point. □

To add one more example, sometimes there is harmony, but to see it, we need operations that are *new* in the literature. For instance, we will now give an evidence counterpart tracked by our earlier plausibility transformation of radical upgrade $\Uparrow\varphi$, again with φ viewed as a subset. Define the following map $up(\varphi, \mathbf{M})$ on evidence models $\mathbf{M}$ with a subset φ:

(i) if $E \in \mathcal{E}$ has $E - \varphi$ non-empty, then replace it by $E \cup \varphi$,
(ii) if $E \in \mathcal{E}$ has $E \cap \varphi$ non-empty, then add $E \cap \varphi$.

Fact 7.3 *For all evidence models* $\boldsymbol{M}$ *and factual propositions* φ, *we have that* $ord(up(\varphi, \boldsymbol{M})) = ord(\boldsymbol{M}) \Uparrow \varphi$.

Proof There are four cases for two points x, y: they can satisfy (a) $\varphi\ \varphi$, (b) $\varphi\ \neg\varphi$, (c) $\neg\varphi\ \varphi$ or (d) $\neg\varphi\ \neg\varphi$. In case (a), if $x \leq y$ initially, then all new evidence satisfied by x holds for y, as the modifications do not affect what happened inside the φ-area. If not $x \leq y$ initially, then new evidence of type (i) no longer distinguishes, but new evidence of type (ii) will. In case (b), the new evidence of type (ii) rules out more plausible $\neg\varphi$-worlds. In case (c), all new evidence true for $\neg\varphi$-worlds also holds for all φ-worlds. Finally, in case (d), the order stays the same among the $\neg\varphi$-worlds: only new evidence of type (i) is relevant, and this has not changed within this zone. □

In the background of these three individual tracking diagrams lie a number of more general observations.

Fact 7.4 *For every map* f *between plausibility models, there exists a map* g *between evidence models that is tracked by* f. *In particular, one can define* g *on models* $\boldsymbol{M}$ *as follows:* $evi(f(ord(\boldsymbol{M})))$.

[20] If we think of updates just as generic maps, tracking really applies to *families* of maps on a whole current level of models. We will leave this matter of uniformity in diagrams implicit, but it is the way one should really think of our discussion in this section.

The choice of g is not unique here, any value $g(\mathbf{M})$ that induces the same plausibility order will do. If we want to enforce uniqueness, we need to sharpen up the definition of our levels and mappings between them, say, in a category-theoretic framework.

Our observations are no coincidence: there is a general construction behind the preceding fact. Consider operations on plausibility models defined in the "flat *PDL* program format" of van Benthem and Liu (2007), being unions of simple relational expressions of the forms

$$?(\neg)\varphi; \leq; ?(\neg)\varphi, \quad ?(\neg)\varphi; T; ?(\neg)\varphi$$

Fact 7.5 *There is an algorithm that takes any plausibility transformation in flat program format and returns an evidence transformation tracked by it.*

The proof of this result involves some tedious combinatorics which we omit.

One reason why tracking is so appealing is that we can consider the coarser-level update as giving a sort of "best approximate content" for the information provided by the finer-level update. For instance, think of the visible observable content of some partly hidden higher-order operation.

Another perspective on tracking diagrams is in terms of *translation*. They allow us to extend the earlier translation from a static evidence language for agent attitudes to the matching plausibility language with dynamic modalities for definable update operations. However, this technical perspective is not our main concern here, and we continue with tracking per se.

7.2 Non-trackable Operations

Perhaps the more interesting question runs in the opposite direction from the preceding one. Given a map g between two evidence models, is it "*trackable*," in the sense that there exists a map f that tracks their plausibility projections? It is not obvious that such a companion map always exists, so let us first state a precise criterion.

Fact 7.6 *A map g is trackable iff it has this property: on evidence models with the same* ord *projection, it yields values with the same* ord *projections.*

Proof First consider necessity of the criterion. If g is tracked by some map f, then, $ord(g(\mathbf{M})) = f(ord(\mathbf{M}))$. Now let $ord(\mathbf{N}) = ord(\mathbf{M})$. Then $ord(g(\mathbf{N})) = f(ord(\mathbf{N})) = f(ord(\mathbf{M})) = ord(g(\mathbf{M}))$. Conversely, if the stated invariance holds, it is easy to see that it is well-defined to let a tracking function $f(\mathbf{M})$ on plausibility models just output the model $ord(g(evi(\mathbf{M}))$. □

Using this criterion we show that some natural dynamic update operations on evidence models are not trackable.

Fact 7.7 *The deletion operator $-\varphi$ as defined earlier is not trackable.*

Proof Let the domain of $\mathbf{M}$ be $\{1, 2, 3, 4\}$, with evidence sets $\{1, 2\}, \{2, 3\}, \{3, 4\}, \{4, 1\}$, and φ holding only at 1 and 2. The induced plausibility model *ord*($\mathbf{M}$) consists of four incomparable points. The operation $-\varphi$ turns $\mathbf{M}$ into an evidence model with only the evidence sets $\{2, 3\}, \{3, 4\}, \{4, 1\}$. In its induced plausibility model *ord*($\mathbf{M} - \varphi$), the points 2 and 4 remain incomparable, but now $1 \leq 4$ and $2 \leq 3$. Now let the evidence model $\mathbf{M}^+$ have the same domain and valuation as $\mathbf{M}$, but with additional evidence sets $\{1, 3\}$, $\{2, 4\}$. This induces the same plausibility model as $\mathbf{M}$. However, when we apply $-\varphi$ to $\mathbf{M}^+$, deleting $\{1, 2\}$, due to the additional evidence sets remaining, the induced plausibility model *ord*($\mathbf{M}^+$) still consists of four incomparable points. This refutes trackability by the above criterion. □

Here is another example. Consider the above evidence operation tracked by radical upgrade. Now consider only one half, namely the following map:

shift($\varphi, \mathbf{M}$) sends $E \in \mathcal{E}$ with $E - \varphi$ non-empty to $E \cup \varphi$.

Fact 7.8 *The operator shift*($\varphi, \boldsymbol{M}$) *is not trackable.*

Proof Let $\mathbf{M}$ consist of $\{1, 2, 3\}$, with evidence sets $\{1, 2\}, \{2\}, \{1, 3\}$, and φ only true at 1 and 2. The plausibility model *ord*($\mathbf{M}$) has 1 and 2 incomparable, while $3 \leq 1$. The map *shift*($\varphi, -$) turns $\mathbf{M}$ into a new evidence model with evidence sets $\{1, 2\}$, $\{2\}$ and $\{1, 2, 3\}$. Its induced plausibility model has $1 \leq 2$, $3 \leq 1$, $3 \leq 2$. Let $\mathbf{M}^+$ have the same domain and valuation, but with one more evidence set $\{1\}$. This induces the same plausibility model as $\mathbf{M}$. However, applying $-\varphi$ to $\mathbf{M}^+$ yields evidence sets $\{1, 2\}$, $\{1\}$, $\{2\}$, $\{1, 2, 3\}$, and the induced plausibility model has 1 and 2 incomparable. □

The preceding results suggest that deletion of evidence is a delicate operation lacking an obvious plausibility counterpart, essentially, since we have lost the specific sets generating the ordering.

Still things depend on how we perform the deletion exactly. Here is one more interesting option, showing the richness of natural operations on evidence that come to light in the realm of our models.

Excursion One more version of deletion.

Instead of, as before, removing all evidence E implying the proposition φ, we now make a weaker move:

$\sim\varphi$ "dilutes" such evidence E by replacing it with $E \cup \neg\varphi$.

This time, the effect can be described faithfully in terms of plausibility order. If a point x does not satisfy φ, then its evidence is not affected by while seeing a more plausible world with $\neg\varphi$, then it satisfies no pure φ-evidence, and nothing changes for it.[21]

[21] Another test on a natural deletion operator $\sim\varphi$ is that its dynamic logic can be axiomatized completely by means of recursion laws for evidence and belief in the style of van Benthem and Pacuit (2011). Here is one such law for deletion: $[\sim\varphi]\Box\psi \leftrightarrow \Box^{\neg\varphi}\psi$, where $\Box^{\alpha}\psi$ says that there is evidence for ψ that is consistent with α.

One can see non-trackability as a problem, but in our view the opposite is the case. Absence of tracking shows that the richer world of evidence models suggests natural operations that are sui generis.

7.3 Internal Operations

Next, trackability is not the same as having externally measurable effects. Consider the earlier "internal" operations on evidence models that did not change the induced plausibility model. Our prime example was intersection of evidence sets. Its tracking map is extremely simple.

Fact 7.9 *Evidence combination is tracked by identity of plausibility models.*

This may be considered a disappointment, but we can also view this trackability by the identity map as a defining feature of what we mean by internal operations in the first place.

What this still leaves open is the issue of what internal operations are good for. We can view them partly as inferences (the combination rule is a sort of conjunction inference), and in that sense, their value may become apparent only at yet higher syntactic levels of representing information where acts of inference can be modeled explicitly (van Benthem and Martinez 2008) and (van Benthem and Quesada 2010).

But we can also think of internal operations as ways of rearranging or streamlining the current evidence without disturbing the plausibility order. For instance, in finite models, we can prune a given evidence family to a smallest one inducing the same order. Or, we can choose new evidence sets by combination that induce the same order in more perspicuous ways.

8 Logical Aspects I: Languages and Invariance Relations

In this paper so far, the term "levels" has been used quite loosely. But really, a level of structure does not just arise by specifying a similarity type of models, but also by giving transformations or invariance relations between what one takes to be "the same" structures (van Benthem 1996, 2011). This standard methodology of mathematics also applies to logic, and it has several interesting consequences for any systematic theory that aims at incorporating our earlier observations.

8.1 Invariance Relations

As for invariances between evidence models, (van Benthem and Pacuit 2011) study a standard brand of *neighborhood bisimulations*, close to topo-bisimulations (van

Benthem and Bezhanishvili 2007). But they also point out that richer languages of evidence models need more discriminating structural bisimulations. One recent example of the latter are the two-way bisimulations of van Benthem et al. (2015), which have a strengthened symmetric clause where the cross-model relation has to be total between the two neighborhoods compared in the back-and-forth step.

The same variety in defining notions of bisimulation can be found with plausibility models. See van Benthem (2011) and Andersen et al. (2013) for some examples, depending on how much structure one wants to preserve at this coarser level of doxastic representation.

8.2 Languages

Next, it is well-known that there is another side to this same coin. Invariance relations suggest introducing *languages* that can define the properties appropriate to a given invariance level. The model theory of modal logic or first-order logic provides key instances of this harmony: cf. van Benthem (2002) and van Benthem and Bonnay (2008) for general discussion.

Thus, there is also an issue of which languages we have in mind when discussing evidence models or plausibility models. In the latter realm, candidates considered included modalities for absolute and conditional belief, but also for safe belief, and so on. On evidence models, we had the basic evidence modality $\Box\varphi$, but also others suggested by the dynamics. A recent stronger candidate is the "instantial modality" (van Benthem et al. 2015):

> $[](\varphi_1, \ldots, \varphi_n; \psi)$: there exists an evidence set all of whose points satisfy ψ, while that set also contains φ_i-points for each i with $1 \leq i \leq n$.

8.3 Invariance and Dynamics

The choice of an invariance relation also has consequences for the dynamic update actions that are appropriate to the models for a given structure level. This issue has been studied extensively in the literature on process algebra (Bergstra et al. 2001), dynamic-epistemic logic (van Benthem 2011), or logics of games (van Benthem 2014). Generally speaking, model transformations need to respect a given structural invariance in the following sense: invariant input models lead to invariant output models.[22]

In a general treatment of our notion of tracking, all these issues will have to find their place. We conclude with one comment on this.[23]

---

[22]We can make this requirement even stricter in terms of definability (Hollenberg 1998).

[23]**Caveat.** One might think that by choosing the right invariance, one can also cross between levels, thereby undermining the whole intuitive picture of different levels that we started with. For instance,

Remark Category theory.

Our discussion of levels and tracking in this paper has been progressively more sketchy and programmatic. A full logical treatment of the landscape of information levels should probably involve a category-theoretic setting, perhaps of a sort already used for dynamic-epistemic logics (Baltag and Moss 2004). For a first attempt at such an uniform presentation, we refer to the follow-up study (Baltag et al. 2015).

9 Logical Aspects II: Looking Across Levels

We conclude our exploration with two aspects of logics and languages that occur when we compare different levels of the sort we had so far.

9.1 Tracking and Translation

Tracking diagrams naturally complement earlier translations between static languages for models at different levels. For instance, what the commuting diagrams of Sect. 7 said in linguistic terms is that the given translation t from the language of plausibility models to that of evidence models can be extended with clauses such as the following:

$[\#\varphi]\psi$ matches $[+t(\varphi)]t(\psi)$

The earlier issue of updates respecting notions of bisimulation then returns concretely in the inductive proof of the invariance of dynamic formulas for bisimulation. The upshot of such an analysis is this:

Fact 9.1 *The earlier translation results between levels for the static language of knowledge, belief and evidence extend to the extended dynamic languages, at least, for pairs of operators that track each other.*

9.2 Logics for Translation, Representation, and Change

Our final theme adds a bit of speculation. We can go yet further than the preceding dynamics. The perspective on information in this paper has two dimensions. In a

(Footnote 23 continued)
one could take "inducing the same plausibility order" as a strong notion of behavioral equivalence between evidence models. See the formal representation results in van Benthem et al. (2014) for some concrete examples, in the spirit of Andréka et al. (2002). While such a rough simulation may indeed blur our level distinctions, we believe that our earlier intuitions will stand up.

"horizontal" sense, we had update actions along models at the same level. But there is also a natural "vertical" dynamics, that of moving back and forth between different levels of representing information.

Perhaps a bit outrageously, this suggests a two-level dynamic logic combining updates at one level with acts of level change.[24] Its language would have the earlier syntax of dynamic logics of attitudes and updates at various levels, say two for simplicity. But to connect these, we add explicit *notation for level-connecting operations*, both up and down. The resulting dynamic logic for the new level-crossing modalities will revolve around recursion laws, as earlier in this paper for definable transformations, but now these laws will reflect the recursive clauses of the above translations.

Additional principles of such a two-dimensional logic of update and level change can be found as reflections of our earlier tracking diagrams in Sect. 7, that will now return in the form of commutation axioms. The resulting system can be seen as formalizing some of the elementary meta-theory of intra-level update and cross-level switching, just as dynamic-epistemic logics formalized basic properties of model-changing operations.

As a final example of this way of thinking, here is an issue of *complexity*. Commutative diagrams with tracking are what logicians like, and in a more general setting, what category theorists love.[25] On the other hand, we also know from modal logic (Blackburn et al. 2000) that complete logics of frames with commuting relations tend to be of high complexity, as their grid structure can encode complex geometrical tiling problems. Which force wins out in two-dimensional logics of the sort considered here?

10 Further Information Levels

A question that often come up when the preceding ideas are presented to audiences is the grand view of the total reprsesentation system for information. There is much more to information than the above three levels of epistemics, plausibility, and evidence—and there are several major directions to go.

Ordered Evidence. A good test on the issues raised so far is how evidence models fare with an obvious next level of structure, with a binary ordering that can stand for entrenchment, trust, or probabilistic weight. There is some concrete logical theory of such models: cf. Girard (2008) and Liu (2011), based on the "priority graphs" of Andréka et al. (2002). One can define new modalities of knowledge and belief based on ordered evidence, depending on how we view the order intuitively. This then leads to an extended theory of representation results and translation between the

---

[24]For a static perspective on level shifts in the concrete case of time, see (Montanari 1996).

[25]Tracking diagrams are not exactly commuting diagrams, but we ignore this finesse.

enriched logics, as well as a richer dynamics on priority graphs, including tracking and non-tracking results extending our earlier analysis.

Probability. One view is that our preceding levels represent relative plausibility as a form of qualitative probability, comparable to what was pursued by De Finetti and other pioneers in the 1930s. Finding precise connections here would involve a fresh look at logics for probability (Holliday and Icard 2013), while doing justice to the very different intuitions underlying plausibility models and probability models where many low scorers can team up to form a high-score area (van Benthem 2013). Also relevant in this comparative setting is Kelly and Lin (2012) on the impossibility of tracking quantitative Bayesian update in qualitative plausibility terms. Mierzewski (2015) proposes a new approach fixing the recent Leitgeb acceptance rule plus Bayesian conditioning to derive a tracking belief revision rule that can be studied in the spirit of nonmonotonic logics.

Syntax and Proof. One can also strike out in a different direction and think of evidence sets as *reasons* put forward to ground beliefs, or even knowledge, as in reason-based epistemology. Then the earlier "internal operations" become crucial, and they suggest a richer habitat with matching more fine-grained notions of information where, for instance, acts of inference have real update effects. In this case, the appropriate setting would seem to be real syntactic proof structure, where detailed formulation can be crucial to information flow consuming coded resources (van Benthem and Martinez 2008).[26] While there are some dynamic-epistemic logics at this level (van Benthem and Velazquez-Quesada 2010), the connection between syntactic and semantic approaches to information remains far from clear.

Interpolating Levels. Once we have the general picture, new questions of their own will arise. E.g., our semantic models and pure syntax seem very far apart. But then, the question is what natural intermediate levels of information structure exist in between the two. One such level is that of *algebraic logic*, where structures range widely, from rather syntactic ones like free algebras to algebras directly associated with set-theoretic models. In particular, we believe that the representation theory of modal algebras in terms of possible-worlds models (Blackburn et al. 2000; Andréka et al. 2014) may have a natural connection with the above uses of plausibility models and evidence models.

We conclude this picture of information levels with a caveat. It goes without saying that we are not claiming that there is a linear hierarchy of information levels. There can well be a more graph-like family with many forks.

---

[26]For a concrete illustration, in science, one well-chosen syntactic notation may be much more informative than a semantically equivalent one.

11 Further Issues and Directions

Expansion and Compression. Many further actions make sense in our two-dimensional perspective of levels and zooming in on richer levels, or zooming out to coarser ones. At the level of evidence, or reasons or proofs, these include giving reasons to others, say, when answering doubts.

But perhaps the major issue in our grand picture is the interplay of storing a lot of information about the past versus *compressing information*. The latter may be thought of as a necessity due to memory limitations, or intelligent resource constraints in solving tasks. It can also be viewed as *forgetting*, perhaps a human defect, but also a basic feature of civilization.

Cognitive Agency. The preceding topics also raise an issue of cognitive realism for our picture. We can think of all our information levels as mathematical structures with eternal connections, waiting to be used by human agents, but equally serene if no one ever comes to visit. But it is also tempting to think of this paper as exploring a world where human cognition takes place, and many of our topics make sense from the viewpoint of agents solving informational tasks. This involves at least two further features:

(a) the computational nature of agents (say, in an automata hierarchy),
(b) the nature of the specific issues or tasks that trigger level change.

In particular, on a task-and agent-oriented view, we may be making local excursions in our landscape of information levels to answer specific questions, rather than engage in dramatic migrations from one level to another.[27]

The Temporal Long-Term. The dynamics in this paper consisted of local steps, whether horizontal as update at one level of information, or as a vertical step toward a richer or poorer level. But informational inquiry also involves global patterns over time, witness the protocols that regulate learning (Hoshi 2009; van Benthem 2011; Kelly 1996; Gierasimczuk 2010). Our style of analyzing information structure is not in conflict with this long-term view: but it still needs to be added to our picture.

Mathematical Frameworks. What would be a best framework for placing the considerations and observations of this paper? One option are Chu spaces for abstract information, whose general treatment can be found in Barwise and Seligman (1995). Another option would be a general category-theoretic framework, for which a first attempt is found in Baltag et al. (2015).

However, there is yet another alternative, namely, to work at a suitable algebraic abstraction level, and explore the laws of information change there. While writing this paper, I increasingly felt that Mike Dunn's *Gaggle Theory* (Dunn 1991) may

[27] The point about local issues was made by Fenrong Liu (p.c.). For logicians, a concrete technical illustration of its utility is the widespread use of *filtration* in modal logic where models get coarsened using just a small finite set of "relevant formulas".

well be an ideal stance from which to study the themes explored in this paper, for its austerity, elegance and broad sweep.[28]

12 Conclusion

We have argued that information is a multi-faceted notion that is best studied at a variety of levels, each supporting their own intuitions of invariance and their own best languages for bringing out structure. In doing so, we found many new notions and distinctions, such as the contrast between internal operations that merely rearrange or elucidate, and update operations that are visible, and non-trivially trackable at other levels. We have also emphasized the further cross-level dimension of a "dynamics of zoom," with both information expansion and information reduction.

In all this, we have shown how logical notions and methods apply, making our approach a conscious extension of the logical approach to information in earlier traditions such as relevance logic, resource logics, or situation theory (van Benthem and Martinez 2008). But we are acutely aware that our current presentation is not the end stage, since we need more general mathematical perspectives to do justice to our landscape. And as we have stated in the preceding section, one great source for doing that is Mike Dunn's work.[29]

Acknowledgments I thank audiences at seminars and workshops in Amsterdam, Beijing, and Stanford that have listened to various versions of this talk, and I thank Katalin Bimbó and the referee for this book for their help. Finally, I want to thank Alexandru Baltag, Giovanni Cina, and Fenrong Liu for their feedback.

References

Adriaans, P., & van Benthem, J. (Eds.). (2008). *Handbook of the philosophy of information*. Amsterdam: Elsevier.

Andersen, M., Bolander, T., van Ditmarsch, H., & Jensen, M. (2013). Bisimulation for single-agent plausibility models. *Proceedings of Advances in Artificial Intelligence: 26th Australasian Joint Conference*, number 827 in *Lecture Notes in Computer Science* (pp. 277–288). New York, NY: Springer

Andréka, H., van Benthem, J., Bezhanishvili, N., & Németi, I. (2014). Changing a semantics: Opportunism or courage? In M. Manzano, et al. (Eds.), *The life and work of Leon Henkin* (pp. 307–337). Heidelberg: Birkhäuser.

Andréka, H., Ryan, M., & Schobbens, P.-Y. (2002). Operators and laws for combining preference relations. *Journal of Logic and Computation*, *12*, 13–53.

[28]This impression has been reinforced by recent intriguing work tying dynamic-epistemic logics to Gaggle Theory and general ternary semantics for relevant logics (Aucher 2014)—see also (Conradie et al. 2014), (Dunn and Meyer 1997) and (Kamide and Wansing 2011).

[29]Compared to Mike Dunn's outreach toward my own interests in Dunn (2014) this may still be too little, but I see this programmatic paper as a serious promissory note.

Aucher, G. (2014). Dynamic epistemic logic as a substructural logic. In A. Baltag & S. Smets (Eds.), *Johan van Benthem on logic and information dynamics*, Outstanding Contributions to Logic (pp. 855–880). Springer

Baltag, A., & Smets, S. (2006). Dynamic belief revision over multi-agent plausibility models. In G. Bonanno, W. van der Hoek, & M. Wooldridge (Eds.), *Proceedings LOFT '06* (pp. 11–24). Department of Computing, University of Liverpool

Baltag, A., Gierasimczuk, N., & Smets, S. (2011). Belief revision as a truth tracking process. In K. Apt (Ed.), *Proceedings of TARK 2011, Groningen* (pp. 187–191)

Baltag, A., van Benthem, J., & Cina, G. (2015). Category-theoretic perspectives on information structure, invariance, logic, and translation. ILLC, University of Amsterdam.

Baltag, A., & Moss, L. (2004). Logics for epistemic programs. *Synthese: Knowledge, Rationality, and Action*, *2*, 165–224.

Barwise, J., & Seligman, J. (1995). *Information flow, logic and distributed systems*. Cambridge, UK: Cambridge University Press.

Bergstra, J., Ponse, A., & Smolka, S. (2001). *Handbook of process algebra*. Amsterdam: Elsevier.

Bimbó, K. (2015). *Proof Theory: Sequent Calculi and Related Formalisms*. Discrete Mathematics and its Applications. Boca Raton, FL: CRC Press

Blackburn, P., de Rijke, M., & Venema, Y. (2000). *Modal logic*. Cambridge, UK: Cambridge University Press.

Conradie, W., Ghilardi, S., & Palmigiano, A. (2014). Unified correspondence. In A. Baltag & S. Smets (Eds.), *Johan van Benthem on Logic and Information Dynamics* (pp. 933–975). Outstanding Contributions to Logic. Springer

Dunn, J. M. (1991). Gaggle theory: An abstraction of Galois connections and residuation with applications to negation, implication, and various logical operators. In J. van Eijck (Ed.), *Logics in AI: European Workshop JELIA '90*, number 478 in *Lecture Notes in Computer Science* (pp. 31–51). Berlin: Springer

Dunn, J. M. (2014). Arrows pointing at arrows: Arrow logic, relevance logic and relation algebras. In A. Baltag & S. Smets (Eds.), *Johan van Benthem on Logic and Information Dynamics* (pp. 881–894). Outstanding Contributions to Logic. New York, NY: Springer

Dunn, J. M. (1971). An intuitive semantics for first degree relevant implications (abstract). *Journal of Symbolic Logic*, *36*, 362–363.

Dunn, J. M. (1976). Intuitive semantics for first-degree entailments and 'coupled trees'. *Philosophical Studies*, *29*, 149–168.

Dunn, J. M., & Meyer, R. K. (1997). Combinators and structurally free logic. *Logic Journal of the IGPL*, *5*(4), 505–537.

Gierasimczuk, N. (2010). *Knowing One's Limits: Logical Analysis of Inductive Inference*. PhD thesis, ILLC, University of Amsterdam.

Girard, P. (2008). *Modal Logic for Belief and Preference Change*. PhD thesis, Department of Philosophy, Stanford University and ILLC, University of Amsterdam.

Girard, P., Liu, F., & Seligman, J. (2012). General dynamic dynamic logic. In T. Bolander, T. Brauner, S. Ghilardi, & L. Moss (Eds.), *Advances in modal logic* (Vol. 9, pp. 239–260). London: College Publications.

Hollenberg, M. (1998). *Logic and Bisimulation*. PhD thesis, Philosophical Institute, University of Utrecht.

Holliday, W. (2012). *Knowing What Follows. Epistemic Closure and Epistemic Logic*. PhD thesis, Department of Philosophy, Stanford University and ILLC, University of Amsterdam.

Holliday, W., & Icard, T. (2013). Measure semantics and qualitative semantics for epistemic modals. In *Proceedings SALT 2013 Semantics and Linguistic Theory*. Santa Cruz, CA: UC Santa Cruz.

Hoshi, T. (2009). *Epistemic Dynamics and Protocol Information*. PhD thesis, Department of Philosophy, Stanford University and ILLC University of Amsterdam.

Kamide, N., & Wansing, H. (2011). Connexive modal logic based on positive S4. In J.-Y. Béziau & M. Coniglio (Eds.), *Logic without Frontiers. Festschrift for Walter Alexandre Carnielli on the Occasion of his 60th Birthday* (pp. 389–409). College Publications, London

Kelly, K. (1996). *The logic of reliable inquiry*. Oxford, UK: Oxford University Press.
Kelly, K., & Lin, H. (2012). Propositional reasoning that tracks probabilistic reasoning. *Journal of Philosophical Logic, 41*(6), 957–981.
Liu, F. (2011). *Reasoning about preference dynamics*. Dordrecht: Springer.
Mierzewski, K. (2015). *The logic of stable belief under Bayesian conditioning*. Department of Philosophy, Stanford University
Montanari, A. (1996). *Metric and Layered Logic for Temporal Granularity*. PhD thesis, University of Amsterdam.
van Benthem, J. (1996). *Exploring logical dynamics*. Stanford, CA: CSLI Publications.
van Benthem, J. (2000). Information transfer across Chu spaces. *Logic Journal of the IGPL, 8*(6), 719–731
van Benthem, J. (2002). Invariance and definability: Two faces of logical constants. In W. Sieg, R. Sommer & C. Talcott (Eds.), *Reflections on the Foundations of Mathematics. Essays in Honor of Sol Feferman*, number 15 in *ASL Lecture Notes in Logic*, ASL (pp. 426–446)
van Benthem, J. (2013). A problem concerning qualitative probabilistic update. *Research report*, ILLC, University of Amsterdam
van Benthem, J. (2015). Talking about knowledge. In C. Baskent, L. Moss, & E. Pacuit (Eds.), *Rohit Parikh*. Outstanding Contributions to Logic, Springer
van Benthem, J., Bezhanishvili, N. & Yu, J. (2015). Instantial neighborhood logic, *working paper*, ILLC, University of Amsterdam and Joint Research Center in Logic, Tsinghua University.
van Benthem, J., Fernandez, D. & Pacuit, E. (2012). Evidence logic: A new look at neighborhood structures. In T. B. et al. (Ed.), *Advances in Modal Logic, Copenhagen 2012* (pp. 97–118). London, UK: College Publications
van Benthem, J. (2011). *Logical dynamics of information and interaction*. Cambridge, UK: Cambridge University Press.
van Benthem, J. (2014). *Logic in games*. Cambridge, MA: MIT Press.
van Benthem, J., & Bezhanishvili, G. (2007). Modal logics of space. In M. Aiello, I. Pratt-Hartmann, & J. van Benthem (Eds.), *Handbook of spatial logics* (pp. 217–298). Dordrecht: Springer.
van Benthem, J., & Bonnay, D. (2008). Modal logic and invariance. *Journal of Applied Non-Classical Logics, 18*(2–3), 153–173.
van Benthem, J., Grossi, D., & Liu, F. (2014). Priority structures in deontic logic. *Theoria, 80*(2), 116–152.
van Benthem, J., & Liu, F. (2007). Dynamic logic of preference upgrade. *Journal of Applied Non-Classical Logics, 17*(2), 157–182.
van Benthem, J., & Martinez, M. (2008). The stories of logic and information. In P. Adriaans & J. van Benthem (Eds.), *Handbook of the philosophy of information* (pp. 217–280). Amsterdam: Elsevier.
van Benthem, J., & Pacuit, E. (2011). Dynamic logic of evidence-based beliefs. *Studia Logica, 99*(1), 61–92.
van Benthem, J., & Quesada, F. V. (2010). The dynamics of awareness. *Synthese, 177*(1), 5–27.
van Benthem, J., & Smets, S. (2015). Dynamic logics of belief change. In H. van Ditmarsch, J. Halpern, W. van der Hoek, & B. Kooi (Eds.), *Handbook of logics for knowledge and belief*. London, UK: College Publications.
van Ditmarsch, H., van der Hoek, W., & Kooi, B. (2007). *Dynamic epistemic logic*. Cambridge, UK: Cambridge University Press.

Syllogistic Logic with Cardinality Comparisons

Lawrence S. Moss

Abstract This paper enlarges classical syllogistic logic with assertions having to do with comparisons between the sizes of sets. So in addition to assertions like All x are y and Some x are y, we also have There are at least as many x as y, and There are more x than y. Our work also allows all nouns to be complemented. We thus obtain sentences equivalent to No x are y and At least half of the universe are x. We work on finite models exclusively. We formulate a syllogistic logic for our language. The main result is a soundness/completeness theorem. The logic has a rule of *ex falso quodlibet*, and *reductio ad absurdum* is admissible. There are efficient algorithms for proof search and model construction, and the logic has been implemented.

Keywords Cardinality · Completeness · Natural logic and quantification · Syllogistic logic

Dedication

It is a pleasure to dedicate this paper to J. Michael Dunn. Ever since his arrival at Indiana University in 1969, Mike has been a central part of IU's logic group. He played the key role in Jon Barwise's move to IU in 1990. My own move to IU came at the same time. I credit Mike more than anyone else for nurturing logic at IU, for serious interdisciplinarity, and for organizational and intellectual leadership. As he "graduated" from Philosophy to Informatics, he showed by his example that becoming a founding dean of a School need not entail checking one's research like a bag at the door. I wish Mike many more productive years in logic, and in all of his other pursuits.

I feel fortunate to have written two papers with Mike (so far), papers on quantum logic re-worked from the point of view of quantum computation. The paper which you are reading is not related to that work, and indeed it is not directly connected to the Dunn *oeuvre*. But I believe that it has a thematic connection with his work. To explain the matter, I comment on a passage from an unpublished draft of his.

L.S. Moss (✉)
Department of Mathematics, Indiana University, Bloomington, IN, USA
e-mail: lsm@cs.indiana.edu
URL: http://www.indiana.edu/~iulg/moss

K. Bimbó (ed.), *J. Michael Dunn on Information Based Logics*,
Outstanding Contributions to Logic 8, DOI 10.1007/978-3-319-29300-4_18

> I would no longer describe myself so much as a lawyer of logics, but more as an engineer of logics – a maker of tools. Man has been defined in the Aristotelean tradition as a rational animal. Benjamin Franklin defined man as a tool making animal. ... Primitive man had primitive tools, and primitive man also had primitive rationality. As humanity developed, it developed more sophisticated tools, and these included more sophisticated tools for reasoning. That logical reasoning is far from an inherent skill is obvious to anyone who has taught elementary logic. It often starts with learning to use simple tools (informal reasoning, fallacies, syllogisms) and builds to the use of more sophisticated tools (propositional calculus, then first-order logic). And this progression shows itself not just in the use of these tools but in their actual construction over the history of logic. It was only in the twentieth century that non-classical logics came into their own. This, at least originally, had nothing to do with the invention of computers. But their connection with computing has become more and more established. — Dunn (2015)

This offers some insight into his own philosophy of logic and computation. The line of work in this paper, *natural logic*, could be described as an "engineering" approach to logic. It is strongly connected to computational issues. And it builds tools. Those tools are supposed to have something to do with human reasoning. Where natural logic deviates from the quote is that it is not connected to the main "progression" in the history of logic. In fact, the topic of this paper is more connected to syllogistic logic than to propositional and first-order logic.

1 Introduction

Syllogistic logic is one of the most historically important logical systems. For fifteen hundred years Aristotle was the *magister*, and his writings nourished syllogistic logic as a living tree. But with the advent of first-order logic, the tree of syllogistic logic withered. The general topic of this paper is a recent resuscitation of the tree. One aims for logical systems which are "big enough" to cover interesting phenomena, and yet "small enough" to be decidable, or even efficiently so. This is rather close to what the logic engineer is doing. Perhaps the difference is that I want to describe logical tools that seem to be implicit in the human reasoning facility. It is much less important to me to connect to the mainstream of the post-Frege logical tradition. Put another way, human reasoning makes use of "simple tools" which we aim to formalize. They are buds on the branches of the tree. This paper is about one of those buds.

The main logical issue in this paper is *reasoning about the sizes of discrete sets*. To get to the issues quickly, consider the following argument:

$$\frac{\begin{array}{l}\textsf{There are more students than professors at the party}\\ \textsf{There are more professors than deans at the party}\end{array}}{\textsf{There are more students than deans at the party}} \qquad (1)$$

I take it as clear that the conclusion follows from the premises. More controversially, my intuition is that the transitivity of more ... than ... is a basic feature of human reasoning, on a par with the transitivity of all ... are ... that we see in the syllogistic rule (BARBARA). One should not formalize the argument in (1) by translating it into another logic (for example, logical systems which incorporate natural numbers); the point is that the general logical principles of the target systems are likely to be much more complicated than necessary for that task. So my aim in this paper is to ask what the syllogistic would look like if we re-engineered it to look as close as possible to the original syllogistic (or to modern reconstructions such as Łukasiewicz 1957; Corcoran 1972; Martin 1997), but also with additional sentences expressing cardinality comparison. In addition to the syllogistic rules of All, Some, and No what other rules would be needed?

Before discussing the actual content of the paper, let us widen the discussion a little. In addition to more ... than ..., we also find in language the weaker assertion there are at least as many ... as .... Here is another argument which we take to be valid:

$$\frac{\begin{array}{l}\text{There are at least as many rabbits as deer}\\ \text{There are more deer than goats}\end{array}}{\text{There are more rabbits than goats}} \qquad (2)$$

Here is an argument of a different character:

$$\frac{\begin{array}{l}\text{All violas are stringed instruments}\\ \text{There are at least as many violas as stringed instruments}\end{array}}{\text{All stringed instruments are violas}} \qquad (3)$$

A moment's thought will convince the reader that this is valid, provided that we are speaking of *finite* situations. We shall restrict attention to finite universes, in order to obtain a logical system that we think if of greater "human interest" than the weaker logic that would result if we allowed infinite structures and thus denied the validity of (3).

We are aiming at a logic which is capable of representing (1)–(3). In addition, we shall make our logical language more expressive yet by allowing *complementation of nouns*. Here are some examples:

$$\frac{\text{There are at least as many } x \text{ as } y}{\text{There are at least as many non-}y \text{ as non-}x}$$

$$\frac{\begin{array}{l}\text{There are at least as many } x \text{ as non-}x\\ \text{There are at least as many } y \text{ as non-}y\end{array}}{\text{There are at least as many } x \text{ as non-}y} \qquad (4)$$

The first example just above shows an inference whose soundness depends on the fact that we are looking at a finite universe. The second uses a property of "half": if the universe has N objects, the premises tell us that the xs are at least $\frac{N}{2}$ in number. The y's number at least $\frac{N}{2}$, and so the non-y's number at most $\frac{N}{2}$. Thus the x's number at least as much as the non-y's. The fact that we can do all of this with cardinality comparison and complement makes this work interesting and non-trivial. But there is no real combinatorial component to the work, unlike other papers in the area (Endrullis and Moss to appear; Lai et al. to appear).

Contents. The main result in this paper is a sound and complete logical system whose sentences are of the form All x are y, Some x are y, There are at least as many x as y, and There are more x than y. Our logic does not involve translating the cardinality assertions into any other language. The proof system is sound and strongly complete: for a finite set $\Gamma \cup \{\varphi\}$ of sentences, φ is true in every model of Γ if and only if there is a derivation of φ from Γ. The language and proof system are discussed in Sect. 2, and the remaining sections discuss aspects of the completeness proof.

Implementation. Our quote from Dunn's (2015) remarks on the connection between non-classical logics and computing. My sense that he could be thinking of logics motivated by computational practice, such as logics motivated by his own work in relevance logic, quantum logic, or many-valued logic. The point of connection between computing and the work of this paper is somewhat different. The point is that much of the practice in fields like artificial intelligence, cognitive science, and linguistics, one finds a stated or unstated constraint that theories and proposals of all types be computational. If a proposal calls for computational or logical systems that are "Turing complete," then this is evidence against it. Conversely, the more feasible, the better. The logic in this paper is about as computationally "light" as one can get. I do not wish to discuss its computational complexity in this paper, or even the specific algorithm that comes from the work we do. But I do want to advertise the fact that syllogistic logic with cardinality comparisons has been implemented in the Sage programming language, and the implementation is currently available on https://cloud.sagemath.com. (That is, this author can share it.) For example, one may enter

```
assumptions= ['All non-a are b',
'There are more c than non-b',
'There are more non-c than non-b',
'There are at least as many non-d as d',
'There are at least as many c as non-c',
'There are at least as many non-d as non-a']
conclusion = 'All a are non-c'
follows(assumptions,conclusion)
```

The last line indicates that we are asking if a given conclusion follows from a given list of six assumptions. Then the program returns, telling us that the conclusion does not follow. Additionally, it produces a *counter-model*, a model where all of the assumptions are true and the conclusion false.

```
Here is a counter-model.
We take the universe of the model to be {0, 1, 2, 3, 4, 5}

  noun   semantics        complement
+------+-------------+--------------------+
  a      {2, 3}           {0, 1, 4, 5}
  b      {0, 1, 4, 5}     {2, 3}
  c      {0, 2, 3}        {1, 4, 5}
  d      {}               {0, 1, 2, 3, 4, 5}
```

So it gives the semantics of a, b, c, and d as subsets of $\{0, \ldots, 5\}$. Notice that the assumptions are true in the model, but the conclusion is false. In the cases that the conclusion did follow, the system would output a proof in our system. See Example 2.4 for an illustration of a formal proof found by a computer.

The reader should also try to construct a counter-model by hand in order to get a feeling for the issues in this paper. Queries which feature negation are especially difficult for people to work with.

Remark 1.1 One reviewer noticed that it is possible to get a counter-model with just two elements and asked "why the size of the model [produced by a program] is so 'large.'" Here is a quick answer. Suppose one is building a first-order model of a finite sequence $\varphi_1, \ldots, \varphi_n$ of sentences in first-order logic, and suppose that φ_1 is $(\exists x)R(x)$ and that φ_2 is $(\exists x)S(x)$. It is natural to build a model by taking new objects, say c and d, and declaring that $R(c)$ and $S(d)$. When one does this, it is natural to make d different than c, because φ_3 might be $(\forall x)(R(x) \rightarrow \neg S(x))$. The point is that in building a model step-by-step, the natural steps will not result in a minimal model. The same thing happens with the logic in this paper.

The advantage of working with a syllogistic system formulated using *ex falso quodlibet* rather than *reductio ad absurdum* is that proof search and counter-model generation are closely related. In a sense, they are both results of the same algorithm. Moreover, the algorithm is efficient. That is, the question of whether a sentence follows from a list of assumptions is in polynomial time. I am not going to discuss the algorithmic aspects of the logic in this paper, except to mention examples and some of the general issues.

Related Work. There is a large body of work on generalized quantifiers in logic, including quantifiers coming from natural language. Two papers to mention are Mostowski (1957) and Lindström (1966). The logical systems in this paper would constitute a small fragment of what we find in those papers. This is because traditionally, logical systems *extend* first-order logic. A fortiori, their decision problems would be undecidable. For more on such logics, see (Herre et al. 1991). Closer to our topic would be generalized quantifiers on finite structures, as in Kolaitis and Väänänen (1995). But I am not aware of any work that studies logical completeness and decidability theorems. This is again related to the issue of the base logic. In what I am calling "natural logic," one does not begin with first-order logic or even propositional logic, and this is what enables the results in this paper. For work on natural logic more broadly, see (Moss 2015; Pratt-Hartmann and Moss 2009; van Benthem 2008). However, this paper stands on its own in the sense that all of the work is self-contained (and elementary). Work on the numerical syllogistic is close to what we are doing; see (Pratt-Hartmann 2009) for a negative result which contrasts with the positive results here.

2 Syntax and Semantics

The logic $\mathcal{S}^\dagger(\text{card})$ of this paper starts with a set of *raw variables* $\mathbf{P}_0$ which we denote by $p, q, \ldots$. It also has *complement variables* $\overline{p}, \overline{q}, \ldots$ which correspond to these. Our set $\mathbf{P}$ of *nouns* is the disjoint union of two copies of $\mathbf{P}_0$:

$$\mathbf{P} = \mathbf{P}_0 \cup \{\,\overline{p} \colon p \in \mathbf{P}_0\,\}.$$

However, we shall use the same letters $p, q, \ldots$ to denote nouns which might be complemented, and we also extend the complement notation "classically," so that $\overline{\overline{p}}$ is always identified with p.

$\mathcal{S}^\dagger(\text{card})$ has sentences of the form $\forall(p,q)$ and $\exists(p,q)$, $\exists^{\geq}(p,q)$, $\exists^{>}(p,q)$. We read "$\exists^{\geq}(p,q)$" as "there are at least as many p as q," and we read "$\exists^{>}(p,q)$" as "there are more p than q." There are no connectives.

Our raw variables are denoted with letters like p and q. But in examples and in proofs we frequently use x, y, and z. We hope that this is not confusing. Late in the paper we also see other letters, chosen to make various arguments easier to follow. Throughout the paper, lower case Roman letters will be used for raw variables.

We interpret this language $\mathcal{S}^\dagger(\text{card})$ on finite models $\mathcal{M}$ which are sets M together with interpretations $[\![p]\!]$ for all nouns p. The set M is sometimes called the *universe* of the model $\mathcal{M}$. We require that the complement operation work classically in the semantics: $[\![\overline{p}]\!] = M \setminus [\![p]\!]$, for all p. The sentences $\exists^{\geq}(p,q)$ and $\exists^{>}(p,q)$ have the following semantics:

$$\begin{array}{lll}
\mathcal{M} \models \forall(p,q) & \text{iff} & [\![p]\!] \subseteq [\![q]\!] \\
\mathcal{M} \models \exists(p,q) & \text{iff} & [\![p]\!] \cap [\![q]\!] \neq \emptyset \\
\mathcal{M} \models \exists^{\geq}(p,q) & \text{iff} & \text{card}([\![p]\!]) \geq \text{card}([\![q]\!]) \\
\mathcal{M} \models \exists^{>}(p,q) & \text{iff} & \text{card}([\![p]\!]) > \text{card}([\![q]\!])
\end{array} \tag{5}$$

On the right in (5), the symbol $\text{card}(S)$ stands for the cardinality (number of elements) of a given set S.

We allow the empty model. Nothing much hinges on this. We could have disallowed the empty model and only small changes would result in what we do.

It should be noted that in (5), the letters p and q range over all nouns, not just over the raw variables. In other words, we have saved a lot of needless repetition by allowing p and q to be either raw variables or complemented variables.

Special attention should be given to sentences where $q = \overline{p}$. So $\exists^{\geq}(p, \overline{p})$ says that there are at least as many p's as non-p's. This sentence $\exists^{\geq}(p, \overline{p})$ might therefore be read as "the p's are at least half of the objects in the universe." Similarly, $\exists^{\geq}(\overline{p}, p)$ might be read as "the p's are at most half of the objects in the universe." We can also read $\exists^{>}(p, \overline{p})$ as "the p's are more than half of the objects in the universe," and $\exists^{>}(\overline{p}, p)$ as "the p's are less than half of the objects in the universe."

Remark 2.1 As we mentioned, the logical language $\mathcal{S}^{\dagger}$(card) does not have boolean connectives, and it also does not have a negation symbol. It does have a *semantic negation*: for every sentence φ, there is a sentence $\overline{\varphi}$ such that $\mathcal{M} \models \overline{\varphi}$ iff $\mathcal{M} \not\models \varphi$. Here is how this works:

φ	$\overline{\varphi}$
$\forall(p, q)$	$\exists(p, \overline{q})$
$\exists(p, q)$	$\forall(p, \overline{q})$
$\exists^{\geq}(p, q)$	$\exists^{>}(q, p)$
$\exists^{>}(p, q)$	$\exists^{\geq}(q, p)$

We are interested in working with this semantics only on *finite universes*. This is because the logic is stronger this way. That is, some of the rules which we shall see shortly are not sound for infinite universes.

2.1 $\mathcal{S}$(card) and $\mathcal{S}^{\dagger}$(card)

We are mainly interested in $\mathcal{S}^{\dagger}$(card) in this paper. But we are also interested in a smaller language which we call $\mathcal{S}$(card). This language $\mathcal{S}$(card) is the same as $\mathcal{S}^{\dagger}$(card), but it lacks complemented variables. The semantics is the same.

Incidentally, the notations $\mathcal{S}$ and $\mathcal{S}^{\dagger}$ come from Pratt-Hartmann and Moss (2009). $\mathcal{S}$ was used there for the classical syllogistic, and so it seems appropriate to use the notation $\mathcal{S}$(card) for the logical system that extends $\mathcal{S}$ with cardinality assertions. In Pratt-Hartmann and Moss (2009), the dagger notation was used in connection with logics with full complementation on all nouns. So the difference between $\mathcal{S}$ and $\mathcal{S}^{\dagger}$ is that the latter system allows sentences $\forall(\overline{p}, \overline{q})$ and $\exists(\overline{p}, \overline{q})$; in $\mathcal{S}$ one cannot even say $\forall(\overline{p}, \overline{q})$. It turns out that allowing complementation on all nouns has a significant effect on the logical system.

2.2 Proof System

Let Γ be a set of sentences in $\mathcal{S}^{\dagger}$(card). A *proof tree over* Γ is a finite tree $\mathcal{T}$ whose nodes are labeled with sentences, and each node is either a leaf node labeled with an element of Γ, or else matches one of the rules in the proof system in Fig. 1. $\Gamma \vdash \varphi$ means that there is a proof tree $\mathcal{T}$ for over Γ whose root is labeled φ. See Sect. 2.3 for examples.

Recall that we are also working with the smaller language $\mathcal{S}$(card) which does not have complemented variables. The proof rules for $\mathcal{S}$(card) are the rules in Fig. 1 above the line. One rule in this logic for $\mathcal{S}$(card) is derivable in the bigger logic for $\mathcal{S}^{\dagger}$(card). This is (MORE-RIGHT). So in the larger logic for $\mathcal{S}^{\dagger}$(card), this rule could be dropped. (Actually, (MORE-RIGHT) and (MORE-LEFT) are inter-derivable. So we only need one of these.) But we need it to get a complete system for $\mathcal{S}$(card).

One rule which uses the finiteness assertion is (CARD MIX). It says that if all y are x, and there are at least as many elements in the bigger set y as in x, then the sets have to be the same.

We turn to the rules at the bottom of Fig. 1, since they show the interaction of the different sentence types and also involve the "half" interpretation from above.

The logic has an *Ex falso quodlibet rule*, listed with an (X) at the bottom of the top half of Fig. 1. In addition, there is a second *Ex falso* rule which is derivable from the first. See Example 2.5 for this.

We next mention the soundness of the systems for $\mathcal{S}$(card) and $\mathcal{S}^{\dagger}$(card). Proposition 2.1 is stated for $\mathcal{S}^{\dagger}$(card), but the same work holds for $\mathcal{S}$(card), mutatis mutandis. We write $\Gamma \models \varphi$ to mean that every model of all sentences in Γ is a model of φ.

Proposition 2.1 (Soundness) *If* $\Gamma \vdash \varphi$, *then* $\Gamma \models \varphi$.

Proof The proof is by induction on the heights of proof trees. The proof reduces to showing that all of the rules are individually sound.

The rules without $\exists^{\geq}$ and $\exists^{>}$ are well-known syllogistic rules, and they are easily seen to be sound. We discussed (CARD-MIX) in the Introduction; its soundness depends on the fact that we restrict to finite universes. The soundness of (SUBSET-SIZE) does not need this restriction; it just says that if $[\![p]\!] \subseteq [\![q]\!]$, then the size of $[\![q]\!]$ is at least as large as the size of $[\![p]\!]$. The (X) rule just says that if in a given model $\mathcal{M}$ there are at least as many q's as p's, and if in that same model $\mathcal{M}$ there are more p's than q's, then $\mathcal{M}$ satisfies *every* sentence. Of course, this is because the assumptions cannot both hold in one and the same model.

Turning to the rules below the line in Fig. 1, (ZERO) and (ONE) are from Moss (2010). (In fact, using the purely syllogistic rules above the line and the *ex falso quodlibet* rule that we present in Example 2.5, we have a complete logic for the language $\mathcal{S}^{\dagger}$ that lacks cardinality assertions.) The finiteness assumption is needed in the soundness of (MORE-ANTI): if there are more q's than p's, then there are more non-p's than non-q's. The rule (INT) says that if there is some p (so that the universe is not empty), and the q's constitute at least half of the universe, then there is at least one q. This is because the size of the universe must be an integer ≥ 1.

The rules (HALF), (STRICT HALF), and (MAJ) are the least obvious and hence the most interesting rules in the system. If there are at least as many p as non-p, then the p's number at least half the size of the universe. If in addition the non-q's number at least half the size of the universe, there must be at least as many p's as q's. (This is because "half the size of the universe" is a well-defined number. It does not depend on p or q.) The justification of (STRICT HALF) is similar.

$$\frac{}{\forall(p,p)}\ \text{(AXIOM)} \qquad \frac{\forall(n,p)\quad \forall(p,q)}{\forall(n,q)}\ \text{(BARBARA)}$$

$$\frac{\exists(p,q)}{\exists(p,p)}\ \text{(SOME)} \qquad \frac{\exists(q,p)}{\exists(p,q)}\ \text{(CONVERSION)}$$

$$\frac{\exists(p,n)\quad \forall(n,q)}{\exists(p,q)}\ \text{(DARII)} \qquad \frac{\forall(p,q)\quad \exists^{\geq}(p,q)}{\forall(q,p)}\ \text{(CARD-MIX)}$$

$$\frac{\forall(p,q)}{\exists^{\geq}(q,p)}\ \text{(SUBSET-SIZE)} \qquad \frac{\exists^{\geq}(n,p)\quad \exists^{\geq}(p,q)}{\exists^{\geq}(n,q)}\ \text{(CARD-TRANS)}$$

$$\frac{\exists(p,p)\quad \exists^{\geq}(q,p)}{\exists(q,q)}\ \text{(CARD-}\exists\text{)} \qquad \frac{\exists^{>}(p,q)}{\exists^{\geq}(p,q)}\ \text{(MORE-AT LEAST)}$$

$$\frac{\exists^{>}(n,p)\quad \exists^{\geq}(p,q)}{\exists^{>}(n,q)}\ \text{(MORE-LEFT)} \qquad \frac{\exists^{\geq}(n,p)\quad \exists^{>}(p,q)}{\exists^{>}(n,q)}\ \text{(MORE-RIGHT)}$$

$$\frac{\exists^{\geq}(p,q)\quad \exists^{>}(q,p)}{\varphi}\ \text{(X)}$$

$$\frac{\forall(p,\overline{p})}{\forall(p,q)}\ \text{(ZERO)} \qquad \frac{\forall(\overline{p},p)}{\forall(q,p)}\ \text{(ONE)}$$

$$\frac{\forall(q,p)\quad \exists(p,\overline{q})}{\exists^{>}(p,q)}\ \text{(MORE)} \qquad \frac{\exists^{>}(p,q)}{\exists(p,\overline{q})}\ \text{(MORE-SOME)}$$

$$\frac{\exists^{>}(q,p)}{\exists^{>}(\overline{p},\overline{q})}\ \text{(MORE-ANTI)} \qquad \frac{\forall(p,q)}{\forall(\overline{q},\overline{p})}\ \text{(ANTI)}$$

$$\frac{\exists^{\geq}(p,q)}{\exists^{\geq}(\overline{q},\overline{p})}\ \text{(CARD-ANTI)} \qquad \frac{\exists(p,p)\quad \exists^{\geq}(q,\overline{q})}{\exists(q,q)}\ \text{(INT)}$$

$$\frac{\exists^{\geq}(p,\overline{p})\quad \exists^{\geq}(\overline{q},q)}{\exists^{\geq}(p,q)}\ \text{(HALF)} \qquad \frac{\exists^{>}(p,\overline{p})\quad \exists^{\geq}(\overline{q},q)}{\exists^{>}(p,q)}\ \text{(STRICT HALF)}$$

$$\frac{\exists^{\geq}(p,\overline{p})\quad \exists^{\geq}(q,\overline{q})\quad \exists(\overline{p},\overline{q})}{\exists(p,q)}\ \text{(MAJ)}$$

Fig. 1 Rules for $\mathcal{S}^{\dagger}$(card), above and below the line. The rules for the smaller system $\mathcal{S}$(card) are found above the line

Turning to the (MAJ) rule, assume about a model $\mathcal{M}$ that there are at least as many p as non-p, and there are at least as many q as non-q. Then the p's and q's have at least half of the elements of M. Also assume towards a contradiction that the p's and q's are disjoint. Then they must be complement sets, each with exactly half of the elements of M. From this, it follows that $[\![\overline{p}]\!] = [\![q]\!]$, and $[\![\overline{q}]\!] = [\![p]\!]$. Also, $[\![\overline{p}]\!] \cap [\![\overline{q}]\!] = [\![q]\!] \cap [\![p]\!] = \emptyset$. This contradicts the third premise.

2.3 Examples

This section contains a few examples of derivations.

Example 2.2 Here is an argument in English:

$$\frac{\begin{array}{l}\textsf{There are more } p \textsf{ than non-}p\\ \textsf{There are at least as many } q \textsf{ as non-}q\end{array}}{\textsf{Some } p \textsf{ is a } q}$$

The conclusion does follow. The p's are a strict majority, and the q's are either a strict majority or have exactly half of the objects in the universe. Some p must also be a q, for if not, all p's would be non-q's, and the p's would number at most half of the universe. Our logic does not have proofs by *reductio ad absurdum*, but nevertheless we do have a derivation that the conclusion follows from the premises:

$$\dfrac{\dfrac{\exists^{>}(p,\overline{p}) \quad \exists^{\geq}(q,\overline{q})}{\exists^{>}(p,\overline{q})}\ \text{(STRICT HALF)}}{\exists(p,q)}\ \text{(MORE–SOME)}$$

Example 2.3 All x are non-y follows from the list of assumptions below:

1. There are at least as many non-y as y
2. There are at least as many non-z as z
3. All x are z
4. All non-y are z

Here is a formal proof in our system:

$$\dfrac{\forall(x,z) \qquad \dfrac{\forall(\overline{y},z) \qquad \dfrac{\exists^{\geq}(\overline{y},y) \quad \exists^{\geq}(\overline{z},z)}{\exists^{\geq}(\overline{y},z)}\ \text{(HALF)}}{\forall(z,\overline{y})}\ \text{(CARD MIX)}}{\forall(x,\overline{y})}\ \text{(BARBARA)}$$

Example 2.4 Here is an example of a derivation found by our implementation. We ask whether the putative conclusion below really follows:

$$\frac{\begin{array}{l}\textsf{All non-x are x}\\ \textsf{Some non-y are z}\end{array}}{\textsf{There are more x than y}}$$

The program returns:

```
Here is a formal proof in our system:

1   All               non-x   are    x       Assumption
2   All               y       are    x       One    1
3   All               non-x   are    x       Assumption
4   All               non-y   are    x       One    3
5   Some              non-y   are    z       Assumption
6   Some              non-y   are    non-y   Some   5
7   Some              non-y   are    x       Darii    4   6
8   Some              x       are    non-y   Conversion   7
9   There are more    x       than   y       More    2   8
```

Note that the proof is displayed as a list rather than a tree. But this is merely a cosmetic difference.

Example 2.5 The *ex falso quodlibet* rule of syllogistic logic (Pratt-Hartmann and Moss 2009) allows one to derive an arbitrary sentence from No p are q and Some p are q. This rule is derivable in $\mathcal{S}^{\dagger}$(card):

$$\dfrac{\dfrac{\dfrac{}{\forall(p,p)}\ (\text{AXIOM})}{\exists^{\geq}(p,p)}\ (\text{SUBSET-SIZE}) \qquad \dfrac{\dfrac{}{\forall(p,p)}\ (\text{AXIOM}) \qquad \dfrac{\exists(p,q) \quad \dfrac{\forall(p,\overline{q})}{\forall(q,\overline{p})}\ (\text{ANTI})}{\exists(p,\overline{p})}\ (\text{DARII})}{\exists^{>}(p,p)}\ (\text{MORE})}{\varphi}\ (\text{X})$$

Remark 2.6 Reductio ad absurdum (RAA) turns out to be admissible in the system: If $\Gamma \cup \{\varphi\} \vdash \psi$ and $\Gamma \cup \{\varphi\} \vdash \overline{\psi}$, then $\Gamma \vdash \overline{\varphi}$. But this admissibility fact is not obvious, and indeed the only argument which we know for it follows from the completeness of the system without reductio. That is, it seems difficult to establish the admissibility proof-theoretically.

As it happens, much of the technical work in this paper is needed simply because the proof system does not have (RAA). If we were to replace the (X) rule with (RAA), the set of rules would be much smaller and the completeness proof would be much easier. The point is that completeness would reduce to showing that if Γ is consistent in the logic, then it has a model. We are going to show this as a lemma in our completeness proof. But in order to show completeness, we must show that if $\Gamma \nvdash \varphi$, then $\Gamma \cup \{\overline{\varphi}\}$ has a model. And we need to do this without knowing that this last set $\Gamma \cup \{\overline{\varphi}\}$ is consistent in the logic.

In connection with (RAA) and the (X) rules, we also should mention that our system is much more algorithmically manageable because it does *not* have (RAA). This is because (RAA) complicates the proof search in the logic.

Remark 2.7 In the remainder of this paper, Γ denotes a *finite* set of sentences. The reason for this restriction is that the logic is not compact.

3 Notation and Preliminaries

The main results in this paper are the completeness of $\mathcal{S}$(card) and $\mathcal{S}^{\dagger}$(card). The method of proof is *model-construction*, so this section contains some preliminary work that will be useful in building models.

A set of sentences Γ in either $\mathcal{S}$(card) or $\mathcal{S}^{\dagger}$(card) is *consistent* if it is not the case that $\Gamma \vdash \varphi$ for all φ. Equivalently, there are no derivations from Γ which use (X).

In both completeness proofs, we are going to fix a consistent set Γ and show that if $\Gamma \nvdash \varphi$, then there is a model of Γ where φ is false.

In building this model, Γ is fixed throughout, and so it is convenient to suppress Γ from the notation. We also will adopt suggestive notation for various assertions in the logic.

Definition 3.1 Let Γ be a (finite) set of sentences. We write $x \leq y$ for $\Gamma \vdash \forall(x, y)$. Note that Γ is left off the notation. We write $x \equiv y$ for $x \leq y \leq x$.

We write $x \leq_c y$ for $\Gamma \vdash \exists^{\geq}(y, x)$. We also write $x \equiv_c y$ for $x \leq_c y \leq_c x$, and $x <_c y$ for $x \leq_c y$ but $x \not\equiv_c y$.

Finally, we write $x <_{more} y$ if $\Gamma \vdash \exists^{>}(y, x)$.

Proposition 3.2 *Let $\mathcal{V}$ be the set of variables in a set Γ.*

1. *If $x \leq y$, then $x \leq_c y$.*
2. *$(\mathcal{V}, \leq_c)$ is a preorder: a reflexive and transitive relation.*
3. *$(\mathcal{V}, <_c)$ is a strict preorder.*
4. *If $x \leq_c y \leq x$, then $x \leq y$.*
5. *If $x \leq_c y$, $x \equiv x'$, and $y \equiv y'$, then $x' \leq_c y'$.*
6. *If $w \leq_c x <_{more} y \leq_c z$, then $w \leq_{more} z$.*

Proof Part (1) uses the (SUBSET-SIZE) rule. In part (2), the reflexivity of $\leq_c$ comes from that of $\leq$ and part (1); the transitivity is by (CARD-TRANS). Part (3) follows from the previous part. Part (4) is by (CARD-MIX). Part (5) uses part (1) and transitivity. Part (6) uses (MORE-LEFT) and (MORE-RIGHT).

Remark 3.1 Let us emphasize that there is a difference between $<_c$ and $<_{more}$. When we write $p <_c q$, we mean that

$$\Gamma \vdash \exists^{\geq}(q, p) \quad \text{and} \quad \Gamma \nvdash \exists^{\geq}(p, q).$$

This is weaker than $p <_{more} q$; recall that this last assertion means that $\Gamma \vdash \exists^{>}(q, p)$. For example, if Γ contains just the sentence $\exists^{\geq}(q, p)$ (and nothing else), then $p <_c q$ but not $p <_{more} q$.

3.1 Preliminary: Listings of Finite Transitive Relations

A *listing* of a set X is a sequence $x_1, \ldots, x_n$ from X so that if $i \neq j$, then $x_i \neq x_j$. Let $(T, <)$ be a finite set with a transitive, irreflexive relation. A *proper listing of*

$(T, <)$ is a listing of the set T with the property that if $t_i < t_j$, then $i < j$. In words, the $<$-predecessors of each point are listed before it. This is also called a *topological sort*.

Lemma 3.3 *Let* $(T, <)$ *be a finite set with a transitive, irreflexive relation. Then* $(T, <)$ *has a proper listing.*

Proof By induction on the size of T. If T has 0 or 1 element, the result is trivial. Assume the result when T has size n, and let $(T, \leq)$ be of size $n + 1$. Let x be such that there is no $y < x$. Such x must exist since T is finite. (Here is the argument in more detail: Suppose towards a contradiction that for every z there were some $w < z$, we would have an infinite sequence $z_0 > z_1 > \cdots > z_n > \cdots$. By finiteness there is $m < n$ so that $z_m = z_n$. But by transitivity we have $z_m > z_n$. This contradicts the irreflexivity of $<$.) Let $T' = T \setminus \{x\}$, and consider T' with the restriction $<'$ of $<$. This order $(T', <')$ is again transitive and irreflexive, and T' has size n. By induction hypothesis, there exists a listing of T', say $t_1, t_2, \ldots, t_n$. Then we take for the listing of T the list $x, t_1, t_2, \ldots, t_n$.

We also need to refine Lemma 3.3.

Lemma 3.4 *Let* $(T, <)$ *be a finite set with a transitive, irreflexive relation. Let* $y \in T$. *Then there is a proper listing of* $(T, <)$ *in which every* x *such that* $y \not\leq x$ *comes before* y *in the listing.*

Proof Start with a proper listing of T as in Lemma 3.3. Let X be the set of points x which come after T in the listing and such that $y \not\leq x$. Note that X might be empty, and $y \notin X$. Move all points in X to immediately before y in the listing, in their order. We check that this new listing is proper. The only way this could fail is that there are some $x \in X$ and some $z < x$ such that z comes after x in the new listing. In this case, we have $z \notin X$, for if z were in X, we would have moved it to before x in the new listing. Thus $y \leq z$. And since $z < x$, we have $y < x$. This contradicts $x \in X$.

4 $\mathcal{S}$(card) and the Construction Lemma

The Completeness Theorem for $\mathcal{S}^\dagger$(card) takes a fair amount of work, and it makes sense to study the smaller system first. (Recall that $\mathcal{S}$(card) is the same as $\mathcal{S}^\dagger$(card), but it lacks complemented variables.) We work with $\mathcal{S}$(card) as a stepping stone towards results for $\mathcal{S}^\dagger$(card).

Definition 4.1 Let Γ be a finite consistent set in $\mathcal{S}$(card). Let $\mathcal{V}$ be a finite set of variables which include all the variables occurring in Γ. Let $\mathcal{V}/\equiv_c$ be the set of equivalence classes of variables in Γ under $\equiv_c$. Let $[u_0], [u_2], \ldots, [u_k]$ be a proper listing of $(\mathcal{V}/\equiv_c, <_c)$. We write $x \prec y$ to mean that for some $i < j$, $x \in [u_i]$ and $y \in [u_j]$. We call $\prec$ the *construction preorder*.

Definition 4.2 We define sets K_v for $v \in V$ as follows:

$$K_v = \{\varphi \in \Gamma \colon \varphi \text{ is either } \exists(w, u) \text{ or } \exists(u, w), \text{ for some } u \text{ and some } w \equiv v\} \quad (6)$$

Lemma 4.3 *Concerning the sets* K_v*:*

1. *If* $v_1 \equiv v_2$, *then* $K_{v_1} = K_{v_2}$.
2. *If* $K_v \neq \emptyset$, *then* $\Gamma \vdash \exists(v, v)$.
3. *If* $K_u \cap K_v \neq \emptyset$, *then* $\Gamma \vdash \exists(u, v)$.

Lemma 4.4 *Let* Γ *be a finite consistent set in* $\mathcal{S}(\text{card})$. *Let* $\prec$ *be the construction preorder from Definition 4.1, and let the sets* K_v *be from Definition 4.2. Then there is a model* $\mathcal{M} = \mathcal{M}_\Gamma$ *such that for all* $a, b \in \mathcal{V}$ *and* $0 \leq i, j \leq k$,

$$K_v \subseteq [\![v]\!]. \quad (7)$$

$$\textit{If } a \leq b, \textit{ then } [\![a]\!] \subseteq [\![b]\!]. \quad (8)$$

$$\textit{If } a \leq b, \textit{ then } \text{card}([\![a]\!]) \leq \text{card}([\![b]\!]). \quad (9)$$

$$\textit{If } i < j \textit{ and } \exists^{\geq}(u_j, u_i), \textit{ then } \text{card}([\![u_i]\!]) \leq \text{card}([\![u_j]\!]). \quad (10)$$

$$\textit{If } i < j \textit{ and } \exists^{>}(u_j, u_i), \textit{ then } \text{card}([\![u_i]\!]) < \text{card}([\![u_j]\!]). \quad (11)$$

$$\textit{If } [\![v]\!] \neq \emptyset, \textit{ then } \Gamma \vdash \exists(v, v). \quad (12)$$

$$\textit{If } [\![u]\!] \cap [\![v]\!] \neq \emptyset, \textit{ then } \Gamma \vdash \exists(u, v). \quad (13)$$

We construct the sets $[\![u]\!]$ *in a step-by-step fashion. Moreover suppose that in the course of the construction, we add extra fresh points to the interpretations of some or all of the variables. Suppose that we do this in such a way that if* $u \equiv y$, *then the same additional points are added to* $[\![u]\!]$ *and* $[\![v]\!]$. *Then parts (8)–(11) continue to hold. Suppose that in addition, whenever we add a point to* $[\![v]\!]$, *then* $\Gamma \vdash \exists(v, v)$. *Then the parts (12) and (13) also hold.*

Proof We define by recursion on $i \leq k$ the interpretation $[\![v]\!]$ of all $v \in [u_i]$. Suppose that for all $j < i$ and all $w \equiv_c u_j$, we have an interpretation $[\![w]\!]$. Recall that $[u_i]$ is the equivalence class of u_i under $\equiv_c$. We work on the elements of $[u_i]$ according to their class under the finer equivalence relation $\equiv$. (That is, we insure that if $a \equiv b$, then $[\![a]\!]$ and $[\![b]\!]$ are the same set.) So at this point we take representatives of the $\equiv$-classes inside of $[u_i]$.

For $v \in [u_i]$, let

$$B_v = \bigcup\{ [\![x]\!] : x \leq v \text{ and } x \prec v\} \quad \text{and} \quad C_v = K_v \cup B_v \tag{14}$$

We start by setting $[\![v]\!]$ to be C_v, but we might need to add more points in order to satisfy some of the requirements. Let

$$n = \max\{ \text{card}(C_v) : v \in [u_i] \}.$$

For each v, add fresh points to $[\![v]\!]$ in order that they all have the same size. That is, add $n - \text{card}(C_v)$ points to $[\![v]\!]$. Finally, if there is some $i < j$ such that $\exists^{>}(u_j, u_i)$, and yet $\text{card}([\![u_j]\!]) = \text{card}([\![u_i]\!])$, then add one fresh point to $[\![v]\!]$ for all $v \in [u_i]$.

It is not hard to see that (8)–(11) hold. Here is the proof of (9): Let $[u_i]$ and $[u_j]$ be such that $x \in [u_i]$ and $y \in [u_j]$. Either $x \equiv_c y$, or else $y <_c x$. In the first case, $[u_i] = [u_j]$ and thus $\text{card}([\![u_i]\!]) = \text{card}(]\!]u_j]\!])$. In the second case, we have $[u_i] < [u_j]$, and so in the proper listing, $[u_j]$ comes before $[u_i]$. Thus, $j < i$. And so by (10), we see that in $\mathcal{M}$, $\text{card}([\![u_j]\!]) = \text{card}([\![u_i]\!])$.

For (11), note that if $u_i <_{more} u_j$, then $u_i <_c u_j$. (This is where we use (X) in the logic, and also the assumption that Γ is consistent.) $[u_i]$ must come before $[u_j]$ in the listing. And so the model construction indeed arranges that $\text{card}([\![u_i]\!]) < \text{card}([\![u_j]\!])$, as desired.

The proof of (12) is by induction on $\prec$. Fix v, and assume (12) for all $u \prec v$. Assume that $[\![v]\!] \neq \emptyset$. If $K_v \neq \emptyset$, we are done by Lemma 4.3, part (2). So we assume that $K_v = \emptyset$. Suppose that $B_v \neq \emptyset$. Then there is some $x \prec v$ such that $x \leq v$ and $[\![x]\!] \neq \emptyset$. By induction hypothesis, $\Gamma \vdash \exists(x, x)$. By (DARII), $\Gamma \vdash \exists(v, v)$. We are left with the case that $B_v = \emptyset$, and thus $[\![v]\!]$ consists entirely of the points added to it in the construction, points not in $K_v \cup B_v$. Then for some $a \prec v$ such that $[\![a]\!] \neq \emptyset$, we have $\Gamma \vdash \exists(a, a)$ (by induction hypothesis), and also $\exists^{\geq}(u_i, a)$. So using (CARD-$\exists$) and other rules of the logic, we see that $\exists(v, v)$.

Equation (13) is a generalization of part (12), and the proof is also a generalization. We show by induction on $\prec$ that for all v and all $u \preceq v$, that if $[\![u]\!] \cap [\![v]\!] \neq \emptyset$, then also $\Gamma \vdash \exists(u, v)$. Assume that $[\![u]\!] \cap [\![v]\!] \neq \emptyset$. If $u = v$, or even if $u \equiv v$, then we are done by the result in (12). Otherwise, we have $u \prec v$. If $K_u \cap K_v \neq \emptyset$, we are done by Lemma 4.3, part (3). So we may assume that $K_u \cap K_v = \emptyset$. Thus, $[\![v]\!]$ is the union of some sets $[\![w]\!]$ (where $w \leq v$ and $w \prec v$), together with some fresh points. Those fresh points are not in $[\![u]\!]$, by construction. So since $[\![u]\!] \cap [\![v]\!] \neq \emptyset$, there is some $w \prec v$ such that $w \leq v$ and $[\![u]\!] \cap [\![w]\!] \neq \emptyset$. We apply our induction hypothesis to u or to w, whichever came later in construction preorder. We see that $\Gamma \vdash \exists(u, w)$. Since $u \leq v$, we also have $\Gamma \vdash \exists(u, v)$, as desired.

Lemma 4.5 *Every set Γ which is consistent in the logic for $\mathcal{S}$(card) has a model.*

Proof Let $\mathcal{M}$ be the model from Lemma 4.4. If the sentence $\exists(u, v)$ belongs to Γ, this sentence itself belongs to $K_u \cap K_v$, hence to $[\![u]\!] \cap [\![v]\!]$ in $\mathcal{M}$. Hence $\mathcal{M}$ satisfies all $\exists$ sentences in Γ. Parts (8)–(11) in Lemma 4.4 insure that the rest of the sentences in Γ also hold.

5 The Completeness Theorem for $\mathcal{S}$(card)

Theorem 5.1 *The logic for* $\mathcal{S}$(card) *found at the top of Fig. 1 is complete: for all sentences* φ*, if* $\Gamma \models \varphi$*, then* $\Gamma \vdash \varphi$*.*

Proof We may assume that Γ is *consistent*, that is, no derivation from Γ uses (X). For if Γ were inconsistent, then for all φ, $\Gamma \vdash \varphi$. For consistent Γ, we prove that if $\Gamma \nvdash \varphi$, then there is a model of Γ where φ fails. We break into cases according to the shape of φ.

The first case: φ is of the form $\exists(x, y)$. We use the model $\mathcal{M}$ of Γ from the Construction Lemma 4.4. In $\mathcal{M}$, $[\![x]\!] \cap [\![y]\!] = \emptyset$. So by part (13) in the lemma, $\Gamma \vdash \exists(x, y)$.

The next case: φ is of the form $\exists^{\geq}(x, y)$. Assuming that $\Gamma \nvdash \varphi$, we see that $y \not\leq_c x$. We use Lemma 3.4 to start off with a listing of $\mathcal{V}/\equiv_c$ which puts $[x]$ before $[y]$. When we define $[\![y]\!]$ we add additional fresh elements to insure that $\text{card}([\![y]\!]) > \text{card}([\![x]\!])$. (We also add the same fresh elements to $[\![z]\!]$ whenever $z \equiv y$.) Incidentally, the hypothesis at the very end of Lemma 4.4 might *not* hold for this model, since it might be the case that $\Gamma \nvdash \exists(y, y)$. But this is not a problem because we do not need to know (13) for this model.

The next case: φ is of the form $\forall(x, y)$. We build our model of Γ using Lemma 4.4, taking any listing of $\mathcal{V}/\equiv_c$. In addition to all of the steps in the Construction Lemma, we add a fresh point $*$ to K_u whenever $u \equiv x$. The point $*$ never gets into K_y: an easy induction shows that $* \in K_z$ iff $x \leq z$.

The final case: φ is of the form $\exists^{>}(x, y)$. We use Lemma 3.4 to start with a proper listing of $(\mathcal{V}/\equiv_c, <_c)$ which puts before $[y]$ all $[z]$ such that $[y] \not\leq_c [z]$. If $[x]$ is one of those $[z]$'s, then in the listing, $[x]$ comes before $[y]$. We can use use the Construction Lemma (with a modification) to arrange that $[\![y]\!]$ be at least as large as $[\![x]\!]$. We do this simply by adding more points to $[\![y]\!]$. In the resulting model $\exists^{>}(x, y)$ will fail. So we assume that $y \leq_c x$. Then we use the proof of Lemma 3.3 again, but in dual form, to further modify the listing so that all $[z]$ with $[z] \not\leq_c [x]$ come after $[x]$. The upshot is that $[y]$ comes before $[x]$, and all $[z]$ which come in between $[y]$ and $[x]$ in the listing satisfy $[y] \leq_c [z] \leq_c [x]$.

Build the model as in the Construction Lemma, up until $[y]$. Change $[\![y]\!]$ to

$$[\![y]\!] = \bigcup_{v} K_v \cup \bigcup \{ [\![x]\!] : x \prec y \}$$

So $[\![y]\!]$ includes the union of everything already defined, and also all $\exists$ sentences in Γ. By adding one point (if need be), we also may arrange that $\text{card}([\![y]\!])$ is strictly larger than $\text{card}([\![a]\!])$ whenever $a \prec y$. We can also arrange that whenever $y \equiv y'$, $[\![y]\!] = [\![y']\!]$.

We shall modify the construction to insure that $[\![x]\!] = [\![y]\!]$. Suppose that z is such that $y \prec z \preceq x$, and that for w such that $y \preceq w \prec z$, we have arranged that $[\![w]\!] = [\![y]\!]$. Then $K_z \cup B_z$ from (14) is a subset of $[\![y]\!]$. For K_z, this is by definition of $[\![y]\!]$. For B_z, we have some cases. Let $w \leq z$ and $w \prec z$. If $w \prec y$, then $[\![w]\!] \subseteq [\![y]\!]$ by definition of $[\![y]\!]$ again. And if $y \prec w$, then $[\![w]\!] \subseteq [\![y]\!]$ by our assumption on z. So at this point we know that $K_z \cup B_z \subseteq [\![y]\!]$. We thus need only consider the two ways that fresh points are added to $[\![z]\!]$.

No fresh points are needed in order to make $\text{card}([\![u]\!]) \leq \text{card}([\![z]\!])$ for any u such that $u \prec z$ and $\exists^{\geq}(z, u)$. This is because whenever $u \prec z$, we have $[\![u]\!] \subseteq [\![y]\!]$.

Let us check that no fresh points are needed in order to make $\text{card}([\![u]\!]) < \text{card}([\![z]\!])$ for any u such that $u \prec z$ and $\exists^{>}(z, u)$. If $u \prec y$, then we already have $\text{card}([\![u]\!]) < \text{card}([\![y]\!])$, and we can arrange that $\text{card}([\![u]\!]) < \text{card}([\![z]\!])$ by taking $[\![z]\!] = [\![y]\!]$. Further, if $y \preceq u$, then the main feature of our listing tells us that $[y] \leq_c [u] \leq_c [x]$. So we do not have $\exists^{>}(z, u)$. For if we did, then together with $\exists^{\geq}(x, z)$ and $\exists^{>}(u, y)$, we would have $\exists^{>}(x, y)$. This would contradict our statement of this final case in our theorem. In this way, we build a model where $\text{card}([\![x]\!]) = \text{card}([\![y]\!])$. This concludes the proof of Theorem 5.1.

6 The Completeness Theorem for $\mathcal{S}^{\dagger}(\text{card})$

This section proves the following result, the last in our paper.

Theorem 6.1 *The logic of Fig. 1 is complete for* $\mathcal{S}^{\dagger}(\text{card})$.

Let us first motivate the work in Sect. 6.1. At this point in the paper we have completeness for the logical system in this paper provided that no complemented variables are used. We want to use this as a preliminary result in dealing with the complemented variables in the syntax and additional rules in the logic.

As before, we first show how to build models of consistent sets Γ, but this time in the full logic, and then we prove completeness by showing that if $\Gamma \nvdash \varphi$, then there is a model where φ fails. The basic idea is to start out by saying which variables *must* denote sets which are half the size of the universe. We call the set of such variables half. These are the variables p such that $\Gamma \vdash \exists^{\geq}(p, \overline{p})$ and also $\Gamma \vdash \exists^{\geq}(\overline{p}, p)$. For the remaining ones, we somehow will partition them into two groups: those which are going to be interpreted by sets smaller than half the size of the universe, and those interpreted by larger sets. Naturally we call these sets of atoms small and large. We are going to use Γ to help with this division, but it is not uniquely determined. For example, if p does not appear in Γ, then we could take $p \in$ small and $\overline{p} \in$ large. Or, we could take $\overline{p} \in$ small and $p \in$ large. (The way we do things, we will never put both p and $\overline{p}$ into half.)

Sections 6.1 and 6.2 have results on partitions of our nouns. Once we make such a partition, the idea is to focus on the small nouns. These include no noun and its complement. Temporarily forget the complement symbols, or rather forget that

$[\![x]\!]$ and $[\![\overline{x}]\!]$ must be interpreted as complementary sets, and build a model of Γ that "otherwise" was a model of Γ. (We could hope to get such a model using the techniques which we have already seen.) Then to rectify matters, we would like to be sure that the atoms in half denote sets which are half the size of the universe, and also that $[\![x]\!]$ and $[\![\overline{x}]\!]$ must be complements. The work on this is done in Sect. 6.3.

6.1 *Small, Large, and Half*

Lemma 6.2 *Let Γ be consistent in $\mathcal{S}^{\dagger}$(card). There is a partition of the nouns into three classes,* small, half, *and* large, *with the following properties:*

(i) $p \in$ half *iff* $(p \leq_c \overline{p}$ *and* $\overline{p} \leq_c p)$.
(ii) $p \in$ large *iff* $\overline{p} \in$ small.
(iii) *If* $p \in$ small *and* $q \leq_c p$, *then* $q \in$ small.

Moreover, we get two additional properties:

(iv) *If* $p \leq_c \overline{p}$, *then either* $p \in$ small *or* $p \in$ half.
(v) *If* $p \in$ half *and* $q \leq_c p$, *then either* $q \in$ small *or* $q \in$ half.

Proof We first check that (i)–(iii) imply (iv) and (v). Suppose towards a contradiction in (iv) that $p \leq_c \overline{p}$ but $p \in$ large. The $\overline{p} \in$ small by (ii), and so $p \in$ small by (iii). This contradicts the pairwise disjointness, and so proves (iv). For (v), suppose that $p \in$ half, $q \in$ large, and $q \leq_c p$. Then by (CARD-ANTI), $\overline{p} \leq_c \overline{q}$. By (ii), $\overline{q} \in$ small. We have just proved (iv), and from this it follows that $\overline{p} \in$ small as well. This again contradicts the pairwise disjointness.

Recall that our nouns are either raw variables p or complemented variables $\overline{p}$. We have been working all along in this paper with the simplified notation that allows us to use p to denote a noun, in particular a complemented variable. But in the current discussion, we need to drop this convention. So for the rest of this proof, p denotes a raw variable, and $\overline{p}$ its associated complemented variable.

The existence of small, half, and large is proved by induction on the number of raw variables in the language. For $n = 0$, the result is trivial.

Assume our result for n, and let the raw variables be $p_0, \ldots, p_n, q$. Let small, half, and large be the sets for $p_0, \ldots, p_n$. We need to see where q and $\overline{q}$ belong.

(a) If $q \leq_c \overline{q} \leq_c q$, then put both q and $\overline{q}$ in half.
(b) Otherwise, if $q \leq_c \overline{q}$ but $\neg(\overline{q} \leq_c q)$, put $q \in$ small and $\overline{q}$ in large.
(c) Otherwise, if $\overline{q} \leq_c q$ but $\neg(q \leq_c \overline{q})$, put $q \in$ large and $\overline{q} \in$ small.
(d) Otherwise, if neither $q \leq_c \overline{q}$ nor $\overline{q} \leq_c q$, but there is some $p \in$ small such that $q \leq_c p$, then put $q \in$ small and $\overline{q} \in$ large.
(e) In all other cases, put $\overline{q} \in$ small and $q \in$ large.

We must check that (iii) holds. There are cases depending on which of (a)–(e) is responsible for putting q and $\overline{q}$ into one of our sets. In case (a), there is nothing to show, since the hypothesis of (iii) will not apply. For (b), suppose that $q \leq_c \overline{q}$ but not conversely, so that $q \in \mathsf{small}$. Suppose also that $p_i \leq_c q$. We must check that $p_i \in \mathsf{small}$ as well. If $p_i \in \mathsf{large}$, then since $p_i \leq_c q$, we also have $\overline{q} \leq_c \overline{p}_i$. Thus $p_i \leq_c q \leq_c \overline{q} \leq_c \overline{p}_i$. We thus have a contradiction, using (iv). And if $p_i \in \mathsf{half}$, then $\overline{p}_i \leq p_i$. Since $\overline{q} \leq_c \overline{p}_i$, we have $\overline{q} \leq_c \overline{p}_i \leq p_i \leq q$. This contradicts the hypothesis of case (b). So we are left with $p_i \in \mathsf{small}$, as desired.

We omit the details on case (c). Next, suppose that we put $q \in \mathsf{small}$ due to (d). So there is some $p \in \mathsf{small}$ and $q \leq_c p$; and also that $p_i \leq_c q$. We need to see that $p_i \in \mathsf{small}$ as well. This is due to $p_i \leq_c p$ and the assumption that our partition of $p_0, \ldots, p_n$ satisfies the conditions of our lemma.

The last case is when $\overline{q} \in \mathsf{small}$ due to (e): there is no p such that both $p \in \mathsf{small}$ and $q \leq_c p$; and also p_i is such that $p_i \leq_c \overline{q}$. Again we must verify that $p_i \in \mathsf{small}$. If $p_i \in \mathsf{large}$, then $\overline{p}_i \in \mathsf{small}$. Hence $\overline{p}_i \leq_c p_i$. And so $q \leq_c \overline{p}_i$, a contradiction. If $p_i \in \mathsf{half}$, then also $\overline{p}_i \in \mathsf{half}$. By the rule (HALF), $p_i \equiv_c \overline{p}_i$. And so $q \leq_c \overline{p}_i \equiv_c p_i \leq_c \overline{q}$. This contradicts the assumption that we are in case (e).

6.2 *Refinements*

We are going to need several refinements to Lemma 6.2. Recall that the proof of Lemma 6.2 was by induction on the number of raw variables in the language. The proof shows that if we have a partition of a subset of nouns which is closed under complement, and if that partition satisfied the conditions (1)–(3) in Lemma 6.2, then this partition extends to a partition of all the of the nouns in the language. This observation makes the construction in the lemma quite flexible, as the results below show.

We say that the small class is *smaller than* the half class, and both of these are *smaller than* the large class.

Lemma 6.3 *Suppose that* $\Gamma \nvdash \exists^{\geq}(p, q)$. *Then there are sets* small, half, *and* large *as in Lemma 6.2 such that one of the following holds:*

1. p *and* q *are both in* small.
2. $\overline{p}$ *and* $\overline{q}$ *are both in* small.
3. p *and* q *are in different classes, and the class of* p *is smaller than the class of* q.

Proof By the remark just above, we need only show that $\{ p, \overline{p}, q, \overline{q} \}$ may be partitioned into classes small, half, and large satisfying (1)–(3) in Lemma 6.2 and also one of the conditions (1)–(3) in the present lemma.

If $p \equiv_c \overline{p}$, then we put $p \in \mathsf{half}$. In this case, we cannot have $q \leq_c \overline{q}$, by (HALF). Indeed, we put $q \in \mathsf{large}$, and then we handle the rest to insure that (1)–(3) in Lemma 6.2 hold. If $p <_c \overline{p}$, put $p \in \mathsf{small}$. If $\overline{p} <_c p$, then put $p \in \mathsf{large}$. In this case, we cannot have have $q \leq_c \overline{q}$. So we may put $q \in \mathsf{large}$ as well.

We say that the small class is *smaller than* the half class, and both of these are *smaller than* the large class.

Lemma 6.4 *Suppose that* $\Gamma \nvdash \exists^{>}(p, q)$. *Then there are sets* small, half, *and* large *as in Lemma 6.2 such that one of the following holds:*

1. p *and* q *are both in* small.
2. $\overline{p}$ *and* $\overline{q}$ *are both in* small.
3. $\overline{p}$ *and* $\overline{q}$ *are both in* half.
4. p *and* q *are in different classes, and the class of* p *is smaller than the class of* q.

Proof This is similar to the proof of Lemma 6.3. The only thing that changes is that now it is possible to have p and q both in half.

Lemma 6.5 *Suppose that* $\Gamma \nvdash \exists(p, q)$. *Then there is a partition of the nouns as in Lemma 6.2 such that (a) either* p *or* q *does not belong to* large, *and (b) if one of them does belong to* large, *then the other belongs to* small.

Proof By the remark just above, we need only show that $\{ p, \overline{p}, q, \overline{q} \}$ may be partitioned into classes small, half, and large satisfying (1)–(3) in Lemma 6.2 and also the assertions in the present lemma.

If $\overline{p} <_c p$, then we are forced to put $p \in$ large. In all other cases, it will be possible to put p in either small or half (following what we did in Lemma 6.2). Let us check that if we are forced to put $p \in$ large, then we cannot also be forced to put $q \in$ large (by having $\overline{q} <_c q$) or even $q \in$ half ($\overline{q} \leq_c q$): we must be able to put q in small.

For this, suppose towards a contradiction that $\overline{p} <_c p$ and $\overline{q} \leq_c q$. Looking back at Example 2.2, we see that $\Gamma \vdash \exists(p, q)$. This contradicts the consistency of Γ.

This also covers a related point: if we are forced to put $p \in$ large, we cannot be forced to have $q \in$ large by having $p \leq_c q$. The reason is that in this case, $\overline{q} \leq_c \overline{p} <_c p \leq_c q$.

The conclusion here is that if we are forced to put $p \in$ large, then we cannot also be forced to put $q \in$ large. The same argument works when we interchange p and q, of course, since $\exists(p, q)$ and $\exists(q, p)$ are equivalent in the logic, due to (CONVERSION). The argument also shows that if we are forced to put one of these in large, we cannot be forced to put the other in half.

6.3 Building a Model of a Consistent Set Γ *of* $\mathcal{S}^{\dagger}$*(card)*

Lemma 6.6 *Every consistent set* Γ *has a model.*

We prove this lemma in stages in this section. Fix a consistent set Γ.

The preliminary model $\mathcal{P}$. Our set Γ is consistent in the logic for $\mathcal{S}^{\dagger}$(card), hence it is consistent in the smaller logic for $\mathcal{S}$(card). However, the smaller language does not have complemented variables, and so we have to change our way of thinking.

Note 6.1 The language $\mathcal{S}$(card) may be formulated on top of any set of "raw variables." When we originally discussed it, we had in mind that it would be formulated on top of the set $\mathbf{P}_0$ of raw variables. It might suggest things to write $\mathcal{S}^\dagger(\mathbf{P}_0)$. At this point, we want to consider $\mathcal{S}^\dagger(\mathbf{P})$. That is, we want to formulate $\mathcal{S}$(card) on top of *all* of the variables, including the complemented ones. Thus, we temporarily regard p and $\overline{p}$ as *completely unrelated variables*, and then we consider the language of $\mathcal{S}$(card) formulated over this set. Γ is consistent in this language.

By the Construction Lemma 4.4, Γ has a model, say $\mathcal{P}$. This model will *almost* be a model of the kind we seek, an $\mathcal{S}^\dagger$(card)-model of Γ. But there is a problem: for each p, $[\![p]\!] \cup [\![\overline{p}]\!]$ will probably not equal the universe P. It *will* be true that $[\![p]\!] \cap [\![\overline{p}]\!] = \emptyset$, by (13) in the Construction Lemma. It even will be true that if $p \in \mathsf{half}$, then $[\![p]\!]$ and $[\![\overline{p}]\!]$ will be sets of the same size. (This follows from the proof of the Construction Lemma 4.4, when we use K_v as in (6). That is, since the construction will not demand that one of the two sets $[\![p]\!]$ and $[\![\overline{p}]\!]$ is strictly larger than the other, the two sets will come out with the same cardinality.) But again, there is no reason to think that $[\![\overline{p}]\!] = \overline{[\![p]\!]}$. Thus, we need a few more steps to arrange this.

Add a point, if necessary. If the size of the universe P of $\mathcal{P}$ is an odd number, add a point $*$ so that the size is even. This point $*$ does not get added to the interpretation of any atom. Call the resulting model $\mathcal{O}$. (If card(P) is even, set $\mathcal{O} = \mathcal{P}$.) The construction arranges that $\mathcal{O} \models \Gamma$.

Make the half variables have size half the universe, and the large variables be the complements of the small ones. For each $p \in \mathsf{half}$, let $S_p = P \setminus ([\![p]\!] \cup [\![\overline{p}]\!])$. The cardinality of S_p is even. Take half the points in this set and add them to $[\![p]\!]$, and then add the other half to $[\![\overline{p}]\!]$. We need to do this carefully, so that if $p \leq q$ belong to half, then the same points are added to these sets. (In other words, we must carry out this step for successive variables in half.)

Also, if $p \in \mathsf{small}$, then change $[\![\overline{p}]\!]$ to be $\overline{[\![p]\!]}$. That is, replace $[\![\overline{p}]\!]$ by the complement of $[\![p]\!]$. Doing this makes for a larger set.

We call this model $\mathcal{N}$. Now we would like it to be the case that $\mathcal{N}$ still satisfies Γ. We can verify that $\mathcal{N} \models \varphi$ for all $\varphi \in \Gamma$ except those of the form $\exists^{\geq}(l, h)$ and $\exists^{>}(l, h)$ with $l \in \mathsf{large}$ and $h \in \mathsf{half}$. The most interesting verification has to do with sentences $\forall(p, \overline{p})$ for $p \in \mathsf{half}$. Suppose such a sentence belongs to Γ. If $N = \emptyset$, then $\mathcal{N} \models \forall(p, \overline{p})$. If $N \neq \emptyset$, then $\Gamma \vdash \exists(q, q)$ for some q, by (13). (In more detail, if $\Gamma \nvdash \exists(p, q)$ for any p and q, then $\mathcal{P}$ is empty. We check this by induction on the construction in Lemma 4.4. So $\mathcal{O}$ is also empty, and so is $\mathcal{N}$.) Fix q so that $\Gamma \vdash \exists(q, q)$. By (INT), $\Gamma \vdash \exists(p, p)$. Since Γ contains $\forall(p, \overline{p})$, we get a contradiction, using (DARII). This contradicts the assumption that Γ is consistent.

Make the large variables have a larger size than the half variables. If need be, add some number of fresh points to $[\![l]\!]$ for all large l, so that whenever $l \in \mathsf{large}$ and $h \in \mathsf{half}$, $\exists^{>}(l, h)$ holds in the resulting model, which we call $\mathcal{M}$.

Doing this preserves all of the good things about $\mathcal{N}$ and in addition makes $\mathcal{M}$ satisfy whichever of the sentences in Γ which do not hold in $\mathcal{N}$. So $\mathcal{M} \models \Gamma$.

This completes the proof of Lemma 6.6.

6.4 The Completeness Theorem for $\mathcal{S}^\dagger(card)$

The rest of this section completes the proof of Theorem 6.1. As with Theorem 5.1, we argue the contrapositive, showing that if $\Gamma \nvdash \varphi$, then there is a model of Γ where φ fails. Again, we split into cases according to φ.

The first case: φ is of the form $\exists^{\geq}(x, y)$. We invoke Lemma 6.3 to obtain small, half, and large with one of the three options stated in the lemma. If x and y both belong to small, then by Theorem 5.1 we may assume that in $\mathcal{P}$, $\text{card}([\![x]\!]) < \text{card}([\![y]\!])$. This fact persists to the other models, and so we get a model of Γ where it holds. If x and y both belong to large, then we consider $\overline{x}$ and $\overline{y}$ and apply what we just saw. If the class of y is larger than the class of x, then in $\mathcal{M}$ we again have $\text{card}([\![x]\!]) < \text{card}([\![y]\!])$.

The next case: φ is of the form $\exists^{>}(x, y)$. This is similar to the first case, except that we call on Lemma 6.4 instead of Lemma 6.3. If our partition has x and y both in half, then in the models $\mathcal{N}$ and $\mathcal{M}$, $[\![x]\!]$ and $[\![y]\!]$ will have the same size.

The next case: φ is of the form $\forall(x, y)$. Note that we cannot have $\Gamma \vdash \forall(x, \overline{x})$, by (ZERO). And using (ONE), we cannot have $\Gamma \vdash \forall(\overline{y}, y)$. Fix a partition of the variables as in Lemma 6.2. Let $*$ be any object. We use the Construction Lemma 4.4, taking a new point $*$ and putting it into $[\![x]\!]$ and also $[\![\overline{y}]\!]$. This point is the only additional point added, beyond what is in Lemma 4.4. We check that the resulting model $\mathcal{P}$ has $[\![z]\!] \cap [\![\overline{z}]\!] = \emptyset$ for all z. (We need to do this in order to build $\mathcal{N}$.) To begin, an induction on the construction shows that $* \in [\![z]\!]$ iff $x \leq z$ or $\overline{y} \leq z$. This applies to $\overline{z}$, of course. Assume towards a contradiction that $* \in [\![z]\!] \cap [\![\overline{z}]\!]$. We get four cases, and all of them contradict $\Gamma \nvdash \varphi$. We use (ANTI), (ZERO), and (ONE) from the logic. This shows that $* \notin [\![z]\!] \cap [\![\overline{z}]\!]$. We also need to consider points $**$ which the Construction Lemma added to $\mathcal{P}$ after $*$. Let $**$ be such a point. Then $\{w\colon ** \in [\![w]\!]\}$ is of the form $\{v\colon w \leq v\}$, where w is a $\prec$-minimal atom whose interpretation contains $**$. (Indeed, w is the "first" atom in the construction whose interpretation contains $**$.) Suppose towards a contradiction that $w \leq z$ and also $w \leq \overline{z}$. By our logic, $w \leq \overline{w}$. A final induction shows that every w with $w \leq \overline{w}$ has $[\![w]\!] = \emptyset$, and this is our contradiction. So we now know that $\mathcal{P}$ has the required disjointness property. We define $\mathcal{O}$, $\mathcal{N}$, and $\mathcal{M}$ as before, and we indeed obtain a model of Γ where $[\![x]\!] \cap [\![\overline{y}]\!]$ contains $*$ and is thus not empty.

The final case: φ is of the form $\exists(x, y)$. In this case, we appeal to Lemma 6.5. We may assume that one of the following holds:

(i) $x \in \mathsf{small}$, $y \in \mathsf{small}$ or $y \in \mathsf{half}$.
(ii) $x \in \mathsf{small}$, $y \in \mathsf{large}$.
(iii) $x, y \in \mathsf{half}$.

In (i), Theorem 5.1 gives $\mathcal{P} \models \Gamma$ where $[\![x]\!] \cap [\![y]\!] = \emptyset$. As we build $\mathcal{M}$ and the models leading up to it, we add no points to $[\![x]\!]$, and all of the points added to $[\![y]\!]$ are out of P. Thus, in $\mathcal{M}$ we also have $[\![x]\!] \cap [\![y]\!] = \emptyset$.

In (ii), we cannot have $\Gamma \vdash \exists^{>}(x, \overline{y})$, due to the rule (MORE-SOME). By what we already know, there is a model of Γ where $\text{card}([\![x]\!]) \leq \text{card}([\![\overline{y}]\!])$. Thus there is a listing where $[x]$ comes before or equal to $[\overline{y}]$. Moreover, we either have $\Gamma \nvdash \forall(x, y)$, or else $\Gamma \nvdash \exists(x, x)$. (For if we could derive both $\forall(x, y)$ and $\exists(x, x)$, then by (DARII) we would have $\exists(x, y)$.) If $\Gamma \nvdash \exists(x, x)$, then we already know that that there is a model of Γ where $[\![x]\!] = \emptyset$. In this model, $\exists(x, y)$ is of course false. So we may assume that $\Gamma \vdash \exists(x, x)$ and $\Gamma \nvdash \forall(x, y)$.

We form the model by the Construction Lemma, except that when we define $[\![\overline{y}]\!]$, we also throw in $[\![x]\!]$. This arranges that $[\![x]\!] \subseteq [\![\overline{y}]\!]$.

When we do this, it is important that for all $w \in \mathsf{half}$, $[\![w]\!]$ and $[\![\overline{w}]\!]$ are disjoint. The only way that this could fail is if $x \leq w$ and also $\overline{y} \leq \overline{w}$. But then we would have $x \leq y$, contrary to what we saw above. It is also important that all of the sentences $\exists(a, b)$ in Γ hold in $\mathcal{P}$. The only way that one of these could fail is if $a \leq x, \overline{y} \leq \overline{b}$. (That is, the sentence $\exists(a, b)$ would be put into $[\![a]\!] \cap [\![\overline{b}]\!]$.) But in this case, we have $\Gamma \vdash \exists(x, y)$, contrary to this case. The upshot is that indeed $\mathcal{P} \models \Gamma$, and in $\mathcal{P}$, each half atom has $[\![w]\!] \cap [\![\overline{w}]\!] = \emptyset$.

We thus can continue to build the models $\mathcal{O}$, $\mathcal{N}$ and (especially) $\mathcal{M}$, the construction still has arranged that $[\![x]\!] \cap [\![y]\!] = \emptyset$.

Finally, we turn to (iii). We assume that $\Gamma \nvdash \exists(x, y)$, and $x, y \in \mathsf{half}$. Consider the definition of $\mathcal{N}$. When we define $[\![h]\!]$ for $h \in \mathsf{half}$, we first note that for our x and y, $[\![x]\!] \cap [\![y]\!] = \emptyset$. This is by (13) and the assumption that $\Gamma \nvdash \exists(x, y)$. But this assumption also implies that $\Gamma \nvdash \exists(\overline{x}, \overline{y})$: this is the one and only place in the proof where we use (MAJ). Thus we also have $[\![\overline{x}]\!] \cap [\![\overline{y}]\!] = \emptyset$. It follows that $([\![x]\!] \cup [\![\overline{y}]\!]) \cap ([\![\overline{x}]\!] \cup [\![y]\!]) = \emptyset$. Before we carry out the definition of $\mathcal{N}$, set $[\![u]\!] = [\![x]\!] \cup [\![\overline{y}]\!]$ whenever $u \equiv x$ or $u \equiv \overline{y}$. Also, set $[\![u]\!] = [\![\overline{x}]\!] \cup [\![y]\!]$ whenever $u \equiv \overline{x}$ or $u \equiv y$. (We cannot have both conditions, lest $\Gamma \vdash \forall(x, \overline{x})$.) Then define $\mathcal{N}$ and $\mathcal{M}$ as before. We may arrange that $[\![x]\!] = [\![\overline{y}]\!]$ in $\mathcal{N}$. The passage from $\mathcal{N}$ to $\mathcal{M}$ does not change $[\![x]\!]$ or $[\![y]\!]$. So $\mathcal{M}$ is as desired.

This concludes the proof of Theorem 6.1.

Additional Remarks. As mentioned, the completeness of the logic implies that *reductio ad absurdum* is admissible in the logic. Adding *reductio* makes the logic more useable by people. For computers, it complicates the proof search.

A final point about the system concerns the (MAJ) rule. This is the only rule with three premises. It can be shown that this rule cannot be eliminated from the system. That is, the (MAJ) rule does not follow from the other rules in the system. At the same time, every valid two-premise rule is derivable in the system. Unfortunately, one verifies these points by exhaustive search rather than by an insightful argument. And so we conclude that this logic $\mathcal{S}^{\dagger}(\text{card})$ is not axiomatizable with only two-premise rules.

7 Conclusion

We have presented a sound and complete logic extending syllogistic logic with There are at least as many x as y, and There are more x than y. The completeness proof is more intricate than in other work in the area, mostly because the constructions are not "canonical": one has lots of flexibility in building models with tools like Lemmas 4.4 and 6.2, but this flexibility comes with a cost in terms of extra details to track in the proofs. As with all proofs in this area, there are many details to check.

We did not discuss details of the implementation. Here is a very brief explanation. One takes a set Γ and a sentence φ and wants to know whether $\Gamma \vdash \varphi$ or not. Since *reductio* is admissible, one can instead ask if $\Delta = \Gamma \cup \{\overline{\varphi}\}$ is consistent or not. To tell, generate all possible proofs in our system. This can be done in polynomial time. If Δ is inconsistent, then we already know that $\Gamma \vdash \varphi$. And if Δ is consistent, one wants to build a model. Here one has to take all the work done in this paper and make it algorithmic.

The model-building part of the paper does not provide models of minimal size. (See Remark 1.1 for a related discussion.) It is open to build minimal models of satisfiable sets of sentences in polynomial time.

I did want to comment that work on the implementation was both helpful and not helpful to work on this topic. While working on the implementation I discovered several of the rules of the system, including (MAJ). I also found that some rules which were originally part of the system were derivable from others. Further, some of the steps in the completeness proof were suggested by work on the algorithm. On the other hand, no amount of testing can actually prove the completeness of any logical system. (It might have helped to have a proof assistant for that.) I found myself debating which I believed more: a thousand randomly generated examples of proofs and counter-models, or a long proof that involves numerous similarly-looking cases that are never written out in full.

The next steps in this line of work would be to expand the logical system. The two papers that come closest to this one are Endrullis and Moss (to appear) and Lai et al. (to appear). Endrullis and Moss (to appear) studies the logic with All, Some, and Most x are y. But it lacks the syntactic constructions in this paper: There are at least as many x as y, and There are more x than y. It also lacks complemented variables. It is open to merge the two systems, or to show that this is impossible. Similar remarks apply to Lai et al. (to appear). That paper studies Most x are y and All x are y on top of propositional logic.

Acknowledgments I thank anonymous reviewers for their comments and corrections to this paper.

References

Corcoran, J. (1972). Completeness of an ancient logic. *Journal of Symbolic Logic*, *37*(4), 696–702.

Dunn, J. M. (2015). Logic(s) as tool(s). unpublished ms.

Endrullis, J., & Moss, L. S. (to appear). Syllogistic logic with "most". In *Proceedings of WoLLIC 2015* (15 pp). LNCS, Springer

Herre, H., Krynicki, M., Pinus, A., & Väänänen, J. (1991). The Härtig quantifier: A survey. *Journal of Symbolic Logic*, *56*(4), 1153–1183.

Kolaitis, P. G., & Väänänen, J. A. (1995). Generalized quantifiers and pebble games on finite structures. *Annals of Pure and Applied Logic*, *74*(1), 23–75.

Lai, T., Endrullis, J., & Moss, L. S. (to appear). Proportionality digraphs. In *Proceedings of the American Mathematical Society* (15 pp)

Lindström, P. (1966). First-order predicate logic with generalized quantifiers. *Theoria*, *32*, 186–195.

Łukasiewicz, J. (1957). *Aristotle's syllogistic* (2nd ed.). Oxford: Clarendon Press.

Martin, J. N. (1997). Aristotle's natural deduction revisited. *History and Philosophy of Logic*, *18*(1), 1–15.

Moss, L. S. (2010). Syllogistic logic with complements, Games, Norms and Reasons. In *Proceedings of the Second Indian Conference on Logic and its Applications* (19 pp). Springer Synthese Library, Mumbai

Moss, L. S. (2015). Natural logic. *Handbook of Contemporary Semantic Theory*, 2nd edn. Wiley, chapter 18

Mostowski, A. (1957). On a generalization of quantifiers. *Fundamenta Mathematicae*, *44*, 12–36.

Pratt-Hartmann, I. (2009). No syllogisms for the numerical syllogistic. *Languages: From Formal to Natural*. Volume 5533 of LNCS (pp. 192–203). Springer

Pratt-Hartmann, I., & Moss, L. S. (2009). Logics for the relational syllogistic. *Review of Symbolic Logic*, *2*(4), 647–683.

van Benthem, J. (2008). A brief history of natural logic. In M. Chakraborty, B. Löwe, M. N. Mitra, & S. Sarukkai (Eds.), *Logic, Navya-Nyaya and Applications, Homage to Bimal Krishna Matilal*. London: College Publications.

A “Reply” to My “Critics”

J. Michael Dunn

Abstract Despite the joking title, this is not really a reply to my critics. Rather it is a response to my fellow researchers in acknowledgment of their expert contributions to this volume on information based logics. Their papers extend my work or their own, in a good way. In my responses, I try to say something interesting, maybe just to set a context, to suggest future work, to clarify something, or to make further connections to my own work.

This is really not an appropriate title, at least without the scare quotes, since the wonderful contributors to this volume are too nice to really criticize me, at least directly. Rather they have found excuses to say nice things about me and my work, and at the same time extend my work, and/or theirs, in a good way. The description of the series *Outstanding Contributions to Logic* says that a typical volume contains a response to the contributions by the logician to whom the volume is devoted. This is that response. But it is not always a “reply” in any strict sense. Do not expect this to be a series of public referee reports. It is too late anyway since I understand that the contributions have already been put through a scrupulous referee process. Moreover there simply is not enough space for a detailed reaction to each and every paper. But I try to say something interesting, maybe just to set a context, sometimes to suggest future work. And occasionally to try to clarify something. Anyone who knows me knows I like things to be clear—this is just me.

I might explain and/or apologize for something in advance. You will often find me citing and explaining my own work. I try to think that I wasn't just being “moi,” but was rather just being me. I was trying to establish reasons why these contributions would appear in this volume and make bridges to my own work.

J.M. Dunn (✉)
Indiana University, Bloomington, IN, USA
e-mail: dunn@indiana.edu
URL: http://www.philosophy.indiana.edu/people/dunn.shtml,
http://www.soic.indiana.edu/all-people/profile.html?profile_id=194

K. Bimbó (ed.), *J. Michael Dunn on Information Based Logics*,
Outstanding Contributions to Logic 8, DOI 10.1007/978-3-319-29300-4_19

I of course want to thank the various authors who have contributed to this volume, many of whom are close friends and colleagues and former students (the classes are far from exclusive). Most of all I want to thank the editor of this volume, Katalin Bimbó, one of my former students, and now a friend and colleague. I of course also want to thank the editor of the series Sven Ove Hansson, and also the editorial board, for choosing to do a volume about me and information based logics. And I want to thank Heinrich Wansing (a member of the editorial board) who first approached me with the idea.

Arnon Avron: RM and its Nice Properties. Arnon always says, and proves, nice things about **RM** (**R**-Mingle). In his abstract he says (p. 41):

> Dunn–McCall logic **RM** is by far the best understood and the most well-behaved logic in the family of logics developed by the school of Anderson and Belnap. However, it is not considered to be a relevant logic by the relevant logicians, since it fails to have the variable-sharing property. Instead **RM** is usually characterized as being a "semi-relevant" logic without explaining what this notion means.

I do not entirely agree with Arnon on this last. I am not sure who first started calling **RM** a "semi-relevant logic," but I think it was likely Bob Meyer. Bob was the one who first showed that in **RM** the negation of a theorem implies any theorem whatsoever, so we get provable implications without the standard requirement for relevance of the Variable Sharing Property, e.g., $(p \wedge \sim p) \rightarrow (q \vee \sim q)$, or $\sim(p \rightarrow p) \rightarrow (q \rightarrow q)$.[1] Bob was the contributing author of Sect. 2.3 in Anderson and Belnap's (1975), and on p. 427 it is proved that **RM** satisfies the principle of "weak relevance," namely "that $A \rightarrow B$ is a theorem only if either A and B share a sentential variable or both $\sim A$ and B are theorems. I think that someone referring to **RM** as "semi-relevant" might well be saying this because it only satisfies "weak relevance." However, putting this aside, Arnon's definition of "semi-relevant" turns out to be stronger than this, and in an interesting way.

There is no doubt that **RM** was, as the saying goes, the red-headed stepchild of the relevance logic family. This was always a disappointment to me, and I feel sure also to its other parent Storrs McCall. I always felt it was similar to the situation of **S5** in the modal logic family—too simple and elegant to be appreciated. But of course I have to be fair and acknowledge that like **S5**, **RM** collapses a lot of formulas into logical equivalents that do not always allow it to make important distinctions. But for many purposes, say as the logic for some paraconsistent theories, it is a good enough tool.[2] So I really appreciate Arnon's contribution to this volume that gives an interesting formal characterization of "semi-relevance." This can be viewed as showing

---

[1]It is interesting that the original formulations of implication and negation for **RM** did have the variable sharing property, but adding conjunction or disjunction (they are interdefinable using De Morgan negation) led to the first of the implications, and the second can then be derived. Arnon has done a lot of interesting work on these implication-negation versions of **RM** without conjunction and disjunction.

[2]The reader who wonders about the qualification "some" is referred to Dunn (1979a), where it is shown that Robinson's Arithmetic blows up, whether with **R** or **RM**, if it has 0 as primitive (but not if its numbers start with 1).

that **RM** is at last, and at least, a semi-legitimate member of the relevance logic family.

Chris Mortensen: Wedge Sum, Merge and Inconsistency. Chris says in the first sentence of his article, that Bob Meyer's and my work on 3-valued model theory proved an inspiration to him to construct inconsistent mathematical theories. Maybe I should take this occasion to apologize to Chris for leading him down the wrong path, but I don't think so. Chris's work on Inconsistent Mathematics (the title of his well-known book) has been an inspiration to many of us, and not just for his technical results but also because of his connecting inconsistent mathematics to the inconsistent art of Escher, Penrose, Reutersvärd, and others.

I was glad to obtain in Dunn (1979b) a quite general way of constructing 3-valued models from homomorphic images of classical models. Bob Meyer (1976) had already constructed 3-valued finite models for his relevant arithmetic and used them to show that his system was absolutely consistent (not every sentence is provable), a result sitting on the edge of Gödel's Second Incompleteness Theorem. His models were based on modular arithmetic, and in particular the natural numbers modulo 2 consists of just two "numbers," 0 (representing the even numbers) and 1 (representing the odd numbers). Bob defined his model so that $0 = 1$ had the value Both. The semi-relevant logic **R**-mingle makes an appearance again, and particularly its 3-valued extension **RM3** determined by the 3-element Sugihara algebra which can be intuitively understood as the set of values, True, False, and Both.[3] My generalization of Bob's somewhat ad hoc construction was to make it more elegant, since the natural numbers mod n are all homomorphic images of the natural numbers. The idea was that an equation $[i] = [j]$ would receive the value True if for every $i' \in [i]$ and $j' \in [j]$, $h(i) = h(j)$. And it would receive the value False if for every $i' \in [i]$ and $j' \in [j]$, $i' \neq j'$. $h(i) \neq h(j)$, and of course it would take the value Both if sometimes its images were the same and other times they were distinct. I used the metaphor of a blurred image. I never really pursued it but I also had a 4-valued account where the value Neither could arise when we had a submodel of a model, a kind of incomplete image. I think mentioning these images is very appropriate given Chris's work on Escher, etc.

But in Chris's contribution to this volume, is going the other direction. He isn't looking at homomorphic images (or submodels), he is rather taking two models and combining them. If I can use some Photoshop metaphors, a homomorphic image is obtained by use of the Blur tool, a submodel is obtained by Cropping, and Chris's Merge is analogous to Photomerge. Chris begins his paper by defining a topological notion: the *wedge sum*. I skip over the technical definition to Chris's saying that the definition "amounts to saying that the wedge sum of two spaces is their union, except for having just one pair of overlapping points identified." As an example Chris asks us to think of the wedge sum of a circle with itself as topologically equivalent to the numeral 8. Chris adds (p. 70):

[3] I don't know why but I used N to denote this third value, even though I clearly explain that it is to be understood as Both. Incidentally these three values function similarly to the three values for Graham Priest's "Logic of Paradox" **LP** except that implication is defined differently.

Clearly, this definition can be extended to more than two spaces, and more than one pair of overlapping points, but we do not need that here. In this paper, we directly construct consistent logical theories which describe wedge sums, and inconsistent theories which extend them. Then we utilize the technique Merge to show how to obtain inconsistent theories of the wedge sum in a different way. Finally, it is seen that this technique can be used to study inconsistent structures other than topological spaces.

The *merge* of two theories is the theory which is the deductive and conjunctive closure of their union, i.e., it is the smallest "theory" that includes their union. Chris uses the logic **RM3** to determine these theories, i.e., sets of sentences closed under deduction ($A \in T$ and $A \to B \in T$ implies $B \in T$) and conjunction ($A \in T$ and $B \in T$ implies $A \wedge B \in T$).

Chris compares the wedge sum, which operates at the level of semantics, and the merge, which operates at the level syntax.

Dolph E. Ulrich: Single Axioms and Axiom-Pairs for the Implicational Fragments of R, R-Mingle, and Some Related Systems. I have known Ted (he goes by the nickname of his middle name Edward) in several roles. Ted and I were actually undergraduate philosophy students together at Oberlin. And then I met him again when he was an advanced graduate student at Wayne State looking for a dissertation advisor in logic, and I was a first year faculty member, lucky to have a graduate student of Ted's quality to work with. I learned at least as much from Ted as he did from me. His dissertation *Characteristic Matrices for Sentential Calculi* was completed I think about a year after my own.

Ted's dissertation contained an appendix translating Jerzy Łoś's *O Matrycach Logicznych* (*On Logical Matrices*). I do not believe Ted knew any Polish but he figured it out word by word so to speak. He told me that once he had figured out what the theorem was he would then proceed to try to figure out its proof, going back and forth between his thoughts and Łoś's, as it were, and trying to construct another proof if they didn't match. Wikipedia tells me that in 1970, 70 % of Hamtramck's population was of Polish origin, and of course many of the younger people came to Wayne State as students and took our logic classes. Ted, was teaching one of these classes at the time and he told me that he thought of announcing at the first class something like, "If I mispronounced any of your names please come up and see me after class—I have a project you might help me with."

Perhaps Ted has been channeling Łoś or some other Pole, but for a long time Ted has been the "go-to guy" for questions regarding shortest axiomatizations of sentential calculi, shortest single axioms, and such as if he is simultaneously channeling Łukasiewicz, Meredith, Prior, Wos, etc. Ted has applied his talents to the logic **R** and **R-Mingle** among many others (even classical logic) over the years. It is very nice to see his contribution to this volume reviewing his and others' work, and showing the existence of one and two axiom bases for both the implicational fragments of **R** and **R-Mingle**.

Larisa Maksimova: LC and its Pretabular Relatives. Larisa Maksimova was working on relevance logic in the late 1960s and early 1970s when relevance logics were just being formulated by Anderson and Belnap. But instead of being in Pittsburgh, close to the action as I was, she was working on her Ph.D. in Novosibirsk. This was before the Internet! One of my earliest published papers connecting algebra and logic was Dunn (1970), where I showed **RM** is "pretabular" (that is, every normal extension closed under its rules has a finite characteristic algebraic model—a "Sugihara algebra"). Shortly thereafter Bob Meyer and I showed the same for Dummett's superintuitionistic logic **LC**. Larisa shortly thereafter (Maksimova 1972) greatly improved on this result showing that there are only 3 pretabular extensions of the intuitionistic propositional logic, and then (Maksimova 1975) showed similarly that there are exactly 5 pretabular extensions of the modal logic **S4**, one of which is **S5**. **S5** was the first logic ever to be shown to be pretabular, by Schiller Joe Scroggs in 1951. It was very nice to see Larisa summarizing her work and adding to it by considering related logics and also adding the dimension of complexity. I might finish my thoughts about Larisa's paper by saying how disappointed I was to learn from her paper that K. Swirydowicz in 2008 proved that the logic **R** has an uncountable number of pretabular extensions. **R** should be better behaved than that.

Alasdair Urquhart: The Story of γ. I appreciate Alasdair's contribution about the admissibility of Ackermann's rule γ. This rule was central to Ackermann's systems which were a stimulus to Anderson and Belnap's systems, but they did not want it as a primitive rule. It is a kind of metatheoretic version of modus ponens for the material conditional and not only I am sure did they think it was ugly, but it got in the way of proving relevant versions of the deduction theorem. Alasdair describes how it has been proved to be "admissible" (redundant) in Anderson and Belnap's systems, in a number of different ways. It is funny how an open problem can sit unsolved for quite a while, then be solved, and then be solved in a different way, etc. Often the solutions do not merely get easier and easier, but they involve connections to more and different results and methods. Alasdair can be seen as documenting this in the case of γ, with the first algebraic proof by Meyer and myself in 1969, followed by a proof by Routley and Meyer in 1973 using their Kripke-style ternary semantics, and thirdly the simplest proof of Meyer (1976) using his ingenious notions of metavaluation and metacompleteness. Alasdair points out there is actually a fourth proof by Saul Kripke that has never been published and is based on analogy with proving cut-elimination for a Gentzen system using semantical methods. In Alasdair's story the logic **R-Mingle** again makes a brief but important appearance in that the first proof of γ derives from a certain algebraic method first used in the context of **R-Mingle** (Dunn 1970). **R-Mingle** might be viewed as the proving ground for relevance logic.

What does the admissibility of γ have to do with information based logics? I might answer this with the following syllogism. The admissibility of γ is important for relevance logics. Relevance logics are information based logics. Therefore, the admissibility of γ is important for (some) information based logics. But this is a bit glib. To probe a bit deeper we might look at $a \models A \wedge ({\sim}A \vee B)$ therefore $a \models B$. Let us suppose that A is supported by the information state a and so is ${\sim}A \vee B$ but not because B is but because ${\sim}A$ is (a is an inconsistent information state). So there is no reason to believe that $a \models B$. But wait! We were not supposed to be considering the derivability of γ, but rather its admissibility, i.e., the preservation of theoremhood. Let us assume that theoremhood is equivalent to being supported by all consistent and complete information states ("possible worlds"). It is straightforward to show that if $A \wedge (\sim A \vee B)$ is supported by all consistent and complete information states a, then so is B. So the admissibility of γ is strongly linked to the thought that theoremhood is equivalent to not just validity in all information states, but particularly to validity in all consistent and complete information states. Clearly this last equivalence implies the admissibility of γ. And in the other direction, the second proof of the admissibility of γ recalled by Alasdair, which uses the Routley–Meyer semantics, involves in effect replacing the complete information state 0 rejecting the non-theorem B with a consistent and complete counterpart $0'$ that continues to reject B.

Edwin Mares: Manipulating Sources of Information: Towards an Interpretation of Linear Logic and Strong Relevance Logic. Ed was an early advocate for some kind of information based interpretation of relevance logic. He correctly points out that first such semantics based on the idea of combining pieces of information stems from Alasdair Urquhart's 1972 paper. It is nice to see in his contribution to this volume that he seems to subscribe to my "logics as tools" approach to logic, even though I do not remember ever directly teaching that to him when he was my student. This actually makes it all the nicer since it means it is not just my view. Thus in the final sentence of the substance of his paper is: "Interpreting relevance logic in terms of the construction of sources of information removes relevance logic from the metaphysically extravagant realm of true contradictions and impossible worlds, and locates it as an epistemological tool." I could not have put it better myself. This was my original intention in creating a 4-valued semantics for relevant first-degree entailments, and I have now upgraded the characteristic 4-valued De Morgan lattice to the "Opinion Tetrahedron" (Dunn 2010).

Ed has given in Mares (2004) a 4-valued semantics for various relevance logics, including **R**, but with "the catch" (his words not mine) that the Routley–Meyer frame semantics is enriched to be a neighborhood semantics. Ed in his contribution emphasizes the construction of sources of information from one another, something that I am in complete sympathy with (see Dunn 2014). The particular ways that Ed considers give a useful framework, as he shows, for understanding the contraction-free logics **MALL** (Multiplicative–Additive Linear Logic) and **RW** (**R** minus contraction), and the relevance logics **R** (Relevant Implication) and **LR** (**R** minus distribution).

Sebastian Sequoiah-Grayson: Epistemic Relevance and Epistemic Actions. Sebastian's contribution pairs nicely with Ed Mares'. Sebastian examines the interpretation of the ternary relation R on a Routley–Meyer frames in terms of information, and particularly in terms of what we might call "epistemic information." Sebastian says (p. 164):

> Both epistemic states may carry atomic information, or one epistemic state may carry information of atomic form and the other of conditional form, or both epistemic states may carry information of conditional form. Following Dunn (2015), we will call the first scenario the

Data Combining (DC) interpretation, the second scenario the *Program Applied to Data* (PD) interpretation, and the third scenario the *Program Combining* (PC) interpretation.

Sebastian cites Dunn (2015), which is a good and intuitive place to look, but I want to call attention to the fact that I first described these three varieties of interpretation in Dunn (2001a, c), and that in effect they were used in Dunn and Meyer (1997) and Dunn (2001b).

Sebastian says "Given that inconsistent propositions may hold at points, that is, given that we may have $x \Vdash \phi \wedge \neg\phi$, understanding as 'true at' is a little too crude. Instead, we may understand $x \Vdash \phi$ as 'x carries/stores the information that ϕ.' "

His point is that $\phi \wedge \neg\phi$, being a contradiction, cannot be true. As an "adialethist" I certainly agree with that. Ed Mares and Ross Brady make similar points. Out of respect I will go along with them, though the picky logician in me cannot help but question whether there is a significant distinction between "x carries/stores the information that ϕ" and say "ϕ is true according to the information x."

Anyway getting back from "semantical quibbling," Sebastian then goes on to given an informational interpretation of the ternary relation, that is the kingpin of the Routley–Meyer semantics for relevant implication, saying: "In this case, $Rxyz$ comes out as 'if you combine the information carried by/stored at x with the information which is carried by/stored at y then you get the information which is carried by/stored at z.' " I see that Sebastian had already read my most recent piece on the ternary relation (Dunn 2015) and that he cites it as he goes on to say: "In other words, given the information carried by states x and y, their combination is relevant to the information carried by state z." I actually spelled this out in terms of "contextual relevance": in the context of information x, information y is relevant to information z. But it comes down to the same since, as Sebastian makes clear, I first introduced the idea of interpreting Routley–Meyer's $a \sqsubseteq b$ ($R0ab$) as absolute (Sebastian's non-contextual) relevance, and then the idea of contextual relevance as combining x with y is absolutely relevant to z ($x \bullet y \sqsubseteq z$).

The interesting new feature of Sebastian's analysis is his saying (p. 161):

> Moreover, the very act of combining x and y is itself informationally relevant to z. This is because it is the operation of combining x and y which bring the information at both states together. Sans such an operation, the information at x and y are separate informational entities, neither of which, either considered independently or non-contextually, are informationally relevant to z.

This is what allows him to build a bridge from epistemology to information, from epistemic states to information states. As he says (p. 159):

> That relevance logics provide a logical framework for epistemic relevance and epistemic actions is at the very least not obvious. Such logics are neither thought of as particularly epistemic, nor as dynamic (and actions, epistemic or otherwise, are dynamic if anything is).

Sebastian (p. 163) relates the properties of the active combining of information to Gentzen's structural rules (Commutation, Weakening, Contraction—and one that Gentzen implicitly assumed, Association).

Given that we are understanding the information states as explicit epistemic states, the composition operation as the epistemic action of combining such states, and the partial order of informational inclusion as epistemic relevance, then the epistemic action contexts in which the structural rules hold or fail become salient. They become salient because they specify the properties that said epistemic actions need to possess with regard to guaranteeing epistemic success.

I will stop here so as to not spoil the interesting ending of his paper.

Ross T. Brady: Comparing Contents with Information. Ross is an early and frequent contributor to relevance logic, even if he does pronounce it in his Aussie way as "relevant logic." I do appreciate his contribution to this volume, since in particular Ross has contributed towards an information based understanding of some of the weaker relevance logics. But Ross prefers to talk of "content" rather than "information." He has been motivated in his work on "content algebras" I believe in large part by his efforts to formulate a naive set theory based on relevance logic, which demands weaker logics than the "gold standard" relevance logic **R**. This need was shown in Meyer et al. (1979).

Ross starts out the body of his present article with "Sect. 2. Carnap on contents and information." And while he goes to some length in presenting Carnap's notion of "content" resulting in the definition of the *content* of A as the class of state descriptions that make A false, he somehow skips Carnap's notion of the *information* of A as the class of state descriptions that make A true.[4] Note that for Carnap, content and information are just duals, as Brady recognizes, and it should be more or less arbitrary which one you prefer.

But when we get to Sect. 4, we find that Ross seems to want use the term "information" in such a way that it implies truth, whereas "content" does not require truth. Now in a way this is just a matter of words. Ross is correct in saying that I believe that "it is all a matter of the interpretation of the words 'true' and 'information' as to whether semantic information has to be true." But that does not mean I think it is entirely arbitrary whether one builds "truth" into the definition of "information." Like many a technical definition in science, say that of the mass of an object, it depends on the role it plays in constructing a good theory. Building weight into the concept of mass would not be conducive to theory building, and I think the same holds of "truth" with respect to "information." I argued this in my 2013 review of Luciano Floridi's book *The Philosophy of Information*. Ross cites this book, so I am not going to go through that again. I obviously didn't convince him. So I will end with the following "argument from authority."

Rudolf Carnap and Yehoshua Bar-Hillel in Carnap and Bar-Hillel (1964) say (p. 229):

It should, however, be emphasized that semantic information is here not meant as implying truth. A false sentence which happens to say much is thereby highly informative in our sense. Whether the information it carries is true or false, scientifically valuable or not, and so forth,

[4]Though Ross does talk of the "range" of A, and that is just a definition away from Carnap's concept of "information." The *information* in A is the same as the *range* of A.

does not concern us. A self-contradictory sentence asserts too much; it is too informative to be true.

I also want to take the opportunity to make clear that I do think that contradictions contain information, but unlike Carnap and Bar-Hillel I do not think that the amount of information is in essence infinite. In relevance logic an arbitrary contradiction does not imply every sentence whatsoever. This can be captured semantically using information states that unlike possible worlds, can be inconsistent and also incomplete. This allows distinctions even among contradictions.

So I end my quibbling over mere words. As my late colleague Hector Casteñeda might have put it: Information, Schminformation, who cares? After all Ross does say (p. 179) "What I propose to do is to simplify the matter by focusing on a normative interpretation of information." I am not quite sure what he means by this but it could be similar to my recognizing in Dunn (2008) that there is pragmatic implication that information is true.

The main logical content in Ross's paper is his Logic of Meaning Containment **MC** with its algebraic semantics, and I surely will not quibble with that. The logic **MC** and its extensions that Ross presents are quite natural, though they do avoid axioms for nested implications such as we have in the logic of relevant implication **R**. Let me end though with a substantive question about Ross's algebraic semantics. Is there a nice way to represent these content algebras, say using a ternary accessibility relation along the lines of the Routley–Meyer semantics for relevance logic?

Heinrich Wansing: On Split Negation, Strong Negation, Information, Falsification, and Verification. It is nice to see Heinrich using my Galois connected negations, even if he switches to calling them "split negations" using the better metaphor of Chrysafis Hartonas. Questions about the nature of information and questions about the nature of negation go hand in hand together. If a sentence A carries a certain amount of information, what amount does its negation $\sim A$ contain? (Shannon). If the information contained in A is understood in such-and-such a way, how is the information in $\sim A$ to be understood (Carnap and Bar-Hillel). What is the information, if any, conveyed by a tautology? A contradiction? (Floridi). Wansing addresses the important question as to whether there are two kinds of information, negative as well as positive. In my dissertation I introduced "proposition surrogates" as pairs of positive information, and negative information. I viewed these as sets of "topics," later upgraded to situations and now information states.

Heinrich, and also another friend Yuri Gurevich, have defended the idea that positive and negative information are in fact different, and they should be treated symmetrically. This leads to a kind of "strong negation" of the Nelson variety. A third friend of mine, Sebastian Sequoiah-Grayson, has claimed that there is a strong asymmetry between positive and negative information. Which side do I choose? This is obviously not the best occasion to take a side, so of course the answer is: none of the above. While a proposition surrogate consists of a set of positive information together with a set of negative information, it is important to note that for me the items (let's call them information states) in the sets do not differ in kind. In that sense there is no intrinsic distinction between positive and negative information. Positive and negative are context relative terms. The same information that provides positive support for the statement "Florida is Democratic" provides negative support with respect to the statement "Florida is not Democratic." The negation of the proposition surrogate (X^+, X^-) is simply (X^-, X^+).

Heinrich seems to agree with this for he says (p. 199): "Suppose that we are working with neutral litmus paper, so that by a change of colour we may test for both acids and bases. What is positive information from the point of view of verification is then negative information in the context of falsification (and vice versa)."

Yaroslav Shramko: Truth, Falsehood, Information and Beyond: The American Plan Generalized. I really appreciate Yaroslav's informed and informative account of the development of what is sometimes called the "Belnap–Dunn 4-valued logic"—and other times called by those brave enough to not follow alphabetical order the "Dunn–Belnap 4-valued logic." I note that Yaroslav was politic enough to avoid either of these labels. The closest he comes is to refer to "Dunn and Belnap's American Plan." As he points out the label "American Plan" was introduced by Bob Meyer, who contrasted the 4-valued approach with what he called the "Australian Plan" approach of Richard and Val Routley (1972) which used a "point-shift" in the evaluation of negation: $a \vdash \sim A$ iff $a^* \nvDash A$. (This evaluation is reminiscent of the earlier Białynicki-Birula and Rasiowa representation of De Morgan lattices that defined De Morgan complement—they called it "quasi-complement"—in an analogous way.)

Yaroslav recounts with great precision 4 different versions of the American Plan, with Version 1 starting in my 1966 dissertation where I did not yet have the nerve to speak of truth values and a sentence being both true and false, but instead spoke of "topics" and how a sentence could give both positive and negative information about a given topic. As Yaroslav next points out with his Versions 2 and 3, I finally got the courage to speak of a sentence being both true and false (also neither) in a given "situation." Situations replaced "topics," and a valuation v became a relation that could relate a given sentence to both True and False in a given situation, with the difference between versions 2 and 3 being whether one just left v as a relation or instead looked at its image, i.e., the set of truth values it relates the sentence to in the given situation. I eventually began talking of "information states" rather than "situations," having converted to "computer scientology."[5]

Finally in Version 4 these sets of truth values are promoted to "generalized truth values," which is just what Belnap did with his famous True, False, Both, and Neither. Belnap also introduced the idea that there can be two different orders on the values, which Shramko refers to as the truth order and the information order. And of course it doesn't stop there. As Shramko carefully explains this, "bilattice" can be extended to a 16 element "trilattice," based on generalized truth values corresponding to all

[5]It might be worth pointing out that Jon Barwise and John Perry developed their "situation semantics" much later (see Barwise and Perry 1983) and while their "real" situations cannot be inconsistent although they can be partial, their "abstract" situations can be both. Also the relevantist Ed Mares still talks of situations.

the subsets of the 4 generalized truth values, and of course, *ad infinitum*.

Chunlai Zhou: Logical Foundations of Evidential Reasoning with Contradictory Information. Chunlai and I did some interesting work together on negation, extending my original "kite of negations" to a "lopsided kite" (Dunn and Zhou 2005). Chunlai has continued to work on negation and the problems of inconsistent information and has developed a theory of "belief functions on distributive lattices." He intends this as a generalization of the widely accepted Dempster–Shafer belief functions which are defined on a Boolean algebra (the powerset of a finite "frame of discernment"). These belief functions when defined on a De Morgan lattice can deal with inconsistent (and incomplete) information in a way that does not fall prey to the classical problems of "Explosion" an inconsistency implying everything.

In Dunn (2010) I myself have recently been trying to extend the original Belnap–Dunn framework for the 4-valued Logic so as to have more subtlety for degrees of belief, or subjective probability, than merely "True, False, Both, Neither." Chunlai cites this paper but makes no comparison of this approach to his. I am not sure there is a direct comparison, but it is an idea to explore.

Chunlai several times speaks of expanding the Carnap universe for consistent information of possible worlds to the what he calls the Belnap–Dunn universe of states where a proposition may be both true and false. He then goes on to precisify this in terms of the Belnap–Dunn 4-valued logic, where the values are "True, False, Both, Neither." This suggests to me that in the Belnap–Dunn universe there may well be incomplete as well as inconsistent states, and Chunlai was just emphasizing the inconsistent states because of their "weirdness."

Chunlai says (p. 239):

> Carnap developed a universe of possible worlds that encompasses all possible states of a real-world system. Information about that system, if precise and certain, identifies its actual state. If imprecise but certain, this information identifies a subset of possible system states. Such kind of subsets are called *truth sets*. If uncertain, then the information induces a probability distribution over system states which is defined on *all* subsets of the universe.

Truth sets are sets of possible worlds, i.e., sets of complete and consistent information states, and are in effect "U.C.L.A. propositions" as Alan Anderson called them.[6] Thus the probability distribution in effects assigns probabilities to the various propositions. Dempster–Shafer use a probability function on sets of information states.

Chunlai gracefully combines the Belnap–Dunn ideas with ideas from Ruspini (1987). As Chunlai explains (p. 239):

> Ruspini (1987) noted that Carnap's characterization does not distinguish degrees of precision when the information is uncertain and Carnap's logical approach, while enabling a clearer understanding of the relations between logical and probabilistic concepts, suffers from a

[6]Carnap who avoided abstractions, actually used the syntactic device of "state descriptions" rather than possible worlds, where a state description can be viewed as a set containing every atomic sentence or its negation, but not both.

major handicap: it assumes that observations of the real world always determine unambiguously probability values for *every* subset in the universe. But uncertain information generates a probability function for all subsets of the universe only if it is precise. When information is imprecise, this probability function is defined on some subsets of possible states, which is not discussed in Carnap's methodology.

Janusz Czelakowski: Probabilistic Interpretations of Predicates. Gary Hardegree's and my book (2001) *Algebraic Methods in Philosophical Logic* benefited from Janusz's work in the 1980s on asymmetric ("single sided") and symmetric consequence relations. Janusz is an acknowledged expert on generalizations and abstractions of Tarski's consequence operator, so it was very nice to see his subsequent positive review of our work (Czelakowski 2003).

I was pleased to see that Janusz's contribution to this volume focuses on probabilistic predicates. As I said in my "autobio" in the early part of my career I eschewed anything having to do with probability. But now I believe that probability, statistics, machine learning, etc., and deductive logics of various forms, all belong together in a kit of tools for reasoning by humans and/or machines. I of course was also pleased to see the relation to De Morgan lattices and to gaggles and their weaker versions distributoids.

Janusz's idea, put quickly, is to interpret vague predicates, such as "good" in the sentence "John is a good writer," in terms of probabilities. He focuses on unary predicates, but does consider the general case of n-ary predicates. These cover his other example: "Mike is a better mathematician than Andrew." (While I appreciate Janusz saying this of me, I feel sorry for poor Andrew, whoever he is. :)). Janusz does not do this in the simplistic "fuzzy logic" way of assigning a degree of "truth" between 0 and 1. This I was glad to see because I have always been concerned about fuzzy logic being overly precise about "fuzziness." As Janusz explains (p. 274):

> While in fuzzy set theory it makes sense to assign a numerical value to each m-tuple $\langle a_1, \ldots, a_m \rangle$ of elements of A as a degree of a relation R 'holding' on $\langle a_1, \ldots, a_m \rangle$, the probabilistic interpretation does not do this. It merely provides a global probability distribution for the set A^m.

I would like to make a wild suggestion. Audung Jøsang has created another multivalued approach to uncertainty, using what he calls an "opinion triangle" to locate degrees of belief (truth), degrees of disbelief (falsehood), and degrees of uncertainty.[7] I wonder whether Janusz's definitions and theorems can be modified so as to use Jøsang's approach. In Dunn (2010) I made an extension of the Jøsang's opinion triangle to the "opinion tetrahedron," introducing two kinds of degrees of uncertainty, one kind having to do with uncertainty in the sense of ignorance (Neither) and the other kind having to do with uncertainty in the sense of conflict (Both). The values Neither and Both of course come from the Belnap–Dunn 4-valued logic. It seems to me that the first might correspond to vagueness, and the second to ambiguity. Or something like that. :)

---

[7] Jøsang first presented this at a conference in 1997, but a more accessible source is Jøsang (2001).

Chrysafis Hartonas: Reasoning with Incomplete Information in Generalized Galois Logics Without Distribution: The Case of Negation and Modal Operators. Chrysafis is widely known as "Takis," and this is how I shall refer to him here. In Hartonas and Dunn (1997) Takis and I gave what I believe to be the most natural representation of general lattices, using ideas that combine older ideas from Birkhoff and Goldblatt. My first development of gaggle theory (Dunn 1991) required an underlying distributive lattice. But I soon regretted this decision since generalized Galois connections can be found on non-distributive lattices, semilattices, and even mere partially ordered sets. In Dunn (1993) I looked at some of these and in particular considered "partial gaggles" built upon posets, and on semilattices, and showed how these could be represented. In my original paper on distributive gaggles I had already generalized how one could give a representation of a pair of Galois connections using a binary accessibility relation $\perp$, but in my 1993 paper I showed how this could be done with partial gaggles and semilattice ordered gaggles.

In particular I looked at the case where the Galois operators were really the original motivating Galois connections, defined though not just on a partial order, but on a semilattices. The representation theorem said: Let $(S, \leq, \wedge)$ be a semilattice with a Galois connection $(\sim, \neg)$. Then there is some set U with a binary relation $\perp$ on U so if we define for $X \subseteq U$, ${}^{\perp}X = \{x : \forall a(a \in X \Rightarrow x \perp a)\}$, $X^{\perp} = \{x : \forall a(a \in X \Rightarrow a \perp x)\}$, then not only do we get a Galois connection on the subsets of S, but in fact the given semilattice with its Galois connection is isomorphic to a collection of subsets of S closed under the two "perp" operations above.

So where is the representation of lattices? Well, a lattice is a semi-lattice when viewed from top to bottom, and also a semi-lattice when viewed from bottom to top (with the two semi-lattices "glued together" by the Absorption Laws). Here is the weird part, the identity map ι is the Galois connection. If $a \leq b$ then of course $\iota b \geq \iota a$. Things get a little more complicated after that, but I hope you get the picture. I believe Takis and I had this revelation at almost the same time. I remember sitting in my office waiting for him to arrive so I could tell him, and his rushing over to my office with the same thought freshly in mind.

It unfortunately seemed difficult to add algebraic operations to this picture so as to accommodate lattice-ordered algebras. So it was never used as the basis for "lattice-ordered gaggles." But Takis in Hartonas (1997) showed how to do this with a modified representation of lattices, and his contribution to this volume nicely presents and extends those results.

Robert Goldblatt and Matt Grice: Mereocompactness and Duality for Mereotopological Spaces. Rob and Matt say: "*Duality* in the semantic analysis of propositional logics has been a significant theme in the research of J. Michael Dunn." I have often somewhat humorously explained my recurring interest in duality by the fact that I am very ambidextrous, which means that I cannot always tell my left hand from my right. I think I can though tell a part from a whole, and mereology, as I am sure every reader knows, or can google, is essentially the study of the part-whole relationship. As a fan of duality it was fascinating for me to read Rob and Matt's application of duality to the preexisting notion of a Boolean contact algebra which had been shown to be representable as an algebra of regular closed subsets of a compact topological space. They construct a dual equivalence between the category of Boolean contact algebras and a category of mereotopological spaces that have a property what they call mereocompactness, which is strictly stronger than ordinary compactness.

Rob's semantics for orthologic, using a binary "orthogonality relation" $\perp$, was one of my inspirations for studying various properties of negations defined using such a binary relation. More generally it was one of my inspirations for gaggle theory. Gaggles were also inspired by Jónsson and Tarski's representation of Boolean algebras with operators. It is worth mentioning that Rob extended Jónsson and Tarski's representation by replacing Boolean algebras with the more general distributive lattices, providing a Priestly-style duality theorem for them; he also relaxed the requirement that the operators distribute over join in each of their places to allow distribution over meet. My notion of a "distributoid" generalizes that even further so as to allow operators sometimes to "co-distribute," i.e., change join to meet and vice versa. So Rob and I have sometimes been "picking in the same berry patch."

Gerard Allwein and William L. Harrison: Distributed Modal Logic. Gerry's and William's contribution on distributive modal logic relates directly to my gaggle theory, and it is also related to my view of logics as tools. This last is perhaps the main theme of my "autobio" in this volume, but I never made it a public theme of my research. This is why it is all the more pleasing to see it pop up in Gerry's and William's paper, and in some in the other contributed papers.

Gerry is one of three students of mine who have ended up working in computer security, so it is not surprising to see that the first application mentioned in the paper is to computer security. That is Gerry's "day job," and he is in effect a "logician in engineer's clothing." Gerry and William give another possible application in terms of testing of systems. I cannot help but mention that it has been fun during my career to see modal logic lifted up from the gutter, so to speak, where it was, according to Willard Van Orman Quine, conceived in the sin of confusing use and mention. Now it is a well accepted tool not just in philosophy, but even in computer science.[8]

According to Gerry and William (p. 357), "a *distributed modal logic* is a collection of *local modal logics* linked together by *distributed modal connectives*, each of which takes formulas in one logic and returns formulas in a different logic." They later say (p. 359) "A distributed logic starts with a directed graph where every node constitutes a *local logic*. Each node is a (possibly null) extension of a classical propositional logic with a set of modal connectives, and any axioms and rules to govern behavior." So we have in effect a meta-Kripke model for distributed modal logic, where the nodes correspond not to sets of truths (as do "possible worlds"), but to

[8]Edmund Clarke won the 2007 Turing Award for his pioneering work in using temporal modal logic in model checking. And while I have this footnote as my podium, let me use it to recommend to the reader who wants to read more on "Sinn" (not a typo but Deutsch) Alasdair Urquhart's recent paper (Urquhart 2010).

whole logics.

Johan van Benthem: Tracking Information. Johan is of course one of the pioneering leaders in looking at logic from both dynamic and informational points of view, and so it is very appropriate that his contribution combine the best of both and be on "the dynamics of information-driven agency over time."

Johan and Yde Venema's "Arrow Logic" is an interesting example of a dynamic approach to logic, and it is not just about arrows on a whiteboard, but is a blueprint for creating logics of transitions. In Dunn (2014) I discussed its relationships to the Routley–Meyer semantics for relevance logics, and also the work I did in Dunn (2001b) in representing relation algebras using Routley–Meyer type frames. Johan starts his paper as follows (p. 387):

> Depending on a relevant task at hand, information can be represented at different levels, less or more detailed, each supporting its own appropriate logical languages. We discuss a few of these levels and their connections, and investigate when and how information growth at one level can be tracked at another. The resulting view has two intertwined forms of logical dynamics for informational agents: one of update and one of representation.

Johan's present paper discusses epistemic logic, which of course involves information (the "proposition" that we know or believe). He points out that information can be represented at various levels of detail. I very much like this approach since it fits well with my "logics as tools" view, recognizing that different and more detailed representations may be appropriate for different purposes, much like architect's drawings, plans or blueprints.[9] He does not mention the crudest (classical logic's True vs. False), but jumps right in with the standard (static) epistemic logic based on the modal logic **S5**. A proposition may be thought of as a set of possible worlds. He then introduces (dynamic) epistemic logic with "hard" updates (meaning learning something certain, say by direct observation). This allows for growth of logic through "announcements." After proving that both of these logics are axiomatizable, Johan retreats from axiomatizing the logic of knowledge, turning to the logic of belief and plausibility, and after characterizing these model-theoretically, shows that the dynamic logic of belief change under hard information is completely axiomatizable. I will stop here, since I do not want to spoil the dynamic nature of Johan's story. But the general picture is that Johan considers finer and finer levels of representation, and to each he then adds the apparatus of dynamic updates. I should add that beside looking at the various level of detail in the modelings of belief, Johan also wants to include a way of tracking, or comparing one level with another. The general idea is that the simpler model might be embedded into the more complex model, and vice versa the more complex model might be homomorphically mapped onto the simpler one.

---

[9] I am reminded of the novel *Kandelman's Krim* by the mathematical physicist J. L. Synge. Nuel Belnap called my attention to this novel many years ago. It involves a philosophical discussion between a goddess, a kea, an orc, a unicorn and a plumber. The Plumber says: "I am of course perfectly well aware of the irrationality of π, but on the job, π is 3 1/7, or 3 if I am in a hurry." Maybe Nuel is to be blamed for my "logics as tools" view.

That is a bit rough, but it should intrigue you to look more closely. I cannot help but mention that Johan's representations do not seem to include models for inconsistent beliefs. This is not at all unusual. As a proponent of paraconsistent logic I have gotten used to it. But let me throw out an idea. In my comments on Chris Mortensen's contribution I mention homomorphic images and submodels as ways of generating inconsistent and incomplete models. Maybe something is lying there in wait.

As I got near the end of Johan's article, I of course found particularly pleasing Johan's statement (p. 410) "While writing this paper, I increasingly felt that Mike Dunn's Gaggle Theory (Dunn 1991) may well be an ideal stance from which to study the themes explored in this paper, for its austerity, elegance and broad sweep." I hope someone can follow through on that as well.

Lawrence S. Moss: Syllogistic Logic with Cardinality Comparisons. I welcome the contribution from my colleague Larry Moss. He has been the fearless leader of the Program in Pure and Applied Logic at Indiana University for many years. I am sure that it can be argued that the syllogistic logic of Aristotle was the first information based logic since many introductory logic classes motivate a valid syllogism as one where the information of the conclusion is included in the information from the premises. And it certainly fits my views of a logic as a tool. Associated with Indiana University Bloomington we have the wonderful Stone Age Institute, and it actually experiments in making stone tools for various purposes. Larry's paper fits my idea about how a tool can be extended for new purposes. In this case the Aristotelean logic of All, Some, and No is extended to include the dyadic quantifiers "there are more" and "at least as many."[10] Larry's is concerned with what he calls "Natural Logic," and that I might call "Post Stone Age Logic." I support the idea of this research because I do think it gets at a more natural (dare I say primitive?) logic than what has been produced by Boole, Frege, Russell and Whitehead, etc. (Dare I add C. I. Lewis, Łukasiewicz, Anderson, Belnap, Girard and others to this list?) And while we are talking of tools, I want to point out that Larry adds computational complexity results which are useful in gauging how efficient or "green" a logic is. This last will become increasingly important as computers do more and more of our reasoning for us.

In the words of Blaise Pascal, "I made this very long, because I did not have the leisure to make it shorter." In the words of Frank Herbert, "There is no real ending. It's just the place where you stop the story." Most importantly, in the words of Porky Pig, "That's all folks!"

[10] Since we all frequently travel by plane, I cannot resist mentioning that I have suggested to Larry, mostly as a joke, that he find a way to formalize "Many bags look alike." How many of us, as we have stood waiting for our bags to arrive on the airport carousel, have wondered how to formalize this statement?

References

Anderson, A. R., & Belnap, N. D. (1975). *Entailment: The logic of relevance and necessity* (Vol. I). Princeton, NJ: Princeton University Press.

Barwise, J., & Perry, J. (1983). *Situations and attitudes*. Cambridge, MA: MIT Press.

Carnap, R., & Bar-Hillel, Y. (1964). An outline of a theory of semantic information. In Y. Bar-Hillel (Ed.), *Language and Information: Selected Essays on their Theory and Application* (pp. 221–274). Reading, MA: Addison-Wesley and The Jerusalem Academic Press. (Reprint of Technical Report No. 247, Research Laboratory of Electronics, MIT, 1952.)

Czelakowski, J. (2003). Review of J. M. Dunn and G. M. Hardegree: Algebraic methods in philosophical logic. *Bulletin of Symbolic Logic, 9*, 231–234.

Dunn, J. M. (1970). Algebraic completeness results for R-Mingle and its extensions. *Journal of Symbolic Logic, 35*, 1–13.

Dunn, J. M. (1979a). Relevant Robinson's arithmetic. *Studia Logica, 38*, 407–418.

Dunn, J. M. (1979b). A theorem in 3-valued model theory with connections to number theory, type theory, and relevant logic. *Studia Logica, 38*, 149–169.

Dunn, J. M. (1991). Gaggle theory: An abstraction of Galois connections and residuation with applications to negation, implication, and various logical operators. In J. van Eijck (Ed.), *Logics in AI: European Workshop JELIA '90*, number 478 in *Lecture Notes in Computer Science* (pp. 31–51). Berlin: Springer.

Dunn, J. M. (1993). Partial gaggles applied to logics with restricted structural rules. In K. Došen & P. Schroeder-Heister (Eds.), *Substructural logics* (pp. 63–108). Oxford, UK: Clarendon.

Dunn, J. M. (2001a). The concept of information and the development of modern logic. In W. Stelzner & M. Stöckler (Eds.), *Zwischen traditioneller und moderner Logik: Nichtklassische Ansatze* (pp. 423–447). Paderborn: Mentis-Verlag.

Dunn, J. M. (2001b). A representation of relation algebras using Routley-Meyer frames. In C. A. Anderson & M. Zelёny (Eds.), *Logic, Meaning and Computation. Essays in Memory of Alonzo Church* (pp. 77–108). Kluwer Academic Publishers.

Dunn, J. M. (2001c). Ternary relational semantics and beyond: programs as arguments (data) and programs as functions (programs). *Logical Studies, 7*, 1–20. (Proceedings of the International Conference Third Smirnov Readings (Moscow, May 24–27, 2001), Part 2; Institute of Logic, Russian Academy of Sciences).

Dunn, J. M. (2008). Information in computer science. In P. Adriaans & J. van Benthem (Eds.), *Philosophy of Information*. Volume 8 of Handbook of the Philosophy of Science (D. M. Gabbay, P. Thagard, J. Woods (eds.)) (pp. 581–608). Elsevier, Amsterdam

Dunn, J. M. (2010). Contradictory information: Too much of a good thing. *Journal of Philosophical Logic, 39*, 425–452.

Dunn, J. M. (2014). Arrows pointing at arrows: Arrow logic, relevance logic and relation algebras. In A. Baltag & S. Smets (Eds.), *Johan van Benthem on Logic and Information Dynamics* (pp. 881–894). Outstanding Contributions to Logic. New York, NY: Springer.

Dunn, J. M. (2015). The relevance of relevance to relevance logic. In M. Banerjee & S. N. Krishna (Eds.), *Logic and its Applications*, number 8923 in *Lecture Notes in Computer Science* (pp. 11–29). Heidelberg: Springer.

Dunn, J. M., & Hardegree, G. M. (2001). *Algebraic Methods in Philosophical Logic*. Volume 41 of Oxford Logic Guides. Oxford, UK: Oxford University Press

Dunn, J. M., & Meyer, R. K. (1997). Combinators and structurally free logic. *Logic Journal of the IGPL, 5*, 505–537.

Dunn, J. M., & Zhou, C. (2005). Negation in the context of gaggle theory. *Studia Logica, 80*, 235–264.

Hartonas, C. (1997). Duality for lattice-ordered algebras and for normal algebraizable logics. *Studia Logica, 58*, 403–450.

Hartonas, C., & Dunn, J. M. (1997). Stone duality for lattices. *Algebra Universalis, 37*, 391–401.

Jøsang, A. (2001). A logic for uncertain probabilities. *International Journal of Uncertainty, Fuzziness and Knowledge-Based Systems*, *9*, 279–311.
Maksimova, L. L. (1972). Pretabular superintuitionistic logics. *Algebra and Logic*, *11*, 308–314.
Maksimova, L. L. (1975). Pretabular extensions of Lewis *S*4. *Algebra and Logic*, *14*, 16–33.
Mares, E. D. (2004). 'Four-valued' semantics for the relevant logic R. *Journal of Philosophical Logic*, *33*, 327–341.
Meyer, R. K., Routley, R., & Dunn, J. M. (1979). Curry's paradox. *Analysis (n.s.)*, *39*, 124–128.
Meyer, R. K. (1976). Relevant arithmetic. *Bulletin of the Section of Logic of the Polish Academy of Science*, *5*, 133–137.
Routley, R., & Routley, V. (1972). The semantics of first degree entailment. *Noûs*, *6*(4), 335–359.
Ruspini, E. H. (1987). Epistemic logics, probability, and the calculus of evidence. In J. P. McDermott (Ed.), *Proceedings of the 10th International Joint Conference on Artificial Intelligence* (pp. 924–931). Milan, Italy: Morgan Kaufmann.
Urquhart, A. (2010). Anderson and Belnap's invitation to sin. *Journal of Philosophical Logic*, *39*, 453–472.

Index

K. Bimbó (ed.), *J. Michael Dunn on Information Based Logics*,
Outstanding Contributions to Logic 8, DOI 10.1007/978-3-319-29300-4

MIX
Papier aus verantwortungsvollen Quellen
Paper from responsible sources
FSC® C105338

If you have any concerns about our products,
you can contact us on
ProductSafety@springernature.com

In case Publisher is established outside the EU,
the EU authorized representative is:
Springer Nature Customer Service Center GmbH
Europaplatz 3, 69115 Heidelberg, Germany

Printed by Libri Plureos GmbH
in Hamburg, Germany